Primate Evolution

Glenn C. Conroy

WASHINGTON UNIVERSITY

W·W· NORTON & COMPANY

NEW YORK · LONDON

Fig. 1.9 and **T1.2** from *Primates in Nature* by Allison F. Richard. Copyright © 1985 by W. H. Freeman & Company. Reprinted by permission.

Fig. 2.2b From *Human Evolution: An Illustrated Introduction* by Roger Lewin. Copyright © 1984 by Blackwell Scientific Publications Ltd. Reprinted by permission of W. H. Freeman & Company.

PRINTED IN THE UNITED STATES OF AMERICA.
FIRST EDITION

The text of this book is composed in Baskerville, with display type set in Optima. Composition and Manufacturing by The Maple-Vail Book Group.

Library of Congress Cataloging-in-Publication Data

Conroy, Glenn C.
Primate evolution / Glenn C. Conroy.
p. cm.
1. Primates—Evolution. 2. Primates, Fossil. I. Title.
QL737.P9C66 1990
599.8'0438—dc20 89-26448
ISBN 0-393-95649-0

W. W. Norton & Company, Inc., 500 Fifth Avenue, New York, N.Y. 10110
W. W. Norton & Company, Ltd., 37 Great Russell Street, London WC1B 3NU
1 2 3 4 5 6 7 8 9 0

Contents

Preface xi

1 Elements of Paleoprimatology 2

WHAT IS A PRIMATE? 3
Early Descriptions 3
Modern Definitions 4
CLASSIFICATION AND EVOLUTION 8
Three Approaches to Classification 8
Character Analysis 10
Comparison of Cladism and Evolutionary Taxonomy 11
Two Models of Speciation 15
Limitations in Taxonomy 18
FOSSIL AGES 19
Geologic Time Scale 19
Radiometric Dating 20
Principles of Radioactive Decay 21 Dating Methods 24
Paleomagnetic Measurement 26
MAKING INFERENCES ABOUT THE BIOLOGY OF EXTINCT PRIMATES 27
Allometry 28
Basic Principles 28 Correlation between Metabolic Rate and Body Weight 29
Relationships between Primate Size and Diet 32
Effects of Diet on Dentition 34
ORIGIN OF THE PRIMATES 38
Arboreal Theory 39

Visual Predation Theory 41
Angiosperm Radiation Theory 41
TAXONOMY USED IN THIS TEXT 43

2 Paleocene Primates 48

PALEOCLIMATES AND BIOGEOGRAPHY 49
Extinctions at the Cretaceous-Tertiary Boundary 49
Greenhouse Model 50 Arctic Ocean Spillover Model 51
Asteroid-Impact Model 51
Summary of Climatic Changes in the Paleocene 52
Movements of Major Landmasses and Dispersal of Species 53
SUMMARY OF THE PALEOCENE FOSSIL RECORD 56
PLESIADAPIFORMES 57
Origins 59
General Morphology 63
Cranial Morphology 63 Dentition 68
Plesiadapidae 68
Dentition 69 Cranial Morphology 73
Postcranial Morphology 75
Carpolestidae 78
Dentition 79
Saxonellidae 79
Microsyopidae 81
Issues in Classifying the Microsyopidae 82
Paromomyidae 84
Dentition 84
Picrodontidae 86
FEEDING ADAPTATIONS OF PALEOCENE PRIMATES 86
PHYLOGENY AND CLASSIFICATION OF PALEOCENE PRIMATES 90

3 Eocene Primates 92

PALEOCLIMATES AND BIOGEOGRAPHY 93
Climates Inferred from Paleobotany 93
Land Movements and Mammal Migration Routes 94
SUMMARY OF THE EOCENE FOSSIL RECORD 97

RISE OF THE EUPRIMATES 99
Comparison of Adapids and Omomyids 101
Adapidae 103
Origins and Distribution 103 General Morphology 105 Comparison with Later Primates 106 Notharctinae 109 Adapinae 115
Omomyidae 117
Origins and Distribution 117 Dentition 119 Cranial Morphology 121 Postcranial Morphology 123 Comparison with Later Primates 124
BODY SIZE, FEEDING ADAPTATIONS, AND BEHAVIOR OF EOCENE PRIMATES 125
PHYLOGENY AND CLASSIFICATION OF EOCENE PRIMATES 128
Plesitarsiiform–Simiolemuriform Classification 128
Lemurophile Hypothesis 128
Strepsirhine–Haplorhine Classifications 131
Omomyophile Hypothesis 132 Tarsiphile Hypothesis 134

4 Oligocene Primates 136

PALEOCLIMATES AND BIOGEOGRAPHY 137
Climatic Trends 137
Land Movements and Dispersal Routes 139
SUMMARY OF THE OLIGOCENE FOSSIL RECORD 142
OLIGOCENE PRIMATES IN THE OLD WORLD 143
History and Stratigraphy of the Fayum Site 145
General Morphology of the Fayum Primates 149
Parapithecidae 149
Apidium 154 *Parapithecus* 157 *Qatrania* 158
Propliopithecidae 158
Propliopithecus 158 *Aegyptopithecus* 160
Other Fayum Primates 164
Oligopithecus 164 *Afrotarsius* 165
Body Size, Feeding Adaptations, and Behavior in the Fayum Primates 166
OLIGOCENE PRIMATES IN THE NEW WORLD 168
Branisella 169
Tremacebus 170
Dolichocebus 172
PHYLOGENY AND CLASSIFICATION OF OLIGOCENE PRIMATES 172

Origins of the Anthropoidea 172
Pondaungia 173 *Amphipithecus* 174
Phylogeny of the Fayum Primates 177
Parapithecid Affinities 177 Propliopithecid Affinities 180
Origin of New World Primates 181
Taxonomy of New World Oligocene Anthropoids 182

5 Miocene Primates 184

PALEOCLIMATES AND BIOGEOGRAPHY 185
East African Rift System 186
Land Movements and Dispersal Routes 191
SUMMARY OF THE MIOCENE FOSSIL RECORD 194
African Fossil Record 194
Early Miocene 194 Middle and Late Miocene 196
Eurasian Fossil Record 197
RISE OF THE HOMINOIDEA 200
Dryomorphs 204
Proconsul 206 *Rangwapithecus* 211 *Limnopithecus* 213 *Dendropithecus* 214 *Micropithecus* 216 Craniodental Comparison of African Dryomorphs and Modern African Hominoids 217 Postcranial Adaptations of African Miocene Dryomorphs 220 *Dryopithecus* 227
Ramamorphs 229
Ramamorphs from Europe and Asia Minor 230 Ramamorphs from the Siwaliks (Pakistan and India) 233 Ramamorphs from China 236 Comparison of Asian Ramamorphs and Dryomorphs 239 Ramamorphs from Africa 240
Pliomorphs 242
Pliopithecus 243 Other Pliomorphs 246
Recently Discovered Miocene Hominoids 246
MIOCENE MONKEYS 248
Old World Miocene Monkeys (Cercopithecoidea) 249
New World Miocene Monkeys (Ceboidea) 253
DIETARY PREFERENCES, BEHAVIOR, AND HABITATS OF MIOCENE PRIMATES 255
PHYLOGENY AND CLASSIFICATION OF MIOCENE PRIMATES 256
Hominoidea 256
Simons and Pilbeam's (1965) Taxonomy 256 Theories in Light of Biomolecular Data and Recent Fossil Evidence 258 Phylogeny of the Ramamorphs 264
Cercopithecoidea and Ceboidea 267

6 Plio-Pleistocene Primates 270

PALEOCLIMATES AND BIOGEOGRAPHY 272
SUMMARY OF THE PLIO-PLEISTOCENE FOSSIL RECORD 273
South African Sites 275
Taung 277 Sterkfontein 280 Kromdraai 282 Makapansgat 284 Swartkrans 286 Summary of the South African Sites 288
East African Sites 288
Sites Older than 4 Million Years 289 Laetoli (Northern Tanzania) 290 Hadar (Ethiopia) 292 Omo Basin (Ethiopia) 294 Lake Turkana (Kenya) 297 Olduvai Gorge (Tanzania) 301 Summary of the East African Sites 304
AUSTRALOPITHECUS 305
General Craniodental Trends in Australopithecines 306
Gracile Australopithecines 308
A. afarensis 308 *A. africanus* 311
Robust Australopithecines 315
A. robustus 315 *A. boisei* 320
Comparison of Gracile and Robust Australopithecines 322
Body Size and Encephalization Quotients 324
HOMO 326
H. habilis 326 *H. erectus* 328 *H. sapiens* 328
MORPHOLOGY OF HOMINID LOCOMOTION 331
Biomechanical Principles of Bipedalism 332
Evidence of Bipedalism in Australopithecines and Archaic Humans 337
A. afarensis 337 Other Australopithecines 342 *Homo* 344 Summary of Trends in Hominid Pelvic Morphology 344
BEHAVIORAL AND CULTURAL TRENDS IN AUSTRALOPITHECINES AND ARCHAIC HUMANS 346
Behavioral Theories for the Evolution of Bipedalism 347
Beginnings of Culture 351
PHYLOGENY AND CLASSIFICATION OF EARLY HOMINIDS 357
Phylogenetic Relationships within *Australopithecus* 357
Phylogenetic Relationships within *Homo* 363
EPILOG 365

Appendices 367

A GEOLOGICAL TIME SCALE 367
B BASIC PRIMATE SKELETAL MORPHOLOGY 373

Cranial Morphology 373 Primate Dentition 378 Postcranial Adaptations 382

C CORRELATES OF FEEDING BEHAVIOR IN MODERN PRIMATES 390

D CLASSIFICATIONS OF THE ORDER PRIMATES 393

REFERENCES 404

CREDITS 439

GLOSSARY 447

INDEX 467

Preface

"The history of the world, my sweet, is who gets eaten and who gets to eat."
from *Sweeney Todd, the Demon Barber of Fleet Street*
by Stephen Sondheim

The story of primate evolution, as we know it in the later days of the twentieth century, begins humbly with small, innocuous quadrupedal creatures scampering across the nighttime forests of ancient continents, and ends with large-brained, ubiquitous bipedal creatures of the nuclear age of modern nation states. It is a story of dramatic discoveries, heated controversies, uncertain evidence, and few verifiable facts. By retelling the story, warts and all, to new students, it is my hope that some of them will share my enthusiasm for the subject, and future books will come to tell the story more fully. At the very least, readers of this book will emerge with an appreciation for the techniques and limitations of modern paleontological research.

While this book assumes some prior exposure to physical anthropology or biology, anyone with an interest in the biological and/or earth sciences can easily follow the major episodes in primate evolution outlined here simply by starting at the beginning. To bring the history of ancient primates to life, I have minimized the sometimes excessive focus on descriptive anatomy and systematics found in journal articles in favor of aspects of modern primate paleobiology that are more interesting to beginning students. These include the impact of recent studies in paleoclimatology, biogeography, and continental drift on our understanding of how extinct primate populations lived and dispersed across the globe, the functional morphology of the different extinct primate groups, and the use of allometric data in interpreting early primate lifestyles. Since the study of the behavior and morphology of living primates plays an important role in forming hypotheses about extinct primates, I have included many of these studies throughout the book where they seem most relevant. Naturally, no professional paleoanthropologist can get very far without a fairly detailed understanding of systematics and evolutionary theory, both of which are outlined in the first chapter and discussed and carefully illustrated in each subsequent chapter.

Because of the nature of its subject matter, paleoanthropology is both a science and an art. While its hypotheses may be falsifiable at any time by fresh data, few hypotheses concerning extinct organisms can be confirmed by direct experimental methods. For this reason, disagreements among knowledgeable anthropologists about the nature (but not the fact) of primate evolution abound. I have tried to convey this state of dynamic tension honestly without promoting my views to the exclusion of others. Every chapter clearly presents what I take to be the consensus (though by no means unanimous) view of the major evolutionary events, as well as the important opposing views. My hope is that students will grasp the major events in primate evolutionary history by the end of this book even though its views on specific details or systematic nuances may sometimes differ from those of their professors. This, then, is a book, not an encyclopedia.

The structure and content of this book is derived from lectures in primate evolution I have given over the past 15 years at New York University, Brown University, and Washington University. Chapter 1 gives students the conceptual framework from which to explore the fossils of the various epochs in primate history which follow. Special attention is given in Chapter 1 to developing an understanding of allometric relationships (especially the relationship between diet and dentition), theories of evolutionary change (especially punctuated equilibrium and phyletic gradualism), the geological time scale, radiometric dating techniques, and systematics (especially cladism and evolutionary classification). Perhaps of supreme importance, Chapter 1 outlines the ways in which paleoanthropologists make inferences from primate fossils, a subject that is returned to in virtually every chapter of the book.

Chapters 2 through 6 are arranged by epoch in chronological order beginning 65 MYA at the Paleocene and ending about 1 MYA in the middle of the Plio-Pleistocene. Within each chapter a standard order of topics is followed: paleoclimatology and biogeography, a morphological analysis of the fossils themselves, and finally, the biological inferences and phylogenetic hypotheses that can be drawn from the fossil evidence. The scope of discussion is the earliest primates to the earliest representatives of the *Homo* lineage. Consequently, the evolution of the genus *Homo* is treated only cursorily in Chapter 6, mostly to provide contrast to the developments evident in *Australopithecus*.

Several appendices containing basic information and summary illustrations on geological and paleomagnetic dating, skeletal morphology, correlations of ecological and biological factors in living primates, and systematics will be indispensable for beginning students. The last appendix on systematics provides several different classifications of the order

Primates proposed over the past two decades. I have decided to be less than dogmatic in adopting a particular classification scheme in the text in order to focus students' attention on the more interesting aspects of primate biology. It's my feeling that an obsession with classification can encourage an unfortunate "stamp collecting" mentality in students early on. The appendix is provided for the student's later reference when the issues of classification become more important.

An especially full glossary is provided as a handy aid to interpreting the sometimes overwhelming language of medical anatomy. All new terms are, of course, defined in the text where they are initially introduced.

A book like this is not written in a vacuum. I consider myself fortunate to have been a graduate student in what I would consider Yale's golden years of paleoanthropology, when at any one time you could walk down the hall and talk to David Pilbeam, Elwyn Simons, Rich Kay, Phil Gingerich, Ian Tattersall, Tom Bown, Ken Rose, or John Fleagle and probably get different opinions from each of them! I learned much from my faculty colleagues Cliff Jolly and Jim Shafland at New York University and from the indefatigable George Erikson at Brown University.

While writing a book is never easy, the task was made so much more pleasant by the congenial atmosphere provided by friends and colleagues in the Departments of Anthropology and Anatomy & Neurobiology here at Washington University. To them I am most grateful. I'm particularly indebted to Gerald Fischbach, Chairman of the Department of Anatomy & Neurobiology, for the support and encouragement he has extended to a physical anthropologist hunkered down in the midst of his department, and to Bob and Linda Sussman in the Anthropology Department, who convinced us to move to St. Louis in the first place.

A number of colleagues read and commented on extensive portions of the text. I am especially indebted to Tom Bown, Matt Cartmill, Laurie Godfrey, Cliff Jolly, Henry McHenry, and Randy Skelton for criticism and encouragement. I wish I could blame them for any errors of omission or commission in the text, but alas that responsibility is mine alone! Other colleagues graciously offered the use of photographs, including Peter Andrews, Eric Delson, Bill Kimbel, David Pilbeam, Ken Rose, Jeff Schwartz, Elywn Simons, Phillip Tobias, Alan Walker, and Tim White.

I must add to this list my gratitude to a number of people who labored with me over the production of the book. Susan Middleton's invaluable efforts brought rhyme and reason to what was sometimes an unwieldy manuscript. Mike Reingold's artistry beautifully captured the nuances of primate anatomy. Joe Wisnovsky revised early drafts of the manuscript, and my assistant Heather Trigg helped with a number of office tasks. And from beginning to end this project was encouraged and supported by my editor at W.W. Norton, Jim Jordan.

Lastly, and most importantly, is the loving support provided at the home front. Molly the wonder dog and Shubel, Jesse, and Chloe, the wonder cats, were always waiting at the door when I came home late at night. And to my wife and colleague, Jane Phillips-Conroy—as Elton John wrote, "how wonderful life is, while you're in the world."

Glenn C. Conroy
Washington University School of Medicine
St. Louis, Missouri
February, 1990

Primate Evolution

CHAPTER 1

Elements of Paleoprimatology

What Is a Primate?

Classification and Evolution

Fossil Ages

Making Inferences about the Biology of Extinct Primates

Origin of the Primates

Taxonomy Used in This Text

WHAT IS A PRIMATE?

Animals such as monkeys and apes have always held a special fascination for humans, presumably because of the many anatomical and behavioral similarities between them and us. Baboons, for example, were venerated by the ancient Egyptians, and langur monkeys have long been considered sacred by orthodox Hindus. The elusive lemurs of Madagascar are still believed by some natives of that island to be reincarnations of their ancestors, and in parts of southeast Asia tarsiers are similarly regarded as spirits.

Early Descriptions

The earliest recorded sighting of an African ape was by the Carthaginian navigator Hanno in the fifth century B.C. He returned to his home city from an exploratory voyage along the southwestern coast of Africa with the pelts of three "wild women" who had allegedly attacked his crew. Judging by his description, they were probably chimpanzees (Oikonomides, 1982).

The first systematic attempt to summarize what the ancient world knew of such creatures was made in the fourth century B.C. by Aristotle, who divided them into two groups, depending on whether they had a tail. (The same criterion was reinvoked 2,000 years later by the 18th-century French naturalist Georges Buffon to distinguish monkeys from apes.) Aristotle's observations were supplemented in the second century A.D. by the Roman physician Galen, whose dissections of the Barbary ape (actually a tailless macaque monkey) served as the basis for all anatomical learning in the medical schools of Europe until the 16th century, when Vesalius's meticulously illustrated descriptions of the human body were published (McCown and Kennedy, 1972).

As the pace of worldwide exploration quickened in the 17th century, evidence of other humanlike animals accumulated. In 1625 the English explorer Andrew Battell described a giant "pongo" (evidently a gorilla) he had seen while he was being held captive by the Portuguese in West Africa. The first detailed anatomical description of one of the great apes, a chimpanzee, was published in 1641 by the Dutch physician Claes Pieterzoon Tulp (the same Dr. Tulp depicted in Rembrandt's famous painting *The Anatomy Lesson*). And in 1658 another Dutch physician, Jakob de Bondt, reported the results of his dissection of an orangutan (a name derived from the Malay words meaning "man of the woods") (McCown and Kennedy, 1972).

Modern Definitions

Based on these findings and others, Carolus Linnaeus, the originator of the modern scientific system of classifying and naming organisms, put forward the first definition of the mammalian order **Primates** in the 10th edition of his great work *Systema Naturae* (1758). The members of the order, according to Linnaeus, are characterized by several distinctive anatomical features: four **incisors** (parallel cutting teeth at the front of the upper jaw; two **clavicles** (collarbones), two **mammae** (mammary glands) on the chest; and at least two extremities that function as hands in the sense of being able to grasp objects by means of an opposable first digit.

Linnaeus, working purely on the basis of anatomical comparisons, divided the order Primates into four genera, to which he gave the Latin names *Homo, Simia, Lemur,* and *Vespertilio.* In the genus *Homo* he placed not only humans but also the orangutan (naming it *Homo sylvestris* in a direct transliteration from the original Malay). Under *Simia* he listed the monkeys and the rest of the apes, including the chimpanzee. Under *Lemur* he grouped the lemurs and other "lower," less humanlike forms, and under *Vespertilio* he put various species of bats. In deciding to classify humans together with apes under the same taxonomic heading, he in effect challenged his fellow naturalists to show him a generic trait by which to distinguish human and ape, for he knew of none himself.

A succession of comparative anatomists in the 18th and 19th centuries rose to the challenge, devoting much effort and ingenuity to the task of devising **anthropometric** criteria that could convincingly distinguish human from ape. They pointed not only to measurable anatomical characters, such as bone structure and dentition, but also to qualitative traits, such as the human capacity for speech, reason, and other "higher" brain functions (Owen, 1858; McCown and Kennedy, 1972). None of their alternative classification schemes made much headway, however, and the Linnaean formulation stood more or less unchanged for more than a century.

The true evolutionary link between ape and human was finally recognized in the mid-19th century by Charles Darwin. Having established the general principles of his theory of natural selection in *The Origin of Species* (1859), he went on to apply them specifically to the question of human evolution in *The Descent of Man* (1871):

> It is notorious that man is constructed on the same generalized type or model with the other mammals. All the bones in his skeleton can be compared with the corresponding bones in a monkey, bat or seal. So it is with his muscles, nerves, blood vessels and internal viscera. The brain, the most important of all the organs, follows the same law. . . . It is, in short, scarcely possible to exaggerate the close correspondence in general structure, in the minute structure of tissues, in chemical composition and in constitution, between man and the higher animals, especially the anthropomorphous apes.

Darwin (1871) even speculated on the likely geographic site where primate evolution led to the emergence of the human species. He noted that

> in each great region of the world the living mammals are closely related to the extinct species of the same region. It is therefore probable that Africa was formerly inhabited by extinct apes closely allied to the gorilla and chimpanzee; and as these two species are now man's closest allies, it is somewhat more probable that our early progenitors lived on the African continent than elsewhere.

The first significant post-Darwinian modification of the Linnaean system of primate classification was proposed in 1873 by the English anatomist St. George Mivart. Removing bats and colugos ("flying lemurs") from the order, he reorganized the remaining members into two suborders: the primitive **Prosimii,** or "premonkeys" (lemurs, lorises, and the like), and the more advanced **Anthropoidea** (monkeys, apes, and humans).

Mivart (1873) also proposed an extended list of traits to further distinguish the primates from other mammals. All the members of the order, he wrote, were

> unguiculate [having nails or claws], claviculate [having clavicles] placental mammals, with orbits [eye sockets] encircled by bone, three kinds of teeth [incisors, canines, and molars], at least at one time of life; brain always with a posterior lobe and a calcarine fissure [a transverse groove along the medial surface of that lobe]; the innermost digit of at least one pair of extremities opposable; hallux [big toe] with a flat nail or none; a well-developed caecum [a pouchlike part of the large intestine]; penis pendulous; testes scrotal; always two pectoral mammae.

Although none of these characters, it turns out, is peculiar to primates, their combination has long been accepted as diagnostic of the order (Martin, 1968; Cartmill, 1974).

An alternative approach to the problem of defining the primates, advocated chiefly by the English anatomist Sir Wilfred Le Gros Clark several decades ago, sought to characterize the order in terms of a complex of evolutionary trends rather than a simple listing of morphological traits. According to this view, the distinctive evolutionary trends that set the early primates apart from the other placental mammals include progressive enlargement of the brain, convergence of the axes of vision, shortening of the snout, atrophy of the olfactory sense, prolongation of the postnatal growth period, and specializations of the extremities for grasping (Le Gros Clark, 1959). Most of these hypothesized trends, it was thought, were related to an **arboreal** (tree-living) way of life.

Recently, Robert Martin (1986*a*) of the University of Zurich has proposed a much more elaborate definition of the living primates that takes into account such diverse factors as geographic distribution, habitat, means of locomotion, influence of major sense organs on the shape of the skull, relative brain size, reproductive biology, and dental patterns. For example, he notes that living primates are typically arboreal animals living mainly in tropical and subtropical ecosystems (there are some obvious exceptions such as the more savanna-dwelling baboons of Africa and the temperate forest–dwelling macaques of Asia). Anatomical features of the hands and feet (**manus** and **pes**) are adapted for **prehension** (grasping). This is clearly evidenced in the foot by the widely divergent **hallux** (big toe) in all primates except humans. In addition, the digits have nails instead of claws, and these serve as supportive structures for the tactile cutaneous ridges on the fingertips that reduce slippage on arboreal supports.

In all modern primates the visual sense is emphasized over the olfactory sense. For this reason the eyes are usually relatively large, and the orbits are protected either by a **postorbital bar** (typical of lemurs and lorises) or by a complete bony cup referred to as **postorbital closure** (typical of monkeys, apes, and humans). The emphasis on vision has other anatomical consequences. For example, the orbits have become enlarged and have moved from a more lateral-facing to a more forward-facing position in the skull. This feature is associated with **binocular vision** (where both eyes focus on the target object, allowing the object to be perceived with greater depth perception). Another result of forward-facing orbits is exposure of the **ethmoid** bone on the medial orbital wall, although there are some exceptions to this in lemurs (Cartmill, 1978) (Fig. 1.1).

Compared with most other mammals the primate brain is enlarged relative to body size. Indeed, primates are unique among living mammals in that the brain constitutes a significantly larger proportion of body weight at all stages of gestation. Modern primates have long gestation periods relative to maternal body size, and both fetal and postnatal growth are characteristically slow in relation to maternal size. Consequently, sexual maturation is attained late, and lifespans are correspondingly long relative to body size. In sum, it takes longer for modern primate populations to reproduce themselves than is the case for populations of most other mammals (Martin, 1986*a*).

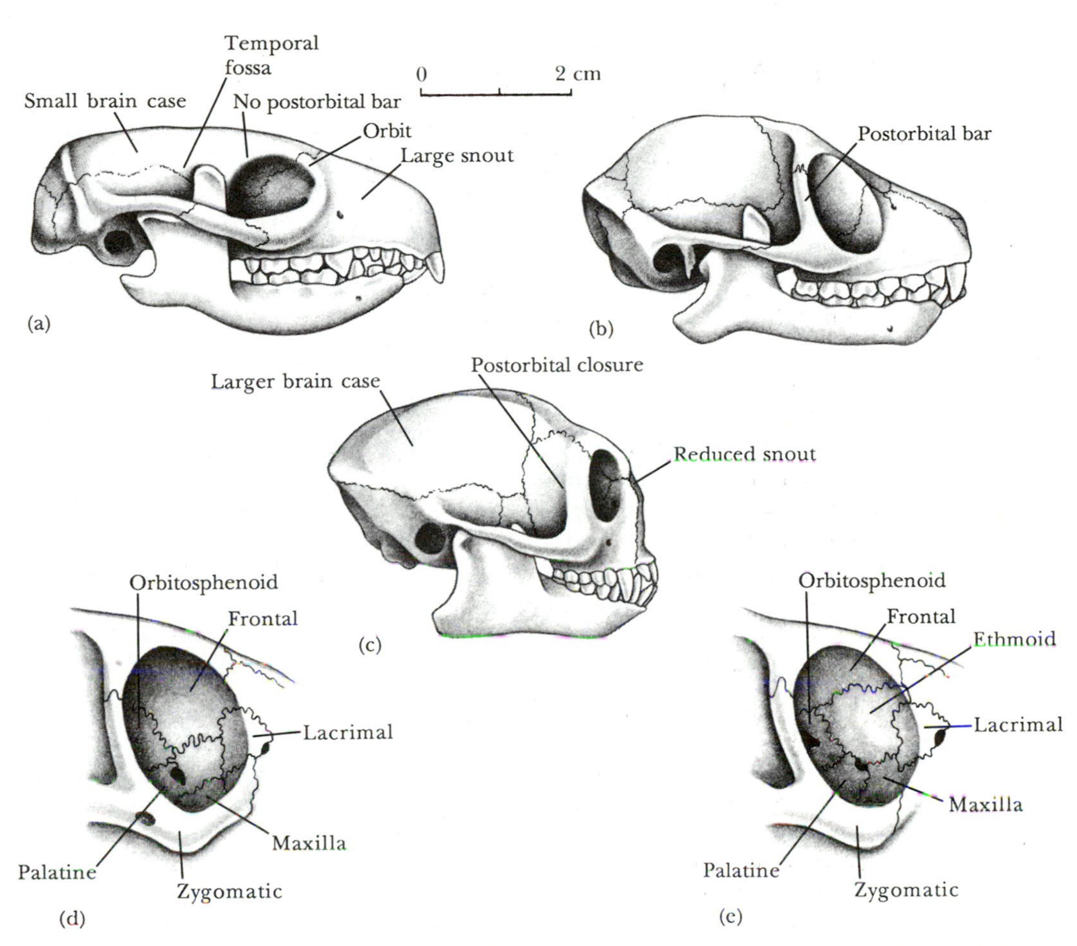

Fig. 1.1 Several Morphological Features Distinguishing Modern Primate and Nonprimate Skulls *(a) Insectivoran (hedgehog); (b) prosimian (Lepilemur);* (c) New World Monkey (*Callithrix*, or marmoset); (d) medial orbital wall showing the absence of the ethmoid typical of many nonprimate mammals and some lemurs (here *Lemur*); (e) medial orbital wall showing the presence of the ethmoid found in most primates (here *Loris*). Note relative size changes of brain and snout, development of the postorbital bar, and position of the eye sockets. Scale in (d,e) about 1½ × natural size. (a–c, Redrawn from Jolly and Plog, *Physical Anthropology and Archaeology,* McGraw-Hill, 1986, reproduced with permission; d,e, redrawn from Cartmill, 1978)

CLASSIFICATION AND EVOLUTION

Three Approaches to Classification

The classification of living and extinct primates is not as straightforward a procedure as it might appear. Implicit in any post-Darwinian system of classifying organisms is a set of underlying evolutionary principles, and the classification of a particular group of organisms depends on which organizing principle one adopts. At present there are three main schools of biological classification: evolutionary taxonomy, phenetic taxonomy, and cladism. In order to grasp some of the central issues in the study of primate evolution today, it is necessary to have a clear understanding of how the three schools differ from one another.

In the classical evolutionary approach to biological classification, called **evolutionary taxonomy** or evolutionary systematics, the observed similarities and differences among groups of organisms are evaluated in light of their presumed **phylogeny** (line of evolutionary descent). Because some **characters** are considered to be better clues than others to evolutionary relationships, they are given greater weight in the classification, a practice known as **character weighting.** (A **character** is simply one particular aspect of an organism chosen for study. The character may evolve through several **character states,** where each state is derived from the one preceding it.) The graphic presentation of this approach is a **phylogram,** or phylogenetic tree, which shows both the branching patterns of ancestor-descendant lineages and the degree of divergence of the descendant branches. In the classical evolutionary approach the taxonomist takes into account not only the branching pattern of phylogeny but also the subsequent evolutionary fate of each branch.

In an effort to avoid the subjectivity inherent in the practice of character weighting, practitioners of **phenetic taxonomy,** also called numerical taxonomy, classify organisms solely on the basis of overall morphological similarity. In the phenetic approach every character is assumed to have an equal weight and is assigned some numerical value. A mathematical operation is then performed on a set of such values to obtain an aggregate value for a particular organism. The resulting value for each organism is finally compared with those of the other organisms in the analysis to arrive at an objective measure of the overall similarity of the group (Sokal, 1974). Because this method does not distinguish between character states that in evolutionary terms are ancestral and those that are derived (see the following discussion), it is seldom used in primate paleontology, and it will not be discussed further here.

The third major school of biological classification, **cladism** (also called phylogenetic systematics), takes a radically different approach. It is concerned solely with the branching patterns of phylogeny, which reflect the distribution of certain characters within a **clade** (a group of related organisms). Members of a clade are so designated because they possess **shared derived characters** (characters they have in common with each other and with their most recent common ancestor) but not with earlier ancestors. To the cladist, phylogeny is nothing more than a sequence of dichotomies, representing at each fork the splitting of a parent **taxon** (taxonomic category) into two daughter taxa (also known as **sister groups**). Trichotomies and even higher multiple branching patterns are ultimately reduced to dichotomies in cladistic analysis (Nelson and Platnick, 1980). In its strictest form, cladism demands that the sister groups have the same **taxonomic rank** (genus, family, and so forth up to phylum) and that the parent taxon ceases to exist after splitting into its two daughter taxa.

Cladistic analysis is represented graphically by branching diagrams known as **cladograms,** or cladistic trees. In the cladogram shown in Fig. 1.2, for example, taxa B and C share a more recent common ancestor than either does with taxon A. Thus B is the sister group of C, and both would have the same taxonomic rank. Similarly, taxa B and C would jointly constitute the sister group of taxon A, and the taxonomic rank of A would have to be the same as that of B and C.

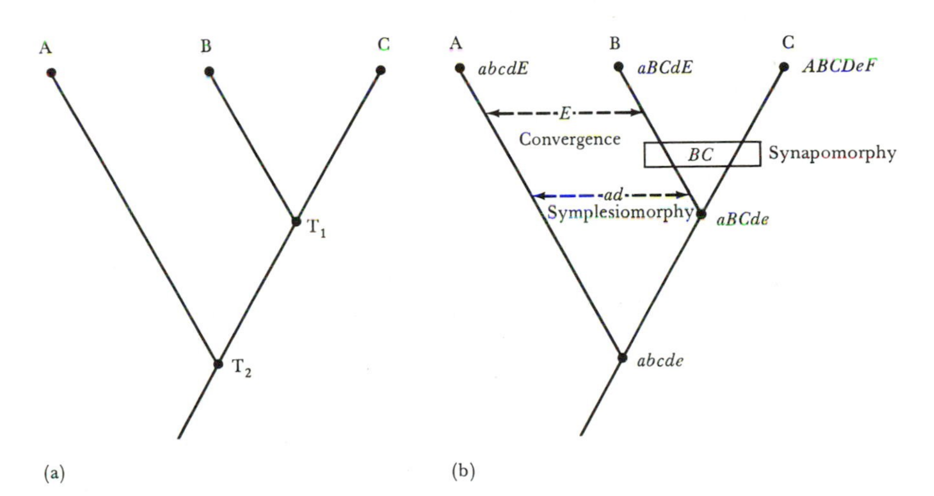

Fig. 1.2 A Cladogram (a) In this simple cladogram, hypothetical taxa B and C are more closely related to one another than either is to taxon A since they share a more recent common ancestor (T_1). Thus taxa B and C together comprise a clade that is the sister group of taxon A. (b) Characters observed in three taxa and inferred in their common ancestors are given beside each group. Lowercase letters signify primitive characters, and capital letters signify derived states of those same characters. Taxa B and C are linked by synapomorphic character state *BC*. Characters *a* and *d* linking taxa A and B are shared primitive (symplesiomorphic) features and thus are not indicative of any special relationship between those two taxa. Character *f* is an autapomorphic character in taxon *C*. (Adapted from Patterson, 1987)

The cladograms in Fig. 1.2 signify that taxa A, B, and C all share derived characters and that B and C have other derived characters in common with each other but not with A (Fig. 1.2b). The phylogenetic implication of these cladograms is simply that B and C share a more recent common ancestor than either one does with A; they carry no further information beyond that simple deduction (Tattersall and Eldredge, 1977; Schwartz et al. 1978).

Character Analysis

The first step in cladistic analysis, as in any other classification scheme, is to cluster together those species that seem morphologically similar to one another. Then it is necessary to categorize the characters that define the similarity into ancestral **(plesiomorphic),** derived **(apomorphic),** or **convergent** states.

More specifically, two groups of organisms may resemble each other in a given character state for any of three reasons:

1. The similar character existed in the ancestry of the two groups before the evolution of their nearest common ancestor. This is said to be a **symplesiomorphic** (shared primitive) character. In Fig. 1.2b, *a* and *d* represent symplesiomorphic characters.
2. The similar character originated in their common ancestor and is shared by all of that ancestor's descendants. Here the character is said to be a **synapomorphic,** or shared derived, character. In Fig. 1.2b, *B* and *C* represent synapomorphic characters. Note that *B* and *C* are *not* present in the ancestor of all three taxa A, B, and C, but only in the immediate ancestor to B and C.
3. The similar character arose independently in several descendant groups. In this case it is said to be a **convergent** character.

A similarity between two species of form or structure shared with their common ancestor is termed **homologous**. An example of a homologous structure is the pentadactyl limb of chimpanzees and humans. Conversely, a similarity between two species of form or structure not shared with their nearest common ancestor is termed **analogous;** that is, it has evolved independently in the two species due to convergent evolution. The wings of birds and bats are analogous structures.

Two groups may differ from each other because one possesses a character that arose for the first time in that group. Such a derived character is called an **autapomorph.** Autapomorphic characters are acquired by a phylogenetic line after it has branched off from its sister group. In Fig. 1.2b, the character *F* is an autapomorph.

In Fig. 1.2, taxa B and C comprise a **monophyletic group.** Monophyletic groups are characterized by synapomorphies (shared derived features) and contain only species that are more closely related to each other than to any species classified outside the group (in this case, taxon A).

One of the main problems in character analysis arises in the case of a **polymorphic** character, one that changes gradually over a **morphocline** (continuum of forms) from one species to another. The problem is to determine the polarity of the morphocline; in other words, which end is primitive and which end is derived? For example, the mammalian forefoot can have anywhere from one digit (in horses) to five digits (in humans). Which of the two conditions is actually the primitive mammalian one?

There are two ways of dealing with this problem. If a given morphological character state is typical of a large number of taxa, and particularly if the state is shared with other closely related taxa of similar rank, then it is reasonable to regard that state as primitive. The alternative is to rely on **ontogenetic** (developmental) data. Although the old notion of ontogeny recapitulating phylogeny is no longer accepted in the literal sense of the terms, morphological characters typical of an early stage of an organism's development are still generally considered more primitive than features appearing at a later stage (Tattersall and Eldredge, 1977). Thus in the case of the mammalian forefoot, five digits are regarded as the primitive condition and any reduction in this number is considered a derived state.

Comparison of Cladism and Evolutionary Taxonomy

Both evolutionary taxonomists and cladists would agree that the analyses of morphological characters discussed in the previous section must be taken into account before a classification can be accepted. Cladists agree with evolutionary taxonomists that some morphological characters are better than others as clues to a biological classification, but there is disagreement over individual characters.

Strictly speaking, cladists do not weight characters; they simply discard from analysis those characters that are not synapomorphies. For example, among prosimians the possession of a shared derived trait, such as a dental comb, is thought to have more taxonomic significance than the possession of a shared primitive trait such as a comparatively small brain. (A **dental comb** is a specialized alignment of the lower canines and incisors for fur grooming; see Fig. 1.3.) In this view the common possession of derived characters proves the common ancestry of a given group of species, whereas the common possession of primitive characters has little if any taxonomic value.

A strict cladist would classify crocodiles and birds, say, as sister groups despite their obvious morphological differences, because they share characters derived from a common reptilian ancestor, even though they subsequently took very different evolutionary pathways. They would also have to be given the same taxonomic rank, even though one sister group (crocodiles) still looks very much like the ancestral group, whereas the

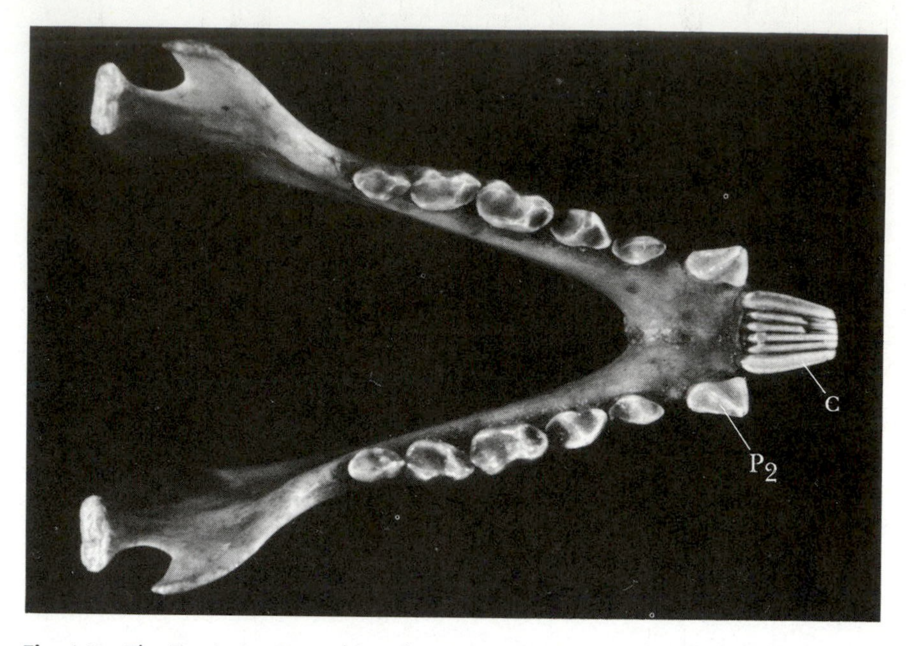

Fig. 1.3 The Prosimian Dental Comb Lower jaw of the prosimian *Varecia variegatus.* Normally made up of the lower four incisors and the two canines, the prosimian dental comb is an adaptation for fur grooming. Since the canines become functional incisors, the first premolars (in this case P_2) become the functional canines. Scale about 1.4 × natural size. Courtesy of J. Schwartz)

other sister group (birds) obviously does not. An evolutionary taxonomist, on the other hand, would take into account the different evolutionary history of each group. Accordingly, the two groups would be assigned to different taxonomic ranks, crocodiles classified as an *order* of the class Reptilia and birds as the altogether separate *class* Aves. In general, the evolutionary taxonomist would hold that only the fossil record can be relied on to determine whether an inferred primitive group is the true one or not. As the eminent paleontologist George Gaylord Simpson once noted, primitiveness and ancientness are not necessarily related, but they usually are (Simpson, 1975).

These contrasting classifications between cladists and evolutionary systematists are illustrated by an example from primate evolution (Fig. 1.4). A cladist would interpret Fig. 1.4a to mean that humans, chimpanzees, and gorillas *(Homo, Pan,* and *Gorilla)* share a more recent common ancestor with one another than they do with orangutans *(Pongo).* Thus *Homo* forms a clade that is the sister group of the clade formed by *Pan* and *Gorilla.* These four genera would in turn comprise the sister group of a third clade consisting of gibbons *(Hylobates)* and siamangs *(Symphalangus).* The classification would reflect this branching pattern. For example, if orangutans are classified at the family level (as Pongidae), then humans, chimpanzees, and gorillas must also be classified together

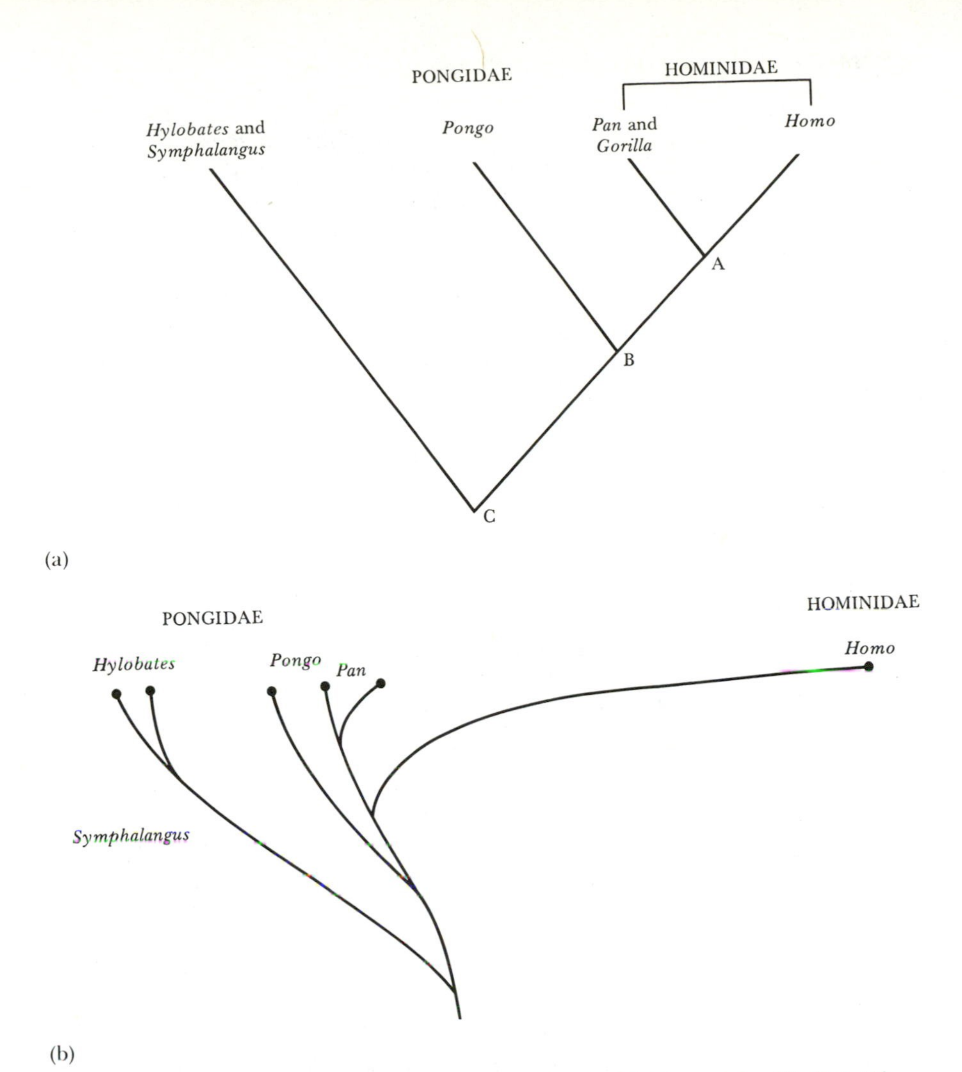

Fig. 1.4 Contrasting Classifications of Living Hominoid Primates by Cladists (Phylogenetic Systematists) and Evolutionary Systematists (a) Cladists classify groups solely according to the branching sequence of phylogeny, and sister groups are given the same taxonomic rank. For example, if *Pongo* is placed in the separate family Pongidae, its sister group (comprised of *Pan, Gorilla,* and *Homo*) must also be included in a separate family, Hominidae. (b) Evolutionary systematists take the same phylogenetic information but classify the taxa differently, by taking into account the unique morphological and behavioral features of *Homo* compared with those of the other great apes. Thus *Homo* is placed in its own family, the Hominidae, and the great apes are placed in a separate family, the Pongidae. (Adapted from Simpson, 1963)

at the family level (Hominidae), since sister groups must have the same taxonomic rank.

An evolutionary systematist could derive a very different classification from the same cladistic relationships depending on how much emphasis is placed on autapomorphic characters. For example, the strict cladist would classify chimpanzees and humans as sister groups and accord them the same taxonomic rank even though chimpanzees still presumably look much like the ancestral group while humans do not. In other words, since their divergence from chimpanzees, humans as a lineage have developed an extensive suite of derived features (autapomorphs)—large brains, bipedal posture, delayed maturation, and elaborate tool use, among others—which is reflected in the classification of the evolutionary systematist (Fig. 1.4b). The result is that all the apes are classified in the family Pongidae, and humans are classified in the family Hominidae, even though chimpanzees share a more recent common ancestor with humans than with some of the other apes such as orangutans and gibbons. In short, evolutionary classifications reflect the adaptive and morphological attributes of a taxon rather than their strict geneological relationships.

Ernst Mayr (1981) has succinctly summed up the distinction between the two taxonomic schools:

> The main difference between cladists and evolutionary taxonomists is in the treatment of autapomorph characters. Instead of automatically giving sister groups the same rank, the evolutionary taxonomist ranks them by considering the relative weight of their autapomorphies as compared to their synapomorphies. For instance, one of the striking autapomorphies of man (in comparison to his sister group, the chimpanzee) is the possession of Broca's center in the brain, a character that is closely correlated with man's speaking ability. This single character is for most taxonomists of greater weight than various synapomorphous similarities or even identities in man and the apes in certain macromolecules such as hemoglobin or cytochrome C. The particular importance of autapomorphies is that they reflect the occupation of new niches and new adaptive zones that may have greater biological significance than synapomorphies in some of the standard molecules.

Even though humans may actually be very similar to apes in certain molecular characters, such as in overall protein and DNA structure, the

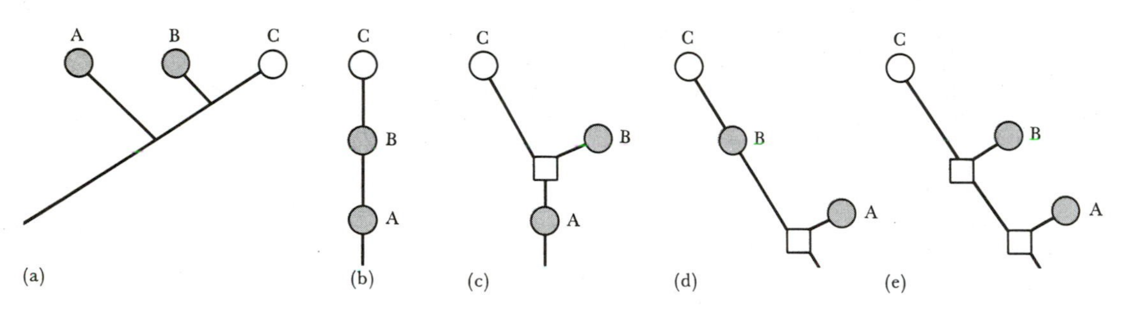

Fig. 1.5 Cladograms and Phylograms Compared for Information Content Open circles represent recent taxa, gray circles represent fossil taxa, and squares represent postulated common ancestors. The four phylograms (b–e) are all consistent with the given cladogram (a) and provide more information about hypothesized ancestor–descendant relationships than does the cladogram. (From Szalay, 1977)

two groups differ so much in mental capacity that no less an authority than Julian Huxley once proposed that humans should be placed in a separate kingdom—Psychozoa.

In graphic terms the main difference between a cladogram and a phylogram is that the former has no time dimension. A phylogram adds to the information contained in a cladogram by specifying not only the nature of the evolutionary relationships postulated but also the temporal sequence of the taxa, provided that reliable information is available on the age of the fossil-bearing rock layers (Tattersall and Eldredge, 1977). The comparative information content of cladograms and phylograms is illustrated in Fig. 1.5.

Two Models of Speciation

Besides the issue of classification, another important goal of paleontology is to obtain evidence concerning what are known as the mode and tempo of **speciation;** in other words, exactly how and at what rate are new species formed? Neither attribute of the evolutionary process readily reveals itself to the paleontologist. As Simpson (1953) once put it,

> Looking more closely into the pattern of evolution, we see that it involves also the organic changes that have occurred in the [branching] sequence and the rates at which these changes have occurred. Trying to see how these arose, were transmitted, and became what they did, we find ourselves grappling finally with every factor and element that is in life or that affects life.

There are at present two principal models to account for the mode and tempo of speciation. The older one, sometimes referred to as **phyletic gradualism,** holds that a daughter species usually originates through a series of small, gradual transformations of a parental species. There are four main attributes of this model:

1. New species arise by the gradual modification of an ancestral population.
2. The transformation is uniform and slow.
3. The transformation usually involves all or at least most of the ancestral population.
4. The transformation takes place over all or at least a large part of the ancestral population's geographic range.

The alternative model, called **punctuated equilibrium** by its originators, Niles Eldredge of the American Museum of Natural History and Stephen Jay Gould of Harvard University, holds that comparatively rapid speciation is much more important than phyletic gradualism as a mode of evolutionary change (Eldredge and Gould, 1972). The punctuated equilibrium model is based on the assumption that the gaps commonly found in the fossil record are attributable to comparatively rapid speciation in small isolated populations, a process known as **allopatric specia-**

tion. Evolution in such populations would appear to be essentially instantaneous in terms of geologic time. A species that originated as a small population and then spread to invade the territory of another population (often that of its parent species) would appear to have arisen abruptly in the fossil record.

What is novel about the punctuated equilibrium model is the premise that most evolutionary change is concentrated in rapid speciation events and that little change occurs before or after the speciation event itself. There are five main attributes of this model:

1. New species arise from the splitting of existing lineages.
2. New species develop rapidly.
3. A small subpopulation of the ancestral species gives rise to the new species.
4. The new species originates in a very small, isolated part of the geographic range of the ancestral species.
5. Species do not change throughout much of their history. (Levinton and Simon, 1980)

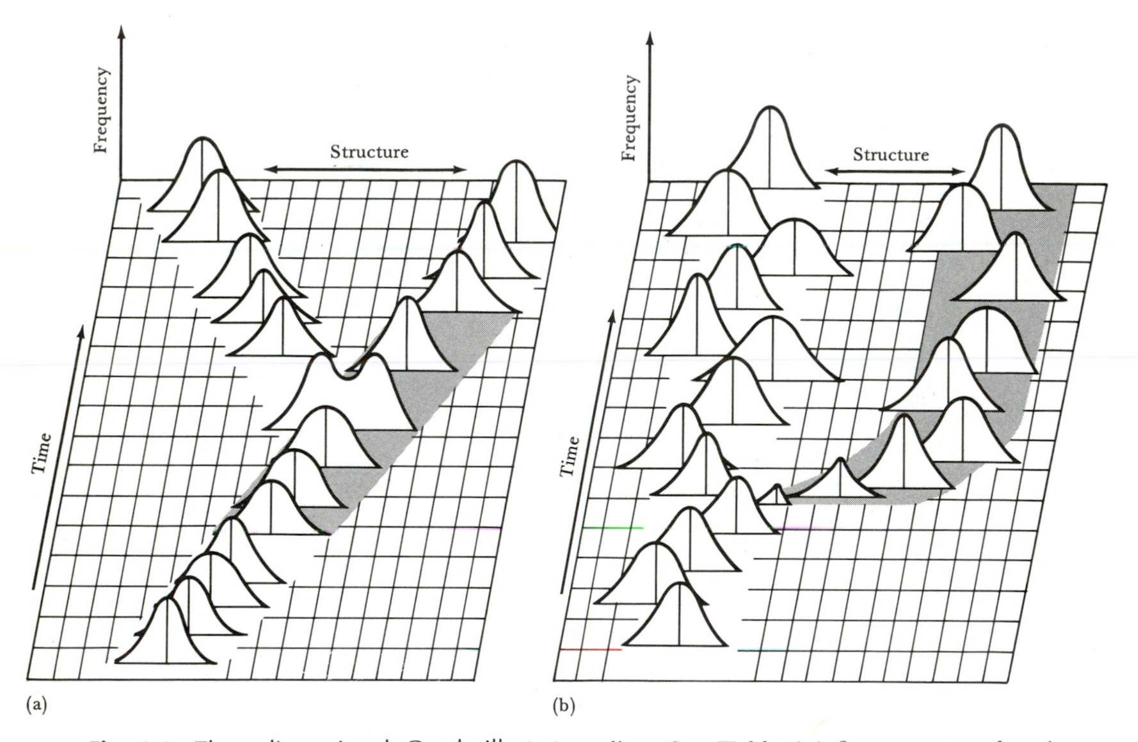

Fig. 1.6 Three-dimensional Graph Illustrating Major Differences between (a) Phyletic Gradualism and (b) Punctuated Equilibrium. Different species are denoted by stippling. See Table 1.1 for summary description of the differences. (Adapted from Vrba, 1980*a*)

Table 1.1 Summary of Differences between Phyletic Gradualism and Punctuated Equilibrium

	Phyletic Gradualism	*Punctuated Equilibrium*
1. Rate of phenotypic change	Uniformly low; no rate increase with splitting	High during speciation; low to absent within species
2. Direction of phenotypic change within a species	Undirectional; evolution is phyletic	Oscillates about a stable mean
3. Population size for speciation	Small or large; can occur in whole species	Only in small isolated populations
4. Whether speciation is purely a function of time	Yes, speeded up by environmental change but does not require it	No, requires environmental change
5. How new species arise	By phyletic speciation or by allopatric speciation in large or small populations	Usually only by allopatric speciation in small, isolated populations
6. Implications for species	Species are arbitrary subdivisions on a continually evolving lineage	Species are real, discrete entities with beginnings and terminations

Source: Adapted from Vrba (1980*a*).

The two models of speciation have important implications for practicing paleontologists. For example, a phyletic gradualist might interpret a morphological discontinuity in a fossil lineage as merely a gap in the fossil record, perhaps using it as a convenient point to divide up the fossil lineage into different species. A punctuationist, on the other hand, would regard the same fossil data as a reflection of the normal evolutionary process: long periods of morphological stasis interrupted by rapid speciation events. Thus, what is no more than missing data to one worker is a critical evolutionary event to the other.

Gradualists, in emphasizing the importance of steady morphological change within fossil lineages, interpret speciation as a special case of phyletic evolution. They see species as arbitrary units, conveniently defined by gaps in the fossil record. Punctuationists, in arguing that morphological change is dominated by abrupt speciation events, see species as discrete units having a beginning and an end in time. Although proponents of the two models agree that both slow and rapid changes are manifested in the fossil record, they interpret the data differently. The major distinctions between the two models are summarized in Fig. 1.6 and Table 1.1.

Most evolutionary biologists today consider the two models of speciation to be at opposite ends of a continuum of possibilities, with some lineages tending toward one end and some toward the other in their pattern of speciation.

Limitations in Taxonomy

In working out plausible classifications and phylogenies of fossil primates, the paleontologist has two basic concerns to keep in mind. First, there is little convincing evidence that fossil primate taxonomy bears any necessary relation to biological reality. For example, there have been few adequate studies of dental variability within natural populations of **extant** (currently existing) species of nonhuman primates, and yet this kind of study can be the only appropriate standard by which to judge dental variability in fossil primates (Phillips, 1978; Phillips-Conroy and Jolly, 1981). The paleontological literature is replete with new species names based solely on dental traits that have never been shown to have any taxonomic significance in living primate populations. The problem of taxonomic placement is difficult for many reasons, not least of which is the lack of taxonomic standards in paleontology that are biologically relevant and universally applied. It is sometimes difficult to distinguish between modern primate species even when such attributes as ecology, behavior, dentition, blood, saliva, palm prints, hair samples, and whole skeletons are known. One cannot help but wonder, then, whether the obsession with taxonomic minutiae in paleontology is a biologically meaningful endeavor or not. Some paleontologists have even gone so far as to suggest that evidence based on soft-tissue anatomy and molecular biology should be ignored in working out phylogenetic relationships since evidence of this kind cannot be ascertained in fossils. Thus one can legitimately question whether phylogenetic conclusions based solely on strict paleontological criteria necessarily have any biological validity.

The second concern involves the preoccupation of many paleoprimatologists with documenting direct fossil lineages. When one considers that all the known primates from Tertiary deposits in Africa have been found within a total geographic area representing only 1/3,000,000 of the present-day land surface of that continent, it stretches credibility to assert that any direct ancestors of later apes, monkeys, or humans have yet been found. However, representatives of various grades of primate evolution have been discovered. A **grade** is a group of species defined by a shared level of adaptation (either primitive or derived) that distinguishes them from other species but does not necessarily define a clade (Ridley, 1986). Recall that a clade is a monophyletic group of species, all of which are descended from a single common ancestor. Fossil primates that show evidence of having attained, for example, the anthropoid grade of evolution, can be considered broadly ancestral to later undisputed anthropoids without necessarily being directly ancestral to them.

The phrase "broadly ancestral" simply refers to the fact that the fossil record samples only a small part of a total population pool. All members of this pool share a basically similar morphological grade from which later forms evolved. The fossils provide a glimpse of this morphological pool without necessarily representing the direct ancestors. For example, Mayr (1981) has stressed that it is not necessary to know the exact species that was the common ancestor of later phyletic lineages. As he states,

> It is of little importance whether *Archaeopteryx* was the first real ancestor of modern birds or some other similar species or genus. What is important to know is whether birds evolved from lizard-like, crocodile-like, or dinosaur-like ancestors. If a reasonably good fossil record is available, it is usually possible, by the backward tracing of evolutionary trends and by the backward projection of divergent phyletic lines, to reconstruct a reasonably convincing facsimile of the representative of a phyletic line at an earlier time.

It is the objective of primate paleontology, and of this book, to construct the most plausible facsimile of primate phylogeny from the available evidence. I will present the more widely accepted scenarios for primate evolution, as well as mention some of the dissenting views.

FOSSILS AGES

Geologic Time Scale

Fossils have been recognized since the time of the ancient Greeks as relics of extinct plants and animals. It was not until the 19th century, however, that geologists were able to demonstrate that the earth was considerably older than anyone had ever realized, and that fossils were in fact documents of a major part of the planet's history. The discovery of the great age of the earth, one of the most significant discoveries in the history of science, was essential to the development of Darwin's theory, since his proposed evolutionary mechanism, natural selection, requires vast amounts of geologic time to operate.

The best current estimate of the age of the earth is about 4.6 billion years. To appreciate the immensity of this number, suppose all geologic time is compressed into one calendar year. On that scale the oldest rock formations on the earth date from mid-March. Living things first appeared in the sea sometime in May. Land plants and animals emerged in late November, and the enormous swamps that formed the worldwide Pennsylvanian coal deposits flourished for four days in early December. Dinosaurs became dominant in mid-December, but they disappeared abruptly on December 26, making way for the Age of Mammals and the emergence of the earliest primates. More advanced, humanlike primates—primitive species of the genus *Homo*—first appeared during the evening of December 31. The most recent continental ice sheets began to recede from the Great Lakes and northern Europe approximately one minute and 15 seconds before midnight. Rome ruled the Western World for five seconds, from 11:59:45 to 11:59:50. Columbus discovered America three seconds before midnight.

The last five days on this scale—roughly the period spanned by primate evolution—correspond in actual time to some 65 million years. Geologists refer to this interval as the **Cenozoic era,** which in turn is subdivided into two **periods,** the **Tertiary** and the **Quaternary.** The principal subdivisions of the Tertiary period are the **Paleocene, Eocene, Oligocene, Miocene,** and **Pliocene epochs.** The **Pleistocene** is the main epoch of the Quaternary. Rocks formed during the first three epochs of the Tertiary (Paleocene, Eocene, and Oligocene) are referred to collectively as the **Paleogene system,** whereas rocks formed during the last two epochs of the Tertiary (Miocene and Pliocene) correspond to the **Neogene system.** Thus the beginning of the Paleocene epoch coincided not only with the advent of the Cenozoic era but also with the start of the Tertiary period and with the initial formation of rocks of the Paleogene system.

This text follows the same chronological sequence, reviewing in successive chapters the major events in primate evolution as revealed in the fossil record from the origins of the primates in the Paleocene through the emergence of hominids in the late Pliocene and early Pleistocene (**Plio-Pleistocene** for short) (see Table 1.2). The aim throughout will be not only to describe and classify some important representatives of the more than 250 known species of extinct primates but also to explore their likely interaction with their environment, paying particular attention to the impact of large-scale climatic and geographic changes on early primate lifestyles and biogeography.

Radiometric Dating

Radiometric dating techniques are key tools in the study of primate paleontology. All such techniques are based on the fact that when the atoms of a radioactive element emit radiation, they decay at a known rate into atoms of another element. By measuring the ratio of parent atoms to daughter atoms in a mineral sample, one can accurately determine the age of the sample. Ideally, when the mineral grains containing the radioactive atoms first crystallize as part of a newly formed rock, they contain none of the daughter atoms, so that the initial daughter–parent ratio is zero.

In 1896 a French physicist, Henri Becquerel, made the serendipitous discovery that uranium emitted mysterious rays that activated photographic plates he had kept wrapped in black paper in a closed drawer. This strange property was named radioactivity by Madame Curie after she discovered the same phenomenon in thorium. Within 10 years of Becquerel's initial discovery, radioactivity was discovered to occur also in radium, rubidium, and potassium.

As early as 1902 the English physicists Ernest Rutherford and Frederick Soddy postulated that radioactive elements actually decayed into other elements during emission of radioactive rays (such as uranium decaying into lead or potassium decaying into argon), and in 1906 Rutherford made the first attempt to measure the ages of minerals from radioactive decay of uranium.

An American chemist, B. Boltwood, proved in 1907 that in unaltered minerals of the same age the lead–uranium ratio was constant, whereas this ratio was different in minerals of different ages: the older the mineral the greater this ratio. This observation formed the basis for most of the subsequent radiometric dating techniques used to date minerals.

Principles of Radioactive Decay In order to understand radiometric dating, we must review some simple concepts in chemistry. Many atoms are unstable and change spontaneously to a lower energy state by radioactive emission. Two atoms that have the same number of protons but a different number of neutrons are called **isotopes.** For example, two common isotopes of uranium (proton number 92) are uranium-235 and uranium-238. (The numbers following the element give the mass number, or total of protons and neutrons. Another way to write these isotopes is $^{235}_{92}U$ and $^{238}_{92}U$.)

Radioactive decay occurs through one or more of the following three processes: alpha decay, beta decay, or electron capture. In **alpha decay** the nucleus of the parent atom loses 2 protons and 2 neutrons; thus the mass number decreases by 4 and the proton number decreases by 2. An example of alpha decay is the decay of uranium to thorium: $^{238}_{92}U \rightarrow ^{234}_{90}Th$.

In beta decay, one of the neutrons in the nucleus turns into a proton; thus the proton number increases by 1 but the mass number remains unchanged. An example of beta decay is the decay of rubidium-87 to strontium, a process used widely in radiometric dating. The nucleus of ^{87}Rb loses a beta particle, which means that a neutron in the nucleus becomes a proton. The mass number remains the same but the atomic number increases from 37 in rubidium to 38 in strontium.

In electron capture, a proton in the nucleus turns into a neutron, thus decreasing the proton number by 1. Again the mass remains unchanged. An example of electron capture is the decay of potassium to argon: $^{40}_{19}K \rightarrow ^{40}_{18}Ar$

Each radioactive element has one particular mode of decay and its unique constant decay rate. Radioactive decay takes place entirely within the atomic nucleus. Thus, if a radioactive element is incorporated into a mineral or rock when it crystallizes (forms), the amount of the radioactive element that decays into its daughter atoms is controlled only by the elapsed time since that crystallization event.

The principle of radiometric dating is often compared to watching the passage of time in an hourglass. When the glass is turned over, sand

Table 1.2 Overview of Primate Evolution

Time (MYA)				Place		
Era	Period	System	Epoch	Africa	Europe	Asia
Cenozoic (65 MYA to present)	Quaternary (1.8 MYA to present)	Quaternary (1.8 MYA to present)	Present Recent 12,000 yr	ESSENTIALLY MODERN FAUNA		
			Pleistocene 1.8 MYA	*Papio* (baboons) replaces *Theropithecus* *Homo* evolves	Cercopithecines present Colobines disappear	Cercopithecines and colobines present
	Tertiary (65–1.8 MYA)	Neogene (22–1.8 MYA)	Pliocene 5 MYA	*Theropithecus* baboon widespread Early hominids appear in east and south	Colobines and cercopithecines present	Colobines and cercopithecines present
			Miocene 22 MYA	Fossils resembling modern prosimians rare but present Colobine monkeys at least 9 MYA Ramamorphs present by 17 MYA Dryomorphs (primitive "apes") appear about 22 MYA	Colobines spread from Africa Ramamorphs present 7–13 MYA Dryomorphs appear about 16 MYA Pliomorphs appear about 16 MYA	Colobines spread from Europe and Africa Ramamorphs present 7–13 MYA Dryomorphs appear about 16 MYA
		Paleogene (65–22 MYA)	Oligocene 38 MYA	Early Old World higher primates (catarrhines) from the Fayum of Egypt	Primates disappear	Primates unknown
			Eocene 55 MYA	A few fragments of possible primates	Adapids and omomyids abundant	Adapids and omomyids present Possible primitive anthropoids in Burma 40 MYA
			Paleocene 65 MYA	No fossil record	Plesiadapiforms present but rare	Plesiadapiforms present but rare

Source: Adapted from Richard (1985).

Table 1.2 (continued)

Time (MYA)				*Place*	
Era	*Period*	*System*	*Epoch*	*North America*	*South America*
Cenozoic (65 MYA to present)	Quaternary (1.8 MYA to present)	Quaternary (1.8 MYA to present)	Present Recent 12,000 yr	ESSENTIALLY MODERN FAUNA	
			Pleistocene 1.8 MYA	No primates	New World monkeys
	Tertiary (65–1.8 MYA)	Neogene (22–1.8 MYA)	Pliocene 5 MYA	No primates	New World monkeys present
			Miocene 22 MYA	No primates	Fragmentary remains tentatively assigned to lineages of living New World monkeys
		Paleogene (65–22 MYA)	Oligocene 38 MYA	Almost all primates disappear	First appearance of primates late in epoch
			Eocene 55 MYA	Adapids and omomyids abundant	Primates not present
			Paleocene 65 MYA	Plesiadapiforms present but rare	No primates

flows from the top to the bottom chamber. The amount of sand in the top chamber relative to the amount accumulated in the bottom one provides a measure of the time that has elapsed. The sand in the top of the hourglass represents decaying radioactive atoms, and the sand in the bottom represents accumulating daughter atoms produced by radioactive decay. Just as the hourglass must be sealed so that extraneous sand cannot leave or enter the system, so too must the atomic lattice structure of the mineral be able to hold both the parent and the daughter atoms without allowing any to escape or enter from an external source (Eicher, 1976).

Each individual atom of a given radioactive isotope has the same probability of decaying within any given year, and this probability remains the same no matter how long the material being dated has been in existence. This probability of decay is termed the **decay constant** *(X),* which simply stipulates the proportion of atoms of that particular element that always decays within any given year. The actual number of atoms that will decay is XN, where N is the number of radioactive parent atoms still present in the system at the beginning of the year. At the beginning of the next year, the number of radioactive parent atoms is obviously smaller, having decreased by XN. The number of radioactive parent atoms decreases with each succeeding year. The time required for half of the atoms of a particular radioactive element to decay is termed its **half-life.** Each radioactive element has its own specific half-life; some have a duration of microseconds and others trillions of years.

The end of one half-life marks the beginning of the next one. Thus, if we let N_0 represent the initial number of atoms, then half that number ($N_0/2$) remain after the first half-life, half of those ($N_0/4$) remain after the end of the second half-life, half of those ($N_0/8$) are left after the third half-life, and so on. If one plots the number of surviving parent atoms as a function of time, the result is a curve like that shown in Fig. 1.7. This relationship is the basis of all radioactive clocks.

As a simple illustration of the technique, suppose one is asked to determine the age of a sample in which the parent–daughter ratio is 1:32 and the half-life of the element is 1 million years. Since 1:32 equals $(\frac{1}{2})^5$, this would signify that the sample has gone through five half-life cycles and is thus 5 million years old.

Dating Methods The fossilized remains of vertebrate animals, including primates, cannot themselves be radiometrically dated, unless they happen to be younger than about 50,000 years old, the outside limit of carbon-14 dating. Since all the fossil primates we will be discussing are millions of years old, carbon-14 dating is not applicable. Consequently, for older Tertiary fossils the usual practice is to measure the age of sediments that can be related stratigraphically to the fossil in question. Ideally one can date the actual layer of sediment the fossil is found in; failing that, one

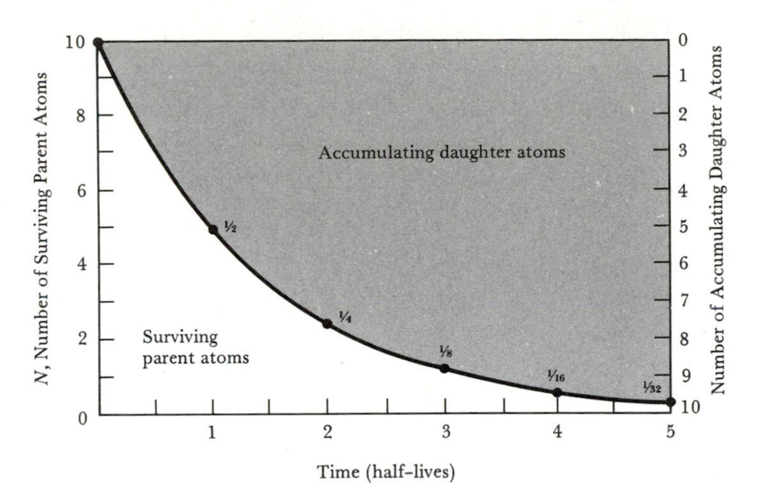

Fig. 1.7 Radioactive Decay Curve The number of parent atoms present when the rock was formed, N_0, is arbitrarily taken here to be 10 units. The fractions along the curve give N/N_0, the ratio of surviving parent atoms to the initial number of parent atoms. Using the right-hand vertical axis, one can also read off the number of daughter atoms present at any given time.

can bracket the fossil by dating the layers of sediment above and below the fossil layer in order to provide minimum and maximum ages.

By far the most important radiometric technique for determining the age of Cenozoic rock samples (that is, samples ranging in age from 65 MYA to the present) is **potassium–argon (K/A) dating,** which relies on the decay of the naturally occurring isotope potassium-40 through a complex sequence of intermediate steps into argon-40 (Miller, 1972). A portion of the potassium-40 decays by electron capture so that one proton becomes a neutron and the atomic number decreases from 19 to 18, the atomic number of argon. **Pyroclastic rocks** (rock fragments or ashes that have been produced by volcanic activity) are the most useful ones for potassium-argon dating. Such rocks may include **tuffs** (consolidated volcanic ash), basalts, obsidian, micas, potash, feldspar, and hornblende. With a half-life of 1.3 billion years, this method can date the oldest rocks on earth as well as rocks as young as 100,000 years in age.

There are several conditions for successful dating using this technique:

1. The parent rock should have negligible amounts of ^{40}Ar.
2. None of the daughter element should be gained or lost over geologic time.
3. There must be a feasible way of measuring parent–daughter ratios. Since only one billionth of potassium decays to argon in 100,000 years, a difference of 100,000 to 300,000 years is hardly perceptible, and a date younger than this is rarely attainable.

There may be two types of dating errors associated with this technique: ages may be discrepantly high or low. Discrepantly high ages occur when radiogenic ^{40}Ar (argon-40 produced from radioactive decay) is somehow included at the time of formation of the rock or mineral to be dated. Discrepantly low ages are usually caused by loss of argon due to diffusion or various chemical reactions. For example, increasing temperatures and stress will accelerate argon loss, while high containing pressures inhibit such loss (Fitch, 1972).

Another useful technique is **uranium–lead dating.** All naturally occurring uranium contains radioactive uranium-238 and uranium-235 in a constant ratio of 138:1. Uranium-238 decays to lead-206 and uranium-235 to lead-207 by emitting an alpha particle. Most uranium minerals also contain radioactive thorium-232, which decays to lead-208. The uranium and thorium decay through a series of intermediate steps before finally arriving at their stable lead daughters.

The most useful uranium-containing mineral for dating purposes is zircon (which typically contains about 0.1% uranium). Small quantities of zircon occur in granitic rocks, and thus uranium dating is widely applicable. The original lead in a uranium-bearing mineral causes the radiometric age to exceed the true age unless it is detected. The decay of uranium-238 to lead-206 has a half-life of 4.5 billion years, and the decay of uranium-235 to lead-207 has a half-life of 713 million years.

Fission track dating differs somewhat from the methods previously discussed. In rock crystals containing ^{238}U, spontaneous **fission** (in which the parent-atom nucleus splits into two or more intermediate-sized fragments with high energy) results in collisions between daughter fragments and surrounding atoms. Instead of measuring parent–daughter ratios, fission track dating counts the number of tracks left behind by the naturally recoiling atoms in the sample; the more tracks per unit volume, the older the sample. This method is thus based on three assumptions: (1) uranium-238 decays at a constant rate; (2) fission tracks are produced and retained with 100% accuracy; and (3) the concentration of uranium in any specimen has remained constant through time.

Given the right kind of sample, fission track dating can be used to date rocks ranging in age from a few decades to 3.4 billion years (the age of the oldest rocks on earth). The technique has been particularly useful in archeology where the clays used for pottery making contain crystals of zircon, which contain uranium. When the pottery was fired, the fission track "clock" was set back to zero, thus dating the pottery (MacDougall, 1976).

Paleomagnetic Measurement Radiometric dating techniques are often employed in conjunction with paleomagnetic measurements that detect the preserved orientation of the magnetic field of the earth in rocks formed at various times in the earth's past. The geomagnetic field is known to have gone through a series of complete reversals in its polarity, and these are recorded in successive rock layers at intervals ranging from less than 40,000 years to more than several million years (Cox, 1972). At various times in the geologic past, the earth's magnetic field has been directed toward the north **(normal polarity)** and at other times towards the south **(reversed polarity);** the earth is presently in a period of normal polarity. What causes these reversals in the magnetic field is not yet clearly understood.

Magnetic minerals in rocks preserve a magnetic vector indicating the orientation and intensity of the global magnetic field within which the rocks formed. The ages of these various magnetic reversals must be dated by other means, usually by radiometric techniques such as those already discussed. If a sequence of fossiliferous sediments can be fitted into a **paleomagnetic column,** or profile, that has been radiometrically dated, then the age of the fossils can be determined within the limits of resolution of the dated paleomagnetic column.

Measurements of past reversals in the earth's magnetic field are important to primate paleontologists for another reason: they document the progress of **continental drift,** a phenomenon that has had a profound impact on the distribution of primate species and hence on their evolution. Continental drift is one of the consequences of **plate tectonics,** in which powerful convective currents in the upper mantle cause the lithosphere (the top layer of the earth to a depth of about 70 km, including oceanic and continental crust) to move slowly over the underlying asthenosphere (the layer some 70 to 200 km below the surface). Volcanic activity at the midoceanic ridges is continually creating new crustal rocks, which, as they cool, become magnetized in the direction of the prevailing geomagnetic field. The new rocks are carried to each side of the ridge by plate-tectonic convection. In effect, the process of seafloor spreading can be likened to a "tape recording" of changes in polarity of the earth's magnetic field on both sides of the ridge. Geologists can "play back" this recording to reconstruct the past positions of the continents (Tarling, 1980).

MAKING INFERENCES ABOUT THE BIOLOGY OF EXTINCT PRIMATES

Many important mammalian traits, particularly those having to do with an organism's behavior or lifestyle, cannot be detected directly from the fossil record. One of the main tasks of the paleontologist is to infer such "nonfossilized" traits from fossilized ones. It turns out that one of the most important mammalian traits, warm-bloodedness, is a key link in a chain of inference that relates fossil evidence of overall body size to another important nonfossilized trait, namely dietary preference. Primates, like all other mammals, are **endothermic,** that is, they maintain a high body temperature by means of internal metabolic processes, regardless of the temperature of the external environment. Living primates range in body size from the tiny mouse lemur, *Microcebus* (some under 100 g), to large male gorillas weighing nearly 2,000 times as much. Extinct primates, as we will see in later chapters, probably had an even greater range of body weights. How can we deduce metabolic rates of living or extinct primates just by knowing their body weights? And how can such information be used to make predictions about certain nonfossilized traits of these animals, such as their dietary preferences?

Let us first ask which animal, *Microcebus* or *Gorilla,* produces more metabolic heat per day in absolute terms. Certainly the gorilla does since there are many more cells in its body. Thus we can say that there is a positive relationship between absolute body size and absolute quantity of metabolic heat produced. But what is the nature of this positive relationship? Is metabolic rate directly proportional (that is, one-to-one) to body size, or does it scale to body size in some nonlinear fashion? To illustrate this question, Kleiber (1961) calculated that if a steer had the same relative metabolic rate as a mouse, its surface temperature would have to be well above the boiling point of water in order to dissipate the amount of heat produced. Conversely, if a mouse had the same relative metabolic rate as the steer, it would need fur 20 cm thick just to keep warm. Clearly, metabolic rate must be scaled by some nonlinear factor to body size. How does metabolic rate change as body weight gets greater or smaller in different primates? To investigate this, we need to understand the concept of allometry.

Allometry

Basic Principles **Allometry** is the study of the relative growth of a part of an organism in relation to the whole. It has come to be most closely associated with those studies that describe quantitatively how the shapes of objects change as they get larger or smaller. In paleontology, as we shall see, allometry is a commonly used tool for establishing lineages when only a limited amount of skeletal evidence is available from the fossil record; for example, one can extrapolate increases in body size from increases in the size of certain teeth. A special case of allometry known as **isometry** (meaning "same measure") describes instances in which the shape of the object remains the same as its physical dimensions increase.

We shall have occasion to refer to allometric relationships throughout this book. A number of principles of animal engineering, feeding habits, and even preferred habitats depend on relationships between area and volume, either of the whole animal or of particular parts such as limbs and organs. Some variables, like metabolic rate, may depend on surface areas; other variables, such as bone strength and muscle power, are determined by cross-sectional areas, while still other variables, such as weight and body size, are determined by volume. Allometric principles can be applied to understanding these relationships.

Let us take the surface-area-to-volume ratio as an example. Allometric and isometric scaling have different effects on this ratio. Consider what happens if we increase all three dimensions of a $1 \times 1 \times 1$ cube by 1 unit. Before the increase, the surface area of the cube is 6 square units (six surfaces measuring 1×1 each), while its volume is 1 cubic unit; the ratio of surface area to volume is thus 6 : 1. If we increase all dimensions by 1 unit (an isometric increase), the new surface area is 24 (6 units measuring 2×2) while its volume is 8 ($2 \times 2 \times 2$); the surface-area-to-volume ratio now is 24:8, or 3:1. Tripling the dimensions ($3 \times 3 \times 3$) gives us a ratio of 2:1, and so forth. Note that as surface area increases, the surface-to-volume ratio decreases; in other words, surface area increases more slowly than does the volume.

Isometric changes involving surface-area-to-volume increases follow a mathematically predictable pattern:

$$A \sim V^{2/3}$$

that is, surface area is proportional to the ⅔ power of the volume. (The general form of this relationship, $Y = aX^b$, is known as the **allometric equation;** a specific form of this equation is discussed in the box on the next page). When a change has occurred and the area-to-volume ratio of the resultant form does not obey the ⅔ power rule, we know that an allometric change has occurred and that the shape of the form is no longer the same.

Correlation between Metabolic Rate and Body Weight How are metabolic rate and body size related? Since basal metabolism can be measured in terms of oxygen consumption, and the surface area of the lungs has been shown to be isometrically related to body size among different animals within a species (intraspecific allometry), there must also be a relationship between metabolic rate and body size. The relationship turns out to be the same as between surface area and volume, that is, an isometric relationship. Within a species, metabolic rate is proportional to the ⅔ power of the body weight (McMahon and Bonner, 1983).

Among animals of different species, however, there is a different relationship between metabolic rate and body weight (a case of interspecific allometry).

Data relating metabolic rate (measured by calculating oxygen consumption) to body weight have been collected for many mammalian species ranging in size from tiny insectivores to elephants (Kleiber, 1961; Schmidt-Nielsen, 1979). The resulting allometric equation is:

$$Y = 0.676\,X^{0.75}$$

Working with Allometric Relationships

In the real world, when we plot the relationship between two characteristics of interest in a group of organisms, there is seldom if ever a linear relationship between them. Take for example, the relationship between the area of the lower first molar and body weight in the series of **frugivorous** (fruit-eating) and **folivorous** (leaf-eating) primates shown in Fig. 1.8a. The best-fit line used to join the cluster of data points in the graph is curved. The relationship is the so-called curvilinear one and is described mathematically by the allometric equation $Y = 0.120\,X^{0.638}$. Taking the natural log of both sides of the equation makes it possible to convert the curvilinear relationship into linear form: $\ln Y = 0.638 \ln X - 2.119$ in which 0.638 is the slope of the line and the Y intercept is -2.119 (Fig. 1.8b). When the slope of the line of an allometric equation is less than 1, there is a **negative allometric relationship,** and the value of Y increases more slowly than does the value of X. In frugivorous and folivorous primates, the surface area of the crown of the lower first molar evidently increases relatively more slowly than does body weight from one primate to the next. A **positive allometric relationship,** in which the value of Y increases more quickly than does X, would be described by a line with a slope in excess of 1. Since in this example we are comparing an area (tooth area) against a volume (body weight), the slope of approximately ⅔ indicates that this tooth maintains essentially the same shape as the animal gets larger; in other words, this is an isometric change (Gingerich and Smith, 1985).

Fig. 1.8 Tooth Size as a Function of Body Weight in Frugivorous and Folivorous Primates Tooth size is given by the area (L × W) of the crown of the lower first molar as seen from above (occlusal view). (a) The curvilinear relationship, $Y = 0.120\,X^{0.638}$, is not useful for analysis because most points are clustered at one end of the distribution. (b) Taking the natural logarithm of both sides of the equation in (a) results in the equation $\ln Y = 0.638 \ln X - 2.119$, which is the equation of a straight line with a slope of 0.638 and an intercept on the $\ln Y$ axis of -2.119 (latter not shown on graph). Note that the slope is less than 1, indicating a negative allometric relationship. (Adapted from Gingerich and Smith, 1985)

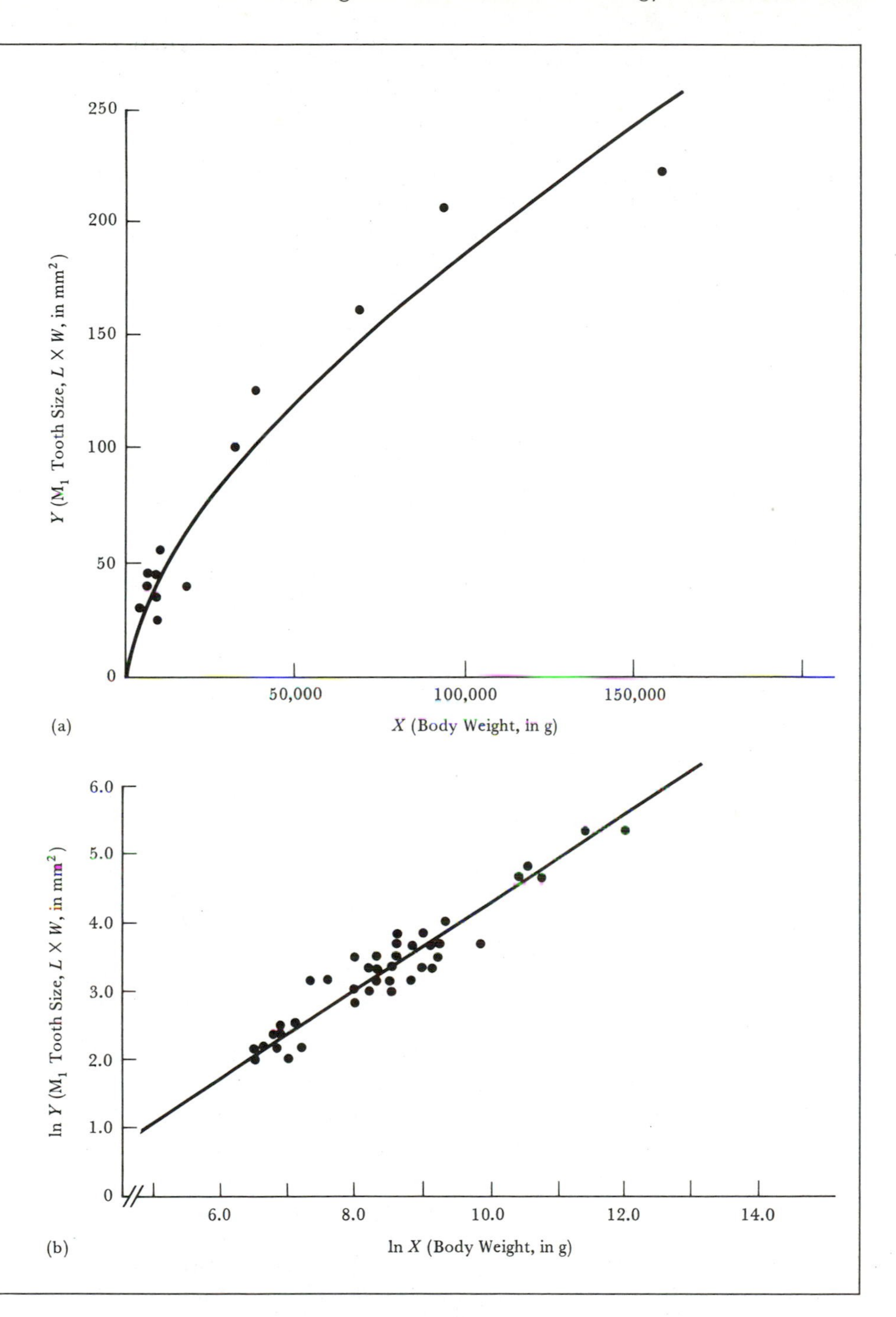
250
200
150
100
50
0
Y (M1 Tooth Size, L × W, in mm²)
50,000
100,000
150,000
X (Body Weight, in g)
(a)
6.0
5.0
4.0
3.0
2.0
1.0
0
ln Y (M1 Tooth Size, L × W, in mm²)
6.0
8.0
10.0
12.0
14.0
ln X (Body Weight, in g)
(b)

where Y equals metabolic rate and X equals body weight. If we change this equation into its logarithmic equivalent, it becomes $\log Y = \log 0.676 + 0.75 \log X$. The Y intercept equals log 0.676, and the slope of the "best-fit" line equals 0.75. Since the slope of the best-fit line for scaling of metabolic rate to body weight is less than 1 (that is, negative allometry), we conclude that in mammals metabolic rate gets relatively lower as body weight increases. Conversely, metabolic rate gets relatively higher as body weight decreases. Thus a smaller mammal has a higher metabolic rate per gram of body tissue than does a larger mammal. In fact, it has been calculated that 1 g of insectivoran tissue consumes oxygen at a rate some 100 times as great as 1 g of an elephant's tissue (Kleiber, 1961; Schmidt-Nielsen, 1979). We will soon see the effect this tremendous oxygen consumption rate (and hence metabolic rate) has for determining dietary preferences in small mammals, including some primates, living and extinct.

As we have just seen, while absolute metabolic requirements are a positive function of body weight over a wide size range of animals, they scale in a negative allometric way to body mass. Thus large animals, while having larger *total* metabolic requirements than small animals, will actually require less energy intake *per unit of body weight* (remember the mouse lemur and the gorilla).

It was recognized in the early 19th century that a smaller mammal must produce more metabolic heat per unit body mass than a larger animal simply to replace the increased amount of body heat lost at its surface due to its relatively large surface-area-to-volume ratio. To accomplish this balance, small mammals must spend relatively more energy, consume more food, and do more work to procure that food than their larger cousins. For example, a human consumes about $\frac{1}{50}$ of his or her own body weight in food per day, whereas a mouse consumes about $\frac{1}{2}$ of its own body weight per day. A mouse is approaching the lower limits in body size for mammals. Much smaller than this, a mammal could neither obtain nor digest the food required to maintain constant body temperature in the face of such heat loss due to its large surface-area-to-volume ratio.

These facts help explain **Bergmann's rule:** animals living in colder climates tend to have larger body sizes than close relatives living in warmer climates as an adaptation to reduce the surface-area-to-volume relationship and thus minimize heat loss (Tracy, 1977).

Relationships between Primate Size and Diet

What do scaling effects contribute to our understanding of primate diet? For example, why is the largest **insectivorous** (insect-eating) primate *(Galago senegalensis)* still about three times smaller than the smallest folivorous primate *(Lepilemur mustelinus)*—250 g versus 700 g?

The answer is readily apparent if we recognize that the nutritional requirements of a mammal are directly related to its metabolic rate: the higher the basal metabolism, the greater the amount of energy the food must be able to deliver per unit volume of body tissue. Recall from our earlier discussion that metabolic rate (and hence energy requirements) of mammals, as viewed over a wide range of different-sized species, scales in a negative allometric way to body mass: while large mammals have greater total metabolic requirements than smaller mammals, they require less energy intake per unit of body weight.

This negative allometric relationship between body weight and nutritional requirements has important implications for the coevolution of diet and body size. These implications comprise the **Jarman-Bell principle** (Jarman, 1968, 1974; Bell, 1971). Large mammals, because of their high daily total food requirements, are usually unable to base their diets primarily on rare, high-energy foods (those high in calories but hard to find and/or catch). On the other hand, large mammals, due to their lower per-unit-weight food requirements, do not need to provide a high rate of nutrient flow to their tissues and are thus able to subsist on more low-quality foods processed in bulk.

Small mammals have the reverse problem: they do not have large total food requirements, and thus larger fractions of their diet can consist of relatively rare foods. However, small mammals must nourish their tissues at high rates and therefore must concentrate on high-quality (highly caloric) foods (Gaulin, 1979; Gaulin and Sailer, 1984; Sailer and Gaulin, 1985; Kurland and Pearson, 1986).

While total metabolic requirements increase as a fractional power of body weight (with a slope of 0.75), gut capacity increases as a liner function of body weight (with a slope of approximately 1.0). The ratio of metabolic rate to gut capacity decreases with body size, so foods can be retained in the gut relatively longer, the larger the mammal is. Since digestibility is related to gut retention time, digestibility of foods will also be a function of body size. Small mammals generally respond by eating rapidly digestible, high-quality foods, whereas large animals usually eat less-digestible, high-fiber foods (Demment, 1983).

Insects are an excellent source of concentrated energy that can be exploited by small mammals (Coe, 1984). Insects are high in energy (protein), relatively easy for small mammals to acquire, and easy to digest. Large mammals usually cannot rely on insects because they cannot capture them fast enough to sustain the larger body size. (The obvious exceptions are certain large mammals like anteaters that can exploit colonies of social insects in large quantities.)

In contrast to insects, leaves are abundant but are usually a much poorer source of energy than animal protein or fruit. Moreover, their structural carbohydrates (such as cellulose) are harder to digest. Studies suggest that cellulose digestion is more complete the longer the food remains in the digestive tract to be broken down by microorganisms and digestive enzymes (Chivers and Hladik, 1984). The increase in digestive assimilation time is directly proportional to body size, so larger mammals can break down more carbohydrates per volume of leaves ingested than can smaller mammals (Kay and Covert, 1984).

In general then, heavier primate species would be expected to eat lower-quality, more-abundant, less-digestible foods, while lighter species would be expected to eat higher-quality, rarer, more easily digested foods.

Primate diets can be described in terms of three principal types of food:

1. Various parts of animals, both vertebrates and invertebrates
2. Reproductive parts of plants (flowers, buds, fruits, nectar, and other resins)
3. Structural parts of plants (leaves, stems, bark, and other plant materials containing a high proportion of structural carbohydrates such as cellulose) (Waterman, 1984)

This way of classifying food types is particularly convenient, since it groups foods by abundance, caloric quality, and digestibility. Foods in the first category are sparsely distributed but rich in nutrients and relatively easy to digest; foods in the third category are abundant but of low nutritional value and are difficult to digest; foods in the second category are intermediate in availability and nutritional quality.

Based on such considerations, Richard Kay and his colleagues at Duke University have concluded that primates weighing more than about 350 g are usually not primarily insectivorous (that is, insects would provide no more than about a third of their energy needs), and that folivory would be difficult to sustain for a primate weighing less than 500 g, a point of demarcation that has come to be known as **Kay's threshold.** Another general observation of Kay and his colleagues concerns two primarily frugivorous groups of primates: (1) those that supplement their diet with protein by eating insects, and (2) those that acquire extra protein from eating leaves. Kay and his colleagues have found that the first group tends to be smaller than the second (Kay, 1984).

In short, if the body weight of fossil primates can be determined with reasonable accuracy, some important inferences can be drawn about dietary preferences (Fig. 1.9). We shall see in later chapters the various ways in which paleontologists determine body weight—and thus dietary preference—from often fragmentary fossil data.

Effects of Diet on Dentition

Dietary specializations also have a strong influence on the form of an animal's teeth. Hence dentition can serve as a basis for inferring dietary preferences in fossil species. For example, insectivores usually have com-

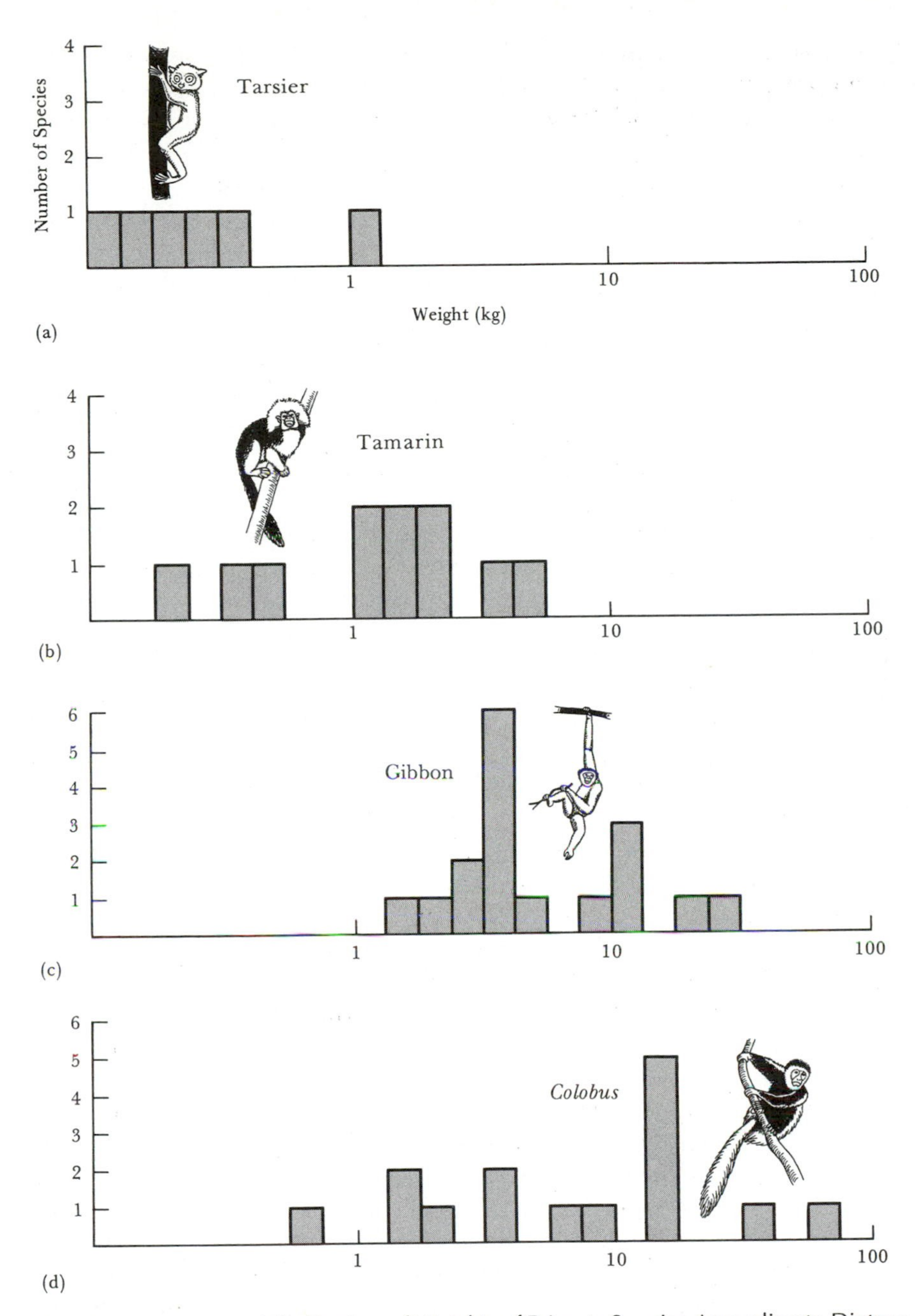

Fig. 1.9 Frequency Distribution of Weights of Primate Species According to Dietary Preferences (a) Insectivores and gummivores (gum eaters); (b) frugivores that also eat insects; (c) frugivores that also eat leaves; (d) granivores (grain or seed eaters), folivores, and herb eaters. The lightest species are insectivores and gummivores, and the heaviest are folivores. Note that the horizontal axis is a logarithmic scale. (Adapted from Richard, 1985)

paratively short, pointed front teeth that are used to kill or immobilize small prey, and their premolars and molars have steep, transversely oriented shearing blades. The premolars are often well differentiated, and the molars tend to have the primitive triangular form, since grinding and crushing surfaces are correspondingly less important.

Frugivores, on the other hand, tend to develop long, protruding front teeth that serve as a mechanical device for opening fruits, nuts, seeds, and the like. In animals that scrape bark, the front teeth are extremely elongated. In both kinds of animals the last premolars tend to become **molariform** (resembling molars) providing additional crushing and grinding surfaces. The cusps and edges of the molars are low and blunt, and the shearing surfaces are flattened and structurally simplified. One or two transverse crests **(lophs)** may develop to guide chewing movements. (Two-crested molars are said to be **bilophodont.**)

In contrast, folivores have comparatively small front teeth, but their **postcanine teeth** (premolars and molars) are equipped with a series of cutting edges (Maier, 1984). Extensive mastication (chewing) is necessary to grind the food particles for adequate digestion, since leaves are high in fiber (structural carbohydrates). Studies by Kay and Sheine (1979) have demonstrated that chewing efficiency among small mammals is important for improving the digestibility of foods that contain high percentages of structural carbohydrates (Fig. 1.10).

A convenient shorthand method of signifying the number of teeth in the upper and lower jaws of an animal is given by the **dental formula.** Primitive mammals, for example, are thought to have a dental formula of 3.1.4.3/3.1.4.3, which means that each side of the upper and lower jaws of these animals has three incisors, one canine, four premolars (bicuspids), and three molars. (The numbers before the slash give the formula for the upper jaw, and those after the slash the lower jaw.) For comparison, an adult human has a dental formula of 2.1.2.3/2.1.2.3, or two incisors, one canine, two premolars, and three molars on each side of the upper and lower jaws. This is interpreted to mean that humans have lost one incisor and two premolars from each side of the upper and lower jaws compared with the ancestral mammalian condition. The teeth that are lost in evolution are considered to be those immediately adjacent to the canine. Thus in the human example it is the upper and lower first two premolars and the upper and lower third incisors that have been lost from the ancestral dental formula. The teeth remaining on each side of the upper and lower human jaw are thus identified as I1, I2, C, P3, P4, M1, M2, and M3 (Fig. 1.11). (When referring to specific teeth on only one jaw, the letter abbreviations are followed by superscript or subscript numbers for the upper or lower jaws, respectively. Hence, P_3 is the lower third premolar and M^2 is the upper second molar.)

There will be much more to say about the dentition of fossil primates in later chapters. Appendix B covers additional dental nomenclature.

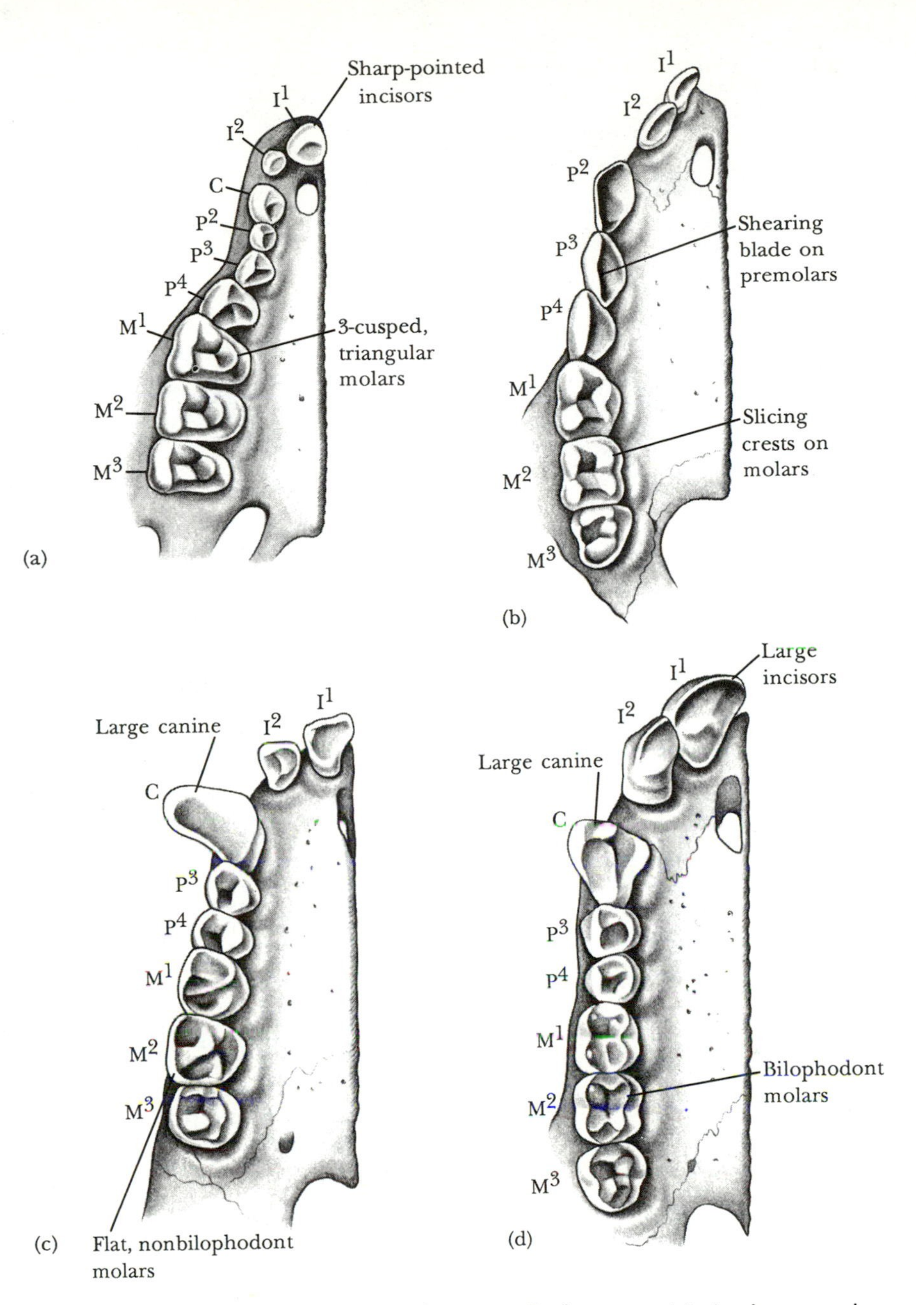

Fig. 1.10 Dental Characteristics and Dietary Preferences (a) An insect-eating prosimian *(Tarsius),* (b) a leaf-eating lemur *(Indri),* (c) a fruit-and-leaf-eating ape *(Hylobates),* and (d) a fruit-eating monkey *(Cercocebus).* Upper right side of the jaw is shown throughout. Insect-eating primates usually have enlarged and sharpened incisors and canines that are used to stab and hold insects. In primates that are primarily leaf eating, the incisors are quite small and the premolars and molars are enlarged with elongated shearing and slicing crests. Mixed fruit-and-leaf eaters like the gibbon have quite broad incisors, but the molars are small and relatively flat. Fruit-eating primates have very large incisors to strip through the fruit rinds, while the molars are relatively small since fruit is usually soft and does not require much grinding. (Redrawn from Jolly and Plog, *Physical Anthropology and Archaeology,* McGraw-Hill, 1986; reproduced with permission.)

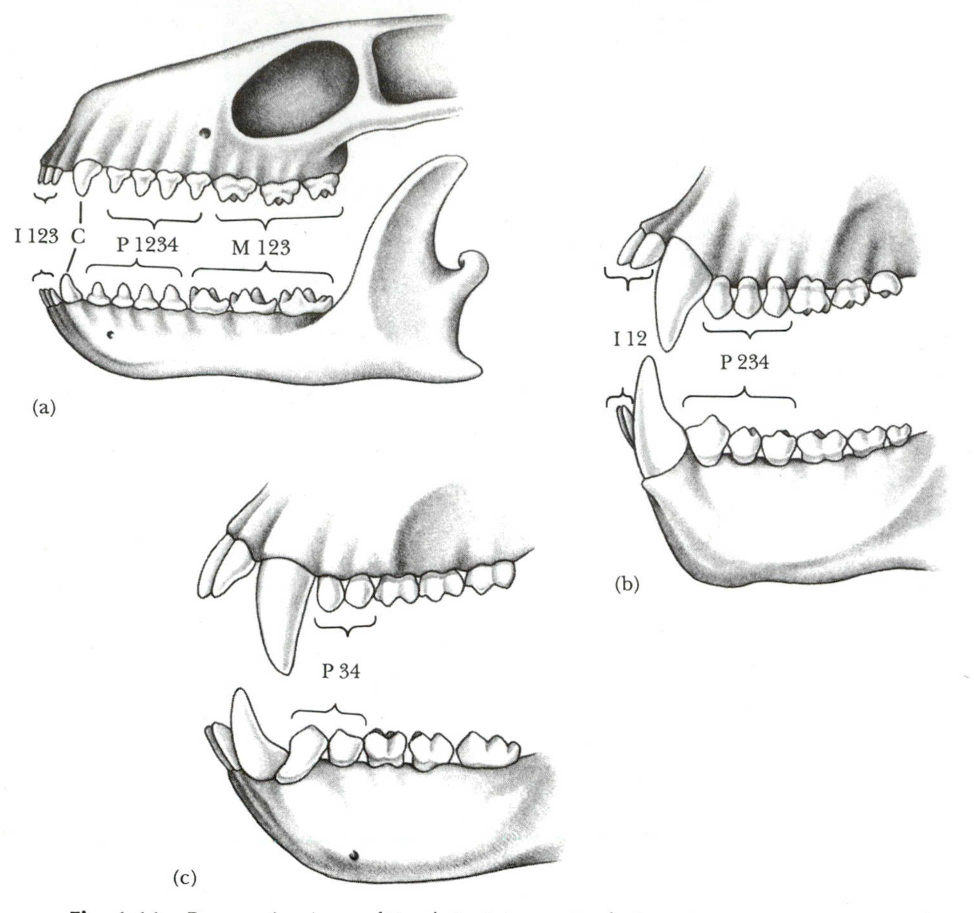

Fig. 1.11 Progressive Loss of Teeth in Primate Evolution (a) Ancestral mammalian condition (3.1.4.3 / 3.1.4.3); (b) condition of New World monkeys (Cebidae) (2.1.3.3 / 2.1.3.3); (c) condition of Old World monkeys, apes, and humans (2.1.2.3 / 2.1.2.3). Notice that New World monkeys have lost the first premolar of the ancestral premolar series and that Old World monkeys, apes, and humans have lost the second premolar as well. All primates have lost the third incisor of the ancestral incisor series. (Redrawn from Le Gros Clark, 1959)

ORIGIN OF THE PRIMATES

The appearance of a new mammalian form in the fossil record is generally believed to represent a successful adaptive shift by the animal to fill an unoccupied niche in its environment. Ideally this adaptive shift is reflected in the animal's morphology and can therefore be inferred by paleontologists and comparative anatomists. Often the behavioral correlates of morphology are obvious, for example, the elongation of the limbs in **cursorial** (fast-running) animals or the accentuated shearing crests in the teeth of folivorous monkeys. However, other morphological characters, such as the distinctive structure of the middle ear in primates or the characteristic prism patterns in the enamel of primate teeth, have so far eluded any widely accepted functional explanation.

Most primatologists find it difficult to define the morphological features and the correlative adaptive shifts that characterize the earliest primates. As Simpson (1961) remarked, "no clear cut diagnostic adaptation or heritage distinguishes the order Primates as a whole, or specifically the primitive primates, from other primitive placental mammals." Indeed, it was because of this widely held belief that Le Gros Clark and others tried to characterize the order in terms of a complex of evolutionary trends, that, taken together, tended to support the view that primitive primates lived in the trees.

Arboreal Theory

The **arboreal theory** of the origin of the primates was originally put forward earlier in this century by two influential English anatomists, G. Eliot Smith and F. Wood Jones (Smith, 1912; Jones, 1916). They attributed many trends in the early evolution of the primates to the presumed arboreal way of life of the remote ancestors of the primates. They reasoned that this tree-living heritage resulted in the elaboration of visual and tactile sense receptors and of their associated control centers in the brain. Increased reliance on vision in an arboreal environment induced a corresponding reduction in olfaction. This led to **orbital convergence,** that is, a shift in the alignment of the visual axes from more laterally directed to more parallel, anteriorly directed lines of sight. Smith and Jones saw this trend as a passive byproduct of the reduction of the olfactory sense and the associated shortening of the snout. Other investigators suggested that convergence of the orbits promoted the development of full stereoscopic vision, an attribute thought to be vital to an arboreal animal; grasping hands and feet, with nails instead of claws, were similarly seen as obvious adaptations to an arboreal way of life (Collins, 1921).

By the 1950s the arboreal theory became dogma in anthropological thought largely through the writings of Le Gros Clark. It seemed to be a unifying theory that could explain the observed morphological adaptations of the order. More recently, Matt Cartmill of Duke University has criticized and modified the arboreal theory (Cartmill, 1972, 1974). The most obvious criticism is that at least nine other orders of mammals include arboreal forms, and none of them shows all of the characteristically primate morphological trends noted by proponents of the arboreal theory. Jones had anticipated this argument, however, by suggesting that most arboreal mammals began to live in the trees only after a prolonged period of adaptation to terrestrial life, and that this adaptation must have included such changes as a reduction in the number of digits, a decrease in the size of the clavicles, and an increased reliance on olfaction.

Cartmill (1974) also raised the following issue concerning the arboreal theory: "If progressive adaptation to living in trees transformed a tree-shrew-like ancestor into a higher primate, then primate-like traits must be better adaptations to arboreal locomotion and foraging than are their antecedents. This expectation is not borne out by studies of arboreal nonprimates." He noted that the representatives of various mammalian taxa, including marsupials (opossums), insectivorans (tree shrews), rodents (squirrels), and carnivorans (raccoons and civets), are found in arboreal niches in present-day forests. He then examined in detail some of the assumed primate trends, such as grasping extremities, convergence of the visual axes and olfactory regression, as they appear in these other arboreal animals, to see whether these trends are causally related to arboreality itself or to other factors.

Grasping hands and feet, for example, are advantageous to animals that habitually forage among small terminal branches. On the other hand, locomotion on sturdier supports is easier for small mammals with clawed, squirrellike hands and feet than it is for small mammals with grasping extremities and reduced claws. Prehensile hands and feet, similar to those of many primates, are most common among shrub-layer insectivores and related herbivorous forms. Cartmill reasoned

> that the distribution of grasping extremities among arboreal mammals suggests that prehensile hands and feet represent an adaptation not to arboreal activity *per se*, but to activity on branches of relatively small diameter, whether in marginal undergrowth or in the forest canopy. For small mammals which habitually climb up and down large vertical trunks, claws apparently provide more effective traction than grasping digits tipped with enlarged pads (Cartmill, 1972).

A second counterexample concerns the presumed causal relationship postulated by the arboreal theory between leaping ability in trees and convergence of the orbits. Acrobatic arboreal mammals do not show significantly increased convergence of the orbits compared with their slower-moving, less acrobatic relatives. In fact, when corrections are made for body size, the slow loris *(Loris tardigradus)* has the most-convergent orbits of any primate.

Convergence of the orbits seems most extreme in vertebrates that depend on vision for detecting prey (for example, owls, hawks, and cats). It seems likely that orbital convergence developed in early primates that were visually directed predators similar to living lorises, galagos, and tarsiers. Precise stereoptic integration of the two visual fields, a trait coordinated with rapid and accurate hand movements, is a necessity for the visually directed, predatory primate (Allman, 1982; Pettigrew, 1986).

Contrary to another principal tenet of the arboreal theory, studies on a variety of arboreal animals show that the sense of smell is extremely important to arboreal animals. Olfactory signals help maintain social organization and reproductive behavior in many arboreal mammals, including tree shrews and the giant tree squirrel *Ratufa* (Cartmill, 1972).

Visual Predation Theory

Cartmill draws two conclusions concerning primate adaptation and differentiation of the order Primates. First, primatelike morphology is not necessarily advantageous to arboreal mammals. Second, the evolution of many ancestral primate adaptations was probably driven by nocturnal, visually directed predation on insects in the terminal branches of the lower layers of tropical forests. His studies suggest that "the close-set eyes, grasping extremities and reduced claws characteristic of most post-Paleocene primates may originally have been adaptations to a way of life like that of [the small arboreal marsupials] *Cercartetus* or *Burramys,* which forage for fruit and insects in the shrub layer of the Australian forests and heaths." According to this interpretation, orbital convergence and correlated neurological specializations were initially predatory adaptations "comparable to the similar specializations seen in cats and owls, and allowing the predator in each case to gauge its victim's distance accurately without having to move its head" (Cartmill, 1974). In addition, olfactory regression is viewed as the necessary result of the coming together of the medial walls of the orbits during orbital convergence.

Is there any evidence in the fossil record to support the visual-predation theory of primate origins? One of the only clues to surface so far is suggestive: at least four of the six families of early Cenozoic primates have molars functionally similar to those of the insectivorous tarsier (Simons, 1974a; Cartmill, 1974).

Angiosperm Radiation Theory

Robert Sussman at Washington University has approached the question of primate origins from an ecologist's point of view. He suggests that early primates were not necessarily visually directed predators that searched for insects in the terminal branches of trees, but rather were omnivores feeding on all kinds of small objects in the same arboreal habitat. He argues that the adaptive radiation of the early primates coincided with the radiation of modern groups of flowering plants (angiosperms), which offered an array of previously unexploited resources such as flowers, fruits, gums, and nectars, and also the insects that fed on these items. According to this ecological argument, visual predation itself

Table 1.3 Traditional Classification of the Living Primates

Order	PRIMATES						
Suborder	PROSIMII (prosimians)						
Infraorder	LEMURIFORMES						TARSIIFORMES
Superfamily	Lemuroidea				Lorisoidea		Tarsioidea
Family	Lemuridae	Indriidae	Daubentoniidae	Cheirogaleidae	Lorisidae		Tarsiidae
Subfamily					Lorisinae	Galaginae	
Tribe							
Genus (Number of species)	*Lemur* (5) *Varecia* (1) *Hapalemur* (2) *Lepilemur* (1)	*Indri* (1) *Propithecus* (1) *Avahi* (1)	*Daubentonia* (1)	*Cheirogaleus* (dwarf lemur) (2) *Microcebus* (2) *Allocebus* (1) *Phaner* (1)	*Loris* (1) } (lorises) *Nycticebus* (1) } (lorises) *Perodicticus* (potto) (1) *Arctocebus* (1)	*Galago* (bush babies) (7)	*Tarsius* (tarsiers) (3)

Source: Adapted from Jolly and Plog, *Anthropology and Archaeology,* McGraw-Hill (1986), reproduced with permission.

is not a sufficient explanation for the visual adaptations of early primates. A more probable explanation, according to Sussman (in press),

> lies in the necessity for increased visual acuity to discriminate between potential food items under the low light conditions of the canopy at night. These nocturnal animals were feeding on and manipulating items of very small size (for example, fruits, flowers, and insects) at very close range. This would necessitate acute powers of discrimination and precise eye-hand coordination.

This new theory has yet to address some obvious problems. For one thing, angiosperms first appear in the fossil record in rocks dating from the Cretaceous Period, millions of years before the appearance of the

Table 1.3 (continued)

PRIMATES												
ANTHROPOIDEA (anthropoids)												
CATARRHINI						PLATYRRHINI (New World monkeys)						
Hominoidea (Apes and humans)			Cercopithecoidea (Old World monkeys)			Ceboidea (New World monkeys)						
Hominidae	Pongidae	Hylobatidae	Cercopithecidae			Cebidae					Callitrichidae	
			Cercopithecinae		Colobinae	Cebinae	Alouattinae	Aotinae	Atelinae	Pitheciinae	Callimiconinae	Callitrichinae
			Cercopithecini	Papionini								
Homo (humans) (1)	*Pan* (chimpanzees) (2) *Gorilla* (gorilla) (1) *Pongo* (orangutan) (1)	*Hylobates* (gibbons) (6)	*Cercopithecus* (17) *Erythrocebus* (1)	*Macaca* (macaques) (11) *Cercocebus* (5) *Papio* (baboons) (1) *Theropithecus* (gelada) (1) *Mandrillus* (2)	*Presbytis* (langurs) (12) *Rhinopithecus* (3) *Nasalis* (2) *Colobus* (4) *Procolobus* (1)	*Cebus* (4) *Saimiri* (squirrel monkeys) (2)	*Alouatta* (howler monkeys) (5)	*Aotus* (1) *Callicebus* (3)	*Ateles* (spider monkeys) (4) *Brachyteles* (1) *Lagothrix* (2)	*Pithecia* (2) *Cacajao* (3) *Chiropotes* (2)	*Callimico* (1)	*Callithrix* (8) *Leontideus* (3) *Cebuella* (1) *Saguinus* (22)

first primates. For another, fossils of the earliest Paleocene primates show little evidence of orbital convergence and the correlated neurological specializations accompanying it. Chapter 2 will document what is known about the morphology of these earliest primates.

TAXONOMY USED IN THIS TEXT

To make the discussion of fossil primates in the following chapters more compatible with traditional notions of primate taxonomy, the classification system of the living primates will be that shown in Table 1.3. The

Table 1.4 The Linnaean Hierarchy of Classification Categories

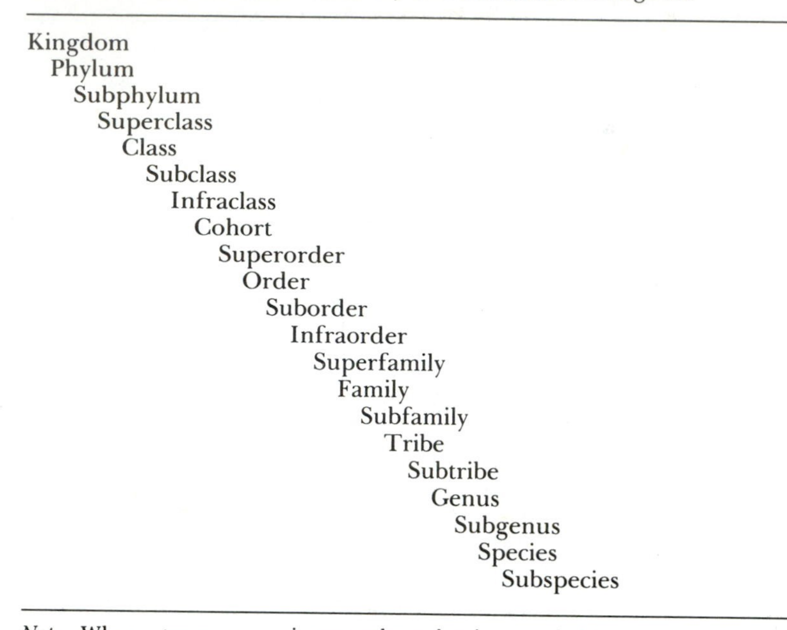

Kingdom
- Phylum
  - Subphylum
    - Superclass
      - Class
        - Subclass
          - Infraclass
            - Cohort
              - Superorder
                - Order
                  - Suborder
                    - Infraorder
                      - Superfamily
                        - Family
                          - Subfamily
                            - Tribe
                              - Subtribe
                                - Genus
                                  - Subgenus
                                    - Species
                                      - Subspecies

Note: When a taxon occurs in parentheses but is preceded by an equals sign, it refers to alternate usage. Examples are: Propliopithecidae (= Pliopithecidae), *Sivapithecus (= Ankarapithecus) meteai,* and *Gigantopithecus giganteus (= bilaspurensis).* Here alternate names for family, genus, and species, respectively, are given. Where subgenera occur, the subgenus name is also given in parentheses but without the equals sign, as in, for example, *Dryopithecus (Proconsul) africanus.*

hierarchical classification of animals used today grew out of the work of the great Swedish naturalist, Carolus Linnaeus, in the 18th century. The major categories in this classification are given in Table 1.4. Throughout the text various animal groupings will be referred to not only by their Latin names, but also by the anglicized versions of these names (for example, Hominidae and hominids).[1] The names applied to some of these higher categories follow certain established rules, as illustrated in Table 1.5.

[1] Members of the order Insectivora are referred to as **insectivorans** to distinguish them from insectivores (insect-eating mammals), which may or may not be insectivorans. Similarly, **carnivores** are meat-eating animals that may also be **carnivorans** (members of the order Carnivora) but need not be.

Table 1.5 Formal and Anglicized Names of Higher Taxa in Primates

Category	*Formal Suffix*	*Genus*	*Stem*	*Formal Name of Taxon*	*Anglicized Name of Taxon*
Infraorder	–iformes	*Lemur*	Lemur–	Lemuriformes	lemuriform
		Tarsius	Tarsi–	Tarsiiformes	tarsiiform
Superfamily	–oidea	*Lemur*	Lemur–	Lemuroidea	lemuroid
		Homo	Homin–	Hominoidea	hominoid
Family	–idae	*Lemur*	Lemur–	Lemuridae	lemurid
		Homo	Homin–	Hominidae	hominid
Subfamily	–inae	*Lemur*	Lemur–	Lemurinae	lemurine
		Homo	Homin–	Homininae	hominine

An alternate classification system divides the living primates into the suborders **Strepsirhini** and **Haplorhini.** The strepsirhines contain only the infraorder **Lemuriformes** (lemurs and lorises),[1] whereas the haplorhines comprise three infraorders: **Tarsiiformes** (tarsiers), **Platyrrhini** (New World monkeys), and **Catarrhini** (Old World monkeys and all apes and humans). In a notable departure from the more traditional classification scheme of Table 1.3, this system, based on recent biochemical evidence, groups gorillas and chimpanzees together with humans as the Hominidae. In the chapters to come, we will examine some of the proposed taxonomies that make use of strepsirhine–haplorhine groupings.

The three major radiations of primate evolution to be discussed in this text reflect the more traditional classification of the order Primates into the suborders **Plesiadapiformes** (archaic, extinct primates), **Prosimii** (lower primates), and **Anthropoidea** (higher primates) (Fig. 1.12). Chapter 2 will cover the plesiadapiforms, Chapter 3 earliest prosimians (Adapidae and Omomyidae), and Chapters 4 through 6 the anthropoids up to the radiation of *Homo* (humans).

[1] In several classifications illustrated later in the text, lorises are accorded separate, equal taxonomic rank (infraorder Lorisiformes) with lemurs. Unless otherwise specified, however, we will assume that lorises and galagos are subsumed under the Lemuriformes.

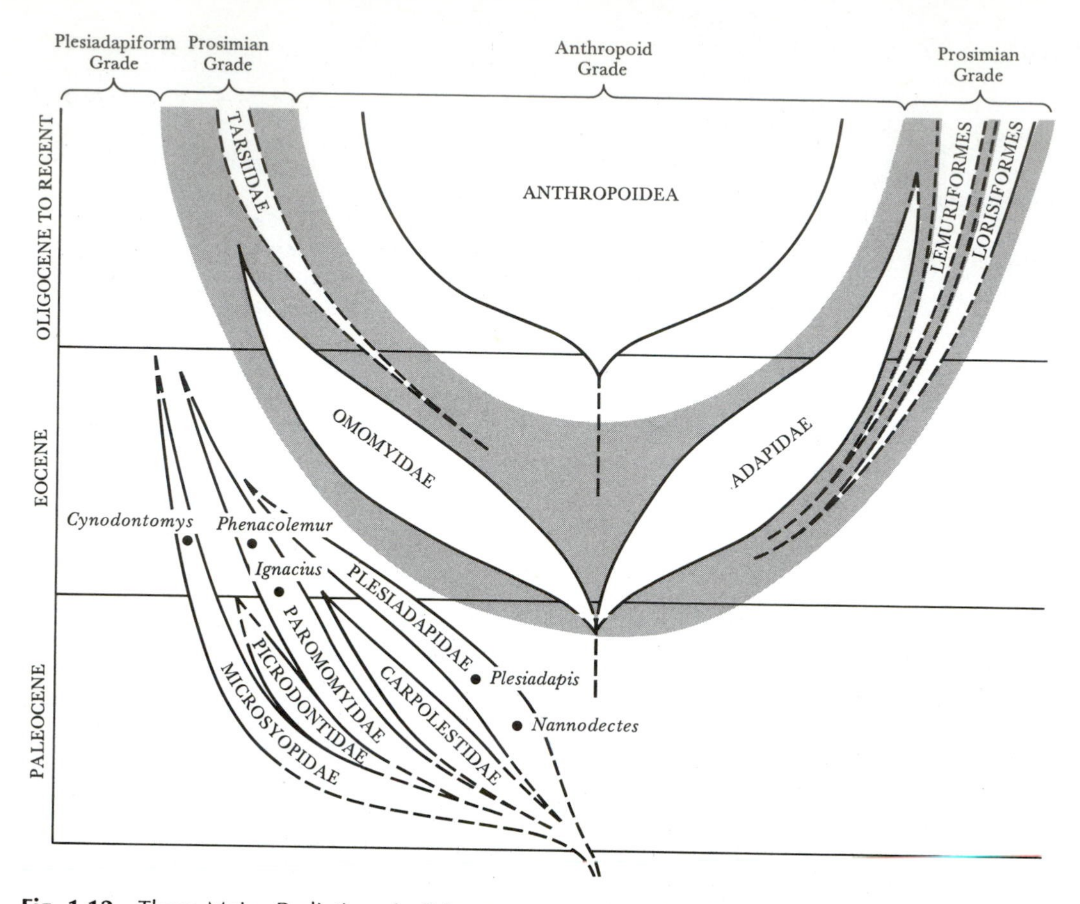

Fig. 1.12 Three Major Radiations in Primate Evolution Solid outlines show distributions of the major primate taxa over time, and dotted lines indicate unresolved phylogenetic relationships. Each portion of the diagram bracketed at the top represents a different grade of primate evolution (prosimian grade in gray for clarity). (Adapted from MacPhee et al., 1983)

CHAPTER 2
Paleocene Primates

Paleoclimates and Biogeography
Summary of the Paleocene Fossil Record
Plesiadapiformes
Feeding Adaptations of Paleocene Primates
Phylogeny and Classification of Paleocene Primates

It was during the Paleocene epoch, beginning some 65 MYA and ending approximately 10 million years later, that archaic primates called plesiadapiforms emerged onto the stage of the evolutionary theater and proliferated into a number of species. Their fossilized remains are known mainly from sites in western North America and in Europe. There are no known primates from the southern continents during this period.

These early primates are called archaic because they lacked many of the morphological features distinctive of modern primates, such as a postorbital bar, nails (instead of claws), and relatively enlarged brains. Indeed, because these traits are not evident, some researchers do not even include plesiadapiforms in the order Primates. In fact, the primate status of plesiadapiforms is based mainly on subtle dental morphology. We shall review these morphological considerations later in the chapter. But first, as will be the pattern in each of the remaining chapters, we begin our discussion by reviewing the important paleoclimatological and biogeographical factors influencing primate evolution during this time.

PALEOCLIMATES AND BIOGEOGRAPHY

Extinctions at the Cretaceous–Tertiary Boundary

The beginning of the Paleocene epoch, some 65 MYA, marked a major transition in geologic time. As noted in the preceding chapter, the Paleocene was the first epoch of the Cenozoic era, the more encompassing unit of geologic time that extends up to the present (see Table 1.2). The Paleocene epoch (65 to 55 MYA) also marks the beginning of the **Age of Mammals.** Fewer than a dozen families of marsupial and placental mammals are known from the preceding late Cretaceous period (Archibald and Clemens, 1984; Archer et al., 1985). By the late Paleocene and early Eocene epochs the number of mammalian families had reached nearly

70. Roughly one-third of all Cenozoic mammalian orders made their first appearance in the Paleocene (Krause, 1984).

Animal and plant life changed dramatically between the end of the Mesozoic and the beginning of the Cenozoic, at the stratigraphic dividing line commonly known as the **Cretaceous–Tertiary (K / T) boundary.** Roughly 75% of all species living in the last part of the Mesozoic vanished by the early Cenozoic. The most famous of these Cretaceous–Tertiary extinction events was the disappearance of the dinosaurs, although mass extinctions were not confined to the large reptiles. Many groups of marine invertebrates also perished (Savage and Russell, 1983; Wolfe and Upchurch, 1986), the most striking being the planktonic Foraminifera (Hallam, 1987). Three of the most widely discussed hypotheses proposed to explain this great extinction event are the greenhouse model, the Arctic Ocean spillover model, and the asteroid-impact model.

Greenhouse Model According to the **greenhouse model,** geological and paleontological evidence dating from the late Mesozoic suggests that a short-term global warming trend (lasting on the order of 100,000 to 1 million years) resulted from greenhouselike conditions in the late Cretaceous, induced by a buildup of carbon dioxide in the earth's atmosphere. During this period a significant reduction in the number of certain carbon dioxide–fixing organisms led to an increase in the amount of free carbon dioxide in the atmosphere. Because carbon dioxide is transparent to incoming solar radiation but absorbs part of the infrared radiation emitted from the earth, increasing quantities of carbon dioxide in the atmosphere would block the escaping radiation and cause a warming of both the lower atmosphere and the surface of the earth (McLean, 1978).

Oxygen-isotope analyses of deep-sea sediments suggest there was a period of lower temperatures during the middle-to-late Mesozoic that preceded the late Mesozoic warming trend. These oceanic sediments consist mainly of **microfossils** (fossilized skeletons of microscopic plants and animals that live in the ocean). Of particular interest for oxygen-isotope analyses are the calcareous shells of foraminiferans. The lime in these shells contains two isotopes of oxygen, the more common oxygen-16 and the rarer oxygen-18. When calcium carbonate is formed in seawater, the ratio of the two oxygen isotopes in the carbonate depends on the water temperature at the time of deposition. For this reason, dated foraminiferal skeletons can be invaluable in reconstructing past ocean temperature patterns.

In addition to the effect of water temperature on isotope ratios, another influence has to be taken into account. When water evaporates from the ocean's surface, molecules containing the lighter isotopes of oxygen are selectively vaporized. During times of extensive ice-sheet formation on the continents, the vaporized water is deposited as snow and is thus removed from recirculation with the ocean. The result is an enrichment of the heavier oxygen-18 isotope in the seawater at the expense of the lighter oxygen-16 (Labeyrie, 1974; Matthews, 1974).

According to the greenhouse theory, animals adapted to such lower temperatures of the late Mesozoic could not readapt to the sudden warming trend in the late Cretaceous. The fact that no land vertebrate larger than about 25 kg survived the K / T extinction event is offered as support for this scenario; in other words, natural selection apparently favored smaller body size. The discussion of body size in Chapter 1 reveals the suggested explanation for this observation. Small animals, because of their higher ratio of body surface to volume, would be able to dissipate heat more effectively than large animals and would therefore be better able to tolerate the modest elevation of temperature at the end of the Cretaceous.

Arctic Ocean Spillover Model The **Arctic Ocean spillover model** postulates that when the rift between Greenland and Norway was initiated some 65 MYA, the colder, less salty, and hence lighter water of the Arctic Ocean spilled out over the North Atlantic and eventually covered the entire ocean surface with a layer of low-temperature, low-salinity water. This surface layer caused the extinction of most of the plankton living at the ocean's surface by depleting the oxygen near the surface. Ultimately the cooling of the ocean affected the earth's climate, resulting in a decline in both global atmospheric temperature and precipitation. According to one interpretation, the change in climate "may have triggered a series of ecological disasters that included the radical change in the distribution of vegetation on the earth as well as the extinction of the dinosaurs" (Gartner and McGuirk, 1979). In this scenario the dinosaurs perished from the effects of drought and reduced temperature rather than from the effects of elevated temperature postulated by the greenhouse model.

Asteroid-Impact Model The third hypothesis to explain Cretaceous–Tertiary extinctions, the **asteroid-impact model,** has received a great amount of publicity since it was first proposed in 1980 (Alvarez et al., 1980). It is based on the observation that certain noble metals (such as iridium, osmium, gold, platinum, rhenium, ruthenium, palladium, nickel, and cobalt) are normally rare in the earth's crust relative to their cosmic abundance. Beginning in 1980 a number of investigators—notably Walter and Luis Alvarez and their coworkers at the University of California at Berkeley—showed that, at a certain stratum in both marine and nonmarine sediments, the concentration of iridium increases by a factor of 5 to 100 over its expected abundance. The iridium-rich layer, which coincides closely with the time of the K / T extinctions, is thought to be extraterrestrial in origin. Calculations show that the impact of a large asteroid or meteorite (10 to 15 km in diameter) could have injected some 60 times the object's mass into the atmosphere as pulverized rock. A fraction of this dust would have stayed in the stratosphere for several years and been distributed worldwide before being precipitated out onto the soil. The resulting darkness would have suppressed photosynthesis and affected the climate dramatically. The expected biological consequences match quite closely the extinctions observed in the paleontological record (Pollack et al., 1983; Alvarez et al., 1984; Tschudy et al., 1984).

The asteroid-impact theory has not gone unchallenged. More recent paleontological evidence reveals that end-of-Cretaceous mass extinctions were not a geologically instantaneous event and were selective in character. To cite one example, a team of investigators headed by Robert Sloan of the University of Minnesota has argued recently that the extinction of the dinosaurs in North America was a gradual process that began 7 million years before the end of the Cretaceous, during an interval of apparent competition with rapidly evolving primitive mammals (Sloan et al., 1986). This interval saw a rapid reduction in both the diversity and population density of dinosaurs. The last appearance of dinosaurs in the fossil record, according to this group's research, coincides with a Paleocene pollen layer that lies 1.3 m above the position of the iridium anomaly at the K / T boundary.

Charles Officer and Charles Drake of Dartmouth College have also questioned the asteroid-impact model on geological grounds (Officer and Drake, 1985). They maintain that iridium and other noble metals were not deposited globally at the K / T boundary but were distributed over a period of 10,000 to 100,000 years by numerous volcanic eruptions. They also stress that there is no physical evidence of an impact crater at the K / T boundary. Thus controversy still surrounds the question of Cretaceous–Tertiary extinctions.

Summary of Climatic Changes in the Paleocene

Temperatures were generally falling toward the end of the Cretaceous period, and lower temperatures prevailed into the early Paleocene of North America. Globally, temperatures were lower during the early and late Paleocene and higher during the middle Paleocene. Jack Wolfe (1980) of the U.S. Geological Survey has summarized the climate of the early Tertiary period as follows:

> During the Paleocene and Eocene, climates were characterized by a low mean annual range of temperature (a maximum of 10–15°C), a moderate to high mean annual temperature (10–20°C) and abundant precipitation; broad-leaved evergreen vegetation extended to almost 60° North during the Paleocene and to well above 61° North during the Eocene. Poleward of the broad-leaved evergreen forests were forests that were broad-leaved deciduous; these deciduous forests, however, were unlike extant broad-leaved deciduous forests in general floristic composition and physiognomy. Coniferous forests probably occupied the northernmost latitudes.

The deciduous broad-leaved forests of the late Paleocene represented by deposits in northwestern Wyoming have been interpreted as indicating a mean annual temperature of about 10°C compared with the 13.5°C estimated for the early Eocene.

Paleocene strata of the western interior of North America contain numerous places where fossils of ferns, palms, conifers, angiosperms, ostracods, diatoms, insects, mollusks, fishes, amphibians, turtles, lizards, crocodilians, and primitive mammals have been found (Savage and Russell, 1983). Forests and savannas thrived in comparatively humid, frost-free subtropical climates at higher latitudes. Much of the Paleocene lowland of the western United States was probably a subtropical savanna. Wolfe and his colleague David Hopkins, analyzing percentages of plant species with leaves indicating a warm, humid climate, concluded: "No known Paleocene flora in North America indicates a climate cooler than warm-temperate, and subtropical climates existed at least as far north as latitude 62° in Alaska during part, if not all, of the Paleocene" (Wolfe and Hopkins, 1967).

Movements of Major Landmasses and Dispersal of Species

In order to understand more fully the evolution of life on this planet, it is necessary to consider the past geographical position of the continents. Paleogeography has played an important role in unraveling some of the mysteries of past and present-day animal distributions. For example, the presence of similar marsupials in such far-flung places as Australia and South America can be understood by knowing that these two continents were once joined with Antarctica to provide a land dispersal route for the early ancestors of these animals. Likewise, we can only understand the distribution of Paleocene mammals, including primates, through knowledge of continental drift, past continental positions, and possible oceanic barriers to animal migrations.

The geography of the Paleocene world about 65 MYA was very different from that of today's world (Fig. 2.1). A review of some of the major plate-tectonic events leading to the configuration of Paleocene landmasses shows that prior to about 200 MYA the continental crust was grouped into a single supercontinent named **Pangaea** ("all earth"). Later in the Mesozoic, Pangaea split into two supercontinents separated from each other by a broad waterway called the **Tethys Sea.** The northern continent, **Laurasia,** consisted of what is now recognized as North America, Europe, and Asia. The southern continent **Gondwanaland,** was made up of South America, Africa, Antarctica, India, Madagascar, and Australia.

Oceanic crust began to form between Australia and Antarctica in the late Cretaceous period, and the separation of their continental margins

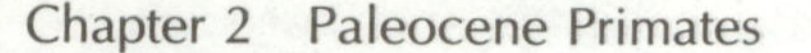

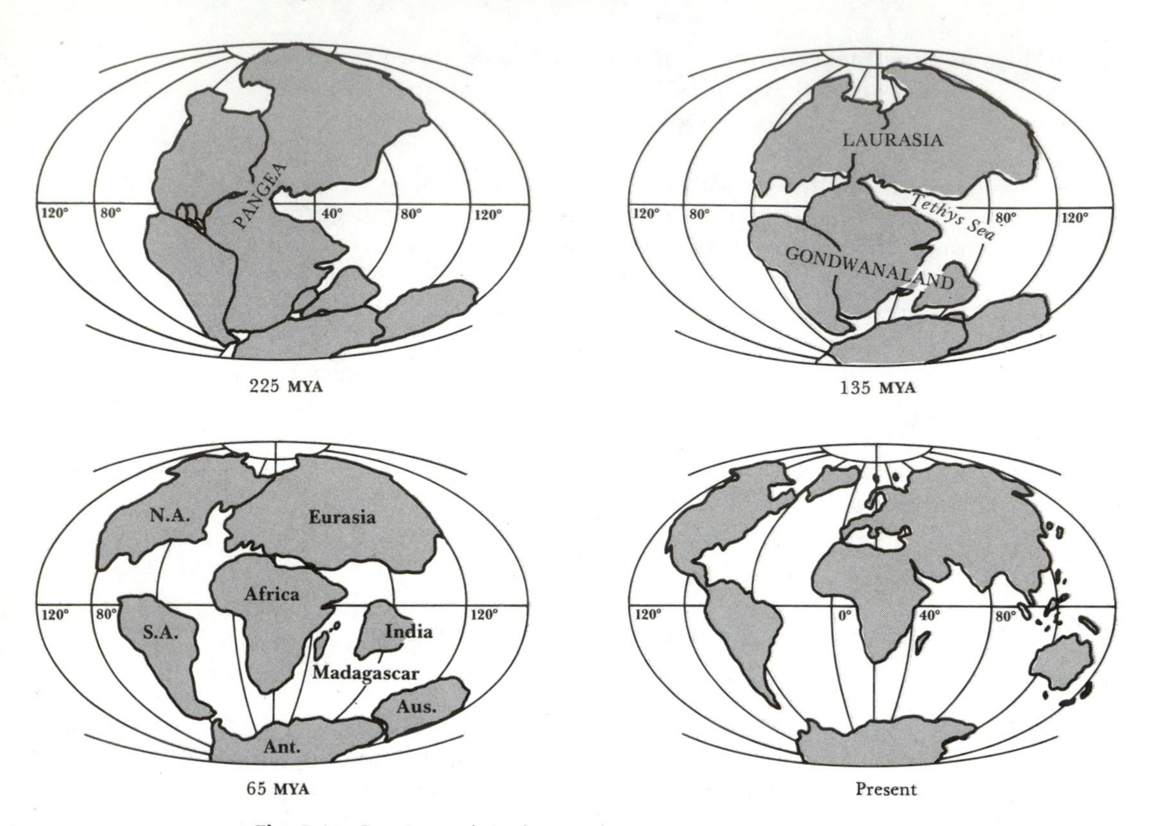

Fig. 2.1 Continental Drift over the Past 225 Million Years

has been dated to the late Paleocene or early Eocene. India probably separated from western Australia sometime in the Cretaceous and drifted 5,000 km northward as an island continent until it collided with Eurasia some 40 to 50 MYA (Markl, 1974).

The Caribbean Sea and the Gulf of Mexico formed about 180 MYA when North America moved away from both Africa and South America, which were then still joined. South America began to drift from Africa in the mid-Cretaceous, about 126 MYA, and the two continents finally separated some 90 MYA. Until as late as the Eocene (55 to 38 MYA), however, South America was closer to Africa, and possibly more accessible to immigration from Africa than from North America (Tarling, 1980). This raises a question concerning the origins of New World monkeys: did they evolve from African ancestors or North American ones? We shall deal with this question more fully in the discussion of Oligocene primates in Chapter 4. Suffice it to say here, that although North and South America were moving farther apart throughout the Paleocene and Eocene, some lizards, mammals, and plants derived from South America seem to have dispersed between these continents before the end of the Eocene (Gingerich, 1985).

North America and Europe were contiguous in the early Tertiary and probably provided a continuously forested avenue for mammalian dispersal until as late as the early Oligocene (McKenna, 1983*a,* 1983*b*). This land bridge was most effective during the early Eocene, when approximately half the mammalian genera were shared between the two continents. By contrast, during the late Paleocene the proportion was only about 10%. Another dispersal route for mammals was over the land bridge **Beringia,** which connected northeastern Asia and northwestern North America across what is now the Bering Strait. Beringia formed in the Cretaceous period and persisted well into the Paleogene.

Large ancient seas imposed on the landmasses of several continents in the Cretaceous and early Tertiary period, impeding the dispersal of species within continents. One continental seaway, an arm of the Arctic Ocean, partially separated northwestern Europe from central and southeastern Europe. Another, the Obik Sea, separated what is now the European part of the U.S.S.R. from the Asian part. The Obik Sea may have reached the Tethys Sea, which separated Africa from Eurasia. In North America a large Cretaceous seaway, the Cannonball Sea, extended from the Gulf of Mexico almost to the present border of Canada (Fig. 2.2).

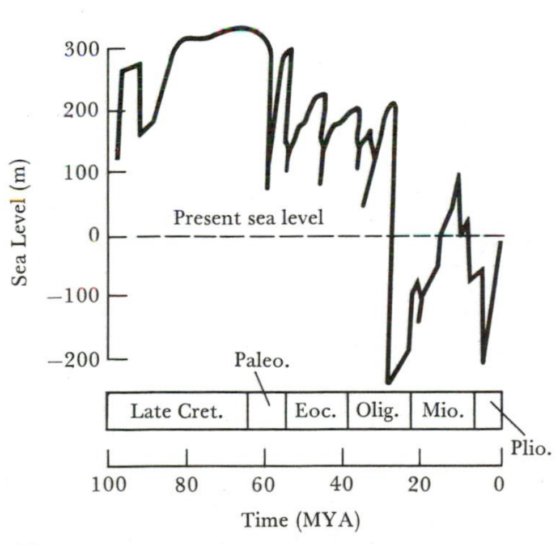

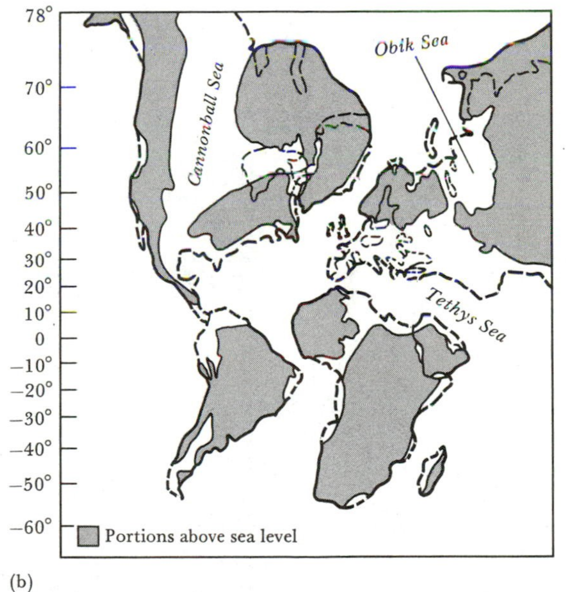

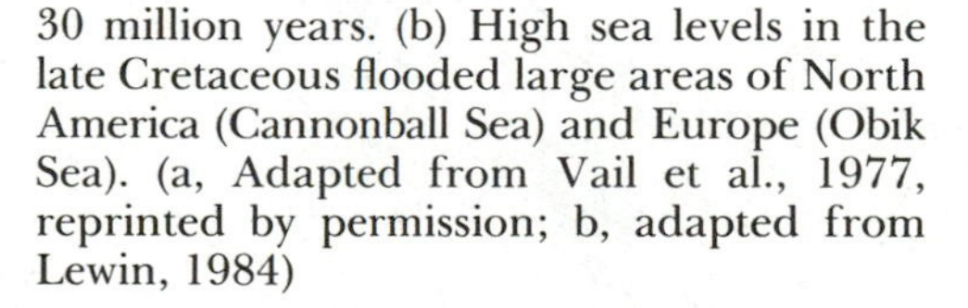

Fig. 2.2 Sea-level Changes through the Late Cretaceous and Tertiary (a) Sea levels have fluctuated dramatically over the past 100 million years. They were highest in the late Cretaceous and dropped to their lowest points in association with glacial onsets over the past 30 million years. (b) High sea levels in the late Cretaceous flooded large areas of North America (Cannonball Sea) and Europe (Obik Sea). (a, Adapted from Vail et al., 1977, reprinted by permission; b, adapted from Lewin, 1984)

SUMMARY OF THE PALEOCENE FOSSIL RECORD

Fossils of Paleocene mammals are known from North America, Europe, Asia, and South America. There is no known terrestrial fossil record for much of Scandinavia, and the early to middle Paleocene record for Europe in general is poor. The most diversified and abundant Paleocene faunas are from North America. Only one Paleocene locality is known from the entire African continent (in Morocco) and none at all from Australia or Antarctica. Undisputed Paleocene primates are known only from North America and Europe. Chinese fossil primates (except for the disputed genera *Petrolemur* and *Decoredon*) are unknown from the Paleocene, and primates appear to have reached eastern Asia only at the beginning of the Eocene (Dashzeveg and McKenna, 1977; Tong, 1979; Szalay and Li, 1986; Szalay et al., 1986). As Fig. 2.3 indicates, primate genera showed much less diversity during the Paleocene than in the succeeding Eocene.

The first Paleocene mammals were found in the 1870s in the San Juan Basin of New Mexico and Colorado by David Baldwin, a collector for the noted paleontologist Othiel C. Marsh of Yale University (Simpson, 1981). This basin contains the localities from which early, middle, and late Paleocene **land-mammal ages** in North America were first defined (Wood et al., 1941; Conroy, 1981; Lucas et al., 1981). (See Appendix A for a delineation of land-mammal ages.) Most of the European Paleocene sites seem to be late Paleocene in age, although the Mons Basin of Belgium may be of middle Paleocene age (Russell, 1975; Savage and Russell, 1983).

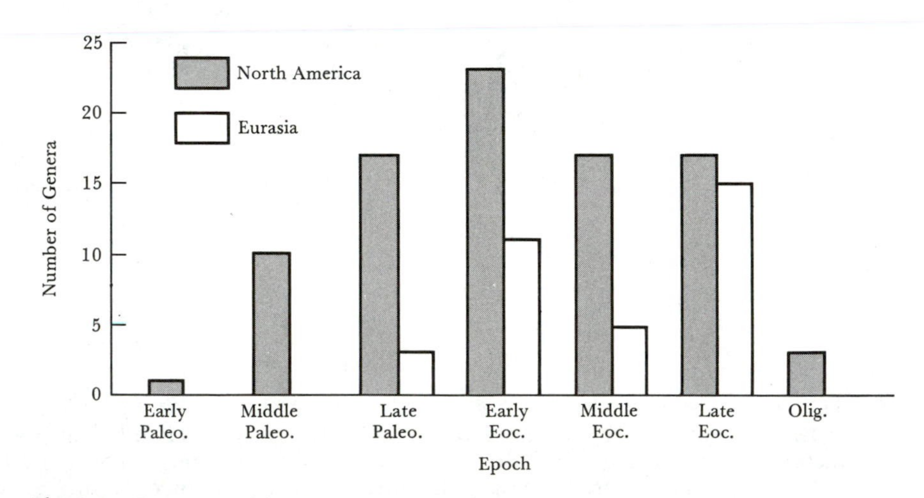

Fig. 2.3 Diversity of the Paleogene Primate Genera

Archaic mammalian orders dominate the Paleocene faunas. These are either orders that first appeared in the Cretaceous but did not change significantly during the Paleocene or orders that became extinct in the Paleocene (Krause, 1984). There were two significant periods of faunal change near the end of the Paleocene in North America. The first occurred at the Tiffanian–Clarkforkian mammal-age boundary, when rodents first appeared in North America. The second occurred at the Clarkforkian–Wasatchian mammal-age boundary, where many modern groups, including modern-looking primates, appeared for the first time in the fossil record. These primates are called modern because by the early Eocene they had evolved some of the structures that characterize living primates, for instance, a postorbital bar, nails instead of claws on the digits, and a widely divergent hallux indicative of arboreal habitats. Many of these features had not yet evolved in the Paleocene primates, as we shall see.

The combined evidence indicates that vertebrates during the Paleocene thrived near water sources in woodland and savanna environments in North America, Eurasia, and South America. The climate was uniformly warm throughout the year in these areas, with no pronounced seasonality, no arid regions, and little or no freezing temperatures. An archaic land-mammal fauna lived in this environment. Between 60 and 55 MYA specialized small archaic primates, all classified as members of the suborder **Plesiadapiformes,** first appeared and were abundant in the forests of the Northern Hemisphere (Savage and Russell, 1983).

PLESIADAPIFORMES

In this section we will discuss the origins of the plesiadapiforms and the general features of their morphology. The following subsections will cover each family separately. Most researchers of early primate evolution recognize six families of plesiadapiforms: **Plesiadapidae, Carpolestidae, Saxonellidae, Microsyopidae, Paromomyidae,** and **Picrodontidae** (Table 2.1).

Of all the faunas of the Cenozoic era, terrestrial Paleocene invertebrates and vertebrates are the least well known (McKenna, 1966). Fortunately for primate paleontologists, the archaic primates commonly known as plesiadapiforms are among the more abundant representatives of the many Paleocene faunas, at least in North America. There the fossilized remains of these archaic primates have been recovered from deposits ranging from Alberta in the north to Texas in the south, with the greatest concentration in the Rocky Mountain area of the western

Table 2.1 Classification of the Plesiadapiformes

Taxa	*Epoch (and Area)*[a]	*Dentition*[b]
Family PLESIADAPIDAE	M. Pal. ≃ E. Eoc.	
Chiromyoides	L. Pal. ≃ E. Eoc. (Eur. and N.A.)	1.?.?3.3/1.0.2.3
Nannodectes	L. Pal.	2.1.3.3/1.0–1.3.3
Platychoerops	E. Eoc. (Eur.)	?/1.0.2.3
Plesiadapis	L. Pal. ≃ E. Eoc. (Eur. and N.A.)	2.0 or 1.3.3/1.0.2–3.3
Pronothodectes	M. Pal.	2.1.3.3/2.1.3.3
Family CARPOLESTIDAE	M. Pal. ≃ E. Eoc.	
Carpodaptes	?M. ≃ L. Pal.	?/2.1.2.3
Carpolestes	L. Pal. ≃ E. Eoc.	2.1.3.3/2.1.2.3
Elphidotarsius	M. Pal.	?/2.1.3.3
Family SAXONELLIDAE		
Saxonella	?L. Pal. (Eur. and N.A.)	?/1.0.2.3
Family MICROSYOPIDAE	M. Pal. ≃ L. Eoc.	
Alsaticopithecus	M. Eoc. (Eur.)	
Berruvius	L. Pal. (Eur.)	
Craseops	L. Eoc.	
Micromomys	L. Pal.	
Microsysops (includes *Cynodontomys*)	E. ≃ L. Eoc.	2.1.3.3/1.0.3.3
Navajovius	L. Pal.	?.1.3.3/1.1.3.3 or 2.1.2.3
Niptomomys	E. Eoc.	?/1.1.3.3
Palaechthon	M. Pal.	2.1.3.3/2.1.3.3
Palenochtha	M. Pal.	2.1.?3.3/2.1.3.3
Plesiolestes (probably = *Palaechthon*)	M. Pal.	
Talpohenach (probably = *Palaechthon*)	M. Pal.	
Tinimomys	E. Eoc.	?/1.0.3.3
Torrejonia	M. Pal.	
Uintasorex	M. ≃ L. Eoc.	?/1.0.3.3
Family PAROMOMYIDAE	M. Pal. ≃ L. Eoc.	
Ignacius	M. Pal. ≃ L. Eoc.	2.1.2.3/1.0.1–2.3
Paromomys	M. Pal.	2.1.3.3/2.1.3.3
Phenocolemur	L. Pal. ≃ M. Eoc. (Eur. and N.A.)	?2.1.3.3/1.0.1.3
Family PICRODONTIDAE	M. ≃ L. Pal.	
Picrodus	M. Pal.	?/2.1.2.3
Zanycteris (probably = *Picrodus*)	L. Pal.	?.1.3.3/?
PLESIADAPIFORMES, *incertae sedis*		
Purgatorius	L. Cret. ≃ E. Pal.	?/?.1.4.3

[a] All genera are known only from North America unless otherwise specified. E. = early, M. = middle, L. = late, Pal. = Paleocene, Eoc. = Eocene, Cret. = Cretaceous, Eur. = Europe, N.A. = North America

[b] Dental formulas are provided where known. However, they may be based on only one species in the genus.

Source: Adapted from Rose and Fleagle (1981).

United States (Gingerich, 1976; Krause, 1978) (Fig. 2.4a). In Europe similar archaic primates are known from France, Belgium, East Germany, and England (Fig. 2.4b). Undoubted Paleocene primate fossils have not been discovered on any other continent. With the possible exception of a single lower molar tooth of the primitive plesiadapiform species *Purgatorius ceratops* found in a late Cretaceous deposit in Montana, no primates have yet been discovered in rocks of Mesozoic age, and only one primate genus (again *Purgatorius*) has been discovered in early Paleocene deposits (Van Valen and Sloan, 1965; Clemens, 1974).

There is an exasperating limitation to our present understanding of the earliest archaic primates, since the identity of both their antecedents and their descendants remains uncertain. Although some of these archaic primates actually survived into the late Eocene, they clearly reached maximum diversity and abundance much earlier in their evolutionary history. After that time they rapidly declined in number and diversity. In fact, the plesiadapiforms were among several archaic mammalian groups that were replaced in the Eocene by more modern mammals. The plesiadapiform primates were probably outcompeted during the Eocene by rodents, a group that evolved in Asia and migrated to North America in the late Paleocene (Krause, 1981).

Origins

It is sometimes difficult to distinguish fossils of the earliest archaic primates from those of other contemporary small mammals. This reflects the obvious fact that related animals are more similar to one another the closer they are in time to their common ancestor. In addition, paleontologists must work almost exclusively with fragmentary bits of jaws and teeth, the structures most likely to be preserved in the fossil record. Fortunately, these are usually the most reliable remains for taxonomic and phylogenetic studies.

There is still no consensus as to which Cretaceous–Paleocene insectivoran family gave rise to the primates. Many candidates have been proposed:

1. The Erinaceomorpha (hedgehoglike animals) (Szalay, 1975*a*)
2. Other members of the suborder Lipotyphla, which includes not only erinaceomorphs but also shrews, moles, tenrecs, golden moles, selenodons, and several fossil taxa
3. The Tupaiidae (tree shrews) (Le Gros Clark, 1959)
4. One of several archaic taxa known as the Leptictidae or the Apatemyidae, which are loosely classified as insectivorans (McKenna, 1966)

The choice is difficult mainly because there is still no consensus on the morphological definition of a primate or on how to recognize one unequivocally in the early fossil record.

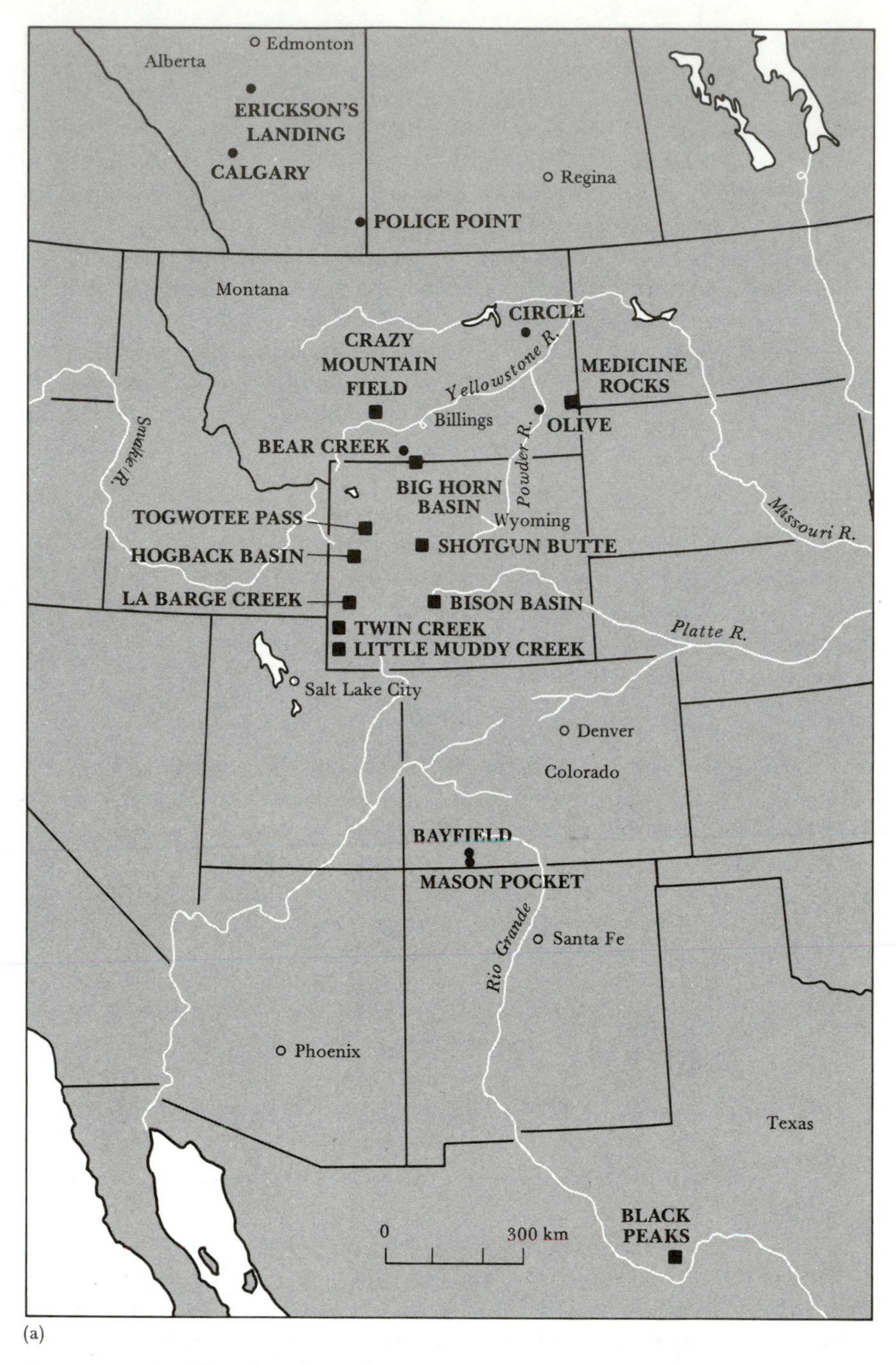

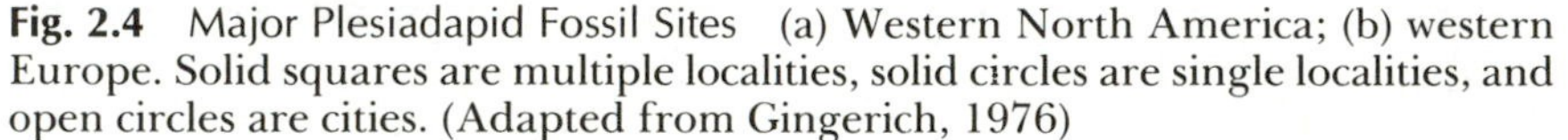
Fig. 2.4 Major Plesiadapid Fossil Sites (a) Western North America; (b) western Europe. Solid squares are multiple localities, solid circles are single localities, and open circles are cities. (Adapted from Gingerich, 1976)

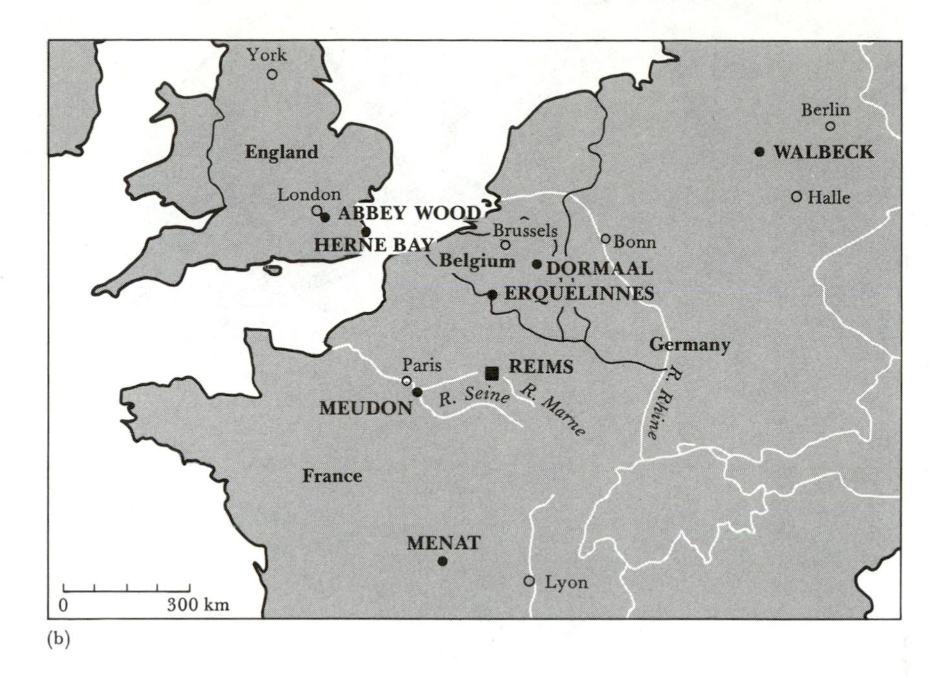

(b)

New fossil evidence (Novacek et al., 1983) has cast doubt on at least one of the competing alternatives, however: that of close ties between erinaceomorphs and primates. The morphological evidence that has been used in the past to promote this relationship was the presumed incorporation of the **petrosal** bone into the **auditory bulla** in both forms (Fig. 2.5). (The auditory bulla, also called the tympanic bulla, is the bony housing

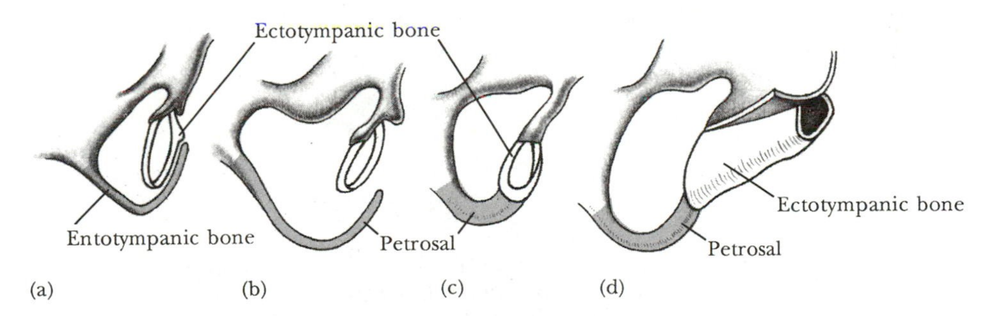

Fig. 2.5 Auditory Bulla in Mammals The eardrum (tympanic membrane) is supported by a ring formed from the ectotympanic bone. In (a) tree shrews and (b) lemurs, it is within the auditory bulla (shaded); in (c) lorises and New World monkeys, it is attached to the edge of the bulla; and (d) in tarsiers, Old World Monkeys, apes, and humans, it is expanded to form a tube. The main bone forming the bulla is the entotympanic in tree shrews and the petrosal (part of the temporal bone) in primates. (Redrawn from Hershkovitz, 1977)

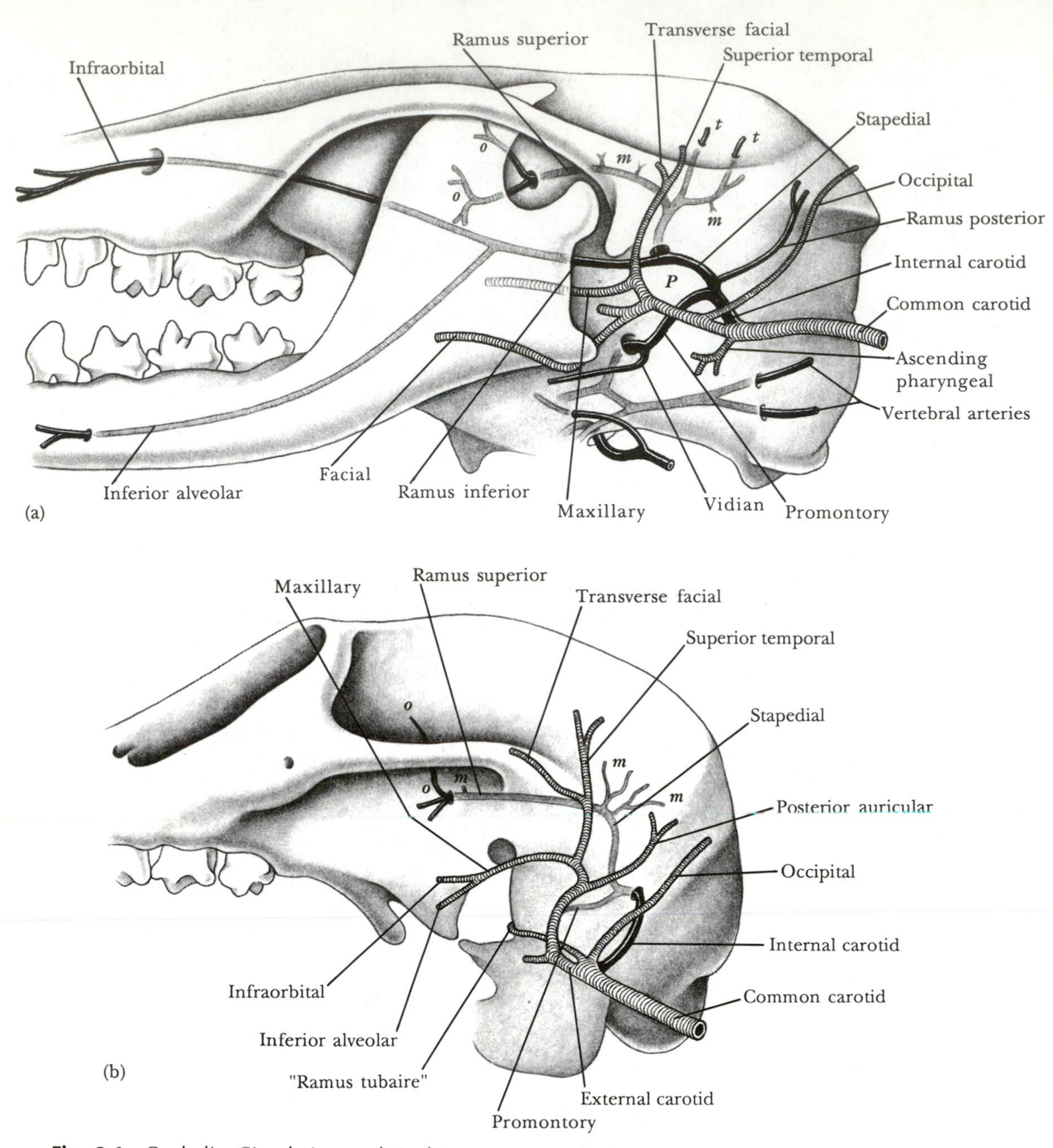

Fig. 2.6 Cephalic Circulation and Auditory Region in Primates and Their Predecessors (a) Schematic representation of the major cephalic arteries in primitive eutherians. Internal carotid, stapedial, and vertebral arteries (and their branches) are black; external carotid artery and its branches are striped. The internal carotid–promontory artery crosses the ventral surface of the petrosal bone's promontory (*P*). Branches of stapedial artery: *o*, orbital; *m*, meningeal; *t*, temporal. (b) Major cephalic arteries of the prosimian *Lemur*, with lettering and shading similar to (a). (c) Right middle-ear region of *Lemur*, a basal view. The auditory bulla formed by the petrous portion of the temporal bone has been partially removed to expose the tympanic ring, internal carotid artery, and carotid foramen. (a,b, From MacPhee and Cartmill, 1986; c, redrawn from Hershkovitz, 1977)

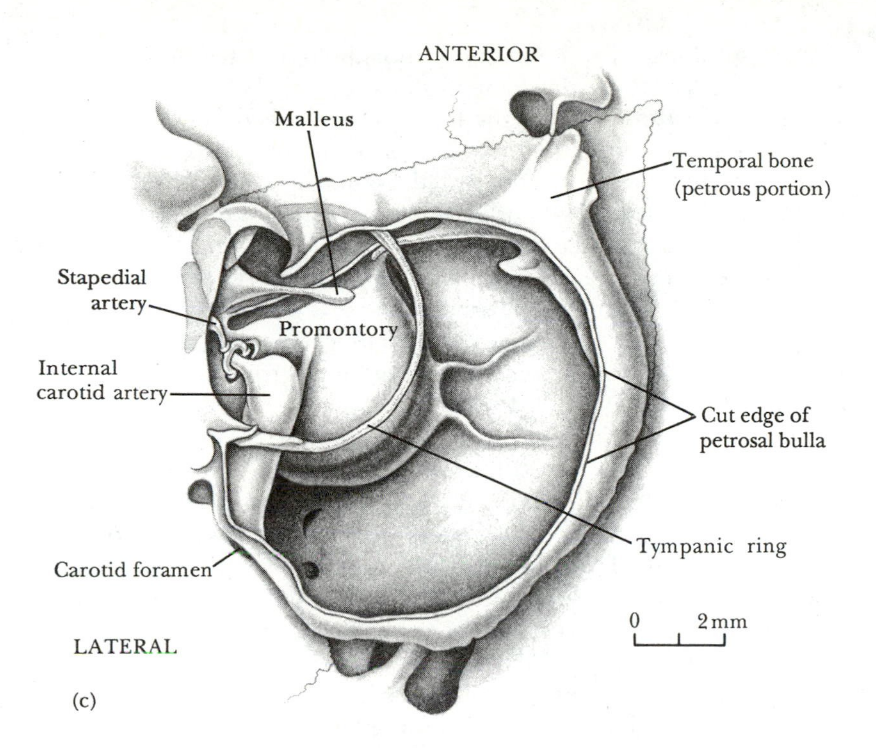

(c)

for the middle-ear cavity and in primates is normally formed by the skull's petrosal bone.) The new evidence, in the form of the fossilized remains of *Diacodon,* an early Eocene erinaceomorph, contradicts this view, since there is no petrosal contribution to the auditory bulla in this genus.

General Morphology

Cranial Morphology In spite of the widespread belief that the order Primates is characterized by morphological trends rather than by specific characters unique to the group, some investigators argue that there are certain anatomical features, particularly those of the auditory region, that can be used to define the earliest members of the order. Such features are said to include the following:

1. The loss of the medial carotid artery and the associated absence of a major vascular foramen on the medial side of the auditory region for this artery;
2. The presence of an ossified auditory bulla formed by the petrosal bone;
3. The presence of a bony canal for the entire intrabullar carotid circulation;
4. The presence of a **fenestra cochleae** (also called the fenestra rotunda, or round window) ventrally shielded by the internal carotid canal (Szalay, 1975*a;* Archibald, 1977; MacPhee et al., 1983) (Fig. 2.6)

The usefulness of some of these morphological features for identifying members of the order Primates has recently been challenged. It is now known, for example, that the first feature—loss of the medial carotid artery—does not apply to all plesiadapiforms; the paromomyid genus *Ignacius,* is a notable exception. In addition, the second feature—presence of an ossified petrosal auditory bulla—cannot be accurately determined without ontogenetic data, and all the known plesiadapiform skulls are adult specimens. To further confuse the issue, some fossils that have been considered archaic primates on the basis of one criterion, such as dentition, totally lack other supposed qualifications, such as the ossified petrosal auditory bulla; a typical example here is the microsyopid genus *Cynodontomys* (Novacek et al., 1983).

The importance attached to carotid circulatory patterns in mammalian systematics can be traced back to the work of William Matthew of the American Museum of Natural History in the early part of the 20th century. He proposed that three separate arteries passed through the middle-ear region of the carnivoran–insectivoran ancestor to supply intracranial structures: separate medial and lateral internal carotid arteries (to supply the brain) and a stapedial artery (to supply the middle ear and adjoining tissues). As stated earlier, it had been previously thought that ancestral primates had lost the medial carotid artery. Recent research, however, suggests that early **eutherians** (placental mammals) never had two separate internal carotid arteries. What were originally thought to be grooves for the medial carotid artery in fossilized skull bones are now interpreted to be grooves for a venous channel, most likely the inferior petrosal sinus (Wible, 1983). Thus the whole question of determining homologous carotid circulatory patterns in fossil and extant primates is currently in a state of considerable flux (MacPhee and Cartmill, 1986).

Modern primates have an artery known as the **promontory artery,** which is the homologue of the human internal carotid artery. In the earliest known primates, the plesiadapiforms, the grooves for the promontory artery are often tiny and/or highly variable in size. In some *(Ignacius)* the arterial grooves are lacking altogether. In others *(Cynodontomys)* the grooves seem well developed. As noted earlier, *Cynodontomys* apparently lacked the petrosal bulla, a supposed hallmark of early primates. Thus any definition of archaic primates that states they were all characterized by having a carotid circulation enclosed in bony tubes within an ossified petrosal bulla is not compatible with the paleontological record as currently understood (MacPhee et al., 1983).

All living and fossil primates, with the exception of the Microsyopidae, have a middle-ear region enclosed by an ossified petrosal bulla. They also have either a ringlike or a tubular **ectotympanic** bone for supporting the **tympanic membrane** (eardrum) (Fig. 2.5). It is not certain which of these two ectotympanic character states is primitive for the order; the ringlike ectotympanic bone characteristic of many lemurs is often considered the primitive character, but the earliest known primate skulls in the fossil record, *Plesiadapis* and *Phenacolemur*, both have tubular ectotympanic bones.

All modern primates also have a complete postorbital bar, whereas none of the archaic Paleocene primate fossils with this region of the skull intact shows evidence of such a feature (Fig. 2.7). (Some of the major osteological features of the mammalian skull are shown in Appendix B.)

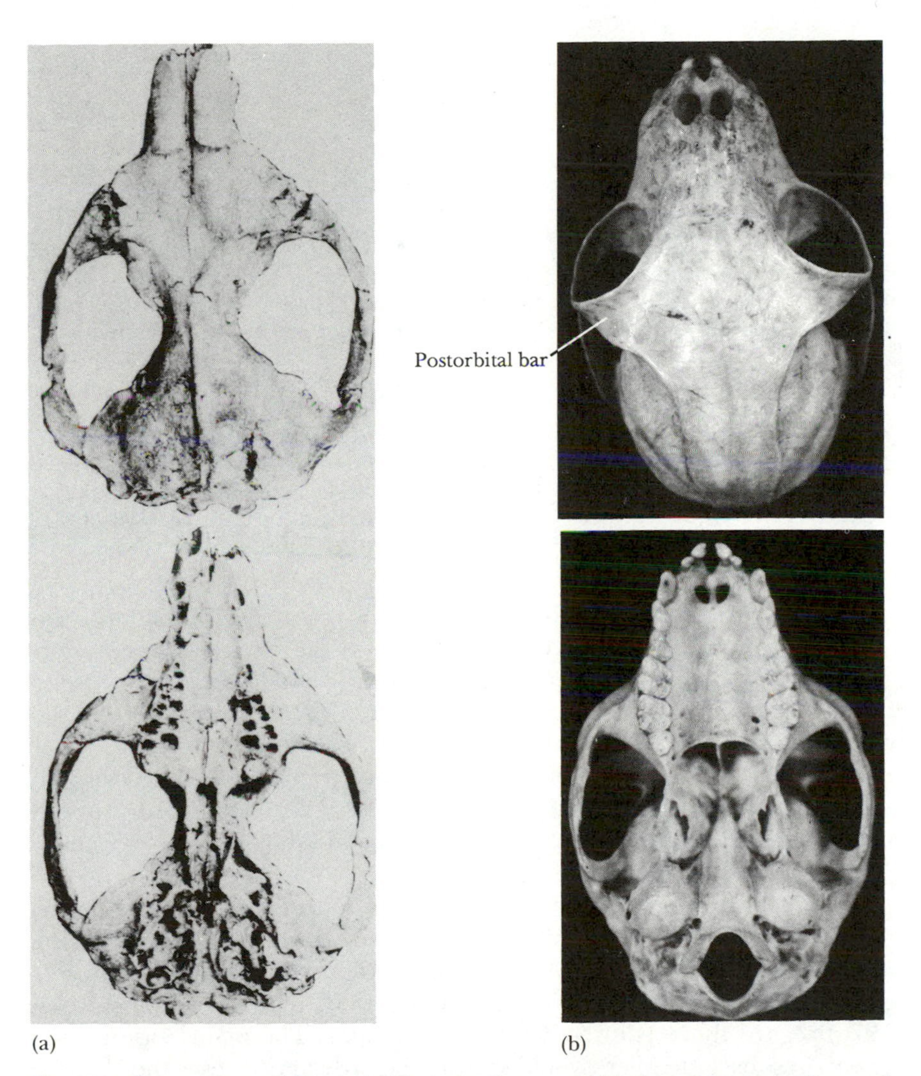

Fig. 2.7 Comparison of Crania in Plesiadapiforms and Modern Primates Top and bottom views of (a) *Plesiadapis*, and (b) *Indri* (a modern prosimian). The postorbital bar, found in all modern primates, is not present in any plesiadapiform. Scale about 6/10 × natural size. (a, from Gingerich, 1976)

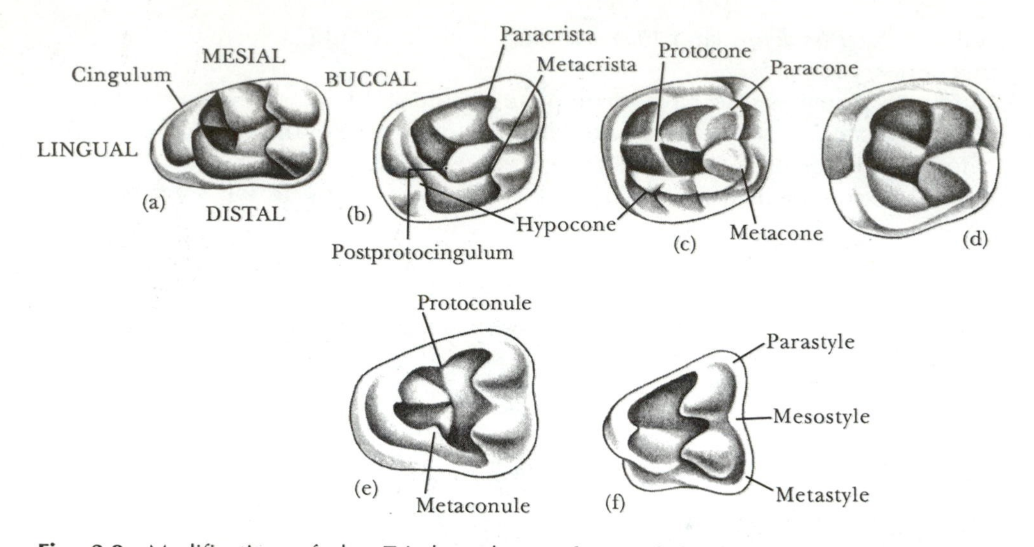

Fig. 2.8 Modification of the Tritubercular Molar Tooth Found in Primates Left upper molar is shown throughout; the four directions are mesial (anterior, or toward the front of the mouth), distal (posterior, or toward the back of the mouth), lingual (medial, or toward the tongue), and buccal (lateral, or toward the cheek). (a) Primitive tritubercular tooth with an internal cingulum in *Tarsius*. (b) Hypocone enlarged to form a quadritubercular tooth in the slow loris *(Nycticebus)*. (c) Quadritubercular tooth with cusps of approximately equal size in the Miocene fossil ape *Proconsul*. (d) Upper molar of the modern human for comparison. (e) Upper molar with a protoconule and a metaconule in the Eocene primate *Pseudoloris*. (f) Upper molar with a pronounced external cingulum bearing three small cuspules in the tree shrew *Tupaia*. (Redrawn from Le Gros Clark, 1959)

It should be apparent from this brief discussion that plesiadapiform primates cannot be unequivocally defined on the basis of cranial morphology alone. The same has been said for the dental evidence: "In spite of the relatively consistent balance of characters in the cheek teeth of the earliest primates . . . it is apparent that . . . Paleocene primates are not readily recognized on dental evidence alone" (Szalay, 1972*a*). Owing to these difficulties, some authors have argued that plesiadapiforms should not be included within the order Primates at all (Martin, 1968; Cartmill, 1972).

Of the six widely recognized families of Paleocene plesiadapiforms (Table 2.1), inclusion of the Microsyopidae is the most controversial. Some authors (Szalay and Delson, 1979) consider them to be primatelike insectivorans, whereas others (Bown and Gingerich, 1973; Bown and Rose, 1976) regard them as insectivoranlike primates. The main reason some researchers exclude Microsyopidae from the order Primates is the already-mentioned fact that some members of this family do not have an ossified petrosal bulla. As each of the six families is reviewed in the following sections, it will become apparent that they are united only by shared

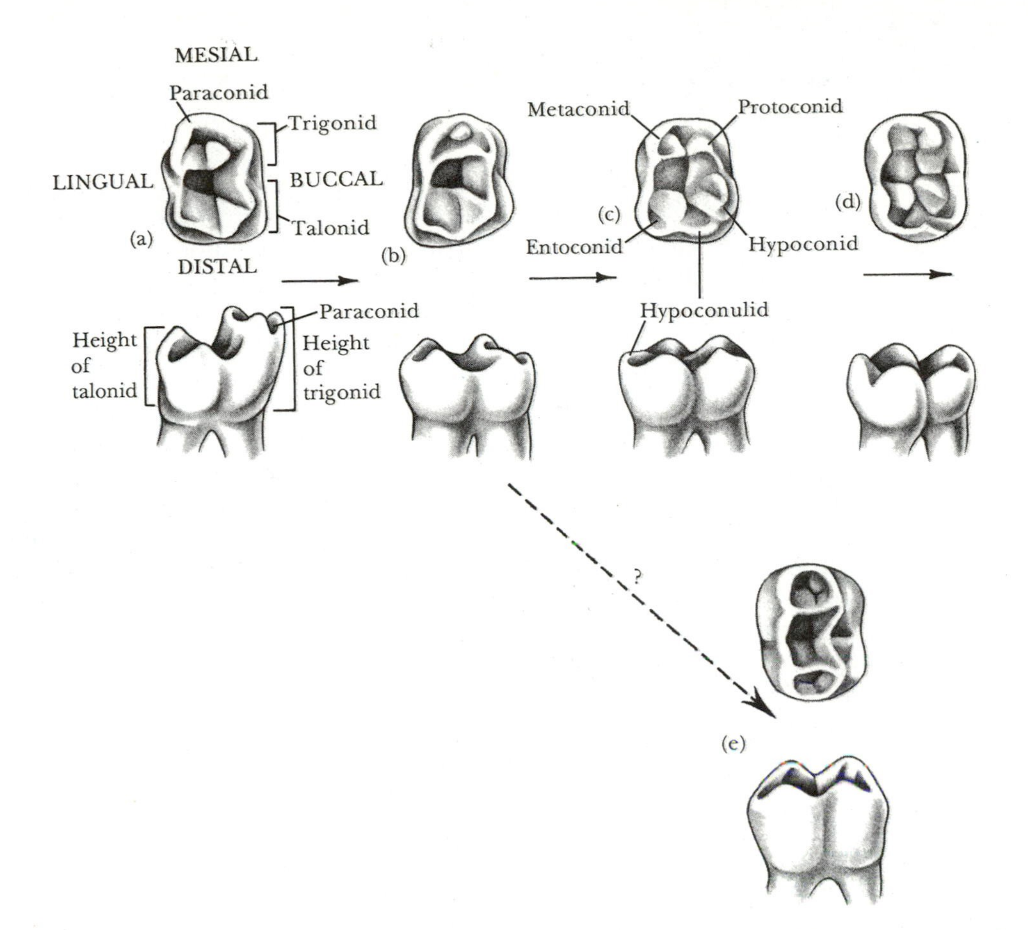

Fig. 2.9 Presumed Evolution in the Cusp Pattern of Primate Molars Top row gives overhead (occlusal) view of a right lower molar with the orientation as indicated; bottom row shows the same molar seen from the buccal (lateral) side, with the mesial end to the right. (a) In the primitive lower molar (still seen in the modern tarsier), the paraconid helps form the trigonid. (b) In the Eocene anthropoid *Amphipithecus* the paraconid is reduced, and the trigonid and talonid occupy roughly the same level. (c) In the Oligocene anthropoid *Parapithecus* the paraconid has disappeared; the trigonid bears the metaconid and protoconid; and the talonid has the hypoconid, entoconid, and a relatively well-developed hypoconulid. (d) In the Miocene ape *Proconsul,* the five cusps have developed about equally and are separated by a characteristic pattern of intervening grooves. (e) In the lower molar of an Old World monkey the characteristic bilophodont pattern is evident. (Redrawn from Le Gros Clark, 1959)

features of their dentition. Other shared features are symplesiomorphic (retained primitive) characters, which, as explained in Chapter 1, do not indicate a special relationship.

This text will follow a taxonomic scheme in which Microsyopidae are included within the order Primates (Bown and Rose, 1976; Rose and Fleagle, 1981). It must be emphasized, however, that other experts on early primate evolution (for example, Szalay and Delson, 1979; Schwartz, 1986) have different taxonomic interpretations of these same early primates and should be consulted for alternative viewpoints.

Dentition For reasons outlined in Chapter 1, dental evidence is crucial to paleontologists. (Dental terminology can be intimidating to the uninitiated; see Appendix B for explanation and illustration of basic terms.) In general, there are several dental trends that characterize the evolution of the primate dentition from their insectivoran ancestors. One trend involves the conversion of the more primitive **tritubercular** (triangular) upper molars to the more **quadritubercular** (rectangular) shapes through the elaboration of the hypocone. In addition, the high, pointed cusps of early insectivorans (typified by *Tupaia,* a genus of tree shrews) become lower and more bulbous in primates. Owing to the reduced emphasis on transverse shearing, the upper cheek teeth of primitive primates are usually narrower than those of insectivorous mammals. The primitive eutherian emphasis on transverse tooth shearing is gradually deemphasized in primate evolution by reducing the long ridgelike structures known as **paracrista** and **metacrista** (Fig. 2.8).

In the lower dentition, as grinding becomes more important, the overall height differential between the anterior **(trigonid)** and posterior **(talonid)** portions of the postcanine dentition is reduced (Szalay, 1968*a*.) The **paraconid** is reduced and finally disappears altogether (Fig. 2.9). These features are particularly evident in the molar teeth of *Plesiadapis* (Fig. 2.10).

Known Paleocene primates share several other dental features: (1) the central incisors are enlarged, and the lower ones are often protruding as well; (2) the **antemolar dentition** (lateral incisors, canines, and premolars) is reduced; and (3) a characteristic structure called the **postprotocingulum,** or ***Nannopithex* fold,** develops on the upper molars (see Fig. 2.8).

Plesiadapidae

Plesiadapids were one of the most successful of early primate families in terms of both diversity of species and abundance of specimens. Five genera *(Pronothodectes, Nannodectes, Plesiadapis, Chiromyoides,* and *Platychoerops)* and approximately two dozen species are currently recognized. They ranged from the middle Paleocene to early Eocene in North America and from the late Paleocene to early Eocene in Europe, a duration of 10 to 12 million years. Their presence in both western North America and Europe (especially the widespread distribution of similar species of *Plesiadapis* and *Chiromyoides*) provides convincing evidence of a continuous land connection between those two landmasses in the early Tertiary.

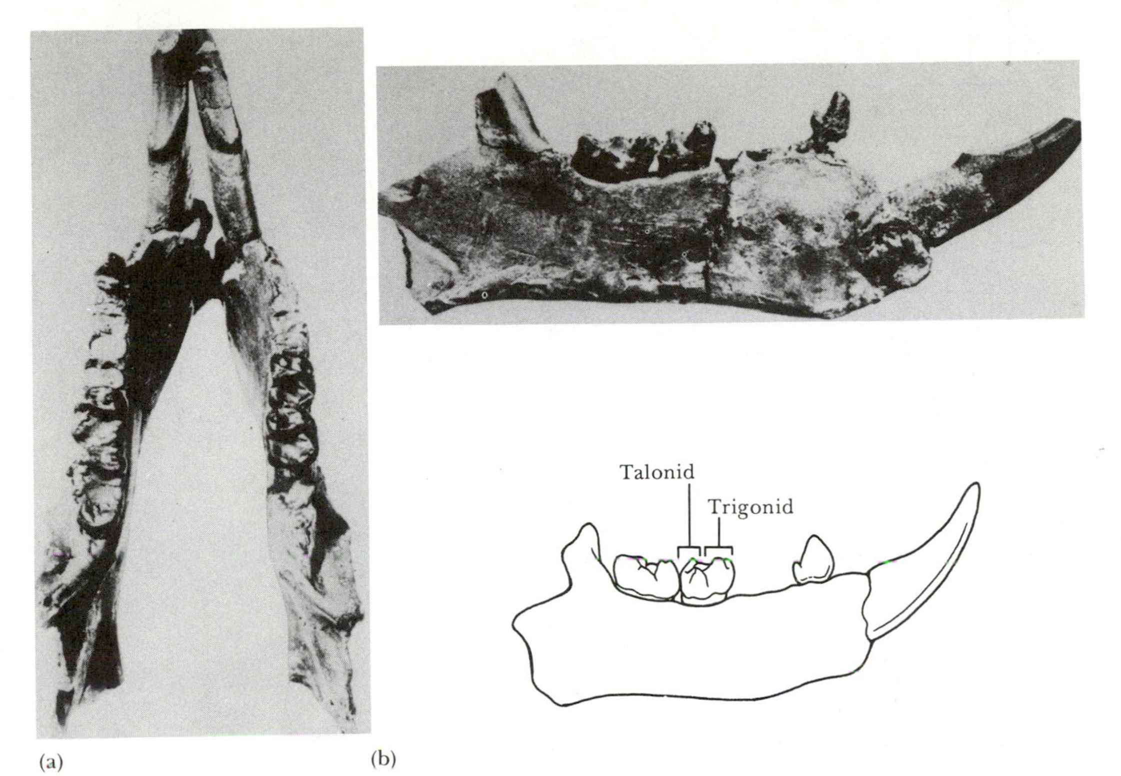

(a) (b)

Fig. 2.10 Lower Jaw of *Plesiadapis gidleyi* (a) Occlusal and (b) side views. Note the reduced difference in height between the trigonid and talonid. Scale about 2.4 × natural size. (Photos from Matthew, 1917)

The most recent comprehensive review of the family Plesiadapidae is by Philip Gingerich of the University of Michigan, and much of the following discussion is based on that study (Gingerich, 1976).

Because these early primates are so common in Paleocene and early Eocene deposits, they have been extremely valuable in Paleogene biostratigraphic studies in the western United States and Europe (Gingerich, 1975*b*). In the United States this zonation has been based mainly on an evolving lineage from *Pronothodectes* to *Plesiadapis,* and in Europe on the evolving lineage from *Plesiadapis* to *Platychoerops* (Fig. 2.11).

Dentition Classification within this family has been based largely on dental evidence. Primitive plesiadapids have nine teeth on each side of the upper and lower jaws, with a dental formula of 2.1.3.3/2.1.3.3. The primitive eutherian dental formula of 3.1.4.3/3.1.4.3 had already been reduced by the middle Paleocene through loss of an incisor and a premolar in both

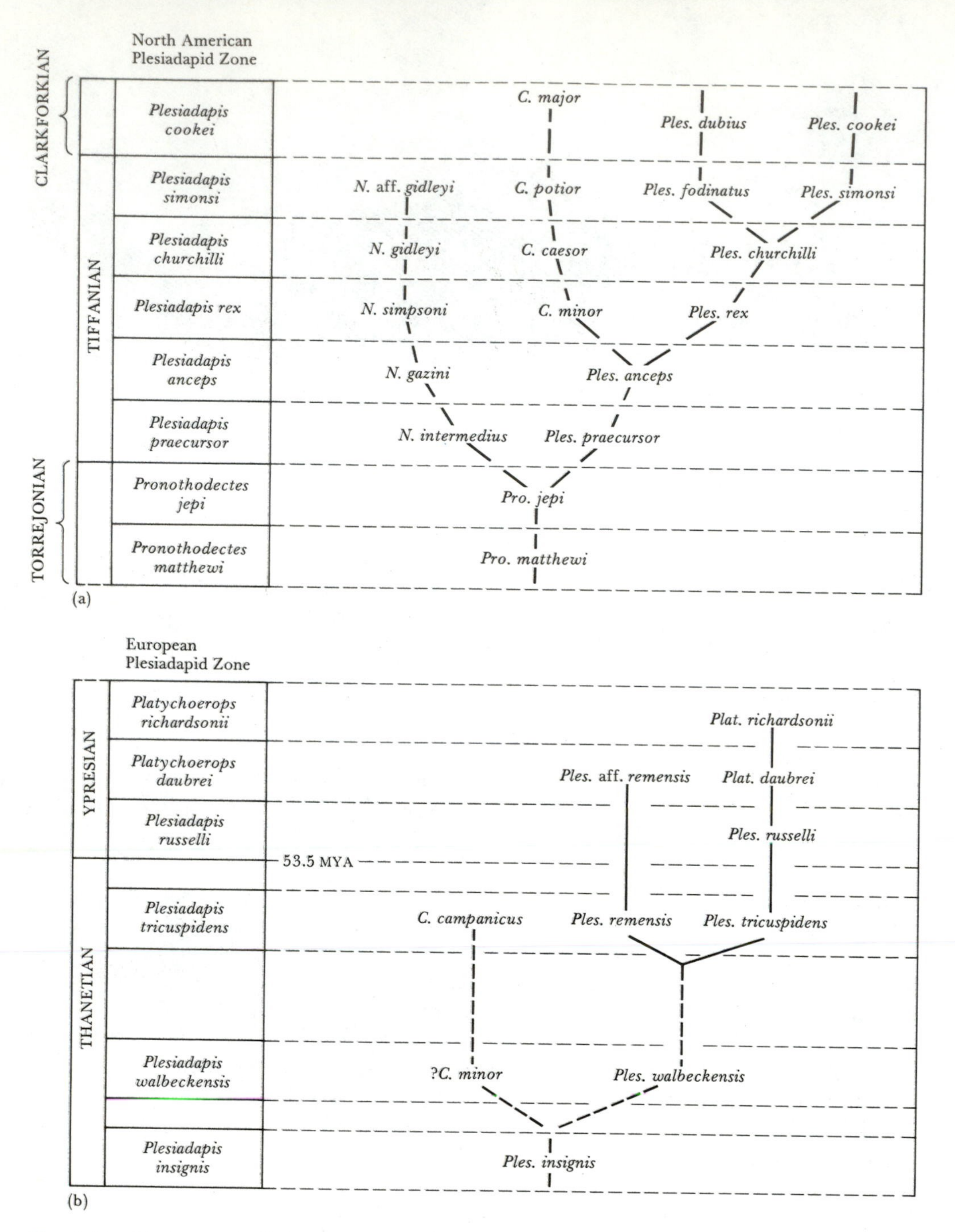

Fig. 2.11 Phylogenetic Relationships of Plesiadapidae (a) North American species of *Pronothodectes, Nannodectes, Chiromyoides,* and *Plesiadapis.* Period covers 8 to 9 million years (approximately 62–53 MYA). (b) European species of *Chiromyoides, Plesiadapis,* and *Platychoerops.* Period covers 7 to 8 million years (approximately 58–50 MYA). (From Gingerich, 1976)

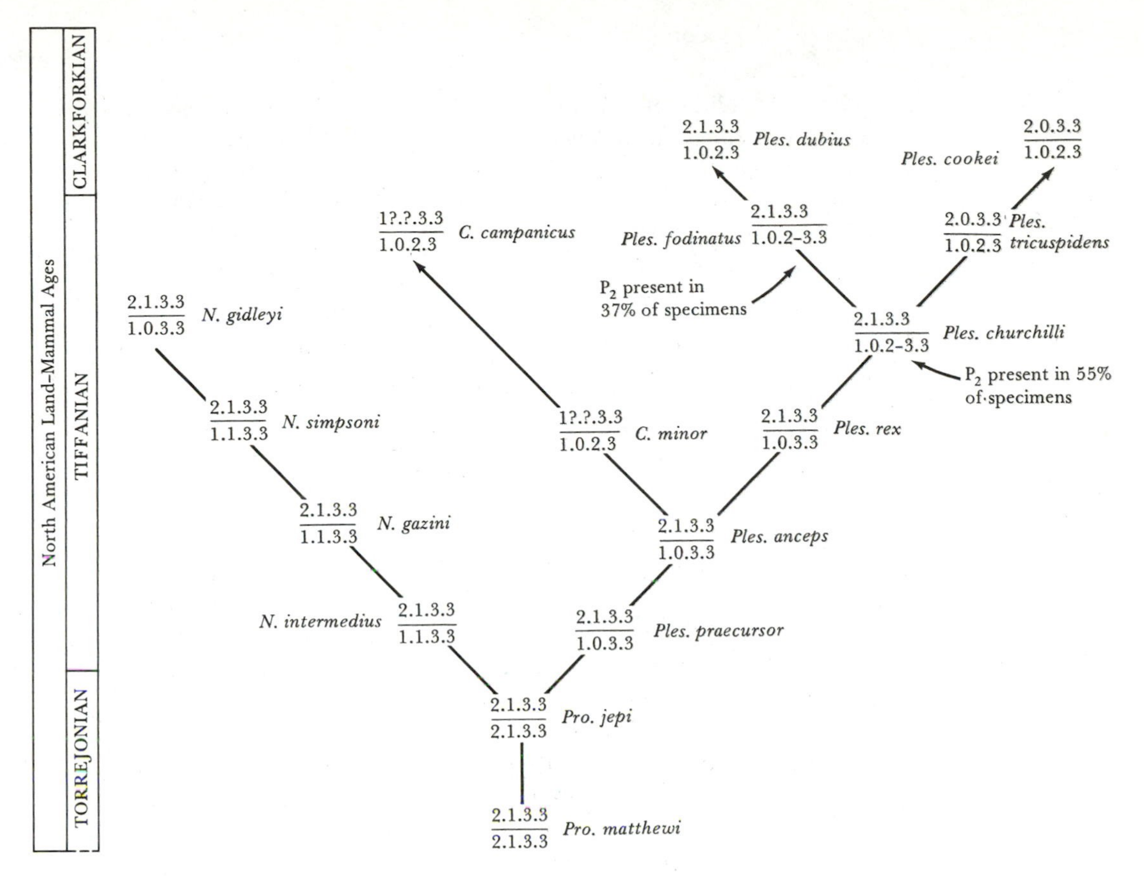

Fig. 2.12 Reduction in Dental Formula of North American Plesiadapids over Time Because they are better known, the European species *Plesiadapis tricuspidens* and *Chiromyoides campanicus* have been substituted for their North American equivalents (*P. simonsi* and *C. caesor* through *C. major*, respectively. (From Gingerich, 1976)

upper and lower jaws. In later plesiadapids the lower dentition was further reduced to 1.0.2.3 (Fig. 2.12).

The earliest plesiadapid genus to appear in the fossil record is the middle Paleocene form *Pronothodectes*. Sometime in the late Torrejonian or early Tiffanian land-mammal age, *Pronothodectes* gave rise to two other plesiadapid lineages, *Nannodectes* and *Plesiadapis*. Although both lineages ultimately lost the lower lateral incisors and canines, the lower canines were retained for a longer time in the *Nannodectes* lineage before being finally lost. In addition, fossils of the *Nannodectes* lineage have narrower, less robust incisors and premolars than those of the *Plesiadapis* lineage. *Nannodectes* species (*N. intermedius, N. gazini, N. simpsoni,* and *N. gidleyi*) seem to have formed an evolving lineage through the late Paleocene of western North America and were generally smaller than contemporaneous species of *Plesiadapis*. The geologically youngest species of *Nannodectes* had a lower dental formula of 1.0.3.3.

The second lineage that evolved from *Pronothodectes* (at least in North America) was the genus *Plesiadapis* itself. It first appears in the late Paleocene of western North America and Europe and ranges into the early Eocene of both continents. Over a dozen species are currently recognized, eight from North America (*P. praecursor, P. anceps, P. rex, P. churchilli, P. fodinatus, P. dubius, P. simonsi,* and *P. cookei*) and five from Europe (*P. insignis, P. walbeckensis, P. remensis, P. tricuspidens,* and *P. russelli*).

Plesiadapis differed from ancestral *Pronothodectes* and most species of *Nannodectes* in lacking lower canines. However, an abundance of new plesiadapid material collected recently in the Paleocene strata of eastern Montana casts some doubt on this distinction. For example, *P. anceps* clearly retained a canine, the loss of which has been traditionally regarded as one of the best diagnostic characters of the genus (Watters and Krause, 1986).

The incisors were very distinctive and rodentlike in this genus (Fig. 2.13). Upper incisors had three pointed cusps toward the front, middle, and side (called the anterocone, the mediocone, and the laterocone respectively). In addition, there was a large posterocone toward the back (a feature also seen in *Pronothodectes, Nannodectes,* and *Chiromyoides*). As mentioned earlier, the second lower incisors, the canines, and the anterior premolars became less functional through time in the Plesiadapidae, and indeed, in many species these teeth were subsequently lost. From analyses of wear facets on upper and lower incisors, it appears that these peculiar teeth were adapted for cutting stems and other soft vegetation. The reduction of canines and anterior premolars in the plesiadapids suggests that these animals ingested food in small pieces, as is typical of many herbivorous species today. The increasing molarization of the premolars, the presence of additional cusps on the cheek teeth, and the increasing development of **crenulations** (wavy surfaces) of the enamel on the **occlusal** (chewing) surfaces of cheek teeth are all considered herbivorous adaptations within this lineage.

Platychoerops represents the final member of one of the European lineages of *Plesiadapis* (Fig. 2.11). The distinctions between *Plesiadapis* and *Platychoerops* are subtle: in *Platychoerops* the upper incisors lack the laterocone, its molars are more molarized, and they have more crenulated enamel than in *Plesiadapis*. *Platychoerops* was the first plesiadapid ever described (Charlesworth, 1854). The type specimen, from Kent in England, is a palate with five teeth that was first diagnosed as being from a primitive ungulate (hoofed mammal); its primate affinities were not recognized until much later (Teilhard de Chardin, 1921). It was once considered to be closely related to the peculiar aye-aye *(Daubentonia)* of Madagascar.

In summary, plesiadapids tended to evolve teeth increasingly specialized for plant eating. Their specializations include the following:

1. Upper and lower rodentlike incisors modified for cropping vegetation

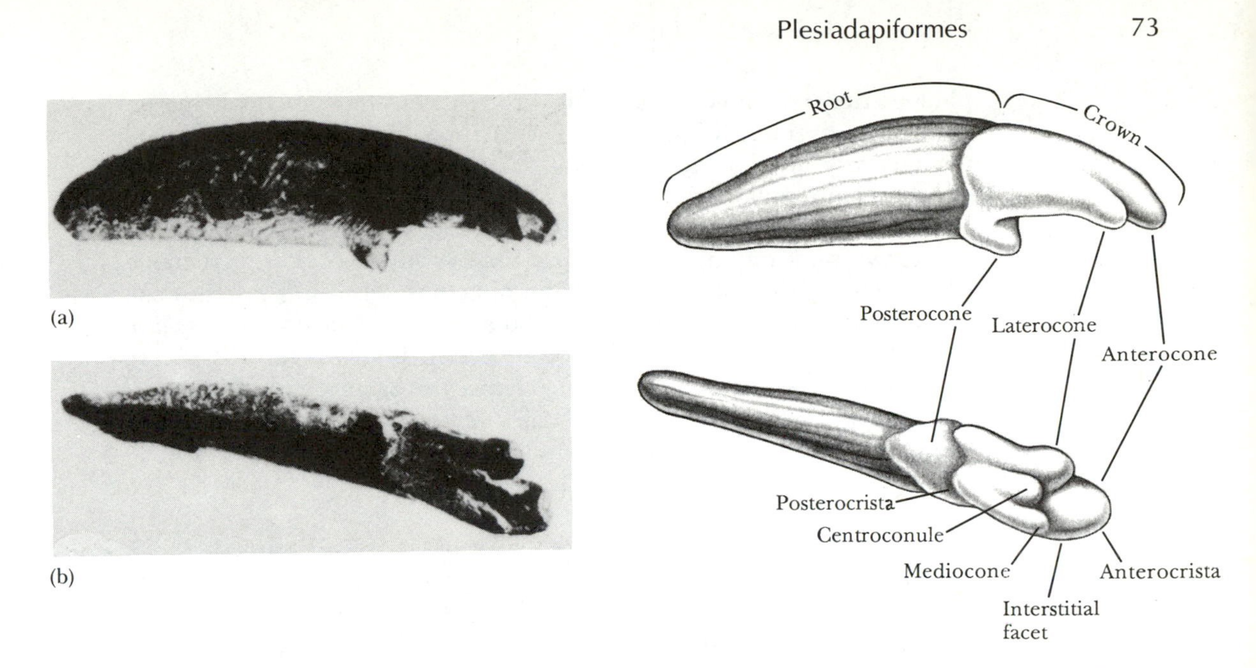

Fig. 2.13 Cusp and Crest Terminology for *Plesiadapis* Upper Incisors (a) Lateral and (b) occlusal views of the upper right incisor. Scale of photos is about 4.5 × natural size. (Photos from Matthew, 1917)

2. Loss of occlusion in the anterior cheek teeth and the subsequent development of **diastemata** (spaces between adjacent teeth) in the mandibles
3. Increasing molarization of premolars
4. Addition of small cusps on the outer surface of the upper molars
5. Additional crenulations on the enamel

Some of these dental specializations were particularly accentuated in another plesiadapid genus present in both North America and Europe, *Chiromyoides*. Five species are currently recognized: *C. minor, C. caesor, C. potior, C. major,* and *C. campanicus*. The European species are *C. campanicus* and *C. minor*. The genus probably evolved from an early *Nannodectes* or *Plesiadapis* species and ranged from the late Paleocene to the early Eocene of North America, but was restricted to the late Paleocene of Europe. Members of this genus have lost both the lower lateral incisor and the lower canine. They have the most robust lower central incisors, the deepest lower jaws, and the most foreshortened faces of any Paleocene primate. The root of the central incisor extends all the way back beneath the lower second molar. In spite of their deep jaws, they, like all other known plesiadapids, lack a fused **mandibular symphysis** (a midline joint between the two halves of the mandible) (Gingerich, 1973*a;* Gingerich and Dorr, 1979).

Cranial Morphology Several skulls or partial skulls of plesiadapids are known (Simpson, 1935; Russell, 1964, Gingerich, 1976). In overall morphology they resemble broad-skulled rodents rather than any living primate. The **premaxilla** (the part of the upper jaw carrying the incisor teeth) is large and contacts the frontal bones of the skull as in many other mammals with enlarged, protruding central incisors. This feature, however, is not present in all early primates.

Known plesiadapids had relatively small brains and elongated snouts, which reflects the importance of the olfactory sense in these animals. Brain volume has been estimated at about 18 cc in *Plesiadapis tricuspidens*.

A useful measure of relative brain size in fossils is termed the **encephalization quotient** (EQ). The EQ expresses brain weight (E_i) for any species i as a proportion of the brain weight expected in an average living mammal of equal body weight (P_i):

$$EQ_i = E_i/0.12\ P_i^{2/3}$$

(For details about how this equation was derived, see Jerison, 1973.) As an illustration, EQs of 1, 0.5, or 2 in fossils simply mean that brain weight is, respectively, equal to, half of, or twice that expected for a living mammal with the same body weight. The average EQ for present-day insectivorans is about 0.5, for prosimians about 1.1, for New and Old World monkeys about 2, and for modern humans about 6.5.

Based on an estimated body weight of 4 kg, the EQ for *Plesiadapis tricuspidens* has been calculated to be 0.62. This is higher than that generally found in archaic ungulates, about the same as that for modern rodents, and smaller than that for living primates. The value is comparable to those determined for Eocene prosimians, using similar methods. Using slightly different parameters, minimum and maximum encephalization quotients of 0.2 to 0.39 have been estimated for *Plesiadapis* (Radinsky, 1977).

As in many other mammals with elongated snouts, the **lacrimal** bone and its internal canal, the **nasolacrimal duct,** are extended out onto the face rather than being restricted to the orbital margin (as it is, for example, in later undoubted Eocene primates such as *Adapis* and *Notharctus*). In plesiadapids the ethmoid bone does not form part of the medial orbital wall (Cartmill, 1971). An ethmoid component of the medial orbital wall, called the **os planum,** was once considered a feature unique to higher primates. An ethmoid contribution to the medial orbital wall has been found, however, in many "lower" primates including the lorisiforms *Microcebus, Indri, Lepilemur, Avahi, Hapalemur,* and *Adapis* (Cartmill and Gingerich, 1978; Cartmill, 1978) (Fig. 2.14). Primates with an os planum are usually either small animals with relatively large eyes or animals with pronounced orbital convergence (Cartmill, 1971).

The auditory bulla of plesiadapids is ossified and appears to be continuous with the petrosal bone. The ectotympanic ring is suspended by bony struts from the lateral wall of the bulla and is expanded laterally to form a tubelike structure. As will be discussed in Chapter 3, this condi-

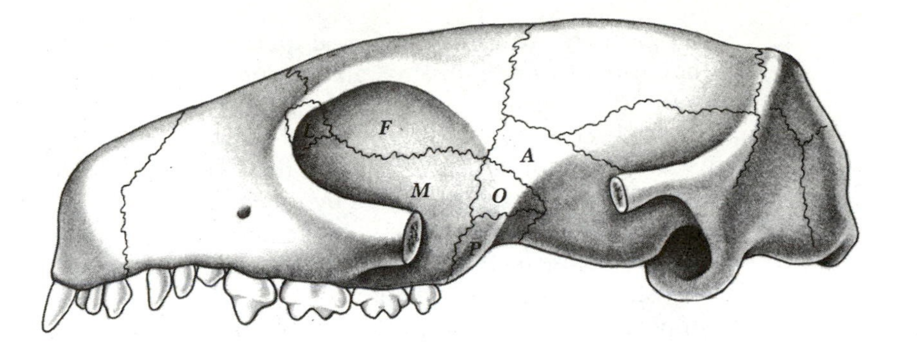

(a)

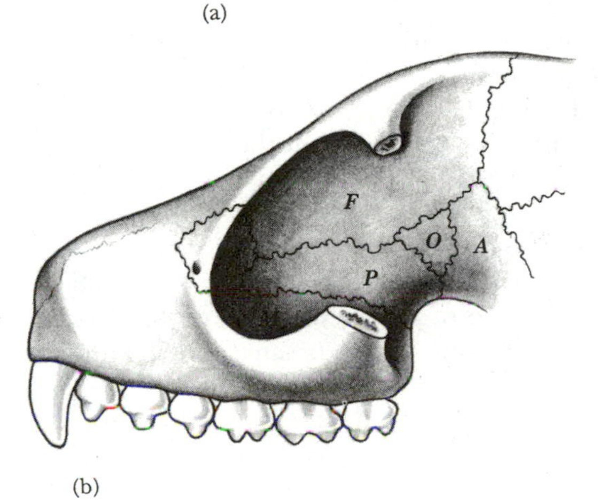

(c)

Fig. 2.14 Cranial Bones Forming the Orbitotemporal Region (a) In the primitive mammal there is wide contact between the frontal bone and maxilla. (b) In recent lemuriforms (including the Tupaioidea) the orbital plate of the palatine extends forward to articulate with the lacrimal. (c) In lorisiforms, *Tarsius*, and anthropoids, the ethmoid separates the frontal bone from the maxilla and the palatine from the lacrimal; the portion of the ethmoid shown here is also called the os planum. Abbreviations: *A*, alisphenoid; *F*, frontal; *L*, lacrimal; *M*, maxilla; *O*, orbitosphenoid; *P*, palatine; shading indicates the os planum.

tion is reminiscent of the configuration seen in Eocene tarsiiforms but is very different from that seen in Eocene lemuriforms. Both the promontory and stapedial arteries of plesiadapids seem reduced in size, judging by the size of the canals presumably housing these vessels in the auditory region of the skull.

Postcranial Morphology Analyses of the **postcranial skeleton** (limbs and vertebral column) are important for determining the locomotor and habitat preferences of both living and fossil mammals. In order to understand such analyses, we must briefly review several important biomechanical aspects of animal locomotion.

Most primates are general **arboreal quadrupeds,** which simply means that they use all four limbs in walking and running among trees and branches. Such primates usually have forelimbs and hindlimbs of approximately equal length. Morphologists have devised a ratio to express relative length of forelimb to hindlimb, the **intermembral index** (length of the humerus and radius divided by the length of the femur and tibia × 100). Thus the intermembral index of most generalized arboreal quadrupeds is close to 100. Moreover, arboreal quadrupeds tend to have relatively short limbs so that their center of gravity lies closer to the arboreal support they are running along. By contrast, arboreal primates that specialize in hindlimb-dominated locomotion such as vertical clinging and leaping (for example, tarsiers) obviously have longer, more powerful hindlimbs, while those that specialize in forelimb-dominated locomotion like **brachiation** (arm swinging; for example, gibbons), have longer forelimbs. The intermembral index would be low in the former (about 55) and high in the latter (about 130).

A few primates have moved away from the more typical arboreal habitats to become more specialized to ground living. These **terrestrial quadrupeds** include such primates as baboons *(Papio)* and patas monkeys *(Erythrocebus)*. The limbs of these primates are adapted for speed and for more fore–aft movements. The speed of any land mammal depends on two main factors: the length of its legs and the frequency with which it can operate them. Thus we would expect terrestrial quadrupeds like baboons to have relatively long fore- and hindlimbs in order to increase stride length. By contrast, the **phalanges** (finger and toe bones) would be short so that they would not "drag" against the ground in terrestrial locomotion.

The frequency of limb movement is aided by the fact that the distal elements of the limb are elongated and light while the bulk of the propulsive muscles are concentrated in the proximal limb segments. This morphological arrangement keeps the center of gravity of the limb near its proximal end and results in a relatively low **moment of inertia** (a measure of resistance to movement) (Smith and Savage, 1956; Gray, 1968). One can get a sense of this phenomenon by noting how much less effort is required to swing a bat by holding it at its "fat" end compared with holding it at its handle.

The morphological ratio expressing the relationship of proximal limb segment length to distal limb segment length is known as the **brachial index** for the upper limb and the **crural index** for the lower limb. The brachial index is the length of the radius divided by the length of the humerus × 100, and the crural index is the length of the tibia divided by the length of the femur × 100.

Some of the main postcranial features associated with different locomotor types in extant primates are shown in Fig. 2.15.

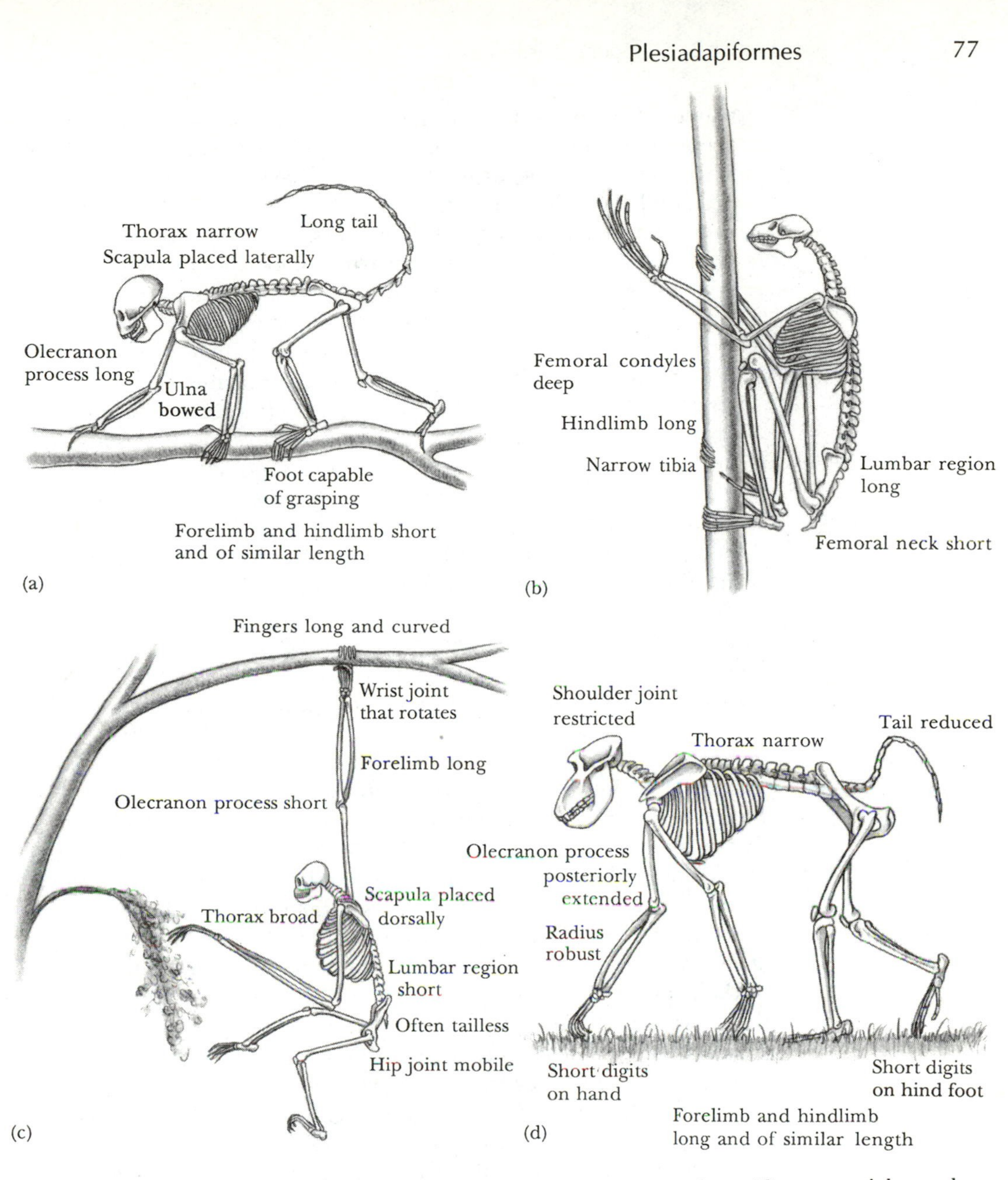

Fig. 2.15 Postcranial Adaptations in Primates of Different Locomotor Types (a) General arboreal quadruped; (b) arboreal primate adapted for leaping and climbing; (c) arboreal primate adapted for suspensory, brachiating locomotion; (d) terrestrial quadruped. Note that in (a) and (d) the intermembral index is about 100, whereas in (b) and (c) it is less and greater than 100 respectively. (Redrawn from Fleagle, 1988)

Analyses of the postcranial skeleton of plesiadapids have yielded conflicting opinions on their mode of locomotion (Conroy and Rose, 1983). Szalay and Decker (1974) argued that the insectivoran–primate transition coincided with the occupation of arboreal niches by early plesiadapiform primates. They suggested that tarsal morphology in Paleocene primates reflects this new emphasis on arboreal locomotion over the more terrestrial quadrupedalism of their insectivoran Cretaceous ancestors. Later studies by Szalay and Dagosto (1980) also concluded that the morphology of the elbow and ankle joints of plesiadapiforms indicates predominantly arboreal habitats for these Paleocene primates.

However, other students of early primates have marshaled at least three major counterarguments to the proposition that plesiadapiforms were primarily adapted to using only arboreal niches:

1. There is compelling dental and cranial evidence to suggest that at least some Paleocene primates were mainly nocturnal, insectivorous, inhabitants of the ground or lower forest canopy, and thus somewhat analogous to modern tree shrews (Jenkins, 1974; Kay and Cartmill, 1977).
2. Limb proportions of *Plesiadapis* are much shorter relative to trunk length than in any living primate and fall within the range of the Sciuridae (rodents such as ground squirrels, marmots, and tree squirrels). *Plesiadapis* also resembles rodents in terms of brachial, crural, and intermembral indices (being most similar to the more terrestrial rodents). Its limbs are more robust than in arboreal squirrels, and it retains claws rather than nails on its digits (Gingerich, 1976).
3. Tarsal morphology can vary tremendously within certain mammalian taxa, thus making it uncertain whether habitat preferences can be accurately deduced from limited fossil material. For example, four different genera of squirrels (groundhogs, chipmunks, fox squirrels, and flying squirrels) all live in different habitats that should be reflected in their tarsal morphology. This is not the case, however. Although there are some minor differences, the **calcaneus** (heel bone) seems remarkably similar in each (Stains, 1959).

It is probably most reasonable to conclude that Paleocene primates utilized both arboreal and terrestrial substrates in a manner analogous to living tree shrews and squirrels. Clearly, plesiadapiforms had not yet evolved the powerful hallucal grasping mechanism or the nails on the digits that were to characterize all later primates.

Carpolestidae

Carpolestids are small, archaic primates that were part of the plesiadapiform radiation. They are known only from dental remains from middle Paleocene–early Eocene deposits in North America. Representatives of the family have not been found in Europe. Three genera *(Elphidotarsius, Carpodaptes,* and *Carpolestes)* and nine species are recognized by Rose (1975), who has made the most thorough study of the group. The three genera form a natural sequence both morphologically and stratigraphically, in that *Elphidotarsius* probably evolved into *Carpodaptes,* which in turn probably evolved into *Carpolestes.* Each genus is temporally restricted: *Elphidotarsius* to the Torrejonian, *Carpodaptes* to the Tiffanian, and *Carpolestes* to the Clarkforkian and early Wasatchian land-mammal ages.

Dentition The most conspicuous dental features of the group are the specialization of the premolars. Through time the lower fourth premolar became progressively enlarged, multicusped, and bladelike, while the lower jaw increased in depth. Accompanying this development was a reduction in the anterior dentition, except for the large, protruding central incisors, which remained prominent throughout the lineage (Fig. 2.16).

As the lower fourth premolar grew, there was a corresponding enlargement of the upper third and fourth premolars. In addition to widening, these premolars developed three longitudinal crests bearing cusps (Fig. 2.16d), somewhat resembling the upper molars of the Multituberculata (early herbivorous mammals that formed a specialized but highly successful and long-lived side branch of the mammalian stock). This particular combination of dental features, which evolved independently in several unrelated mammalian groups, is called the **plagiaulacoid dentition** because of its appearance in members of the multituberculate suborder Plagiaulacoidea. It has evolved in parallel at least twice in primate evolution, in carpolestids and saxonellids. The only extant mammals that share similar dental specializations are some marsupials. The difference, however, is that in carpolestids the upper third and fourth premolars are studded, filelike teeth, not shearing blades as in modern marsupials with plagiaulacoid dentitions.

The dentition suggests that carpolestids were adapted to a diet with a high fiber content, including comparatively large, deformation-resistant food items such as nuts, seeds, and invertebrate animals (Biknevicius, 1986). The detailed convergence between distantly related mammals such as carpolestids of the early Cenozoic, multituberculates of the Mesozoic and early Cenozoic, marsupials of the early and later Cenozoic, and more recent marsupials is probably due to adaptations to similar dietary regimes (Simpson, 1933). Details of incisor and molar morphology reveal that carpolestids and plesiadapids were most likely related through a common ancestor, probably in the early Paleocene.

Saxonellidae

Only one genus, *Saxonella,* is currently recognized in the family Saxonellidae (Russell, 1964; Rose, 1975). *Saxonella* fossils have so far been found in only two late Paleocene deposits, one in Walbeck, East Germany, and

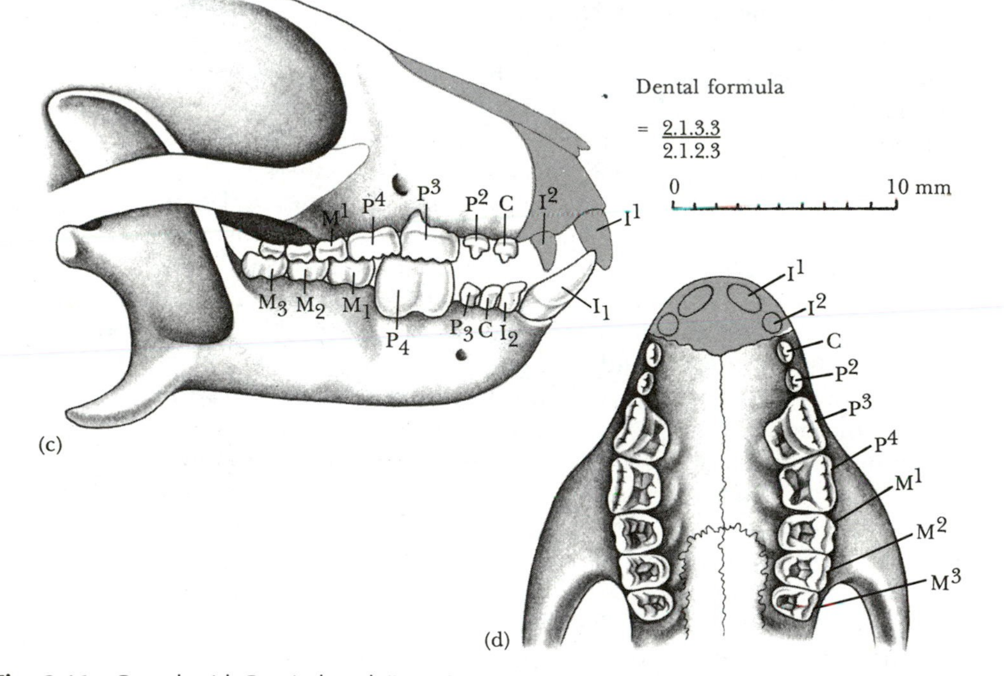

Fig. 2.16 Carpolestid Craniodental Remains Lateral views of the lower jaw in (a) *Elphidotarsius* and (b) *Carpolestes.* (c,d) Lateral and occlusal views of a *Carpolestes* skull and upper jaw. Shading shows reconstructed portions; in (d) one side of the jaw has been reconstructed as the mirror image of the other. In *Carpolestes,* note the greatly enlarged, butcher-block-like upper third and fourth premolars and the enlarged lower fourth premolar, which together form the plagiaulacoid dentition. (From Rose, 1975)

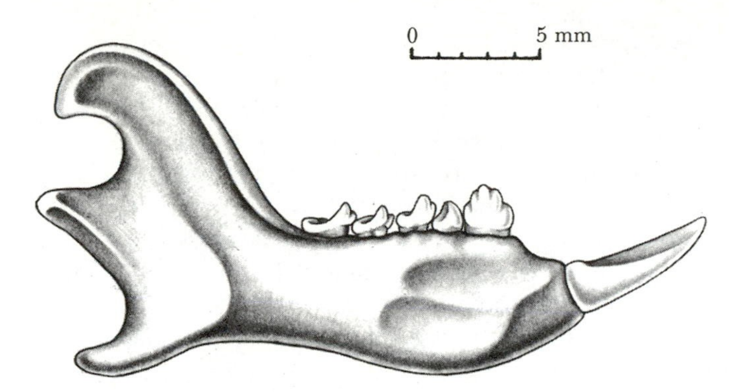

Fig. 2.17 Reconstructed Left Mandible of *Saxonella* Medial view with I_1, P_{3-4}, M_{1-3}. Note the enlarged sharing blade on the lower third premolar; this contrasts with carpolestids, in which the large shearing blade is the lower fourth premolar. (Redrawn from Szalay, 1972*b*)

the other in Alberta, Canada. Only a few specimens of the upper and lower dentition are known, including a mandible, but the evidence is enough to ascertain the specialized nature of the genus (Szalay and Delson, 1979).

When first described, *Saxonella* was placed in the family Carpolestidae (Russell, 1964). Like carpolestids, *Saxonella* has a plagiaulacoid dental complex. Rose (1975), however, discerned that the plagiaulacoid specializations were not identical between the two and therefore recommended placing *Saxonella* in a separate family, a conclusion followed by most subsequent authors. In *Saxonella* the lower third premolar is a bladelike, **trenchant** (cutting) tooth, whereas the lower fourth premolar is comparatively small. In carpolestids, the situation is reversed (Fig. 2.17). In addition, *Saxonella* has fewer teeth in the lower jaw (1.0.2.3) than even the most advanced carpolestids (Rose, 1975). The morphology of both the rodentlike incisors and the molars reveal affinities with the plesiadapids, and for that reason *Saxonella* is included within the suborder Plesiadapiformes.

Microsyopidae

Microsyopids were very primitive but highly successful primates. They were one of only two families of archaic plesiadapiform primates to survive and flourish through the early Eocene of the western interior of North America (Gunnell, 1985). In all, they range from the middle Paleocene to the late Eocene of the Rocky Mountain region and from the late Paleocene to the middle Eocene of France.

Microsyopids were small primates that probably ranged in size from 25 to 2,000 g (roughly the size range of shrews to large rabbits) (Conroy,

1987). The upper dental formula is 2.1.3.3. The lower dental formula is variable within the group. The microsyopids are characterized by progressive reduction of the antemolar dentition in association with an enlarged, laterally compressed, pointed incisor (Fig. 2.18).

Over two dozen genera are currently recognized in the family including: *Palaechthon, Plesiolestes, Torrejonia, Palenochtha, Talpohenach, Microsyops, Craseops, Navajovius, Micromomys, Tinimomys, Niptomomys, Uintasorex, Alsaticopithecus,* and *Berruvius* (see Table 2.1).

The earliest known primate from the fossil record, *Purgatorius,* is quite possibly a microsyopid, although some authors have placed it among the paromomyids (Clemens, 1974; Szalay and Delson, 1979). One species, *P. ceratops,* is known from a single tooth of late Cretaceous age (Van Valen and Sloan, 1965), while a second species, *P. unio,* is known from more-complete dental remains from early Paleocene deposits of Montana. Unlike most other Paleocene primates, *Purgatorius* retained four premolars. Clemens (1974) suggests that *Purgatorius* is ancestral to the microsyopines *Palaechthon* and *Palenochtha.*

Issues in Classifying the Microsyopidae Microsyopids have been studied intensively by several investigators (Szalay, 1969*a,* 1969*b;* Bown and Gingerich, 1973; Bown and Rose, 1976; Gunnell, 1985). As mentioned earlier in the chapter, these authors differ greatly in their taxonomic interpretations of the family.

The taxonomic status of certain genera placed within the Microsyopidae by Bown, Gingerich, and Rose is still debatable. The major reason for this uncertainty concerns the ear region: although the dentition of microsyopids is undoubtedly similar to that of plesiadapiform primates in general, the ear region is most similar to that of certain insectivorans. Because of the nonpetrosal construction of its bulla, Szalay and others would place the Microsyopidae within the Insectivora (McKenna, 1966; Szalay and Delson, 1979). Due to dental similarities, these authors would also consider certain genera that Bown, Gingerich, and Rose include among the Microsyopidae *(Palaechthon, Plesiolestes, Palenochtha, Micromomys, Tinimomys,* and *Navajovius)* to more properly belong within the Paromomyidae.

Bown and Gingerich (1973) deal with this dilemma by concluding that

> the most reasonable interpretation of these relationships is that . . . microsyopids are primates which were derived from leptictid insectivores and retained the entotympanic bulla construction typical of leptictids. It is unlikely that the distinctive molar morphology of early primates was acquired at the same time as the distinctive primate bulla construction. The close relationship of *Plesiolestes* and the Microsyopidae . . . provides evidence that the diagnostic primate molar morphology evolved from that of leptictid insectivores before the characteristically primate petrosal bulla was acquired.

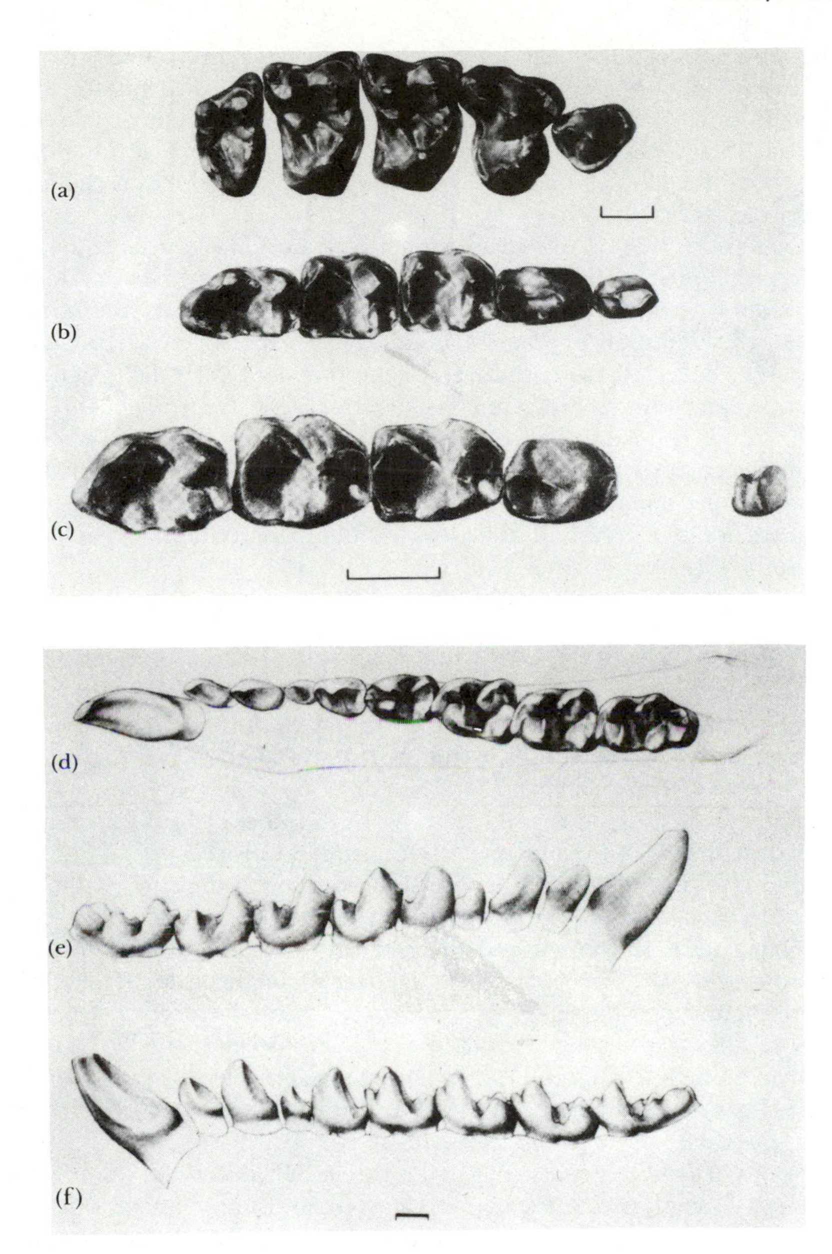

Fig. 2.18 Representative Examples of Microsyopid Dentition (a) Upper right and (b) lower left tooth rows of the middle Paleocene *Palaechthon alticuspis* showing third premolar through third molar. (c) Lower canine and four premolar through third molar of the middle Paleocene *Palenochtha minor.* In (a–c) mesial end is to the right. (d) Occlusal, (e) lateral, and (f) medial views of the lower right dentition of the late Paleocene *Plesiolestes problematicus* showing I_{1-2}, C, P_{2-4} M_{1-3}. Note the large, pointed incisor and the loss of the first premolar. Scale bars equal 1 mm. (a–c, from Simpson, 1955, d–f, after Szalay, 1973)

If this view is accepted, it follows that the presence of an ossified petrosal bulla can no longer be considered a diagnostic feature of primates since known Eocene microsyopids, including *Microsyops,* evidently lacked it. Suffice it to say, the Microsyopidae have been moved back and forth across the insectivoran–primate boundary many times over the past quarter century.

Bown and Rose (1976) consider *Plesiolestes* and *Palaechthon* to be the most typical, or least specialized, microsyopines. Both are North American middle Paleocene (Torrejonian) forms. There are certain similarities between the dentition of *Palaechthon* and the earliest known primate, *Purgatorius,* collected from the late Cretaceous and early Paleocene of Montana. According to Bown and Gingerich (1973), *Palaechthon* and *Plesiolestes* are likely ancestors of the earliest *Microsyops.*

The microsyopids remain a difficult group to classify. However their taxonomic placement is finally resolved, it is clear that the family was one of the longest lived and most successful of the six families included here in the Plesiadapiformes.

Paromomyidae

In Bown and Rose's (1976) revision of the plesiadapiform primates, the Paromomyidae became a tightly knit group composed of just three genera: *Paromomys, Ignacius,* and *Phenacolemur.* The family is known from the middle Paleocene to the late Eocene in the Rocky Mountain region of North America and from the early Eocene in France.

Paromomys, found only in the middle Paleocene, is the earliest-known and least-specialized paromomyid (with the possible exception of early Paleocene species of *Purgatorius*) and is considered ancestral to *Phenacolemur* and *Ignacius.* Two species of *Paromomys* have been recognized, *P. depressidens* and *P. maturus.*

Species of *Ignacius* range from the middle Paleocene to the late Eocene (late Torrejonian to Uintan land-mammal ages) in the Rocky Mountain region (Rose and Gingerich, 1976). The known species include *I. frugivorous, I. hemontensis, I. graybullensis,* and *I. mcgrewi.*

Species of *Phenacolemur* range from the late Paleocene to the middle Eocene (Tiffanian through early Bridgerian land-mammal ages) in the Rocky Mountain area and are also known from early Eocene (Sparnacian land-mammal age) deposits of the Paris basin. *Phenacolemur* ranges possibly into the late Eocene of North America as well (Krishtalka, 1978). Species include *P. praecox, P. pagei, P. simonsi, P. citatus,* and *P. jepseni* from North America, and *P. fuscus* and *P. lapparenti* from France.

Dentition Like many other plesiadapiforms, paromomyids have greatly enlarged, protruding lower incisors. The molars have low crowns with blunt cusps and shallow basins. Both *Ignacius* and *Phenacolemur* show considerable reduction in the antemolar teeth.

Paromomys has the usual lower dental formula of 2.1.3.3 for middle Paleocene plesiadapiform primates. Both *Ignacius* and *Phenacolemur,* however, have lost the small, lower second incisor, the lower canine, and the second premolar. In some species the lower third premolar is lost as well. (The lower dental formula for *Ignacius* is 1.0.1.3 or 1.0.2.3, and for *Phenacolemur* is 1.0.1.3.)

The most striking dental specializations in the family are found in species of *Phenacolemur.* The lower antemolar teeth are greatly reduced and the enlarged lower central incisor is the longest among all the early Tertiary primates (Fig. 2.19).

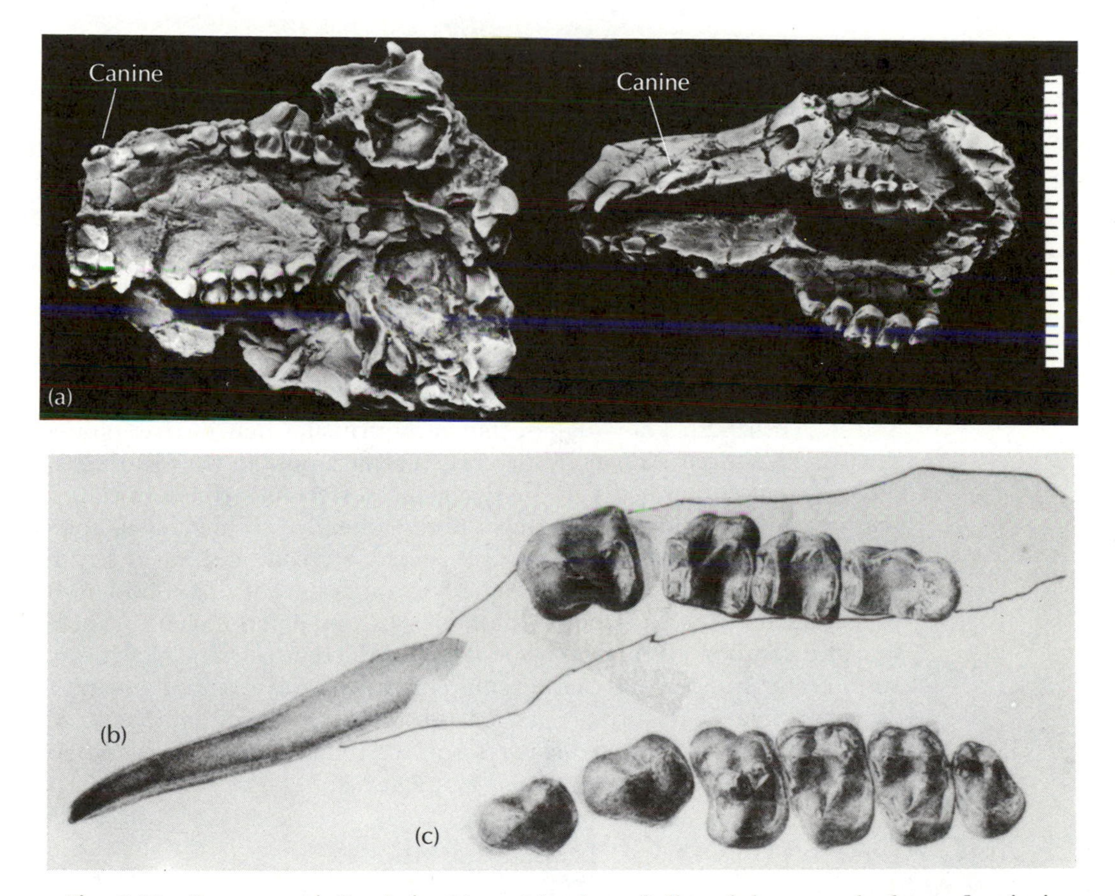

Fig. 2.19 Paromomyid Craniodentition (a) Occlusal views of skulls of *Phenacolemur (left)* and *Ignacius* (right); scale bar in millimeters. (b) Lower right dentition of *Phenacolemur* showing greatly enlarged central incisor and loss of all teeth between the lower first incisor and the last premolar. (c) Upper left dentition of *Phenacolemur.* Scale in (b,c) about 2.7 × natural size. (a, From Rose and Gingerich, 1976; b,c, after Simpson, 1955)

Picrodontidae

The peculiar, small primates classified as the Picrodontidae were often confused with insectivorans or bats by early taxonomists. Two genera are known, *Picrodus* and *Zanycteris.* They range from the middle Paleocene (Torrejonian) to the late Paleocene (Tiffanian) in North America. Their dentition is very similar to that of the nectar- and pollen-eating bats and suggests a similar dietary adaptation for these primates (Williams, 1985). There is a marked reduction of the cusps on the lower molars, and the upper molars show a very enlarged metacone on the first molars (Fig. 2.20). Picrodontids have reduced trigonids and enlarged talonids. Like most Paleocene primates they have enlarged lower incisors. A partial, crushed skull is the type specimen of *Zanycteris* (Szalay, 1968*b*).

In view of the fact that picrodontids have often been confused with bats, it is interesting to note that John Pettigrew of the University of Queensland has recently shown that some *Megachiroptera* (large fruit-eating bats) share a unique visual system with primates. On this basis he suggests that fruit bats and primates shared a specific common ancestry and that bat wings evolved twice, once in megachiropteran "primates" and once in the Microchiroptera (ordinary small bats) (Pettigrew, 1986; Martin, 1986*b*).

The similarities between at least some Paleocene primates and bats have prompted Robert Sussman and Peter Raven of Washington University to advance an interesting hypothesis concerning a reciprocal biogeographic relationship between the two groups in the past (Sussman and Raven, 1978). They suggest that early primates may have exploited the same resources as bats in the early Tertiary period, thus leading to intense competition between the two groups. Certainly the dentition of many Paleocene primates was similar to that of early bats in being adapted to diets consisting predominantly of insects, fruit, nectar, or gum. Although an earlier study (Jepsen, 1970) suggests that this competition precipitated the evolution of flight in bats, Sussman and Raven speculate that the competition brought about the extinction of many Paleocene and Eocene primates and perhaps played a role in the evolution of larger, diurnal primates. In support of the latter idea, they point out that small, nocturnal, nectar-feeding primates survive today only in areas where nectar-eating bats are rare, such as Madagascar.

FEEDING ADAPTATIONS OF PALEOCENE PRIMATES

As previously discussed, to infer dietary preferences in fossil species, one must consider both the dental morphology and the absolute body size of an animal. Large herbivorous primates, such as gorillas, process a great

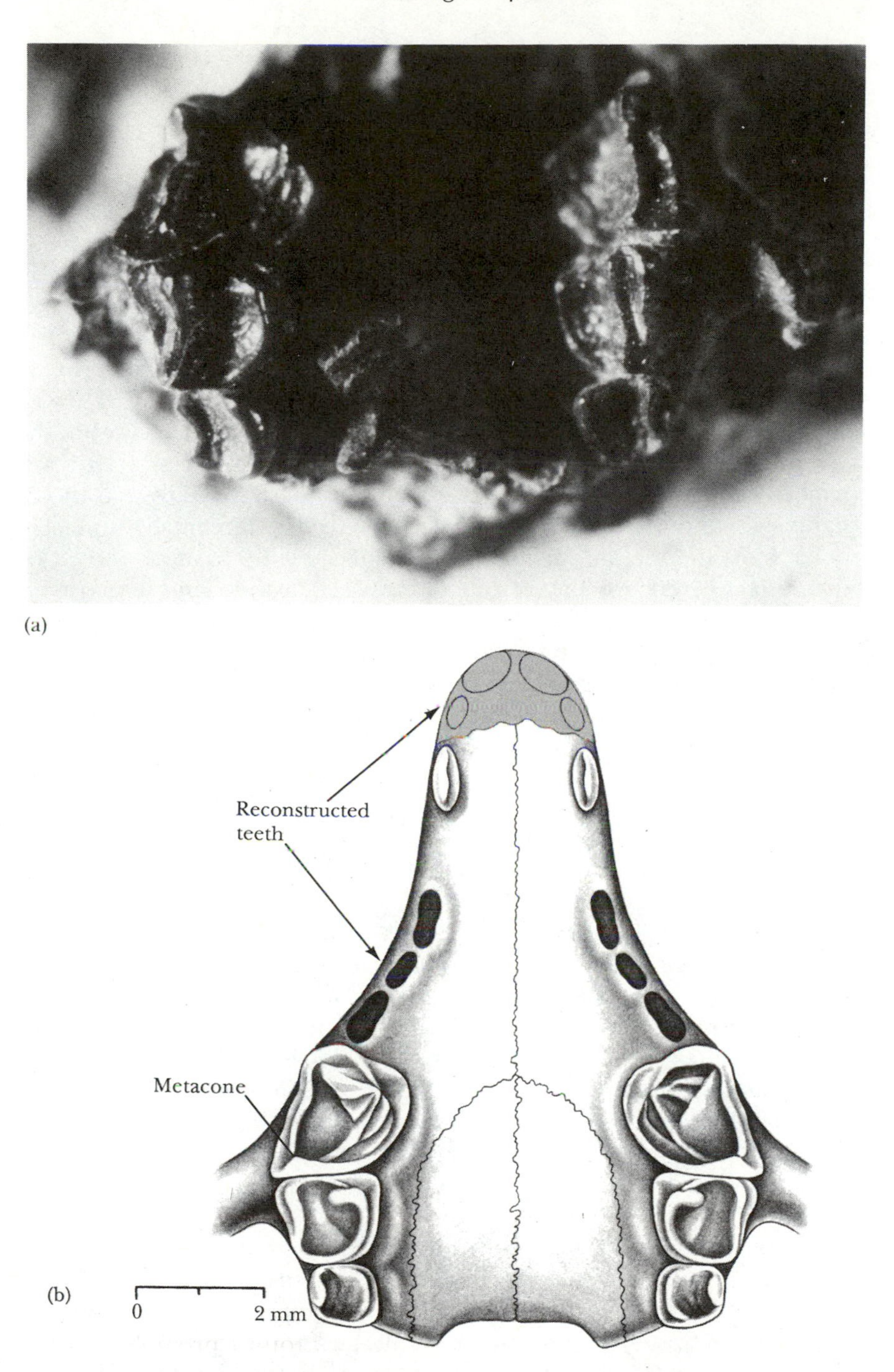

Fig. 2.20 Picrodontid Dentition (a) Occlusal view of the maxilla of *Picrodus* from the late Paleocene of western North America. Scale about 10 × natural size. (b) Reconstructed upper jaw, occlusal view of another picrodontid, *Zanycteris*. One side of the jaw has been reconstructed as the mirror image of the other. Note the enlarged metacone on the first molar. (a, Courtesy of J. Schwartz; b, redrawn from Szalay, 1968*b*).

deal of food in bulk. Normally such food has a low concentration of nutrients. Food processing of this type is possible because larger animals have a low surface-area-to-volume ratio, which allows slow energy assimilation with little loss of energy through radiation of heat. In addition, large herbivores can afford to let plant material digest longer in their gastrointestinal tract, thus increasing the nutrient yield from the bulk food. Small mammals, in contrast, need more concentrated nutrients such as nectar, gum, fruit, or insects to maintain a high body temperature in the face of their comparatively high metabolic rate and surface-area-to-volume ratio (Demment, 1983).

As pointed out in Chapter 1, comparative work on extant primate dentitions by a number of researchers (Hylander, 1975; Kay, 1975; Kay and Covert, 1984) has made it possible to place fossil primates within the very general dietary categories of insectivores (insect eaters), frugivores (fruit eaters), and folivores (leaf eaters). In addition, Kay (1984) has shown that insect- and leaf-eating primate species are fundamentally similar in molar morphology but can be distinguished on the basis of body size, since insect eaters are always smaller than leaf eaters. The similarity in dentition is due to the fact that the chitin of insects and the structural carbohydrates of plants have similar physical and chemical properties: both have to be finely ground to optimize digestibility. As much as half the energy value of insects is contained in chitin, which resembles cellulose in not being easily digestible (Kay and Sheine, 1979). It is broken down by the enzyme chitinase, which apparently is found only in mammals that are predominantly insectivorous (Garber, 1984). Masticatory efficiency is not so critical in frugivorous primates, since most of the constituents of these foods are completely digested regardless of how fine they are ground.

It is also known that living insectivorous primates tend to have larger molars than their fruit- and sap-eating relatives, when adjusted for body size (Kay, 1984; Kay and Hylander, 1978; Gingerich and Smith, 1985). Insect eaters also tend to have higher and more acute cusps, longer shearing blades, and enlarged crushing surfaces on their molars. Folivores, like insectivores, tend to have larger teeth in relation to body size, with well-developed shearing, crushing, and grinding surfaces. In contrast, frugivores have small teeth relative to their adult body size, with poorly developed shearing, crushing, and grinding features on their molars.

Some authors regard early primate teeth as providing the evidence for a major shift in dietary preference from more insectivorous Cretaceous ancestors to more herbivorous Paleocene primates (Szalay, 1968*a*). For example, Szalay suggests that the enlarged fourth premolar in *Phenacolemur* is a seed- and nut-cracking adaptation and that the Paleocene picrodontids fed on fruit pulp, fruit seeds, or nectar. He concludes that the molar patterns of Paleocene primates in general were not adapted for chewing the fibrous materials of animal bodies and that the specialized incisors and the deemphasis of the canines indicates a general lack

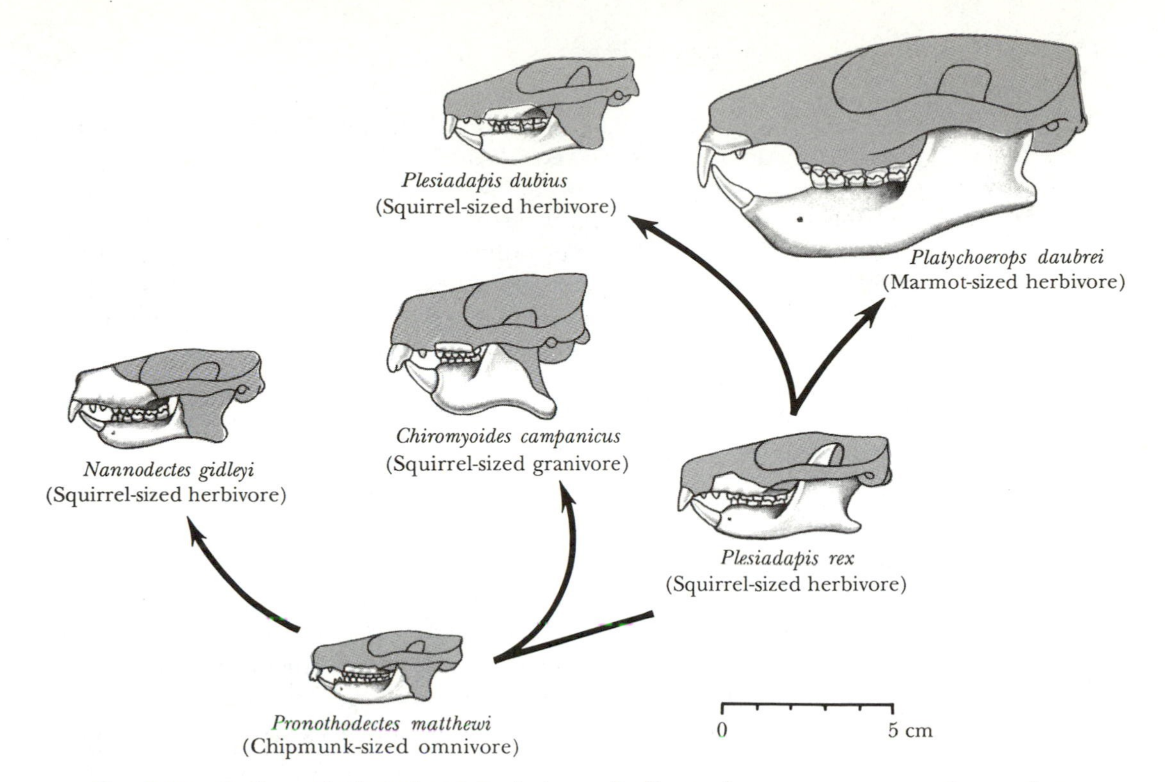

Fig. 2.21 Outline of Plesiadapid Evolution Showing Diversification in Size and Feeding Behavior All skulls drawn to the same scale; shading shows reconstructed portions. (Redrawn from Gingerich, 1976)

of predatory behavior in plesiadapiforms. Other investigators, however, emphasize the similarity of *Phenacolemur's* dentition with that of the gliding marsupial *Petaurus,* which feeds on insects, larvae, small birds, buds, blossoms, nectar, and sap (Gingerich, 1974).

Other small Paleocene primates, such as *Palenochtha, Palaechthon, Elphidotarsius,* and *Navajovius,* have dental characteristics that are similar to *Tarsius,* which is insectivorous. For this reason, John Crook of the University of Bristol and J. S. Gartlan of the University of Wisconsin have postulated that primitive primates were predominantly insectivorous and carnivorous, with diets resembling those of modern mouse lemurs, galagos, and tarsiers (Crook and Gartlan, 1966). Kay and Cartmill (1977) also conclude that *Palaechthon* was predominantly an insect eater. Dental morphology alone suggests that it might have been a leaf eater. However, since the smallest folivorous primates weigh about 700 g and *Palaechthon* is estimated to weigh only about 100 g, there seems little doubt as to its insectivorous, or at least nonfolivorous, dietary regime.

Plesiadapis, however, has an incisor region that probably functioned as a cropping mechanism somewhat analogous to that seen in kangaroos (Gingerich, 1976). Most plesiadapid genera were probably herbivorous (Fig. 2.21).

In summary, considerations of dental structures and body size indicate that many middle Paleocene primates were predominantly insectivorous. A few, such as *Torrejonia* and *Paromomys,* may have had mixed diets including insects, fruit, or gum. By the late Paleocene a range of plant-eating adaptations evolved: *Plesiadapis* had become a folivore, and others, for example *Chiromyoides,* were frugivores. The carpolestids probably fed on fruit and nuts, whereas *Phenacolemur* may have fed largely on tree sap. In short, the long-standing view that the evolution of primates from insectivorans could be traced solely to shifts towards new herbivorous and frugivorous dietary patterns does not seem to be as clear-cut as previously thought.

PHYLOGENY AND CLASSIFICATION OF PALEOCENE PRIMATES

Even though the anatomy of some plesiadapiforms is quite well known, their phyletic relationship to more advanced, modern-looking primates of succeeding epochs is still problematical. It is generally agreed that the anterior dentition of plesiadapiforms is too morphologically specialized (for example, enlarged, rodentlike incisors and reduction in the number of antemolar teeth) to consider them to be directly ancestral to any later primates.

However, as mentioned earlier, a small number of authors have placed plesiadapiforms in the order Primates solely on the basis of dental (particularly molar) similarities to Eocene and modern prosimians like lemurs and *Tarsius* (tarsiers). Some authors have also concluded that plesiadapiform postcranial anatomy reveals adaptations to primatelike arboreal habitats (Szalay and Decker, 1974). A few have argued for a phyletic relationship between plesiadapiforms and the more modern primate group, the Tarsiiformes (tarsiers) (Gingerich, 1976; Krishtalka and Schwartz, 1978). These authors base their reasoning on the grounds that both groups share enlarged, protruding, pointed incisors and remarkably similar configurations of the auditory bulla and tubular ectotympanic bones. This view has been formalized by including plesiadapiforms and tarsiiforms in the suborder **Plesitarsiiformes** as the sister group of the suborder **Simiolemuriformes.** The latter includes the Lemuriformes (lemurs), the Lorisiformes (lorises and galagos), and the Anthropoidea (monkeys, apes, and humans) (Fig. 2.22a).

Several workers have taken another view on plesiadapiform relationships, namely, that plesiadapiforms should not be considered primates at all since they are so distinct from later primates; moreover, they do not possess many of the anatomical features characteristic of modern primates (nails instead of claws, postorbital bars, relatively enlarged brains, reduction of the olfactory sense with elaboration of the visual sense, and, in microsyopids at least, a petrosal bulla) (Martin, 1968; Cartmill, 1974) (Fig. 2.22b).

A third view, adopted here, is that all of the plesiadapiforms, including microsyopids, are part of the earliest primate radiation (Fig. 2.22c).

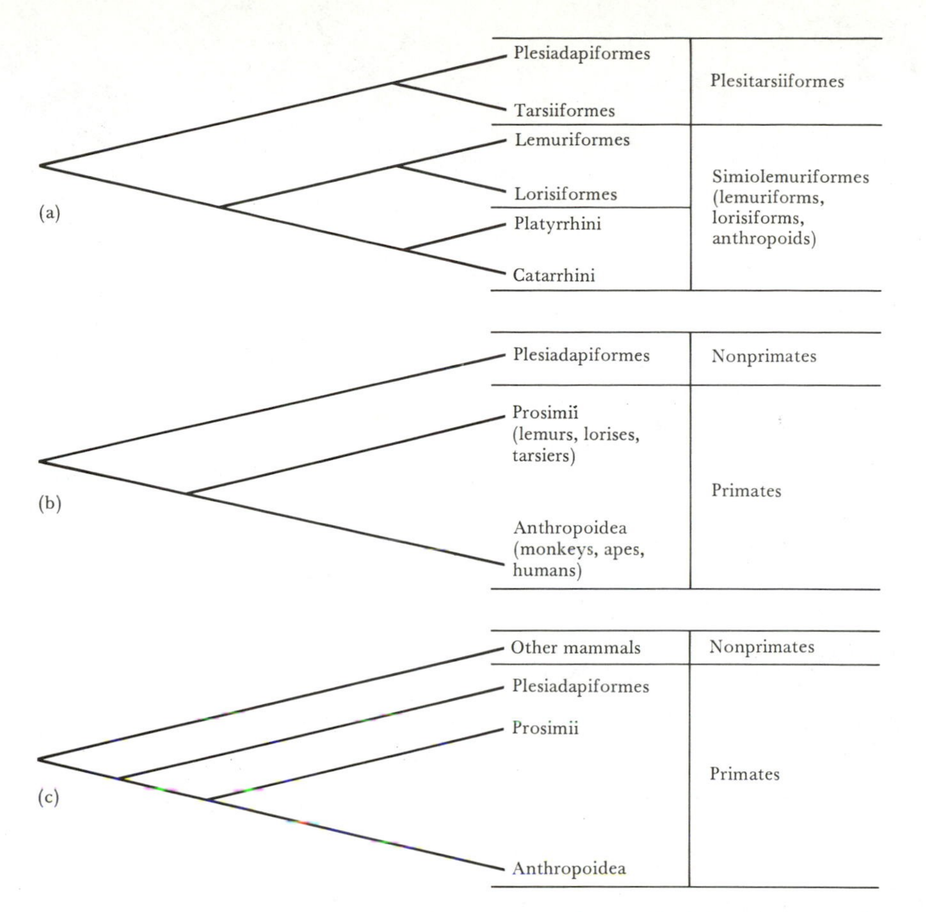

Fig. 2.22 Three Views of Plesiadapiform Relationships Plesiadapiforms as the sister group (a) of tarsiers (proposed, for example, by Gingerich, 1976); (b) of the order Primates (for example, Cartmill, 1974); and (c) of more modern-looking primates (prosimians and anthropoids) (for example, Rose and Fleagle, 1981).

This view has the advantages of grouping all families with primatelike cheek teeth within the same order and of being operable for the classification of fossil species known only by molar teeth in most cases (Rose and Fleagle, 1981). It is fair to say that most students of primate evolution now accept the proposition that plesiadapiforms were nature's first successful experiment with primate evolution, albeit one that ultimately ended in the Eocene without further issue. As we shall see in Chapter 3, the next major episode of primate evolution occurred in the Eocene with the rise of euprimates ("true" or modern primates). Chapter 3 will take a close look at the first two euprimate groups, the Adapidae and the Omomyidae, as well as examine what characters distinguish euprimates from plesiadapiforms. In the meantime, it is important to note, as have some investigators, that the plesiadapiform primates "share euprimate-like features of molar morphology that give them particular significance for understanding primate origins" (MacPhee et al., 1983), even if it is not possible to resolve their precise relationships to one another or to later primates.

CHAPTER 3

Eocene Primates

Paleoclimates and Biogeography

Summary of the Eocene Fossil Record

Rise of the Euprimates

Body Size, Feeding Adaptations, and Behavior of Eocene Primates

Phylogeny and Classification of Eocene Primates

Many significant events in primate evolution took place during the Eocene. During this time, the archaic plesiadapiforms gradually became extinct, and their place in the primate fossil record was taken by primates that had evolved many of the morphological features we associate with primates living today—the so-called "primates of modern aspect," or euprimates. By the end of the epoch, higher primates appeared in the fossil record for the first time. We will discuss these evolutionary trends in this chapter. In addition, we will review the various hypotheses that attempt to link these Eocene primates to primate groups living today.

We begin our examination of this period in primate history with a consideration of the paleogeographical and paleoclimatic forces that were shaping the destiny of primate evolution about 55 to 35 MYA.

PALEOCLIMATES AND BIOGEOGRAPHY

Climates Inferred from Paleobotany

Paleobotanical evidence has been indispensable in reconstructing Eocene climates. Eocene floras have been recovered from many parts of the world, including Asia, Africa, Europe, and both North and South America (Reid and Chandler, 1933). These floras (sometimes referred to as the London Clay floras) include plants characteristic of tropical rain forests and are said to have strong Indo-Malayan affinity since more than 70% of the extant genera live in Southeast Asia.

One of the most complete sequences of Eocene leaf floras in North America is that of the Puget Group in western Washington, ranging in age from about 50 to 34 million years old. The floras all contain numerous leaf species that have driptips (pointed tips to facilitate water runoff) and/or lianalike leaves. The fossil floras represent vegetation that apparently grew under abundant year-round precipitation and would today be classified as broad-leaved evergreen rain forests (Wolfe, 1978).

Paleobotanists agree that the Eocene was the warmest epoch of the Tertiary period. An overall gradual warming took place from the Paleocene into the middle Eocene, which may have been the time of greatest tropicality in the whole Cenozoic era. Eocene floras generally suggest a continuation from the preceding Paleocene of warm, wet climates without great extremes in temperature; there was also a trend toward less seasonality (alternating wet and dry periods) (Wolfe, 1978).

Analyses of early Tertiary floras in North America detail these trends: the mean annual temperature range decreased from the early into the middle Eocene, at which point it was about half that of the present range; the latitudinal temperature gradient along the western coast of North America was also half what it is now. Two cool intervals occurred during this warming trend; in mean annual temperature, the difference between warm intervals was about 7°C and the cool intervals were still about 4 to 5°C warmer than at the present. During the middle Eocene the western coast of North America received abundant precipitation (Wolfe, 1978).

After the middle Eocene, the period of relative tropicality came to an end. Gradual cooling commenced, and the mean annual temperature range increased slightly. In the late Eocene, the climate, at least from evidence in the northern Rocky Mountain area, became drier and more seasonal than at any previous time in the Eocene. This paleobotanical evidence is compatible with the plate-tectonic evidence indicating that world-wide climates cooled rapidly near the end of the Eocene, approximately 35 MYA. The succeeding Oligocene was even cooler, drier, and more arid; never was such a humid, tropical climate reached again in earth's history (Wolfe and Hopkins, 1967; Wolfe, 1971, 1978).

As the climates of the continental areas became drier and more seasonal in the late Eocene, they also became geographically more diverse. This climatic revolution profoundly modified the vegetation and fauna in various parts of the world. The global belt of tropical rain forest was modified, reduced, and interrupted. Within this changing milieu, there were many significant extinctions and originations of mammalian groups. Many modern families of land mammals first appeared during this period, from about 40 to 34 MYA (Savage and Russell, 1983).

Land Movements and Mammal Migration Routes

Eocene paleogeography was still very different from the world of today; many landmasses now separate were connected by land bridges, and landmasses now in contact were isolated. Great plate-tectonic and climatic changes were taking place in the Eocene that had dramatic effects on mammalian biogeography.

Europe and North America remained connected in the early Eocene but began their separation by the middle Eocene. Eocene faunas were greatly influenced by this connection. More than 50% of mammalian genera were common to both continents during this period, a higher percentage than at any other time in the history of mammals. By contrast, the faunal similarity between the two continents in the Paleocene was only about 10% (Russell, 1975).

A land bridge between Asia and Alaska at the Bering Strait existed throughout most of the Cenozoic era; however, its latitude was initially high, so its cooler temperatures probably exerted a partial filtering effect on migrating mammals.

With the final rupture of the land bridge between Europe and North America in the late Eocene, western Europe was left comparatively isolated. Faunal dispersal to Asia was limited by the Ural Mountains and the Obik Sea, a marine arm of the Arctic Ocean that partially bisected the Asiatic landmass (Fig. 3.1). This is evidenced by the fact that late Paleocene faunas of central Asia show little similarity to those of western Europe (Szalay and McKenna, 1971; McKenna, 1983*a*, 1983*b*).

Fig. 3.1 Possible Migration Routes of Eocene Land Mammals in the Northern Hemisphere A land connection existed between North America and Europe in the early Eocene via Iceland, Greenland, and the Faeroes Islands. Migration between Europe and Asia was hindered by the Obik Sea. Land mammal migration between Asia and North America occurred across the Bering Strait. Note that the boundaries of the Tethys and Obik seas are vague. (From Savage and Russell, 1983)

The Eocene was a time of isolation for most continents. Africa, Antarctica, Australia, and South America were all surrounded by sea barriers that restricted the dispersal of land vertebrates. In addition, the developing seasonality of climates discussed earlier contributed to a major reorganization of land-mammal faunas on these isolated continents.

This climatic turnabout was influenced in large measure by plate-tectonic movements involving Antarctica and Australia. During the early Eocene the southern oceans were comparatively warm and Antarctica was largely unglaciated. However, substantial Antarctic sea ice began to form when Australia began its northward drift from Antarctica approximately 55 MYA. This ultimately led to a rapid drop in ocean water temperatures and subsequently to global climatic cooling (Kennett, 1977). Detailed oxygen-isotope studies reveal that both surface and deep ocean waters experienced a series of sharp temperature drops during the early Tertiary, with particularly marked cooling at the Eocene–Oligocene boundary.

Biotic changes occurred that reflect this series of worldwide climatic events marking the transition from warm, ice-free climates to colder climates with more polar ice and stronger temperature gradients between the poles and the equator (Corliss et al., 1984; Keller et al., 1983). The Eocene–Oligocene boundary is of particular paleontological interest because of the major changes in global climate and faunal distributions that occurred about this time. This noticeable discontinuity in mammalian faunas is often referred to as the **Grand Coupure** (French for "great cut") (Stehlin, 1909). At least 60% of the mammalian genera in Europe, including most of the primates, became extinct at this time; the majority of primates disappeared from North America as well. In addition, approximately 65% of the Oligocene mammalian genera in Europe were new (McKenna, 1983*a;* Savage and Russell, 1983). Thus the many distinct faunal changes that occurred from the middle Eocene to the middle Oligocene can be related to the formation of the Antarctic ice sheet and the associated climatic cooling it produced.

Of special significance is the fact that the first undisputed primates, the **euprimates,** appeared in the fossil record at the beginning of the Eocene epoch and the higher primates (anthropoids) made their first appearance at the Eocene–Oligocene boundary.

One of the Eocene euprimate groups (the Adapidae) bears a certain resemblance to lemurs, as we shall examine in some detail later in this chapter. Although in general this text will not discuss the origins of living primates, lemurs are today found only on the island of Madagascar, and their origin is still a mystery. It behooves us therefore to consider briefly here some of the hypotheses for the geological origin of lemurs.

Some investigators have postulated that Madagascar broke away from the eastern coast of Africa about 165 MYA and arrived at its present location with respect to Africa between 75 and 90 MYA (Rabinowitz et al.,

1983). If so, just when and how lemuriform primates first colonized the island is unclear. Several hypotheses have been offered to explain the spread of these early prosimians to Madagascar.

According to the traditional hypothesis, one or more groups of lemurs rafted on floating vegetation across the 400-km-wide Mozambique Channel from Africa to Madagascar (Simpson, 1940*a;* Simons, 1972; Mahe, 1972). Currents produced where the Zambesi and other East African rivers empty into the channel could have helped propel the rafts to Madagascar (Mahe, 1972).

According to a second hypothesis, lemurs originated in India and were rafted to Madagascar by strong southern currents and the southeast trade winds (Gingerich, 1975*a*). Presumably, India was south of the equator during the early Tertiary; such currents could not have propelled rafts once India drifted into the Northern Hemisphere by the end of the Oligocene.

A third hypothesis has lemurs migrating across one or more land connections between Africa and Madagascar (Tattersall, 1982). In this scenario there were several periods between the late Cretaceous and the mid-Oligocene in which the sea level of the Mozambique Channel could have dropped dramatically as a result of increased glaciation (Morner, 1978; Matthews and Poore, 1980).

Ashok Sahni of Punjab University in India has assembled paleontological evidence to show that at the genus and family levels there is a close correspondence among Cretaceous vertebrates of India, Africa, and Madagascar. This suggests to him that a dispersal corridor consisting of currently submerged structures (the Mascarene Plateau and the Chagos-Laccadive Ridge) existed between these landmasses about 80 MYA as India drifted close to eastern Africa.

SUMMARY OF THE EOCENE FOSSIL RECORD

Eocene mammalian faunas were far more diverse and abundant than those of the Paleocene and are known from many more localities throughout the world. For example, there are more than 50 early Eocene fossil-bearing sites on the European continent alone. Eocene mammals have been found on all the continents except Australia, including sites within the Arctic Circle (Dawson et al., 1976).

A notable aspect of these faunas is the small number of taxa that survived through the Paleocene and into the Eocene. For example, during the late Paleocene in Europe approximately 80% of the mammalian genera became extinct. Some of the more archaic Paleocene mammals that did manage to survive into the Eocene include multituberculates, marsupials, plesiadapiforms, the Creodonta (primitive carnivores), and the Condylarthra (primitive ungulates).

The Eocene witnessed the appearance of many new types of animals. Approximately 70% of North American Eocene mammalian genera and 90% of European Eocene genera make their first appearance in the early Eocene—specifically, in the Wasatchian land-mammal age in North America and the Sparnacian land-mammal age in Europe (Savage and Russell, 1983). To account for this phenomenon, one must postulate mass migrations, rapid speciation, and/or mass extinctions since there does not seem to have been enough geologic time between the late Paleocene and the early Eocene to invoke gradual evolutionary change as an explanation for all these faunal changes. In addition, the necessary ancestral taxa are often lacking in known Paleocene faunas. In general, Eocene mammals are so different from those of the late Paleocene that one often has to look back to the Cretaceous or middle Paleocene for possible ancestral stocks. Rapid diversification seems to have occurred in the early Eocene.

The early Eocene faunas of Europe contain relatively few large animals, most being small to medium-sized. These faunas include the first European members of several living mammalian groups such as euprimates, **ungulates** (hooved mammals), and rodents. Rodents are already diverse and numerous, accounting for approximately 25% of all mammalian remains. Because of their abundance and rapidity of evolution, they are important fossils for biostratigraphic correlation. Their evolutionary success may have contributed to the ultimate demise of the archaic, rodentlike plesiadapiforms.

Marsupials are also new arrivals in Europe and show little advance over their Cretaceous ancestors in North America. True insectivorans (members of the suborder Lipotyphla), Dermoptera ("flying lemurs"), and Chiroptera (bats) become more abundant and diverse, whereas multituberculates become rarer and are not known after the Eocene. In Chapter 2, it was noted that the plesiadapiform family Picrodontidae was at one time or another placed in either the Dermoptera or the Chiroptera (Szalay, 1968*b;* Schwartz and Krishtalka, 1977). The discovery of a complete bat skeleton from the Eocene in North America reveals that full flying ability and echolocation have already evolved in the group by this time (Novacek, 1985). A few modern carnivores and ungulates also appear, possibly having spread into Europe from North America. By the later part of the early Eocene larger mammals become more prevalent, evidently reflecting the appearance of drier, more savanna-like environments (Russell, 1975; Krause 1984; Rose, 1984).

Four plesiadapiform families are still represented in the early Eocene: Plesiadapidae, Carpolestidae, Paromomyidae, and Microsyopidae. The former two families disappear by the end of the early Eocene; the latter two survive well into the Eocene (Krishtalka, 1978).

The paleontological significance of Eocene faunas can be summarized as follows: (1) they provide evidence for the existence of a land bridge connecting North America and Europe; (2) they signal the appearance of new, more modern mammalian faunas and document the end of the Cretaceous–Paleocene stage of mammalian evolution; and (3) they reveal the first presence of euprimates in the fossil record.

RISE OF THE EUPRIMATES

By the early Eocene, fossil evidence reveals a reduction in the diversity of archaic plesiadapiforms, thus signaling the evolutionary decline of these earliest primates. In their place euprimates suddenly appear in North America, Europe, and Asia, approximately 53 MYA. (In fact, the early Eocene of North America can be recognized in the field on the basis of the first appearance of the early euprimate genus *Cantius*.)

Elwyn Simons of Duke University introduced the term "euprimates" (literally, true primates) in 1972 to describe what he called "primates of modern aspect." These primates are called modern not because they resemble living primates in all respects, but rather because they have already evolved some structures associated with all modern primates. For instance, nails have replaced claws on the digits; a postorbital bar has developed; and a grasping, opposable hallux (first toe) indicative of arboreal habitats is present.

Eocene euprimates are usually classified into two families, **Adapidae** and **Omomyidae,** which in turn are divided into several subfamilies (Table 3.1). Both families are well known in the fossil record from cranial, dental, and postcranial remains. Some of the better-known adapid genera are *Adapis, Pronycticebus, Notharctus, Cantius, Pelycodus,* and *Smilodectes.* Some of the better-known omomyid genera are *Necrolemur, Tetonius, Hemiacodon,* and *Pseudoloris.* Fairly complete skulls are known for the adapids *Notharctus, Adapis, Smilodectes, Mahgarita,* and *Pronycticebus* and for the omomyids *Shoshonius, Necrolemur, Rooneyia, Tetonius,* and *Nannopithex* (Fig. 3.2).

All primates decline in abundance from northern latitudes toward the end of the Eocene. They disappear entirely from Europe by the Oligocene and become very rare in North America as well. No other primates appear in Europe until the invasion of apelike hominoids (superfamily Hominoidea) in the middle Miocene.

Table 3.1 Classification of Euprimates: Omomyidae and Adapidae

Taxa	*Epoch (and Area)*[a]	*Dentition*[b]
Infraorder LEMURIFORMES		
Family ADAPIDAE	E. Eoc.–E. Olig.	
Subfamily NOTHARCTINAE		
Cantius	E. Eoc. (Eur. and N.A.)	
Copelemur	E. Eoc. (N.A.)	2.1.4.3/2.1.4.3
Notharctus	M. Eoc. (N.A.)	2.1.4.3/2.1.4.3
Pelycodus	E. Eoc. (N.A.)	2.1.4.3/2.1.4.3
Smilodectes	M. Eoc. (N.A.)	2.1.4.3/2.1.4.3
Subfamily ADAPINAE		
Adapis	M. Eoc.–E. Olig. (Eur.)	2.1.4.3/2.1.4.3
Anchomomys	M. Eoc. (Eur.)	2.1.4.3/2.1.4.3
Caenopithecus	M. Eoc. (Eur.)	2.1.4.3/2.1.4.3
Cercamonius	L. Eoc. (Eur.)	
Donrussellia	E. Eoc. (Eur.)	
Europolemur	M. Eoc. (Eur.)	2.1.4.3/2.1.4.3
Indraloris	L. Mio. (India)	
Leptadapis	M.–L. Eoc. (Eur.)	2.1.4.3/2.1.4.3
Lushius	L. Eoc. (China)	
Mahgarita	L. Eoc. (N.A.)	2.1.3.3/2.1.3.3
Microadapis	M. Eoc. (Eur.)	2.1.4.3/2.1.4.3
Periconodon	M. Eoc. (Eur.)	
Pronycticebus	L. Eoc. (Eur.)	2.1.4.3/2.1.4.3
Protoadapis	E.–L. Eoc. (Eur.)	2.1.4.3/2.1.4.3
Infraorder TARSIIFORMES		
Family OMOMYIDAE	E. Eoc.–L. Olig.	
Subfamily ANAPTOMORPHINAE	E. Eoc.–L. Eoc.	
Absarokius	E.–M. Eoc. (N.A.)	2.1.3.3/2.1.2–3.3
Altanius	E. Eoc. (Asia)	
Anaptomorphus	M. Eoc. (N.A.)	2.1.?2.3/2.1.2.3
Anemorhysis	E. Eoc. (N.A.)	
Aycrossia	M. Eoc. (N.A.)	2.1.2.3/2.1.3.3
Chlororhysis	E. Eoc. (N.A.)	
Gazinius	M. Eoc. (N.A.)	2.1.3.3/2.1.2.3
Kohatius	E.–M. Eoc. (Asia)	
Pseudotetonius	E. Eoc. (N.A.)	?/1.1.3.3
Strigorhysis	M. Eoc. (N.A.)	2.1.3.3/2.1.2.3
Teilhardina	E. Eoc. (Eur. and N.A.)	?/2.1.3–4.3
Tetonius	E. Eoc. (N.A.)	2.1.3.3/2.1.3.3
Tetonoides	E. Eoc. (N.A.)	2.1.2.2/2.1.3.3
Trogolemur	M.–L. Eoc. (N.A.)	?/2.1.2.3
Uintonius	M. Eoc. (N.A.)	?/2.1.3.3

[a] E. = early, M. = middle, L. = late, Eoc. = Eocene, Olig. = Oligocene, Mio. = Miocene, Eur. = Europe, N.A. = North America

[b] Dental formulas are provided where known. However, they may be based on only one species in the genus.

Source: Adapted from Rose and Fleagle (1981).

Table 3.1 (continued)

Taxa	*Epoch (and Area)*[a]	*Dentition*[b]
Subfamily OMOMYINAE	E. Eoc.–L. Olig.	
Chumashius	L. Eoc. (N.A.)	?/2.1.3.3
Dyseolemur	L. Eoc. (N.A.)	?/2.1.3.3
Ekgmowechashala	L. Olig.–E. Mio. (N.A.)	?/2.1.3.3
Hemiacodon	M. Eoc. (N.A.)	?/2.1.3.3
Macrotarsius	L. Eoc.–E. Olig. (N.A.)	?/2.1.3.3
Ourayia	L. Eoc. (N.A.)	2.1.3.3/2.1.3.3
Omomys	E.–M. Eoc. (N.A.)	2.1.3.3/2.1.3.3
Rooneyia	E. Olig. (N.A.)	2.1.2.3/?
Shoshonius	E. Eoc. (N.A.)	
Stockia	L. Eoc. (N.A.)	
Utahia	E.–M. Eoc. (N.A.)	
Washakius	M.–L. Eoc. (N.A.)	2.1.3.3/2.1.3.3
Subfamily MICROCHOERINAE	M.–L. Eoc. (Eur.)	
Microchoerus	L. Eoc.–E. Olig. (Eur.)	2.1.3.3/2.1.2.3
Nannopithex	M. Eoc. (Eur.)	2.1.3.3/2.1.2.3
Necrolemur	L. Eoc. (Eur.)	2.1.3.3/2.1.2.3
Pseudoloris	M.–L. Eoc. (Eur.)	
OMOMYIDAE, *incertae sedis*		
Arapahovius	E. Eoc. (N.A.)	2.1.3.3/2.1.3.3
Loveina	L. Eoc. (N.A.)	?/2.1.3.3

Comparison of Adapids and Omomyids

Adapids have been described as lemurlike and omomyids as tarsierlike in morphology. Indeed, the two families are usually subsumed under separate infraorders: Adapidae in the Lemuriformes and Omomyidae in the Tarsiiformes (Table 3.1). Although these labels are somewhat misleading, it is fair to say that already in the early Eocene two distinctive primate morphotypes can be found that resemble in many ways the two extant primate suborders, Strepsirhini (lemurs and lorises) and Haplorhini (tarsiers and anthropoids). (A **morphotype** is comprised of the known character states that are likely to be diagnostic of the ancestor of a species.) The last section in the chapter will discuss various hypotheses concerning the evolutionary relationships of these two families.

Some of the morphological features distinguishing adapids from omomyids include the following:

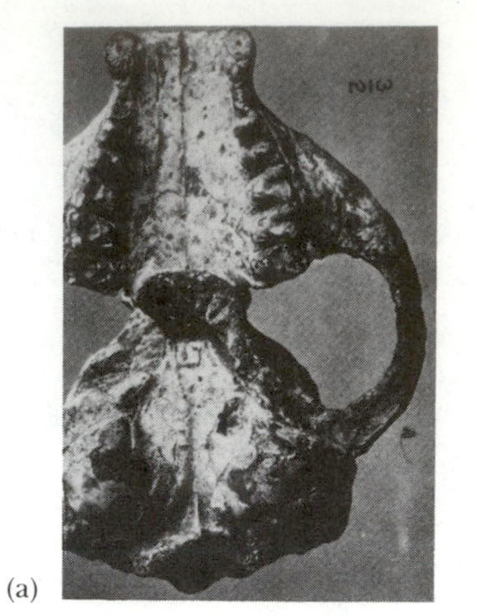

(a)

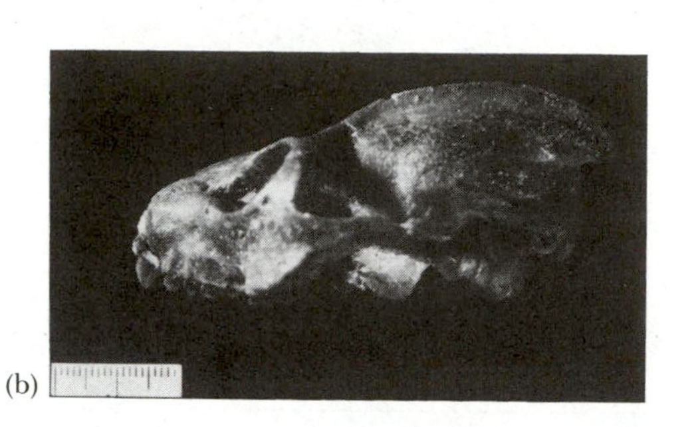

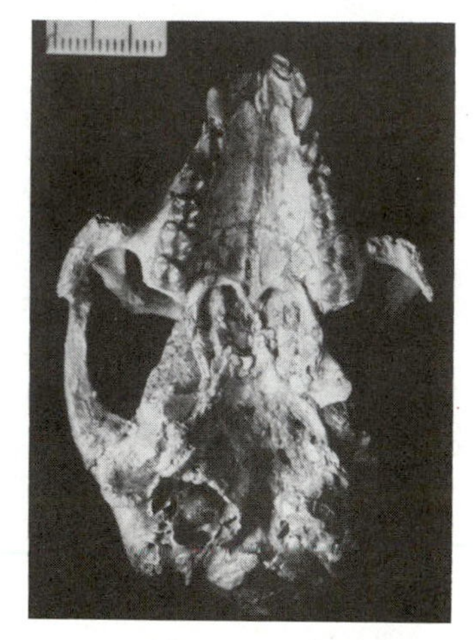

(b)

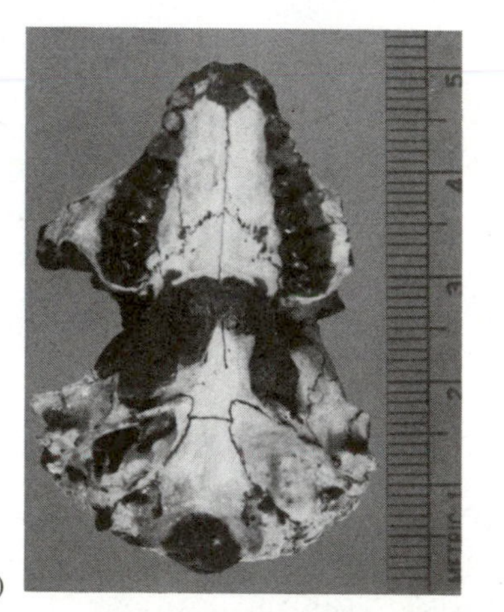

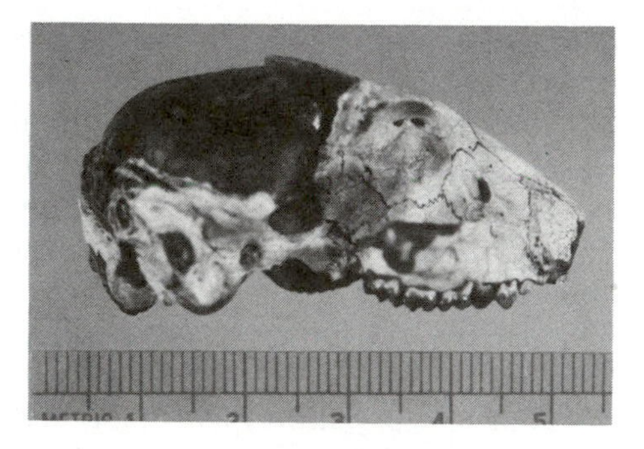

(c)

Fig. 3.2 Examples of (a,b) Adapid and (c) Omomyid Skulls (a) *Pronycticebus gaudryi* (late Eocene, France, about 1 × natural size); (b) *Adapis parisiensis* (middle to late Eocene, France); (c) *Rooneyia viejaensis* (early Oligocene, Texas). (a, From Simpson, 1940b; b, courtesy of J. Schwartz)

1. Adapids have a comparatively elongated snout with jaws that are almost parallel, whereas omomyids have shorter snouts and more divergent, V-shaped, jaws.
2. Except in the earliest forms such as *Cantius* and *Pelycodus,* adapids generally have a fused **mandibular symphysis** (midline joint in the lower jaw), whereas all omomyids have an unfused, mobile mandibular symphysis.
3. Adapids have short, more vertically implanted incisors with **spatulate** (broad, flat) crowns, whereas omomyids have more protruding lower incisors with moderately to sharply pointed crowns.
4. Adapids have a lower central incisor that is smaller than the lateral incisor, whereas omomyids have a lower central incisor that is equal to or larger than the lateral incisor.
5. Adapids have strong, interlocking upper and lower canines, and the most anterior lower premolars are **sectorial,** that is, they have **honing facets** for sharpening the upper canines; in contrast, the canine teeth of omomyids are usually small, **premolariform,** (shaped like a typical premolar), and often not honed by the lower premolars.
6. Adapid canines are **sexually dimorphic** (showing differences in size and/or shape, between the sexes), whereas canine sexual dimorphism is never observed in omomyids.
7. Many Eocene adapids have quadritubercular lower molars with a reduced or absent paraconid, whereas most omomyids retain a paraconid in the lower molars.
8. Adapids have a free ectotympanic ring within the auditory bulla, whereas omomyids have a tubular ectotympanic bone. (Gingerich and Schoeninger, 1977)

These and other features are summarized in Table 3.2.

Adapidae

Origins and Distribution Adapidae are first recognized from the early Eocene of Europe and North America, and some (such as *Indraloris*) apparently survived into the late Miocene of southern Asia. A number of genera are known from Europe and North America but relatively few from Asia and Africa. Adapids are most numerous and diverse during the middle Eocene, particularly during the Bridgerian land-mammal age in North America, from about 49 to 47 MYA.

The presence of several adapid genera in the early Eocene of Europe, compared with only one genus during the same time period in North America, suggests that Europe may have been the group's geographical center of origin. However, Gingerich (1977*c*) has postulated an origin in the Southern Hemisphere for adapids from a *Purgatorius*-like ancestor. He reasons that since Paleocene and Eocene fossil mammals are known mostly from middle to high latitudes of North America, Europe, and Asia, only a portion of the available habitat for mammals during this

Table 3.2 Morphology of Eocene Euprimates and Oligocene Anthropoids

Morphological Characteristics	*Eocene Adapidae*	*Eocene Omomyidae*	*Oligocene Anthropoids*
Body size			
Average body size	Above 500 g	Below 500 g	Above 500 g
Dentition			
Lower dental formula	2.1.4.3 to 2.1.3.3	2.1.4.3 to 2.1.2.3 or 1.1.3.3	2.1.3.3 or 2.1.2.3
Mandibular symphysis	Unfused to fused	Unfused	Fused
Incisor form	Spatulate	Pointed	Spatulate
Incisor size	$I_1 < I_2$	$I_1 \geq I_2$	$I_1 < I_2$
Canine occlusion	Interlocking	Limited	Interlocking
Canine dimorphism	Present	Absent	Present
Canine–premolar hone	Present	Absent	Present
Molar form	Tritubercular to quadrate[a]	Tritubercular	Quadrate[a]
Position of hypocone	On postproto- or basal cingulum	On basal cingulum	On basal cingulum
Cranium			
Encephalization quotient	EQ = 0.39 or 0.41	EQ = 0.42–0.97	EQ = 0.85
Postorbital closure	None	None	Partial to complete
Ectotympanic	Free anulus	Tubular	Free (?) to fused anulus
Stapedial artery	Large or small	Small	Lost
Postcranium			
Calcaneonavicular	Short	Elongated	Short
Tibia–fibula	Unfused	Fused	Unfused

[a]Quadrate = quadritubercular

Source: Adapted from Gingerich (1980*b*).

time has been paleontologically sampled. In addition, the climatic history of the early Tertiary may have had two major effects: (1) forms such as those directly ancestral to adapids may have been limited to equatorial areas (such as Central America or southern Europe and Africa, or perhaps southern Asia), where fossil mammals have not yet been found; and (2) the warming temperatures in the early Eocene may have made land corridors connecting northern and southern continents more accessible to subtropical mammals so that adapids in the Southern Hemisphere could have migrated to more northern latitudes.

Although the taxonomy of the group is still unsettled, approximately two dozen Eocene genera and several Miocene genera are known. Some late Eocene–early Oligocene African and Asian primates such as *Azibius, Oligopithecus,* and *Amphipithecus* have also been considered adapids by some workers (Szalay, 1970; Sudre, 1975; Gingerich, 1980*a*) but these claims have not been universally accepted (Simons, 1972; Szalay and Delson, 1979).

Adapidae are divided into two subfamilies: **Notharctinae** and **Adapinae,** which separate roughly along geographical lines. Notharctines are restricted to North America, with the exception of *Cantius,* which has species in both Europe and North America. Adapines are restricted to Eurasia, with the exception of *Mahgarita,* which is found only in North America. The early Eocene *Cantius* is the most primitive genus and probably gave rise to the subsequent notharctine and adapine genera in North America and Europe respectively. The opening of the North Atlantic in the early Eocene was probably responsible for the geographic separation and independent evolution of post-*Cantius* adapid species. Major adapid taxa are listed in Table 3.1.

It appears that the North American notharctines, at least in the Rocky Mountain area, became extinct about the end of the middle Eocene approximately 47 MYA (Gingerich, 1979). C. Lewis Gazin (1958) of the Smithsonian Institution attributed this extinction and the overall reduction of primates at this time to increasing aridity associated with regional uplift of the western interior of North America. He speculated that primate faunas might have retreated from the interior to more lowland coastal regions during this period. This speculation has recently been confirmed by the discovery of an assortment of late Eocene primate fossils in San Diego County, California, that includes both *Pelycodus* and *Notharctus* (Lillegraven, 1980). The paleogeography of this region is reconstructed as being coastal lowland environment with rivers and deltas.

In Europe, adapines lasted until the late Eocene, although a few genera (such as *Indraloris* and *Sivaladapis*) survived in southern Asia until the late Miocene.

We shall take a look at notharctines and adapines in greater detail later in the chapter. Now we turn to examine the morphology of the Adapidae as a whole.

General Morphology Characteristic of all adapids are the following morphological traits:

1. Fused mandibular symphyses (except in *Cantius* and *Pelycodus*)
2. Retention of four premolars in most forms
3. Projecting, interlocking, and sexually dimorphic canines and vertically implanted incisors in most forms
4. Placement of the lacrimal bone within the orbit instead of extending onto the face as in plesiadapiforms
5. A free, ringlike ectotympanic bone within a petrosal auditory bulla
6. An internal carotid artery that enters the posterolateral margin of the auditory bulla
7. Comparatively well-developed promontory and stapedial arteries, with the stapedial usually larger than the promontory

The small orbital size relative to overall skull length suggests that most adapids were diurnal, although one, *Pronycticebus,* may have been

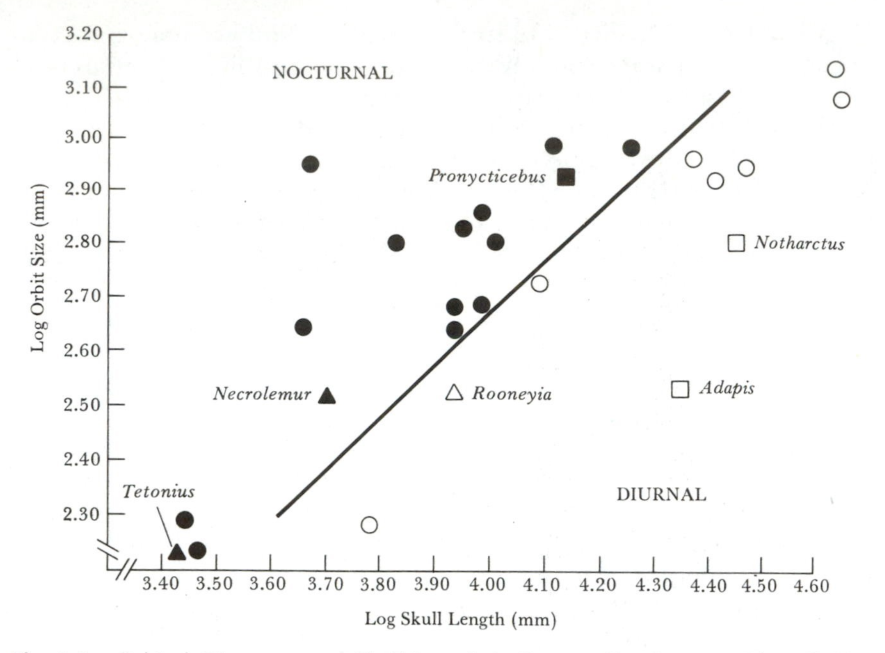

Fig. 3.3 Orbital Diameter and Skull Length in Eocene Euprimates Line divides nocturnal species (solid symbols) from diurnal species (open symbols); squares represent adapids, triangles represent omomyids, and circles represent extant species. (Adapted from Kay and Cartmill, 1977)

nocturnal or **crepuscular** (active at dawn and twilight) judging by its relatively large orbital size (Fig. 3.3). Adapids also have very small infraorbital openings for the maxillary nerve, suggesting that the nose and whiskers were probably of secondary tactile importance (Kay and Cartmill, 1977; Gingerich, 1980*a*).

Comparison with Later Primates As mentioned earlier, adapids have already evolved several advanced primate features, such as a divergent hallux and flattened nails on the digits. Adapids such as *Notharctus* and *Pronycticebus* clearly have a postorbital bar and a petrosal bulla, universal features of all extant primates. Other parts of adapid anatomy, however, are still somewhat primitive. For example, *Adapis* retains a bony configuration of the medial orbital wall in which a wide contact between the frontal and maxillary bones separates the palatine and lacrimal bones (Cartmill, 1971, 1978) (see Fig. 2.14a). In addition, contact between the palatine and frontal bones separates the **orbitosphenoid** and maxillary bones.

Dentally, adapids resemble higher primates in having (1) vertically implanted, spatulate incisors; (2) projecting, sexually dimorphic canines; and (3) rectangular molars with reduced paraconids. Most genera have fused mandibular symphyses as well.

As discussed in Chapter 2, relative brain size (encephalization quotients) can be calculated if brain and body weights are known (Jerison, 1973; Martin 1982). Relative brain size has been determined for some adapids such as *Smilodectes, Notharctus,* and *Adapis* (Gingerich and Martin, 1981; Gurche, 1982) (Table 3.3). These values are within the range of other Eocene mammals but are slightly lower than the encephalization quotients of the Oligocene anthropoid *Aegyptopithecus* (see Chapter 4) and of most living primates (Jerison, 1973; Radinsky, 1982).

Adapids bear some resemblance to lemurs in the middle-ear region. As we have previously noted, the lateral wall of the middle-ear cavity is formed by the tympanic membrane (eardrum), which is supported by

Table 3.3 Encephalization Quotients for Early and Modern Prosimians

Species	*Source of Estimate*[a]	*Mean EQ*	*Minimum EQ*	*Maximum EQ*
Early Prosimians				
Adapis parisiensis	JG	0.39	0.34	0.57
	LR	0.39		
	HJ	0.53		
Smilodectes gracilis	JG	0.47	0.35	0.86
	JG	0.44	0.36	0.64
	LR	0.41		
	HJ	0.53		
Notharctus tenebrosus	JG	0.49	0.36	0.92
Tetonius homunculus[b]	JG	0.43	0.33	0.67
	LR	0.42		
	HJ	0.71		
Necrolemur antiquus	JG	0.56	0.35	0.76
	LR	0.79		
	HJ	0.94		
Rooneyia viejaensis	JG	0.81	0.60	1.07
	LR	0.97		
	HJ	1.23		
Modern Prosimians[c]				
Tarsius spectrum		1.14		
Tarsius syrichta		1.35		

[a]Values calculated from volumetric measurements and body-size estimates by JG (Gurche, 1982), LR (Radinsky, 1978), and HJ (Jerison, 1979).

[b]Based on a brain volume of 1.5 cc, used by all three researchers in their calculations.

[c]Range for modern prosimians: 0.67–1.89; mean: 1.09. Values for modern prosimians were calculated from data presented by Stephan et al. (1970). The mean EQ for modern prosimians was based on mean log body weight and mean log brain weight.

Source: Adapted from Gurche (1982).

the ectotympanic bone (see Fig. 2.5). In lemurs the ectotympanic bone is a free ring inside the bulla and is attached to the side wall of the bulla by a membrane. In lorises the ring is exposed at the lateral surface of the bulla and somewhat expanded to form part of the lateral bulla wall. This lorislike condition is also typical of South American primates, the Miocene loris *Komba,* and early anthropoids such as *Aegyptopithecus.* In contrast, omomyids and other haplorhines (extant tarsiers, Old World monkeys, and hominoids) develop a tubular external auditory tube (outer-ear passage) from the ectotympanic bone. Adapid skulls that preserve this region show some variability in ectotympanic configuration. The ectotympanic bone is clearly a free, ringlike structure in some genera but is closer to the lateral bulla wall in others.

Which of these configurations, the free ectotympanic ring or the tubular ectotympanic bone, is the primitive primate condition? The free ring is found in tree shrews, leptictids, some primitive marsupials, lemurs, and adapids. However, the archaic primates *Plesiadapis* and *Phenacolemur,* and the Pleistocene lemur *Palaeopropithecus,* apparently had tubular ectotympanic bones. Thus the true morphocline polarity of the primate ectotympanic bone is still debatable.

In extant lemurs the internal carotid artery enters the posterolateral part of the bulla and divides into larger stapedial and smaller promontory branches (Fig. 2.6). In lorises, however, the internal carotid artery divides before reaching the base of the skull. The major blood supply to the brain in the latter primates is through the ascending pharyngeal artery, which enters the skull in front of the auditory bulla at the opening called the **foramen lacerum** (Cartmill, 1975). Adapid skulls all seem to resemble the lemurlike condition in this feature.

Even though adapids are now often referred to as being lemurlike in morphology, it should be emphasized that adapids differ craniodentally from modern lemurs in several important respects:

1. The lacrimal bone in adapids does not extend onto the face.
2. The mandibular symphysis in adapids is fused except in *Cantius* and *Pelycodus.*
3. Adapid upper and lower fourth premolars are molariform.
4. Most significantly, there is no hint of the characteristic lemur dental comb.

It might be noted that the position of the lacrimal bones and the morphology of the lower incisors and canines are quite variable among fossil Malagasy lemurs of the Pleistocene epoch. One cannot consider, for example, that Pleistocene lemurs like *Hadropithecus, Palaeopropithecus,* or *Archaeoindris* had tooth combs, although one can argue that their peculiar anterior dental morphologies were derived from tooth combs.

Indeed, there are no unequivocal synapomorphies (shared, derived features) linking adapids to the modern tooth-combed primates (lemurs and lorises). Most of the arguments advanced for such a linkage are based on middle-ear morphology and carotid circulatory patterns. These

Table 3.4 Some Characters Common to Adapids and Anthropoids

1. Body size greater than 500 g
2. Tendency for the mandibular symphysis to fuse
3. Vertical, spatulate incisors
4. I_1 smaller than I_2
5. Interlocking canine occlusion
6. Canines moderately large and projecting
7. Canines sexually dimorphic
8. Canine–premolar honing
9. Molarized P4
10. Tendency toward quadritubercular lower molars
11. Nontubular (partially free) ectotympanic
12. Relatively short calcaneum
13. Unfused tibia–fibula

Source: Adapted from Delson and Rosenberger (1980).

features may be primate symplesiomorphies (shared primitive features), however, and thus of dubious phylogenetic value (Cartmill and Kay, 1978).

The postcranial morphology in some adapids indicates that they approach lemurs in their leaping ability, showing less tendency toward the slow-climbing locomotion of the more rodentlike plesiadapiforms. Notharctines are more lemurlike than adapines in this respect. Details of the postcranial comparison with lemurs will be given shortly as we examine each subfamily separately.

As we shall see later in the chapter, some workers (for example, Gingerich, 1976) have noted that adapids share some morphological features with higher primates (anthropoids) and may be ancestral to them. Table 3.4 summarizes the traits adapids have in common with higher primates.

Having taken a look at the general morphology of adapids and how it compares with that of later primates, we shall now examine the two subfamilies, Notharctinae and Adapinae.

Notharctinae The adapid subfamily Notharctinae is usually divided into five genera: *Cantius, Pelycodus, Notharctus, Smilodectes,* and *Copelemur*. Of these, *Cantius* is found in both North America and Europe. The other genera are known only from North America. The taxonomy of the notharctines has recently been complicated by the wholesale transfer of most species previously considered members of *Pelycodus* to the genus *Cantius* (Gingerich and Haskin, 1981; Gingerich, 1986). The only species left in *Pelycodus* is *P. jarrovii,* the type species of the genus. The early-to-middle Eocene genera *Cantius, Pelycodus,* and *Notharctus* represent a single evolving lineage in North America. The origin of *Smilodectes* is not as clear, but it probably evolved from *Notharctus;* where *Copelemur* fits into the notharctine lineage is also unknown.

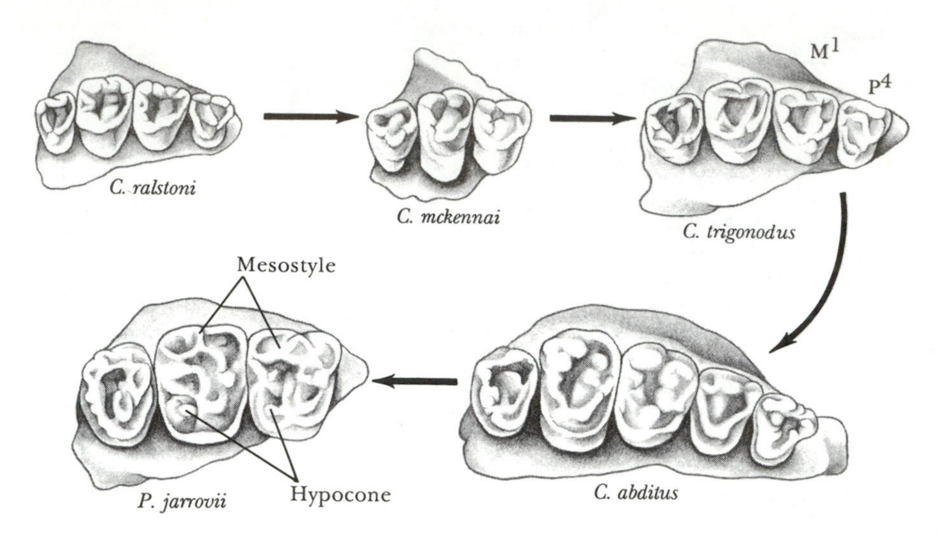

Fig. 3.4 Dental Trends in the Notharctine Lineage of *Cantius* to *Pelycodus* Upper right dentition, all drawn to the same scale, about 2×natural size. Note the increasing tooth size and enlargement of the mesostyle and hypocone. (Redrawn from Gingerich and Simons, 1977)

The transition from *Cantius* to *Notharctus* seems to have been a gradual process, thus making the division between them somewhat arbitrary. There are some distinctions, however. For example, the mesostyle and hypocone become progressively larger through time in *Cantius* (Fig. 3.4), and in *Notharctus* these structures are well developed. The functional significance of the mesostyle is to puncture leaves during the initial stages of mastication (Schoeninger, 1976). Fusion of the mandibular symphysis in this lineage is a convenient demarcation point separating the two genera and fortuitously coincides with the Wasatchian–Bridgerian mammal-age boundary (early to middle Eocene).

Like its contemporary *Cantius*, the early Eocene genus *Copelemur* has an unfused mandibular symphysis and a rudimentary development of the hypocone and mesostyle. The only major difference in molar morphology between these two closely related genera is the deeper notch separating the entoconid from the hypoconulid in the latter.

Excellent postcranial fossils are known for *Cantius, Notharctus* (Fig. 3.5), and *Smilodectes* (Gregory, 1920; Rose and Walker, 1985; Covert, 1985).

In overall skeletal morphology, *Cantius* and *Notharctus* are very similar to the modern lemuriforms *Indri, Lepilemur, Propithecus, Lemur,* and

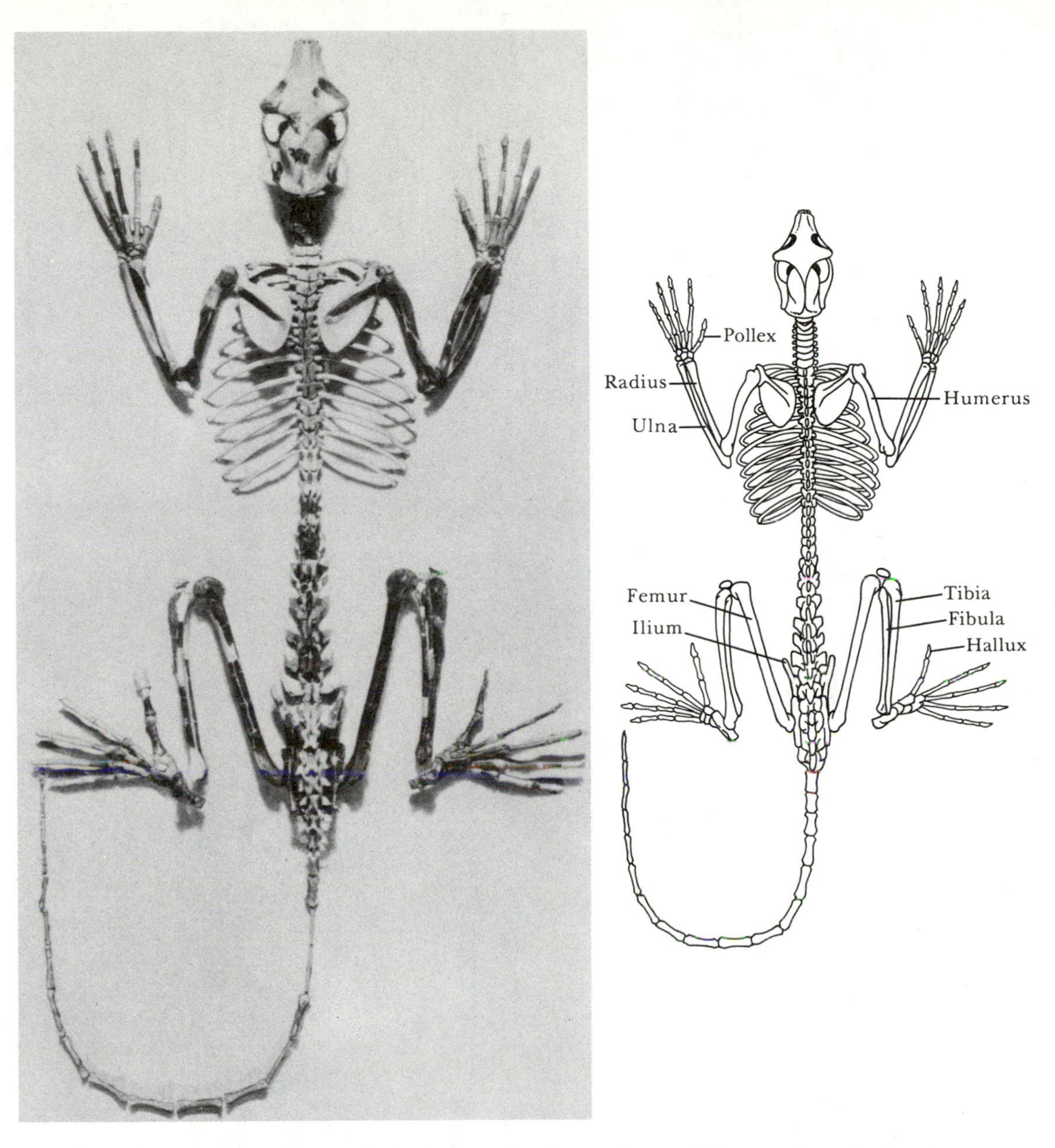

Fig. 3.5 Partially Restored Skeleton of *Notharctus osborni* Reconstructed portions shown in a lighter color. Scale about 1.03 × natural size. (From Gregory, 1920)

Hapalemur. Notharctines have comparatively long hindlimbs and short forelimbs and were clearly arboreal animals capable of leaping and powerful grasping; hands have an opposable **pollex** (thumb) and feet an opposable hallux (Fig. 3.6).

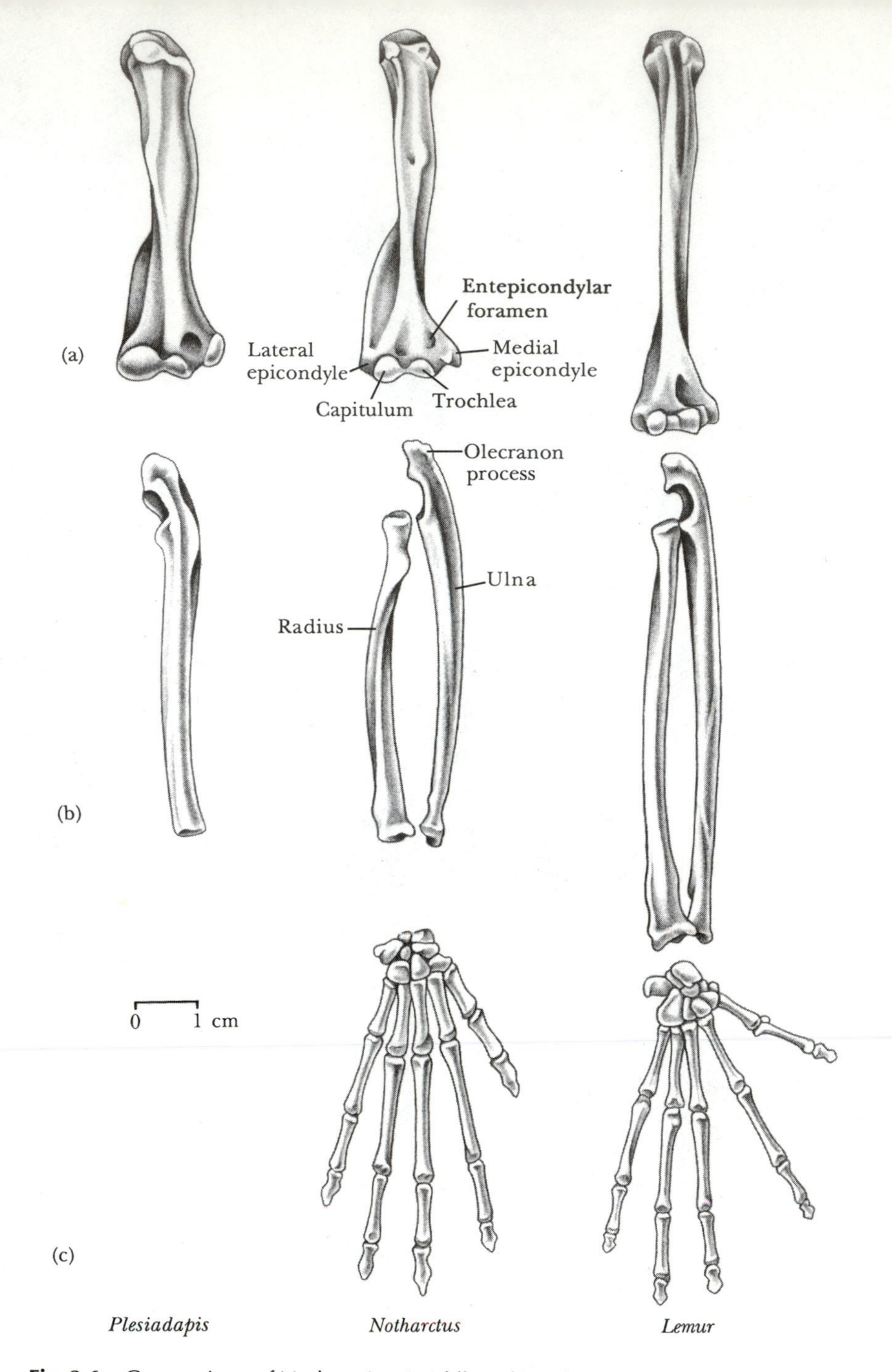

Fig. 3.6 Comparison of Notharctine *(middle)*, Plesiadapiform *(left)*, and Lemuriform *(right)* Forelimb Bones (a) Anterior view of a right humerus, (b) anteromedial view of right radius and ulna (with only ulna shown for *Plesiadapis*), and (c) anterior view of right manus. (The *Plesiadapis* humerus, originally a left humerus in fossil form, has been inverted here for easier comparison.) Note the opposable pollex *Notharctus* shares with later primates; the plesiadapid hand (not shown) does not have an opposable thumb. (Plesiadapis bones redrawn from Szalay et al., 1975; all others redrawn from Gregory, 1920)

The postcranial skeleton of notharctines is characterized by the following features:

1. The brachial index is lower than in all vertical clingers and leapers.
2. Full extension of the elbow (indicative of **suspensory posture**) would have been impossible because of the shallow **olecranon fossa** (the shallow depression on the posterior surface of the distal humerus).
3. The crural index is lower (ranging between 81 and 90) than in nearly all prosimians, but it is closest to the crural index values for vertical clingers and leapers (83 in *Notharctus*).
4. The intermembral index, which is often the best predictor of locomotor habits, is within the range of values for vertical clingers and leapers and lower than the values for slow and active climbing arboreal quadrupeds.

The hindlimb of notharctines more closely resembles the hindlimb of lemurs than that of the more rodentlike plesiadapiforms. The bony pelvis has a shorter ischium, a smaller **obturator foramen,** and a much flatter, more rodlike ilium than the plesiadapiform pelvis. In addition, the **anterior inferior iliac spine** is much more prominent (Fig. 3.7). The

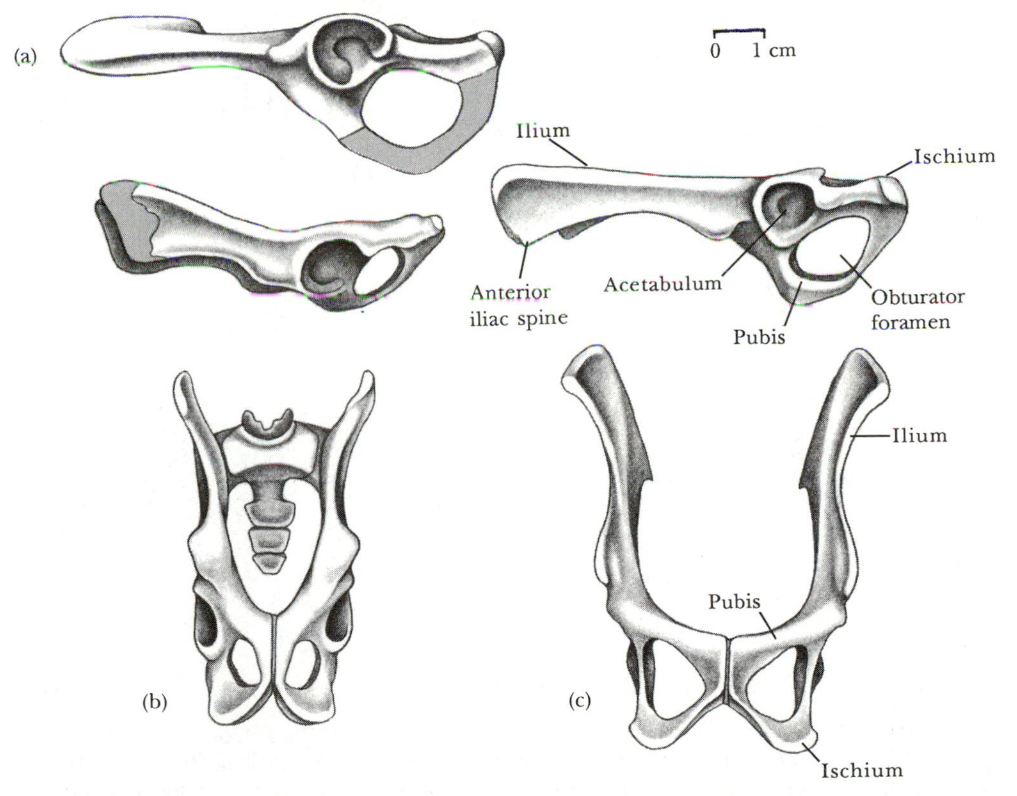

Fig. 3.7 Comparison of the Pelvis in (a) *Plesiadapis,* (b) *Notharctus,* and (c) *Lemur* Left lateral view *(above)* and ventral view *(below)*. (a, Redrawn from Szalay et al., 1975; b,c, redrawn from Gregory, 1920)

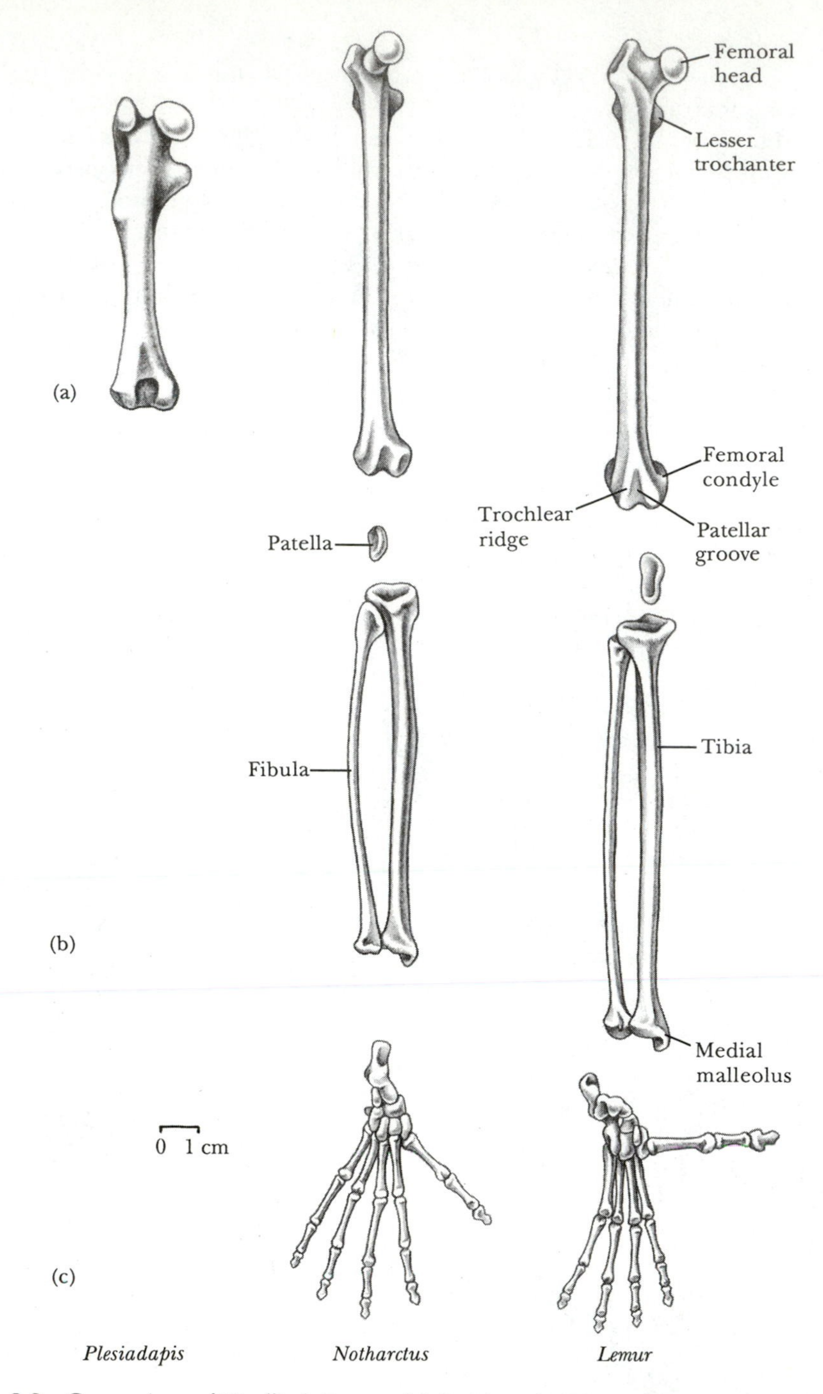

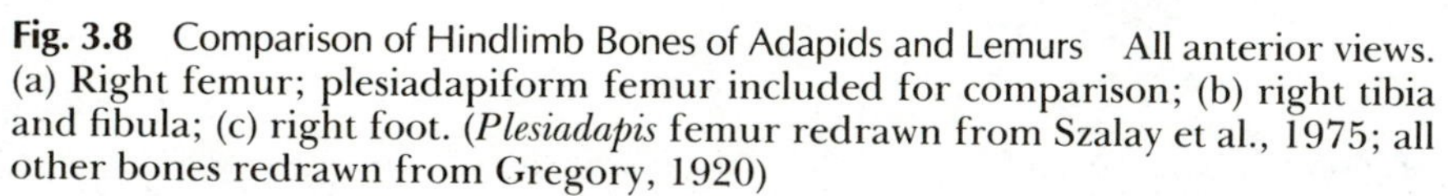

Fig. 3.8 Comparison of Hindlimb Bones of Adapids and Lemurs All anterior views. (a) Right femur; plesiadapiform femur included for comparison; (b) right tibia and fibula; (c) right foot. (*Plesiadapis* femur redrawn from Szalay et al., 1975; all other bones redrawn from Gregory, 1920)

femur is longer and less robust, with a larger **greater trochanter,** a smaller and higher **lesser trochanter,** and a more proximal **third trochanter.** The femoral neck is longer and less vertical. The distal femur is comparatively narrow, with a deeper and narrow **patellar grove** (the depression on the distal end of the femur in which the **patella,** or kneecap, moves). In addition, the **trochlear ridges** are better developed (Rose and Walker, 1985) (Fig. 3.8). (A **trochlea** is any smooth, saddle-shaped bony surface that acts like a pulley as it articulates with other bones; in the femur the trochlea is the distal end in contact with the kneecap. Trochleae have one or more raised margins called trochlear ridges.) These differences in pelvic and femoral morphology point to a greater capacity for deliberate climbing in the plesiadapiforms and more leaping ability in the notharctines.

Adapinae There are at least twice as many Eocene genera of European adapids (adapines) as there are North American ones (notharctines). The European adapines are of particular interest for several reasons:

1. *Adapis* was the first genus of a fossil primate to be described, although it was initially thought to be a transitional form between the Artiodactyla (even-toed hooved mammals) and primates—hence the name *Adapis,* which means literally "toward the sacred bull Apis" (Cuvier, 1821; Simons, 1972) (Fig. 3.9).
2. *Caenopithecus lemuroides* was the first fossil primate to be considered lemurlike (Rutimeyer, 1862).
3. Adapines are common constituents of European Eocene mammal faunas.
4. It has been suggested that anthropoid primates may have evolved directly from an adapine ancestry.

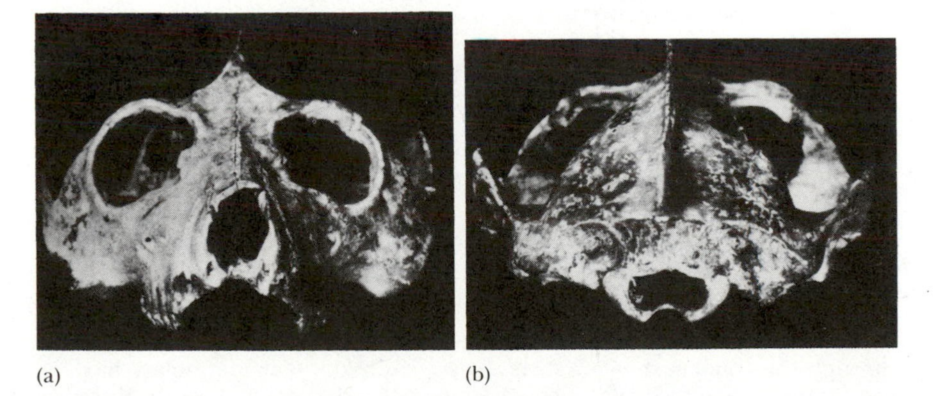

(a) (b)

Fig. 3.9 *Adapis parisiensis* (a) Anterior and (b) posterior views Scale about 1.1 × natural size. (From Gingerich and Martin, 1981)

About a dozen genera of European Adapidae have been described (Szalay and Delson, 1979). Some of these are listed in Table 3.1. The geologically oldest is the notharctine *Cantius eppsi,* known from England and France. *Cantius eppsi* is very similar to the earliest North American notharctine, *C. ralsonti.* Small and medium-sized species of *Protoadapis* next appear in the European fossil record; they differ from *Cantius* in having somewhat higher tooth cusps with sharper interconnecting crests and more strongly developed hypocones on the upper molars.

One of the major dental distinctions between notharctines and adapines is in the formation of the hypocone. In European adapines it arises directly from the **cingulum** (the shelflike extension of enamel around the periphery of the tooth), whereas in notharctines the hypocone develops as an extension of the protocone and as such is really a **pseudohypocone** (Gregory, 1920) (Fig. 3.10).

In three adapine lineages, body weight increases through time *(Protoadapis curvicuspidens–Caenopithecus lemuroides; Adapis priscus–A. magnus;* and *A. sudrei–A. laharpei)* and in three adapine lineages body weight decreases through time *(Protoadapis louisi–Anchomomys gaillardi; Protoad-*

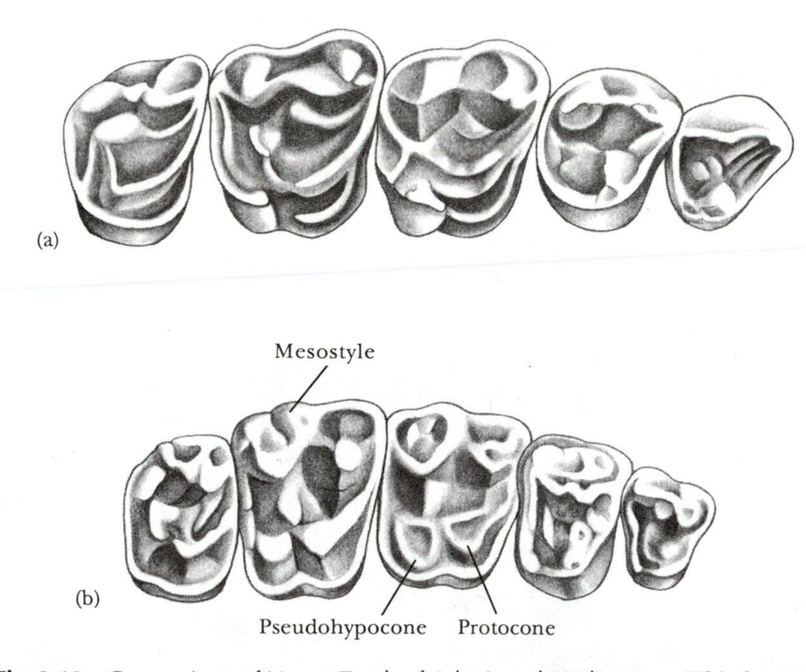

Fig. 3.10 Comparison of Upper Teeth of *Adapis* and *Notharctus* Third premolar through third molar shown. (a) *Adapis magnus.* Note that the molars have no mesostyles and that the hypocones are continuous with the cingulum. (b) *Notharctus crassus.* Note the large mesostyles and pseudohypocones on the molars. Scale about 3.6 × natural size. (Redrawn from Gregory, 1920)

apis recticuspidens–P. klatti; and *Adapis magnus–A. parisiensis)* (Gingerich, 1977*b*). Several examples of parallel evolution are apparently documented in some of these lineages. For example, the lower first premolar was lost and the second premolar reduced in *Protoadapis klatti, Caenopithecus lemuroides,* and *Cercamonius brachyrynchus,* while similar reductions occurred in the North American adapine *Mahgarita;* fusion of the mandibular symphysis occurred in the European *Caenopithecus, Cercamonius,* and *Adapis* and in the North American *Notharctus* and *Mahgarita.*

Adapines seem to have been less agile arboreal quadrupeds than notharctines; some resemble nonleaping prosimians such as *Perodicticus* (the potto), and others resemble more robust forms such as *Varecia* (the ruffed lemur) (Dagosto, 1983). *Adapis,* like recent Lorisinae (lorises but not galagos), has been found to have much shorter hindlimbs relative to forelimbs than do the Lemuridae (lemurs), Indriidae (indri and sifakas), or notharctines. The distal femur of *Adapis* differs from that of notharctines in being broader and flatter with a wide, shallow patellar groove and less accentuated trochlear ridges. The ankle bone is broader and the neck shorter and more medially deviated (more like *Perodicticus*).

Omomyidae

Omomyids are the second family of Eocene euprimates. They are often considered tarsierlike in morphology, and this is reflected in their assignment to the infraorder Tarsiiformes by most authors (Bown and Rose, 1987). Some of the main omomyid genera are listed in Table 3.1.

The first omomyid known to science, *Microchoerus,* was discovered in England in 1843. As the name implies (*choeros* is Greek for pig), its discoverer thought it was a kind of small pig! During the 1860s and 1870s other omomyid fossils were discovered in the western interior of North America *(Omomys, Hemiacodon,* and *Anaptomorphus)* and in Europe *(Necrolemur).* Edward D. Cope (1872) was the first to link omomyids with other primates (Gingerich, 1981).

Origins and Distribution The Omomyidae first appear in the early Eocene of North America, Europe, and Asia. In North America they last through the late Eocene and into the early Oligocene, with one genus, *Ekgmowechashala,* existing into the Miocene. The maximum diversity of omomyids occurs in North America in the late Wasatchian and Bridgerian land-mammal ages. In Europe omomyids last just into the early Oligocene, thus surviving for approximately 16 million years (from about 53 to 37 MYA). The two Asian genera come from the early and middle Eocene of Mongolia *(Altanius)* and Pakistan *(Kohatius)* respectively (Dashzeveg and McKenna, 1977; Gingerich, 1981; Rose and Krause, 1984). One omomyidlike primate from Africa, *Afrotarsius chatrathi,* has recently been described from the Oligocene of Egypt (Simons and Bown, 1985).

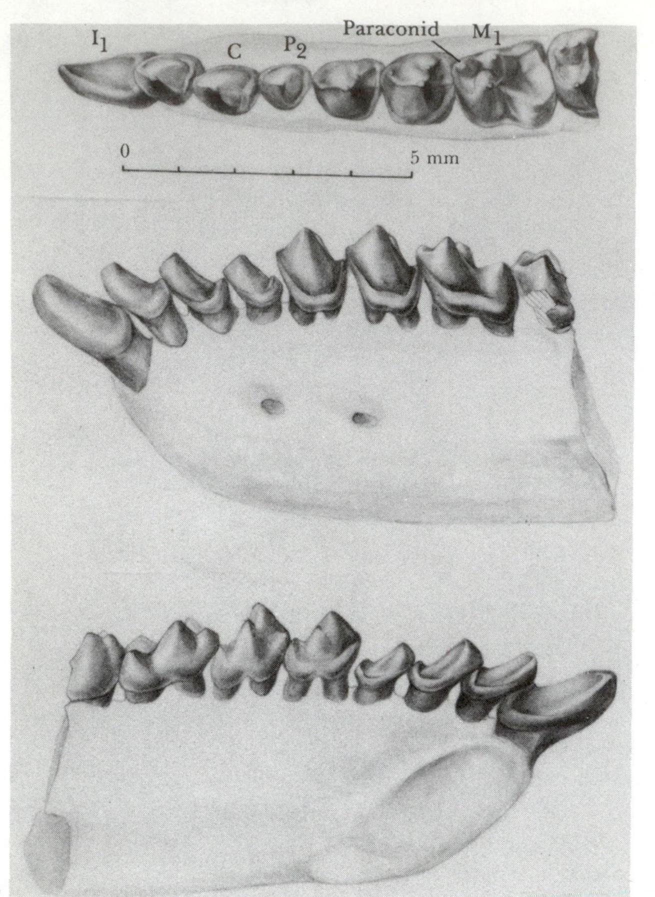

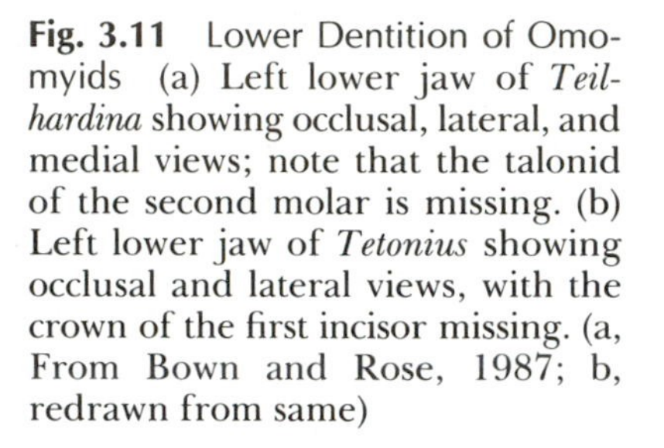
Fig. 3.11 Lower Dentition of Omomyids (a) Left lower jaw of *Teilhardina* showing occlusal, lateral, and medial views; note that the talonid of the second molar is missing. (b) Left lower jaw of *Tetonius* showing occlusal and lateral views, with the crown of the first incisor missing. (a, From Bown and Rose, 1987; b, redrawn from same)

Three subfamilies of the Omomyidae are generally recognized: **Anaptomorphinae, Omomyinae,** and **Microchoerinae** (see Table 3.1). The earliest omomyid in both North America and Europe is the anaptomorphine *Teilhardina. Teilhardina americana* first appears in North America in the Wasatchian land-mammal age (early Eocene) of the Rocky Mountain area and is morphologically similar to *T. belgica* from the European Sparnacian land-mammal age (also early Eocene). Most workers accept that all omomyid primates could have descended from a form similar to *T. belgica* and that all North American omomyids could have had an ancestor similar to *T. americana.* As the North Atlantic opened, this ancestral anaptomorphine stock probably diverged, giving rise to later omomyines in North America and microchoerines in Europe. Anaptomorphines dominated the early Eocene part of the North American omomyid radiation, surviving into the middle Eocene; omomyines dominated the middle and late Eocene radiation in North America; and microchoerines dominated the middle and late Eocene in Europe.

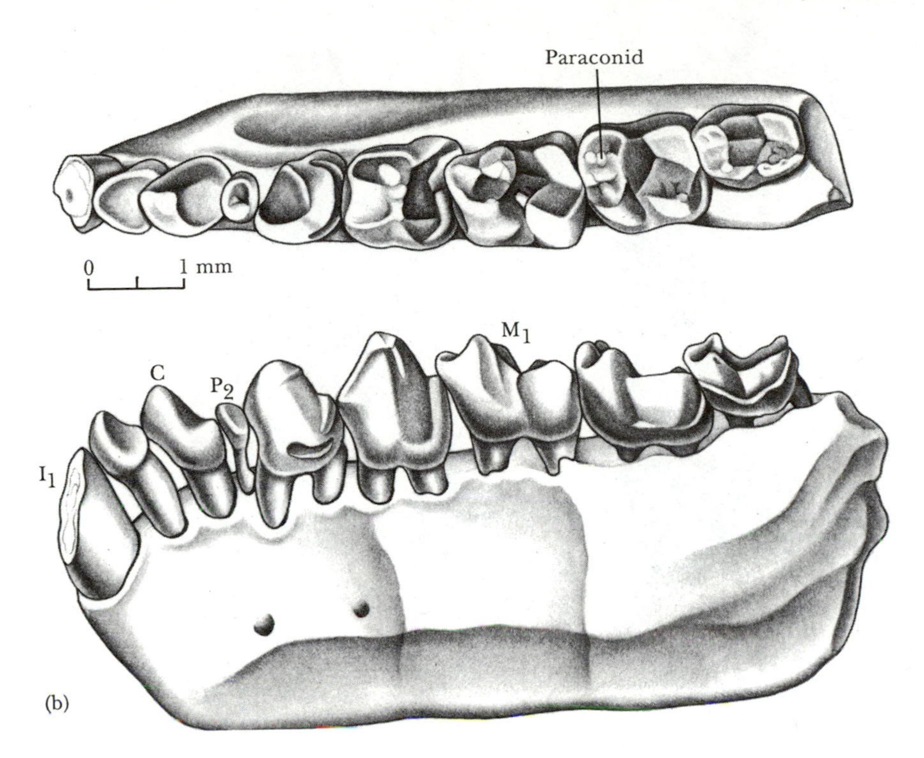

Distinctive characteristics shared by almost all the Omomyidae include the following:
1. Enlarged, pointed, protruding central incisors
2. A reduced dental formula 2.1.3.3 / 2.1.2–3.3 or 1.1.3.3 / 1.1.3.3
3. An ossified postorbital bar with incipient postorbital closure
4. A tubular ectotympanic bone
5. Elongated tarsal bones

Dentition The lower dental formula of omomyids is generally 2.1.2.3 or 2.1.3.3; in a few cases it may be 1.1.3.3. At least some specimens of *Teilhardina,* the most primitive omomyid, probably retained a small lower first premolar, giving it a dental formula of 2.1.4.3 (Bown and Rose, 1987). Characteristics common to the lower dentition of all omomyid genera include the following:
1. A V-shaped tooth row
2. An unfused, mobile mandibular symphysis
3. Pointed incisors with the central pair larger than the lateral pair
4. Comparatively small canines that are not sexually dimorphic
5. Lower molars that retain a three-cusped trigonid and include a distinct paraconid (Fig. 3.11)

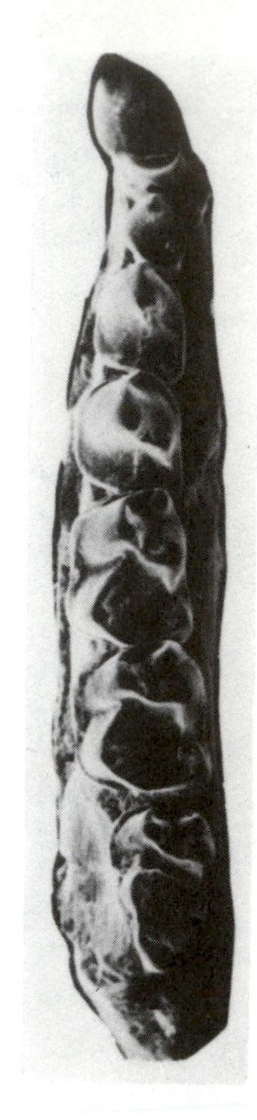
(a)

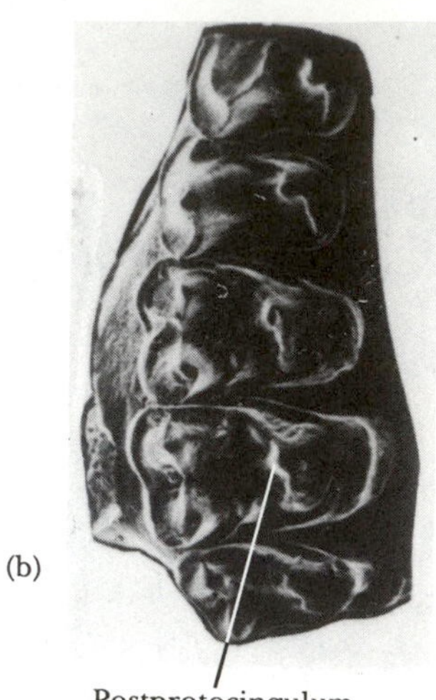

(b)

Fig. 3.12 Scanning Electron Micrographs of Dentition of *Teilhardina* (a) Left lower jaw of *Teilhardina americana* with C, P_2–M_3; (b) upper right jaw of *T. crassidens* with P^3–M^3. Scale about 8.5 × natural size. (Courtesy of K. Rose)

Most omomyids have an upper dental formula of 2.1.3.3. The upper canine is **premolariform** (shaped like a typical premolar) rather than projecting and does not show evidence of sexual dimorphism. The upper molars vary from being simple tritubercular teeth to those having a hypocone and a postprotocingulum (or *Nannopithex* fold) (Fig. 3.12). Omomyids seem to have quite large cheek teeth compared to skull size, as is characteristic of insectivorous mammals generally (Gingerich, 1981). Dental differences between adapids and omomyids are given in Table 3.2.

In anaptomorphines, the most primitive omomyid subfamily, the upper molars retain a postprotocingulum and generally have more-rounded cusps, crests, and basins on their cheek teeth than are typically found in omomyines and microchoerines (Gingerich, 1981). In the latter two subfamilies the postprotocingulum on the upper molars is lost and a hypocone derived from the lingual cingulum is sometimes present. In both anaptomorphines and omomyines the paraconid of the first molar is medially placed. Lower molar cusps are more peripherally situated in omomyines.

Cranial Morphology Skulls of only four omomyids have been identified: *Necrolemur, Shoshonius, Tetonius,* and *Rooneyia.* Unlike adapids, the cranium in each case has a short, tapered snout in which the **lacrimal foramen** (the opening for the tear duct running from the eye to the nasal cavity) opens onto the face. The **infraorbital foramen** (the opening onto the maxilla that allows for passage of the major sensory nerves and blood vessels to the lower face) is comparatively small; this indicates that the nerves and blood vessels supplying the upper lip, nose, and whiskers were reduced in comparison with other mammals (Gingerich, 1984). Omomyids have a postorbital bar, and some show the beginnings of postorbital closure. There is no ethmoid component in the orbital wall. The relative size of the orbits (orbital diameter compared with skull length) suggests that some omomyids were nocturnal (Kay and Cartmill, 1977) (see Fig. 3.3). This combination of features suggests that for omomyids vision was more important than smell in perceiving the world (Fig. 3.13).

Natural **endocasts** of the brain have been preserved for several omomyids, allowing one to learn something of their brain morphology and to calculate encephalization quotients. These endocasts formed when sediment filled the skull and hardened to stone with the passage of time. Because the inner surface of the skull reflects the pattern of convolutions of the brain close to it, the surface morphology of the original brain is duplicated in the cast. (Endocasts can also be made deliberately by filling skulls with latex rubber.)

Tetonius (with a brain volume of 1.5 cc) has a more expanded neocortex, especially in the occipital and temporal regions, than is characteristic of insectivorans and a smaller frontal lobe than is typically found in modern prosimians. The olfactory bulbs also appear to be comparatively small, and the neocortex is smooth except for a faint impression of the **lateral** (or **sylvian**) **sulcus**. (A sulcus is a groove separating convolutions on the surface of the brain; the lateral sulcus separates the temporal from the frontal and parietal lobes.) The brain of *Necrolemur* (2 to 4 cc) is larger than that of *Tetonius,* but otherwise the morphology appears similar. *Rooneyia* has an even larger brain (7 cc) with relatively large frontal lobes and smaller olfactory bulbs (Gurche, 1982).

Radinsky (1982) has calculated encephalization quotients of 0.42 for *Tetonius,* 0.79 for *Necrolemur,* and 0.97 for *Rooneyia,* whereas Jerison (1979) has determined values of 0.71, 0.94, and 1.23 for the same genera (see Table 3.3). Relative brain size in omomyids is less than that of an average living mammal (except possibly for *Rooneyia*) but is moderately larger than that of the average Eocene mammal. Relative brain size in omomyids appears to be slightly larger than in contemporary adapid primates (Gurche, 1982).

Omomyid auditory regions are best known from skulls of *Necrolemur antiquus* (Hurzeler, 1948; Simons, 1961; Szalay, 1975*b,* 1975*c,* 1976) and *Rooneyia viejaensis* (Szalay and Wilson, 1976; Szalay, 1976). Both had an

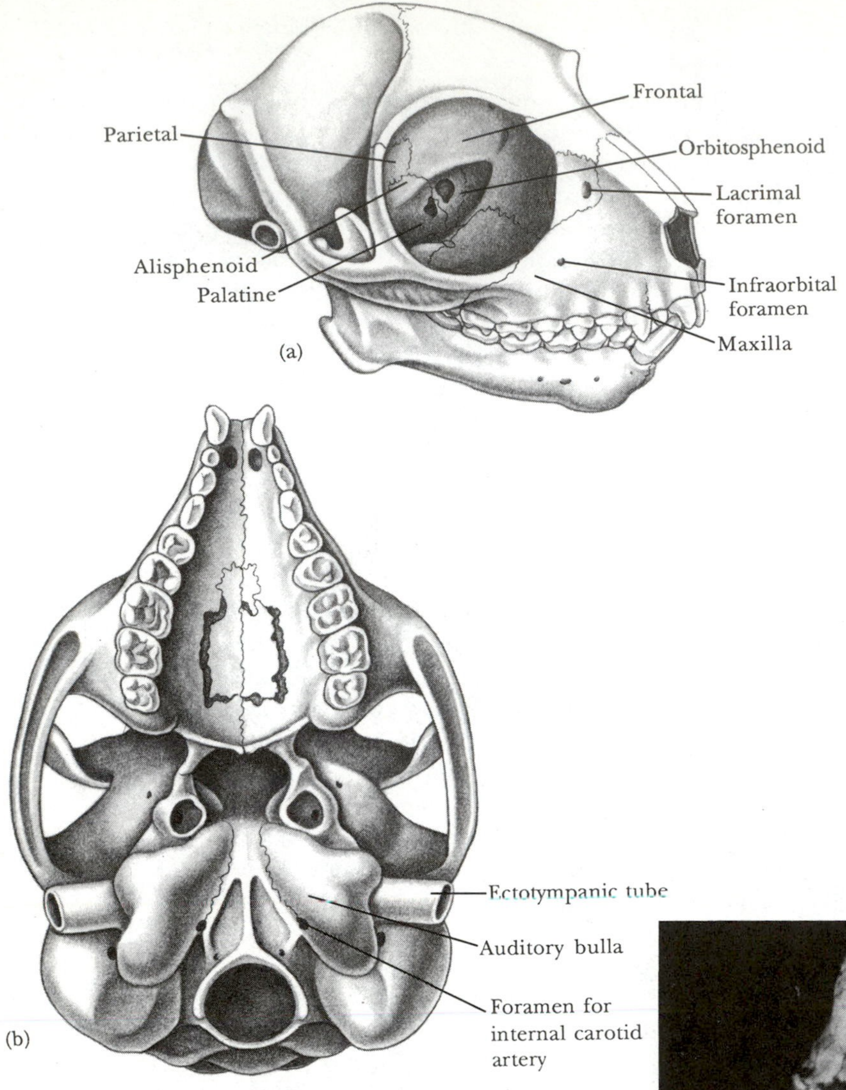

Fig. 3.13 Skull of the Omomyid *Necrolemur* (a) Lateral oblique and (b,c) basal views. Note the tapered snout, lacrimal foramen opening onto the face, large orbits, expanded postorbital bar, and lack of ethmoid in the orbital wall—all features indicating greater dependence on vision than on olfaction. Scale in (a,b) is about 2.4 × natural size; scale bar in (c) is in millimeters. (a,b, Redrawn from Simons and Russell, 1960; c, courtesy of J. Schwartz)

extended tubular auditory meatus like that of *Tarsius* and the plesiadapiforms (Fig. 3.13b).

The carotid circulation in omomyids such as *Necrolemur* and *Rooneyia* has been described by several workers (Simons and Russell, 1960; Szalay and Wilson, 1976). The internal carotid canal enters the auditory bulla posteromedially and then immediately divides into promontory and stapedial canals. Usually the promontory canal is the larger of the two (although in *Necrolemur* the stapedial canal is sometimes only slightly smaller than the promontory canal).

Postcranial Morphology Postcranial remains are known for a number of omomyids such as *Hemiacodon, Necrolemur,* and *Nannopithex.*

Lower hindlimb morphology differs somewhat between omomyines (such as *Hemiacodon*) and microchoerines (such as *Necrolemur*). Microchoerines share several features of the lower leg with *Tarsius,* including distal fusion of the tibia and fibula and an anteriorly–posteriorly compressed distal tibial shaft. These features strongly imply that some microchoerines used leaping as a primary mode of locomotion. Omomyines like *Hemiacodon,* on the other hand, exhibit more primitive features in their lower hindlimb morphology in that the distal tibial shaft is more rounded and tibiofibular fusion is absent (Dagosto, 1985). Presumably, leaping was not as specialized a locomotor activity in this subfamily.

While omomyines may have been less specialized leapers than microchoerines, the fact that the foot is elongated in all omomyids is clear evidence that some leaping ability was an important component of the total locomotor repertoire in the whole family (Godinot and Dagosto, 1983). For example, in *Hemiacodon* the **navicular** bone of the foot is elongated, as it is in primates that habitually leap and cling, while the **entocuneiform** bone and the first metatarsal bone articulate with a broad saddle-shaped joint like that in *Tarsius,* permitting rotation and extreme abduction of the hallux. A greatly enlarged peroneal tubercle on the first metatarsal bone for insertion of the **peroneus longus** muscle (which

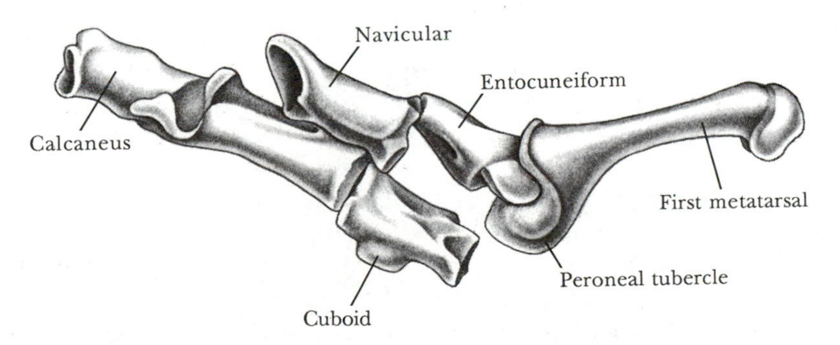

Fig. 3.14 Elements of the Foot Skeleton of the Omomyine *Hemiacodon* Note the elongated navicular and the enlarged peroneal tubercle characteristic of leaping primates. About 2.6 × natural size. (Redrawn from Simpson, 1940b)

adducts the first digit of the foot) indicates that the big toe was also capable of very powerful grasping (Fig. 3.14). In addition, the portion of the foot that is anterior to the ankle joint varies in length from two to three times that found in nonleaping quadrupedal primates living today. This approaches the condition seen in modern tarsiers in which the segment in front of the ankle joint is approximately four times the length of that in a nonleaping primate.

Taken together, the distinctive characteristics of the postcranial skeleton in omomyids suggest that some form of vertical clinging and leaping was an important part of their locomotor repertoire.

Comparison with Later Primates Because omomyids are often placed in the infraorder Tarsiiformes, it behooves us to look more closely at how they resemble and differ from modern tarsiiforms, specifically *Tarsius*.

The dental morphology in omomyids is generally similar to that of *Tarsius* but differs chiefly in having smaller, less projecting lower canine teeth (Gingerich, 1981). The lower dental formula in omomyids is sometimes 1.1.3.3, as in modern tarsiers, but usually is 2.1.2.3 or 2.1.3.3 as mentioned earlier. The shape of the palate also resembles that in *Tarsius* in being constricted across the canine and anterior premolar region.

Like *Tarsius,* the crania of the four identified omomyid skulls have short, tapered snouts with the lacrimal foramen opening onto the face. Unlike *Tarsius,* however, there is no ethmoid component in the orbital wall.

In the ear region omomyids resemble *Tarsius* in having a tubular ectotympanic bone. In this respect, omomyids and tarsiers differ from New World anthropoids and Oligocene Old World anthropoids, in which the bony meatus is limited to a simple ringlike ectotympanic bone attached to the lateral bulla wall. A bony meatal tube may be an adaptation designed to reduce physiological noise produced by motions of the mandible during mastication in species where the jaw joint borders directly on the external ear. The jaw joint does border on the bony meatal tube in tarsiers and in the omomyids *Rooneyia, Necrolemur,* and *Tetonius.* Thus the similarities of the auditory tube between extant anthropoids and tarsiers may represent convergent auditory specializations not necessarily indicative of close phylogenetic relationships (Packer and Sarmiento, 1984).

Several authors consider an omomyid–tarsier–anthropoid clade to be tied together by the posteromedial position of the posterior carotid foramen, the comparatively large promontory artery, and the relatively small stapedial artery (Szalay, 1975*b,* 1976; Szalay and Wilson, 1976; Rosenberger and Szalay, 1980). Others consider such intrabullar pathways of the internal carotid artery to be synapomorphies of a tarsier–anthropoid clade that excludes omomyids and other Paleogene prosimians (Cartmill and Kay, 1978; Cartmill et al., 1981). Other features common to tarsiiforms and anthropoids are given in Table 3.5.

Table 3.5 Some Characters Common to Tarsiiforms and Anthropoids

1. Facial length reduced and face set low on the cranium
2. Nasal region and olfactory components diminished in overall size
3. Closely approximated orbits that fuse posteriorly
4. Expansion of occipital lobes that moderately overlap the cerebellum
5. Absence of a coronolateral sulcus
6. Presence of a central sulcus
7. Tubular ectotympanic bone
8. Enlarged promontory artery (direct continuation of the internal carotid artery), which enters the auditory bulla posteromedially
9. Relatively small stapedial artery
10. Enlarged auditory bulla
11. Morphology of the distal humerus

Source: Adapted from Rosenberger and Szalay (1980).

BODY SIZE, FEEDING ADAPTATIONS, AND BEHAVIOR OF EOCENE PRIMATES

Adapid species range in body weight from 70 to 10,000 g, with most species falling between 350 and 8,000 g (Fig. 3.15). This range of body weights overlaps that found in modern ceboid monkeys. Adapids were quite small when they first appeared in the fossil record (early Eocene), with most species weighing 200 to 1,000 g. Their body weight generally increased through time, so that most later adapids weighed 600 to 8,000 g (Gingerich, 1980*a*).

The correlation between primate body weight and diet discussed in Chapter 1 suggests that the adapids radiated mainly on the folivorous–frugivorous side of Kay's threshold (body weight over 500 g). The dentition indicates a wide range of dietary adaptations, however, including insectivory, folivory, and frugivory. Some genera (for instance, *Protoadapis*) have flatter teeth, which is associated with frugivory, whereas others (for instance, *Adapis*) have more crested teeth indicative of folivory. The European genus *Anchomomys* is the only predominantly insectivorous genus in the family. The North American adapid radiation seems to have taken place entirely within the folivorous–frugivorous dietary regime (Gingerich, 1980*a*).

Several equations have been devised permitting one to estimate omomyid body size from tooth size based on regression analysis of these two quantities in a wide range of living primates (Gingerich et al., 1982; Conroy, 1987). All anaptomorphines fall below Kay's threshold, suggesting that they were predominantly insectivorous or gummivorous rather than

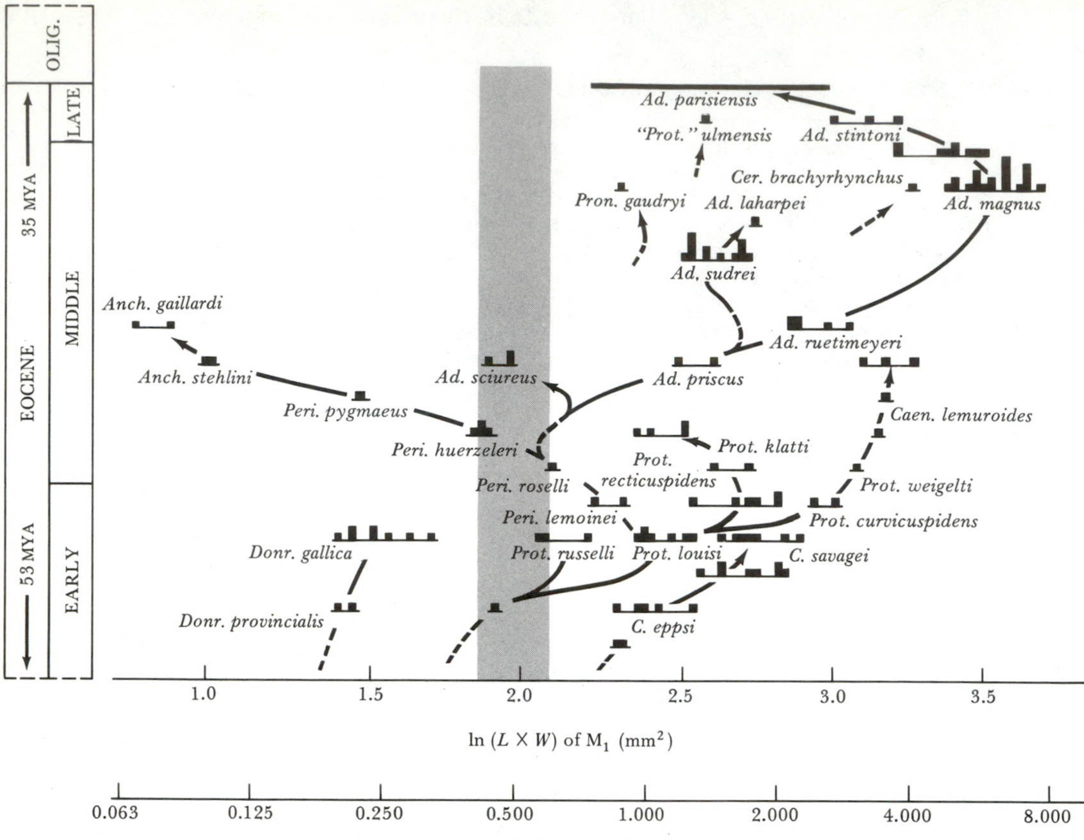

Fig. 3.15 Radiation of Eocene Adapidae in Europe Histograms represent the size distribution of the lower first molar in specimens of European adapids; each black square represents a separate specimen. The horizontal axis shows the natural log of the tooth area of M_1 for each species (length × width in millimeters) and, by inference, body weight in kilograms. Stippling shows Kay's 500-g threshold separating insect-eating (to the left) from leaf-eating and fruit-eating primates. Note that the adapid radiation occurred primarily on the leaf-eating side of this threshold. The line above *Adapis parisiensis* marks the Grand Coupure. Genera: *Ad., Adapis; Anch., Anchomomys; C., Cantius; Caen., Caenopithecus; Cer., Cercamonius; Donr., Donrusselia; Peri., Periconodon; Pron., Pronycticebus; Prot., Protoadapis.* (Adapted from Gingerich, 1980*a*)

folivorous. In these features they are similar to some modern prosimians such as *Phaner, Euoticus,* and *Galago senegalensis* and to some **callitrichids** (marmosets and tamarins) such as *Cebuella* and *Callithrix* (Nash, 1986). Anaptomorphines probably supplemented their diets with fruit as well, judging by the morphology of their broadly basined molar trigonids and talonids (Kay, 1977). Most omomyines also fall on the insectivorous side of the threshold, but a few (for example, *Hemiacodon, Ourayia, Macrotarsius, Rooneyia,* and *Ekgmowechashala*) lie above it (Gingerich, 1981, 1984) (Fig. 3.16).

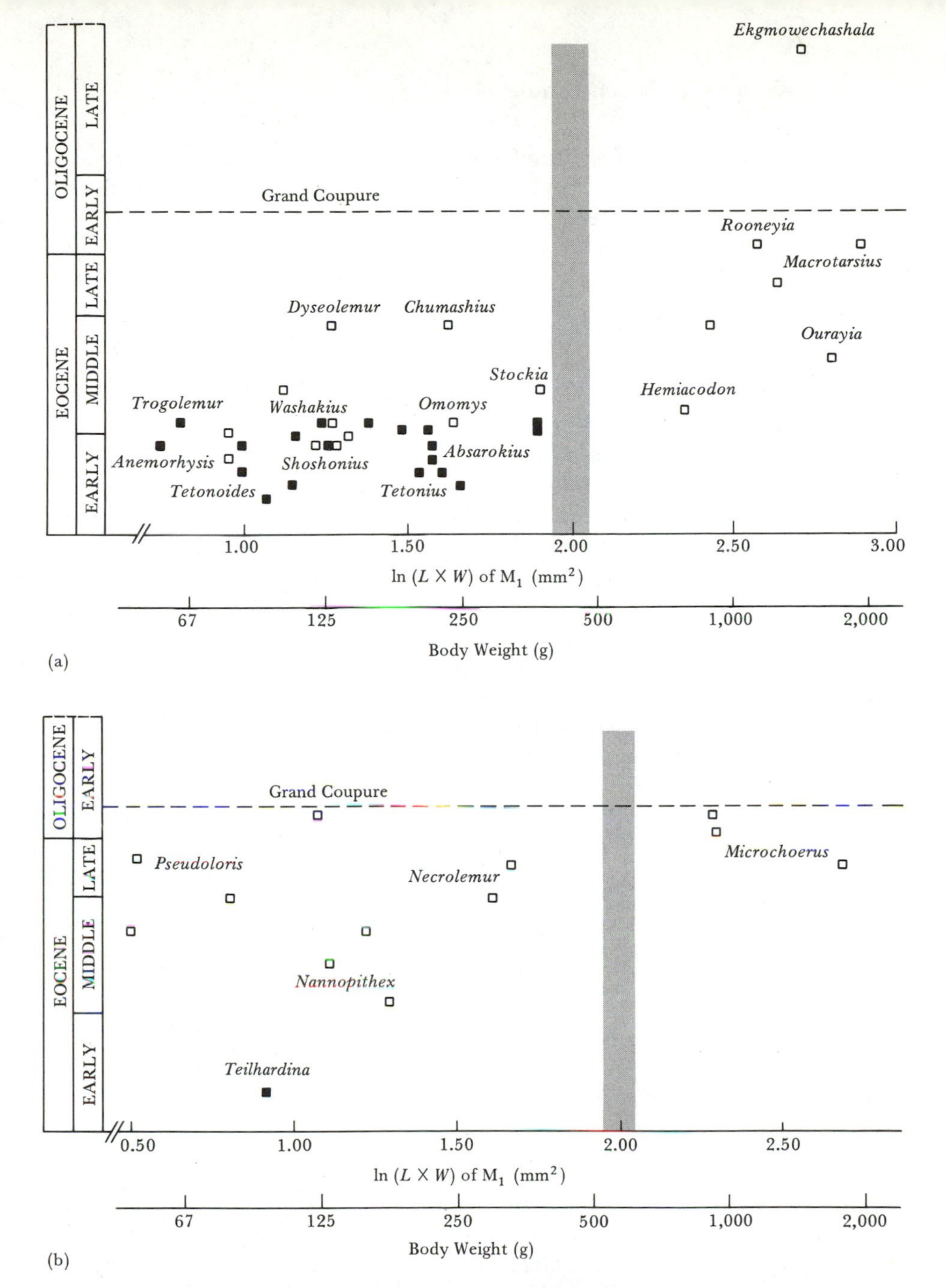

Fig. 3.16 Stratigraphic Distribution of Eocene and Oligocene Omomyidae (a) North America; (b) Europe. The horizontal axis shows the tooth area of M_1 for each species (natural logarithm of length × width in millimeters) and, by inference, body weight in grams. The solid squares represent anaptomorphines and the open squares represent omomyines in (a) and microchoerines in (b). Shaded bars show Kay's threshold. Note that omomyids do not appear in North America until the middle of the early Eocene, and only *Ekgmowechashala* is present after the early Oligocene worldwide cooling event (Grande Coupure); in Europe no omomyids are known before the middle of the early Eocene or after the Grande Coupure. (Adapted from Gingerich, 1981)

Several of the cranial and dental features traditionally used to distinguish omomyids from adapids may be dependent on body size. For example, the retention of four premolars in large adapids, but not in omomyids (except for *Teilhardina*), may simply be a function of tooth crowding in small, short-snouted animals. The specialized molars of the larger adapids are clearly adapted for leaf eating, a diet that is inefficient for small animals. It is probably for this reason that the smaller adapids such as *Anchomomys, Periconodon,* and *Pronycticebus* show somewhat simpler, more omomyidlike occlusal patterns compatible with a diet of insects and fruit. Conversely, the comparatively large omomyid *Macrotarsius* has molars more reminiscent of *Alouatta* (howler monkey) and *Notharctus,* suggesting it may have been a leaf eater.

Adapid skulls, as noted earlier, show greater relative importance of vision over olfaction and indicate that most adapids were active primarily during the day, although *Pronycticebus* may have been crepuscular or nocturnal. Their hindlimb morphology points to climbing and leaping behavior, hallmarks of an arboreal habitat. One can infer from the sexual dimorphism in their canines that adapids may have been sexually dimorphic in body size as well. Sexual dimorphism in mammals is usually associated with living in social groups with differentiated sex roles and polygynous bonding: the larger canines and body size are used by the male to defend his females from appropriation by other males, while rearing of the young is carried out primarily by the female.

Unlike adapids, omomyids are not sexually dimorphic. Judging from their small body size, insectivorous diets, arboreal milieu, and nocturnal habits, omomyids resembled other primates that are typically solitary or live in pairs with little sexual differentiation in social roles (Crook and Gartlan, 1966; Gingerich, 1984).

PHYLOGENY AND CLASSIFICATION OF EOCENE PRIMATES

Various hypotheses regarding adapid and omomyid phylogeny have been proposed that consider the morphological and stratigraphical information outlined in this chapter.

Plesitarsiiform–Simiolemuriform Classification

Lemurophile Hypothesis As discussed earlier, distinctive characteristics shared by almost all the tarsiiform omomyids include (1) enlarged, pointed, protruding central incisors; (2) a reduced dental formula of 2.1.3.3/2.1.2–3.3 or 1.1.3.3/1.1.3.3; (3) an ossified postorbital bar with incipient postorbital closure; (4) a tubular ectotympanic bone; and (5) elongated tarsal bones. Because some of these features are also shared with plesiadapiforms, it was once thought that plesiadapiforms gave rise to omomyids. This hypothesis also proposed that adapids were ancestral to both anthropoids and tooth-combed prosimians. Proponents of these phylogenetic views established two new suborders, Plesitarsiiformes and Simiolemuriformes, to reflect these phyletic hypotheses (Gingerich, 1975*c*, 1976; Schwartz and Krishtalka, 1977) (see Fig. 2.22a). This view has recently been coined the **lemurophile hypothesis** (MacPhee and Cartmill, 1986) (Fig. 3.17).

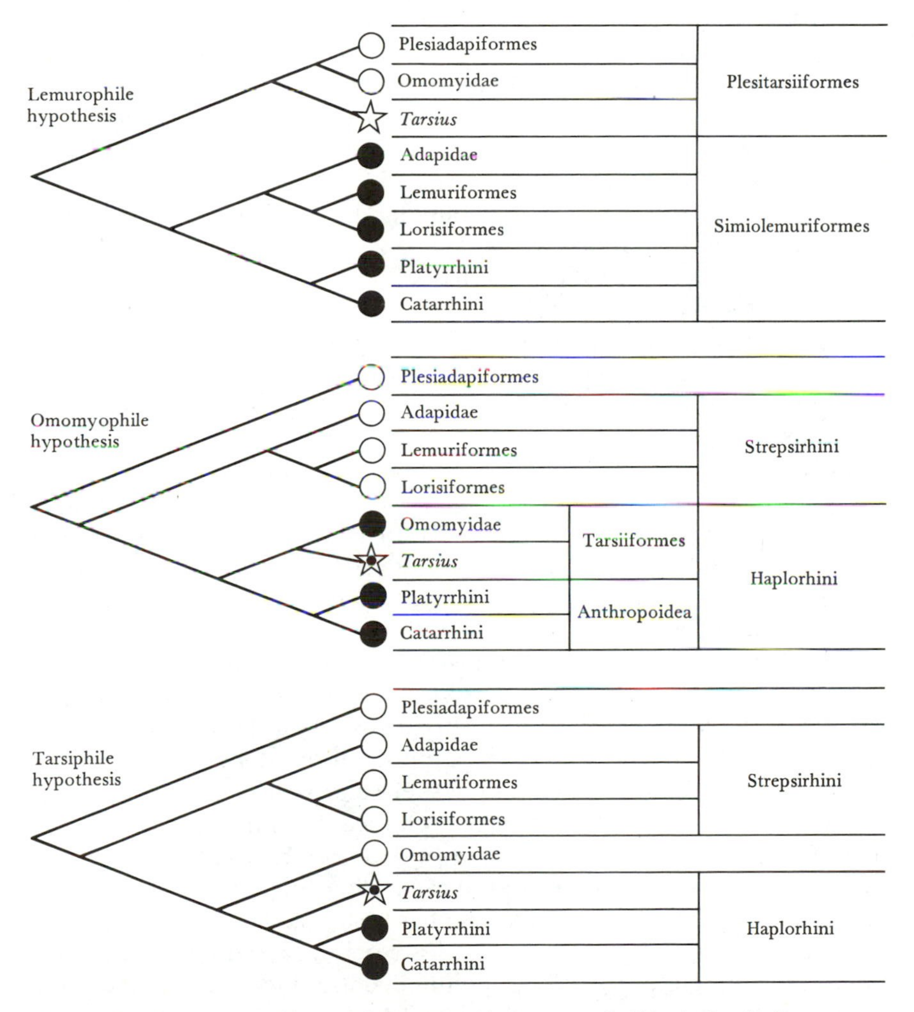

Fig. 3.17 Three Hypotheses of Primate Phylogeny Solid circles indicate taxa composing suggested anthropoid sister groups. The star highlights the different position of *Tarsius* in each hypothesis. (From MacPhee and Cartmill, 1986)

Two dental similarities were submitted as evidence for linking omomyids and plesiadapids: both groups have enlarged central incisors, and both groups have lost the small first premolar. Adapids, on the other hand, retain the full complement of four premolars. Some authors have countered this evidence by pointing out that these features often appear in other mammalian lineages and are of little phyletic value (Cartmill and Kay, 1978).

Another feature used to support the lemurophile hypothesis is the contention that both plesiadapiforms (such as *Plesiadapis*) and omomyids (such as *Necrolemur*) have bony ectotympanic tubes. This is difficult to verify, however, since the tube might be an extension of the petrosal bone as in *Microcebus* or an independent bony element as in *Tupaia.*

The lemurophile hypothesis also proposed a phyletic link between adapids and anthropoids based on the sharing of a number of features:

1. Small, vertically implanted, spatulate incisors with no enlarged central incisor
2. Lower second incisors that are larger than first incisors
3. Upper canines that develop a honing facet to wear against an enlarged anterior lower premolar
4. Fusion of the mandibular symphysis
5. Loss of the paraconid
6. Radiation on the folivorous side of Kay's threshold (Gingerich, 1980*b*)

However, if one accepts the notion that lemurs and anthropoids are more closely related to one another than either group is to tarsiers, then one must assume that the numerous features shared by tarsiiforms and anthropoids result from parallel evolution or retained primitive states.

Recently some advocates of the lemurophile hypothesis (for example, Gingerich, 1981) have moved away from its central tenets for several reasons. The fact that omomyids and adapids first appear together in North America and Europe suggests that these families may have immigrated from a common geographic center of origin. Furthermore, the most primitive omomyid, *Teilhardina belgica,* is not easily distinguished from primitive adapids, particularly in premolar morphology (Gingerich, 1986). Some *Teilhardina belgica* specimens retain the primitive adapid dental formula 2.1.4.3 / 2.1.4.3; judging from the **dental alveoli** (holes in the jaws for the tooth roots), the lower central incisors of this species are not as enlarged relative to the lateral incisors as in other omomyids. Thus most workers would now accept that lemuriform and tarsiiform primates are more closely related to each other than either is to the plesiadapiforms. This realignment implies that the features supposedly linking plesiadapiforms and omomyids (such as a reduced dental formula; the enlarged, protruding, pointed central incisors; and the tubular ectotympanic bone) probably evolved convergently.

As I have stressed previously, there is also little unequivocal evidence phyletically linking adapids to tooth-combed primates. Traditional arguments for this association rest on the lemurlike bulla and carotid patterns, but these are very likely to be primate symplesiomorphies. Other supposed diagnostic strepsirhine features discernible from the adapid fossil record (a large olfactory apparatus and a comparatively small brain) are also likely to be primitive retentions.

Strepsirhine–Haplorhine Classifications

The suborders Strepsirhini and Haplorhini were first defined by R. I. Pocock in 1918. Extant strepsirhines are characterized by the possession of a tooth comb or some variant of it; they also retain the ancestral condition of the muzzle, which includes the following:

1. Laterally split nostrils
2. The presence of a **rhinarium** (patch of skin between nose and upper lip)
3. A **philtrum** (vertical cleft in the rhinarium)
4. A **frenulum** (small fold of mucous skin) immobilizing the upper lip

In contrast, extant haplorhines share such apomorphous characters as the following:

1. Unsplit nostrils
2. A nose with hairy, undifferentiated skin
3. No philtrum
4. A much reduced or absent frenulum (Fig 3.18)

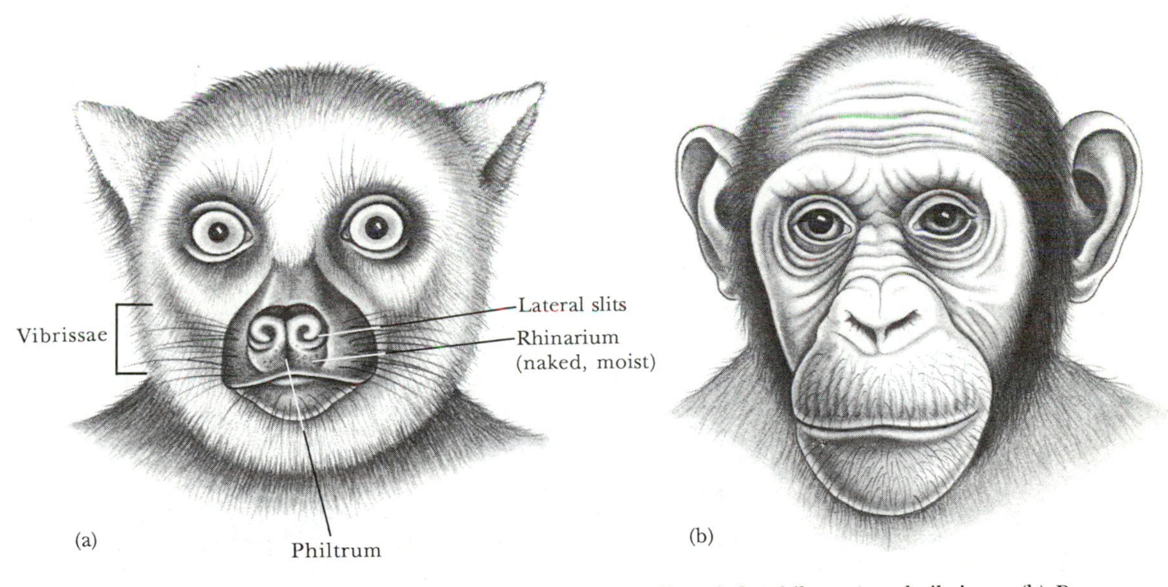

Fig. 3.18 Morphology of the Nasal Region in Living Strepsirhini and Haplorhini (a) *Lemur.* In strepsirhines (lemurs and lorises) the nostrils continue laterally as blind slits, and the entire nasal region is surrounded by the naked, moist, glandular rhinarium with a median cleft (philtrum) and vibrissae. (b) *Pan.* In haplorhines (tarsiers and anthropoids) the nostrils do not form lateral slits, there is no rhinarium, and the philtrum is vestigial or absent. (Redrawn from Schultz, 1969)

Details of fetal membranes and various biochemical studies also support this strepsirhine–haplorhine dichotomy (Luckett, 1975; Goodman, 1975; Dene et al., 1976).

Robert Hoffstetter (1980) of the Institute of Paleontology in Paris has suggested that this dichotomy may correspond to an orientation toward two different activity rhythms: nocturnal for the Strepsirhini and diurnal for the Haplorhini. As we have seen, however, this scenario would not pertain to adapids, which presumably were mainly diurnal strepsirhines.

No known synapomorphies link tarsiers or anthropoids to any tooth-combed prosimians. Indeed, most investigators subscribe to a model in which there was an initial split between plesiadapiforms and euprimates, with the latter then dividing into the Strepsirhini (adapids, lemurs, and lorises) and the Haplorhini (omomyids, anthropoids, and tarsiers). The **omomyophile** and **tarsiphile hypotheses** shown in Fig. 3.17 are both attempts to order taxa within a strepsirhine–haplorhine classification.

Many workers have suggested that the lineage leading to extant tarsiers stems directly from an omomyid or some closely related group; hence the designation Tarsiiformes. A suite of cranial, dental, soft-tissue, and molecular evidence also supports the hypothesis that tarsiiforms and anthropoids are sister taxa that together form the Haplorhini. (Molecular evidence, of course, is lacking for extinct primates such as adapids or omomyids.) This implies that anthropoids are descended from some species that would be classified as a tarsiiform or as an omomyid (see the omomyophile hypothesis of Fig. 3.17).

The strepsirhine–haplorhine dichotomy implies that tarsierlike primates and the anthropoids are sister groups. It does not necessarily mean, however, that anthropoids were derived from a tarsierlike ancestor, as some have suggested. If one abides by the requirement that a taxonomic group be monophyletic, then a strepsirhine–haplorhine dichotomy means only that haplorhines and strepsirhines are sister taxa and that a set of shared derived traits characterizes all members of the Haplorhini, including tarsiers. Thus the Tarsiiformes and Anthropoidea comprise one sister group, and the Lemuriformes (together with the Lorisiformes) are the sister group of the Tarsiiformes and the Anthropoidea. The question of the morphology of the common ancestor of the tarsiiform–anthropoid clade is an entirely separate matter, and it is not necessary that it be tarsierlike, whether or not most champions of the strepsirhine–haplorhine dichotomy reconstruct the common ancestor as such. Discussions and analyses of the two issues should be kept distinct.

Omomyophile Hypothesis The **omomyophile hypothesis** relies heavily on overall cranial similarities between omomyids and anthropoids (see Table 3.5). For example, they share a comparatively enlarged promontory artery in combination with a posteromedial entrance of the internal carotid into the bulla. In lorises, however, the entrance of the carotid artery has also shifted medially. In addition, omomyids and most anthropoids share the following features:

1. A comparatively reduced facial length
2. A reduced nasal fossa (and presumably reduced olfactory apparatus as well)
3. Expanded occipital lobes and reduced olfactory bulbs
4. Absence of the **coronolateral sulcus** typical of the strepsirhine brain (Fig. 3.19)
5. An external auditory tube (ectotympanic)

It should be noted, however, that not all anthropoids share all of these features.

The size of the vascular canals in the middle ear has been invoked in arguments both for and against the notion that anthropoids evolved from omomyids. Gingerich (1973*b*) has noted that a comparatively large canal for the promontory artery can be found in some adapids (for example, *Notharctus*) and thus is not necessarily a sign of omomyid affinities. However, Szalay (1975*c*) has stated that a large canal for the promontory artery relative to canal size for the stapedial artery is a shared derived feature linking tarsiers, omomyids, and anthropoids. The argument is somewhat moot however, since the presence of an arterial groove or canal does not in every case indicate the size or even presence of a corresponding artery (Conroy and Wible, 1978).

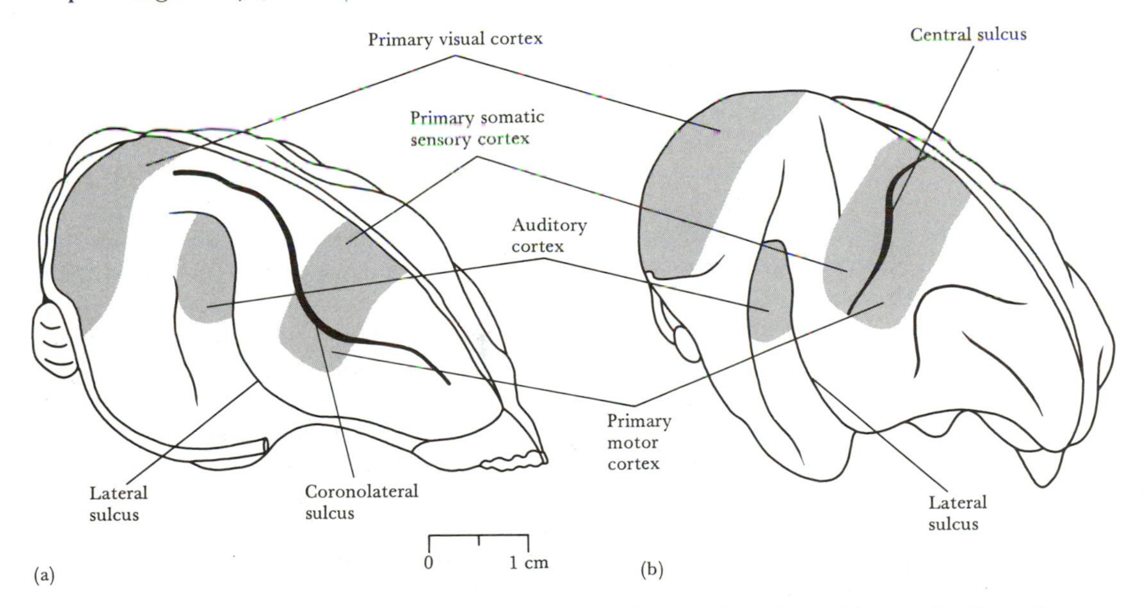

Fig. 3.19 Differences in Cortical Folding of Prosimians and Anthropoids (a) Endocast of *Lemur* showing the coronolateral sulcus running lengthwise along the side of the brain. This sulcus is characteristic of prosimians and many other mammals. (b) Endocast of the Old World monkey *Cercopithecus* showing the central sulcus, which runs over the top of the brain. Shaded areas show major sensory and motor areas of the brain in both prosimians and anthropoids. (Adapted from Radinsky, 1975)

Tarsiphile Hypothesis Cartmill and Kay (1978) have offered an alternative model dealing with the question of how omomyids, anthropoids, and tarsiers are related. According to them, the more traditional hypothesis that anthropoids are derived from omomyids finds little support in the anatomy of the ear region. The only fossil omomyids whose ear regions are adequately known, *Rooneyia* and *Necrolemur,* display no features that unequivocally link them to undisputed anthropoids of the Oligocene (Conroy, 1980; Cartmill et al., 1981). More specifically, early anthropoids of the Oligocene have a ringlike ectotympanic bone attached to the lateral bulla wall, not an elongated tube as in omomyids. For this reason, Cartmill and Kay (1978) do not lump omomyids together with anthropoids and tarsiers in the suborder Haplorhini. Their alternative model, the **tarsiphile hypothesis** shown in Fig. 3.17, proposes that *Tarsius* is the actual phyletic sister group of the Anthropoidea and thus is more closely related to anthropoids than to any other early Tertiary primate group. They emphasize several derived features of the ear region shared by tarsiers and anthropoids but not by omomyids (for example, the prenatal loss of a functional stapedial artery).

Cartmill (1981) has also argued that if postorbital closure was present in the last common ancestor of extant haplorhines, then tarsiers and anthropoids would form a monophyletic clade that would exclude the fossil omomyids of the Eocene, since they lacked this feature.

Tab Rasmussen (1986) has noted the conflict between the neontological and paleontological data used to support different groups for anthropoid ancestry. The evidence from comparative anatomy and biochemistry of living primates indicates that tarsiers are more closely related to anthropoids than are tooth-combed prosimians, thus suggesting an omomyid ancestry for anthropoids. However, paleontological evidence favors adapids as more suitable anthropoid ancestors.

Rasmussen has recently proposed an interesting solution to this apparent paradox (Fig. 3.20). He points out that these seemingly mutually incompatible conclusions are true only if certain assumptions are granted. The neontological argument in favor of an omomyid ancestry depends on the assumption that tooth-combed prosimians are descended from, or are the sister group of, adapids. As previously noted, this phylogenetic link has never been firmly established through identification of unequivocal synapomorphies or through discovery of fossils intermediate in morphology. Perhaps adapids actually form a clade with omomyids, tarsiers, and anthropoids, and this clade shares a common ancestor with tooth-combed prosimians. According to this hypothesis "the comparative study of soft anatomy and biochemistry cannot be used to refute an adapid ancestry of anthropoids, that the haplorhine–strepsirhine dichotomy is of extremely limited value when applied to fossil taxa, and that toothcombed prosimians, rather than tarsiers, may provide the best behavioral and ecological model of an anthropoid ancestor despite their cladistic relationships" (Rasmussen, 1986). In other words, omomyids may be ancestral to tarsiers, adapids may be ancestral to anthropoids,

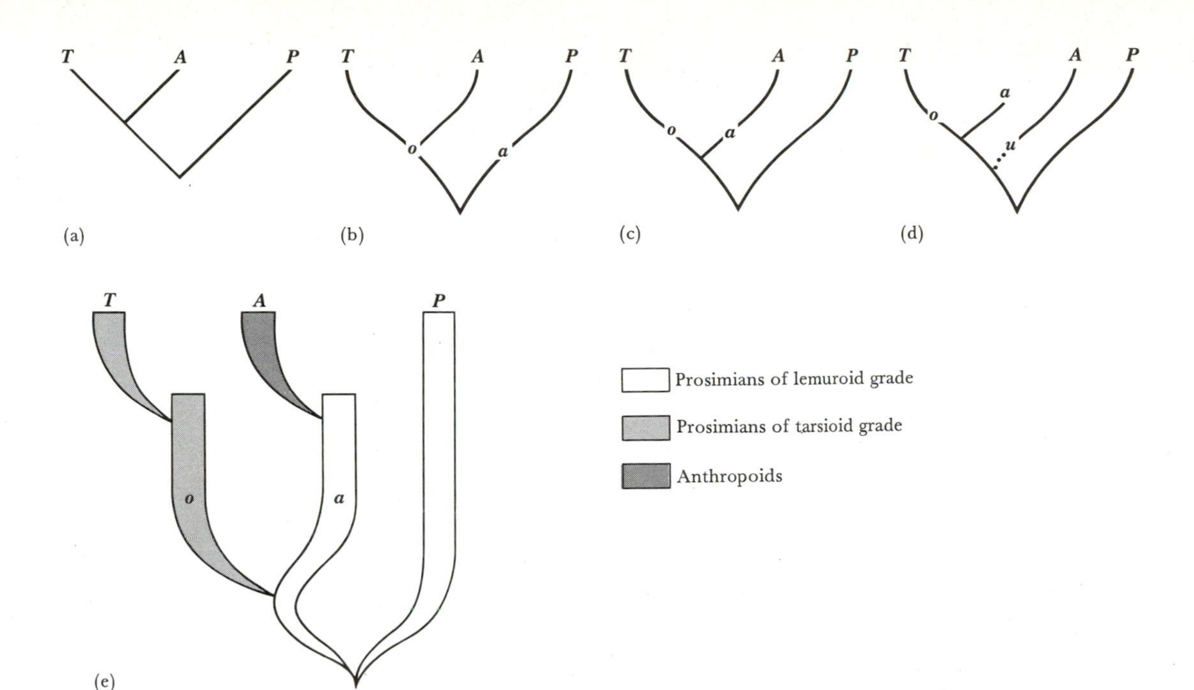

Fig. 3.20 Different Views of Anthropoid Origins *A*, Anthropoidea; *P*, tooth-combed prosimians; *T*, Tarsiidae; *a*, Adapidae; *o*, Omomyidae; *u*, hypothetical unknown group. (a) Cladogram derived from neontological data indicating that tarsiers and anthropoids are more closely related to each other than either group is to tooth-combed prosimians. This cladogram is consistent with postulating (b) an omomyid origin for anthropoids, (c) an adapid origin for anthropoids, or (d) a derivation of anthropoids from an unknown group. The preponderance of fossil evidence supports (c). (e) Rasmussen's suggested phylogeny of the primates. Note that the Tarsiidae and Anthropoidea have a closer cladistic relationship to each other than either does with the tooth-combed prosimians (as supported by neontological evidence), and yet the Adapidae give rise to the Anthropoidea (as supported by paleontological evidence). (Adapted from Rasmussen, 1986)

and together this clade (comprised of these ancestral and modern groups) may be the sister group of the tooth-combed prosimians. However, this "solution" would still necessitate a great amount of parallel evolution to explain the morphological similarities between tarsiers and anthropoids.

As we shall see in the next chapter, the question of anthropoid origins is still an unsolved issue. Cartmill et al. (1981) have concluded:

> It seems likely that anthropoids were derived from a group of Eocene prosimians that is currently unknown (or at least not known from basicranial remains). We find it impossible to believe that the anthropoid otic [ear] complex would have evolved independently in the Old and New Worlds, no matter whether one postulates a last common ancestor that resembled known adapids or known omomyids in the morphology of the ear region. The hypothesis that separate lineages of New and Old World prosimians, whether omomyids, . . . adapids . . . or unspecified but presumably different prosimian stocks . . . gave rise independently to Platyrrhini and Catarrhini seems wholly untenable. Platyrrhines and catarrhines evidently had a last common ancestor that could not have been ancestral to any other known primate. When and where did that common ancestor live?

The next chapter will address this question.

CHAPTER 4

Oligocene Primates

Paleoclimates and Biogeography

Summary of the Oligocene Fossil Record

Oligocene Primates in the Old World

Oligocene Primates in the New World

Phylogeny and Classification of Oligocene Primates

The Oligocene was a most significant time in primate evolution. At the beginning of the epoch, primate groups that flourished during the Eocene disappeared almost completely from the fossil record of the Northern Hemisphere, lingering on in only a few places in North America and Asia. More importantly, higher primates—those having achieved an anthropoid grade of evolution—appeared suddenly for the first time, in the early Oligocene of Africa; by the mid-to-late Oligocene they reached the shores of South America.

In this chapter the morphological criteria used to identify an anthropoid grade of evolution will be defined and the fossil evidence for anthropoid origins will be reviewed. Unfortunately there is no unanimity of opinion about the geographical origin of these higher primates, but most probably it was in Africa. Once again, paleoclimates and biogeography strongly influenced the course of events in primate evolution.

PALEOCLIMATES AND BIOGEOGRAPHY

Climatic Trends

The Oligocene continued to show the trends toward cooling, drying, and alternating seasons that had begun in the late Eocene. As discussed in Chapter 3, global climatic cooling toward the end of the Eocene was associated with increased Antarctic glaciation. The cooling and glaciation were probably both triggered by fundamental changes in ocean circulation resulting from plate-tectonic movements between Antarctica and Australia (Kennett et al., 1985). It has also been suggested that lower sea levels (resulting from increased glaciation) lowered the continental water table, thus contributing to drought and to the faunal and floral discontinuities (the Grand Coupure) seen in the fossil record at the Eocene–Oligocene boundary (Morner, 1978).

The cooler surface waters of the Oligocene restricted the high-latitude tropical wet belts that had been characteristic of the Eocene (Frakes and Kemp, 1972; Kennett, 1977). Consequently, the vegetation changed drastically in the middle to high latitudes of the Northern Hemisphere. Within a geologically short time span, areas that had been occupied by broad-leafed evergreen forests were replaced by more temperate, broad-leaved deciduous forests signifying a major decline in mean annual temperature (about 12° to 13°C at latitude 60° in Alaska). Just as profound was the shift in temperature variability. The mean annual range of temperature in the Pacific Northwest, which had been as little as 3° to 5°C in the middle Eocene, must have been at least 21°C in the Oligocene (Wolfe, 1978; Hubbard and Boulter, 1983).

Leaf floras, which are well represented in the Oligocene record of Europe, show about the same trend in climate inferred for North America. Basically, there was a dramatic change to more "cool tropical" and "warm temperate" plants compared with the more tropical plants of the Eocene (Savage and Russell, 1983).

In earlier chapters the evidence for mass faunal extinction events at various points in the early Tertiary was noted. There is now some evidence of a mid-Oligocene extinction event at about 32 MYA. This event, spanning perhaps 20,000 years or less, appearing in the fossil record of North American land mammals, resulted in the selective disappearance of archaic members of the fauna and the later diversification of other taxa. The selective nature of the extinctions suggests climatic and ecological causes rather than extraterrestrial ones such as meteorite impacts. Such causes would likely include (1) increased mid-Oligocene glaciation; (2) worldwide cooling; (3) a major retreat of the oceans; and (4) abrupt changes in the flora resulting from the changes in global oceanic circulation (Prothero, 1985). Few primates survived this climatic change in northern latitudes.

Primates are virtually absent from the Oligocene faunas of the Northern Hemisphere and are well represented in only one place in the world, the Fayum Depression of Egypt. In the Oligocene, this area was a subtropical to tropical lowland coastal plain with damp soils and seasonal rainfall patterns that supported a variety of plants (including lianas, tall trees, and possibly mangroves) and animals. The southern border of the Tethys Sea was probably close by, and large, meandering brackish streams flowed through the region (Butzer and Hansen, 1968). Most of the floodplain fossiliferous sediments were laid down at a time of high water from channels that had eroded through their levees. Large accumulations of petrified logs suggest some type of riverine forest along the stream edges (Bown et al., 1982).

This paleoenvironmental scenario differs markedly from that suggested by Kortlandt (1980) in which the Fayum is pictured as nearly treeless, sparsely vegetated, and semiarid. These two different paleoenvironmental scenarios have important implications for whether Fayum primates were more arboreal or more terrestrial. While the latter would support a view of the Fayum primates as being primarily ground dwellers, the former scenario is consistent with fossil morphology, which indicates that the Fayum primates spent most of their time in trees. We shall review the morphological evidence concerning this point later in the chapter.

Sedimentological evidence suggests a monsoonlike climate, since ancient Fayum soils show traces of periodic wetting and drying episodes (Simons, 1985*b;* Bown et al, 1982).

In South America, where a small number of Oligocene primates appear, the mid-Tertiary paleoclimate can be summarized as being mostly tropical in nature. The environment, as suggested by mammalian faunas from the southern parts of the continent, consisted of woodland and savanna grading northward into rain forests that were undoubtedly more extensive than at present. It was not until later in the Neogene that the Andes began to rise; this produced a marked ecological effect, mainly by acting as a barrier to moisture-laden Pacific winds. As a result, the pampas that characterize much of Argentina today probably came into prominence during the Neogene, and the rain forests began to recede in the southern part of the continent (Patterson and Pascual, 1972). The important point to keep in mind is that South America was an island continent throughout most of the Tertiary.

Land Movements and Dispersal Routes

Africa was isolated from Eurasia by the Tethys Sea throughout most of the Oligocene (Fig. 4.1). As we shall see in Chapter 5, the movement of animals between these two landmasses would not be reestablished until the early Miocene, some 19 MYA (Van Couvering, 1972). It was during this time of continental separation that higher primates presumably differentiated in Africa. What is still not clear, however, is whether these primates arose in Africa or migrated there from Eurasia in response to the climatic cooling in northern latitudes toward the end of the Eocene. The question of Oligocene primate origins will be addressed later in the chapter.

Estimates of the extent of geographical separation between Africa and South America in the Oligocene and late Eocene are based on various calculations for the initial breakup of western Gondwanaland (Maxwell et al. 1970; Tarling and Tarling, 1971; Keast, 1972; Cachel, 1981). If we assume a seafloor-spreading rate of approximately 2 cm per year, Africa and South America must have been some 3,000 km apart in the south and at least 500 km apart in the north by the end of the Eocene. This distance has bearing on one of the hypotheses for the origin of New World monkeys, namely, that they rafted to South America from Africa

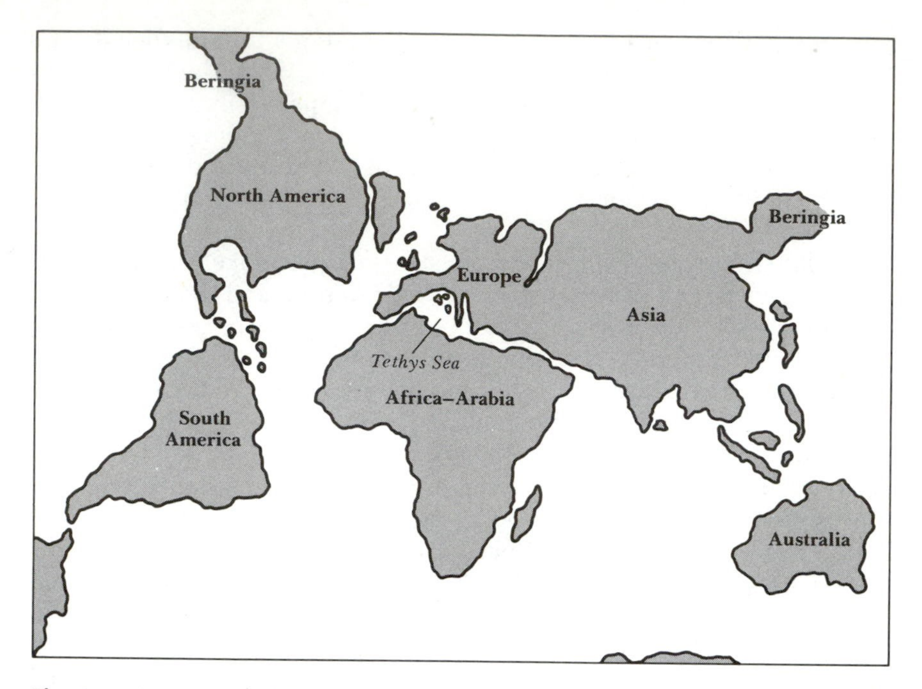

Fig. 4.1 Approximate Distribution of Landmasses in the Early Oligocene (about 35 MYA) (Adapted from Savage and Russell, 1983)

sometime in the early Tertiary. Would a migration route between Africa and South America have been possible at this time?

According to geologist D. H. Tarling (1980) of the University of Newcastle upon Tyne, such a migration route was not only possible but probable. He notes that falling sea levels in the early Oligocene may have exposed some parts of the mid-Atlantic Rise. Consequently, a series of "stepping stones" could have been provided by exposed portions of the Walvis Ridge and the Rio Grande Rise. Further to the north, parts of the oceanic Ceara and Sierra Leone rises may also have been exposed. If this were the case, deep oceanic waters may never have exceeded 200 km in width. In addition, postulated oceanic currents would clearly have assisted the migration of floating organisms from Africa toward South America in equatorial regions. A rise in sea level after the early Oligocene would have submerged most of these postulated trans-Atlantic links, and (except for very isolated islands such as Tristan de Cunha) they would have remained below sea level for the remainder of the Cenozoic (Fig. 4.2).

Another proposed dispersal route during the early Tertiary is between North and South America. During the Cretaceous, North America was moving away from South America (and Africa) in a northwesterly direction. By the mid-Cretaceous this movement resulted in a latitudinal separation between these two continental blocks of some 1,500 km (Tarling, 1980). Apparently there was some kind of land bridge or island chain in the early Tertiary between the two continents (see Fig. 4.1). However, oceanic currents, which in that region flowed in a northerly direction, most likely hindered migration from North to South America. It seems probable that oceanic currents continued to inhibit trans-Caribbean migration until just before a continuous land bridge formed in the late Miocene.

However, there is fossil evidence for at least some migration between the two continents during this time. Late Cretaceous dinosaurs of South America are more similar to North American ones than to those in Africa and Eurasia. The fact that certain characteristic North American dinosaurs are absent from South America, however, suggests a filtering effect of some sort across Central America (Keast, 1972). The presence of *Arctostylops* (a member of the Notoungulata, an order of extinct South American herbivorous mammals) in late Paleocene deposits of North America and some enigmatic fused cervical vertebrae similar to those of the Edentata (toothless mammals such as anteaters, armadillos, and sloths) from the middle Eocene of Wyoming (McKenna, 1975) also provide evidence for limited mammal migrations between South America and North America.

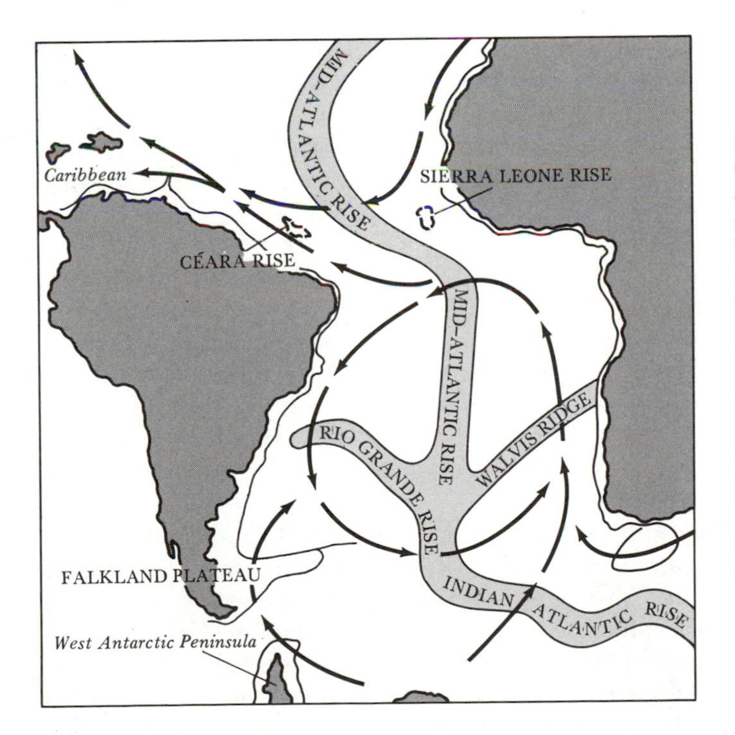

Fig. 4.2 South Atlantic in the Early Oligocene Shown are several oceanic ridges that may have been used by primates during their hypothesized dispersal from Africa to South America in the Oligocene during periods of lowered sea levels. Mode of dispersal may have been via island hopping, most probably across the more northern portion of the South Atlantic. Note the direction of prevailing oceanic circulation, indicated by arrows. (Adapted from Tarling, 1980), with current arrows from Ciochon and Chiarelli, 1980)

SUMMARY OF THE OLIGOCENE FOSSIL RECORD

There is a major gap in our understanding of primate evolution during the Oligocene epoch, between approximately 35 and 24 MYA. As Chapter 3 relates, fossil primates were quite common during the Eocene in Europe, Asia, and North America. However, only a few primates have yet been recovered from the overlying Oligocene sediments on these continents, such as the omomyids *Rooneyia* and *Macrotarsius* from North America and the New World monkeys *Branisella, Dolichocebus,* and *Tremacebus* from South America (discussed later in this chapter). One omomyid *(Ekgmowechashala)* even survived into the early Miocene of North America (MacDonald, 1963; Rose and Rensberger, 1983), although there has recently been a suggestion that it was really a dermopteran ("flying lemur") and not a primate at all (McKenna, personal communication).

In Africa the fossil record is also limited. Early Tertiary rock sediments yielding mammalian fossils of any kind are rare, and none of them has yet yielded undisputed fossil primates. (The primate status of one such North African fossil, *Azibius,* is unclear.) There is only one Oligocene locality on the entire African continent that contains an abundant, diverse primate fauna: the Fayum Depression (or badlands) of Egypt, an area about 150 km southwest of Cairo. This site is significant for two other reasons: it contains the earliest undisputed fossils of the Anthropoidea anywhere in the world, and these are the only anthropoid primates known from the Oligocene in the Old World. These anthropoid primates are probably between 35 and 40 million years in age. In addition, several nonanthropoid primates have also been recovered from these deposits, thus testifying to the diversity of primates present at this location in the Oligocene.

Although the Fayum mammalian fauna has some elements in common with the faunas of Eurasia, it also has many peculiar animals such as the giant herbivore *Arsinoitherium,* early mastodons, diverse Hyracoidea (rock hyraxes) and many Anthracotheridae (primitive hippolike mammals). Conspicuous by their absence are the Perissodactyla (odd-toed ungulates), many of the Eurasian Artiodactyla (even-toed ungulates), the Carnivora (true carnivores), and several rodent groups (for example, the Cricetidae).

Several families of birds make their earliest appearance in the fossil record in the Fayum deposits: lily trotters, ospreys, turacos, grebes, and shoebill storks. Other birds present include hawks and eagles, rails, storks, and herons. Nearly all these birds now live and feed in swamps or near running water, and all but one of the turacos are forest-dwelling species (Simons, 1985*b*). An avian fauna closely analogous to that of the Fayum is found today only in a limited area of Uganda around Lake Victoria, a region of swampland bordered by forest and grasslands that presents marked faunal similarities to the environment inferred for the Egyptian Oligocene (Olson and Rasmussen, 1986).

Petrified fruits, nuts, and seed pods are also known from the Fayum deposits. The plants include mangroves, water lilies, aquatic ferns, figs, palms, cinnamon, and a variety of hardwoods. These plants have tropical Indo-Malayan affinities. Nests of subterranean termites have also been found; such termites only occur in tropical regions (Bown, 1982).

We have no idea how widespread primates were in the Oligocene of Africa, since no other fossiliferous sites of that age have been discovered in sub-Saharan Africa. Except for the Fayum, the few Oligocene sites in northern Africa have not yielded any primates.

Primate fossils appear in South America for the first time in the Oligocene; however, they do not appear there much before 25 MYA, toward the end of the epoch (Hoffstetter, 1969, 1980; MacFadden et al., 1985). Unlike the situation in Africa, the absence of fossil primates from the Paleocene and Eocene in South America is probably not an artifact of a poor paleontological record since sites yielding other fossil mammals are known for every epoch of the Cenozoic in South America. However, these faunas are mainly restricted to the southern part of the continent (Patterson and Pascual, 1972; McKenna, 1980).

The fossil mammals of South America are the result of three major colonizations. Mammals first appeared in the Cretaceous and consisted mainly of condylarths and marsupials. From the late Paleocene to the middle Eocene, the fauna consisted of a diverse group of marsupials, ungulates, and edentates (McKenna, 1980). Of the seven orders present in the latest Paleocene, four are thought to have come from the north. The absence of insectivorans and placental carnivorans, however, suggests an obvious filtering effect in this presumed land route. Similarities with African or Eurasian mammals in the early Tertiary are minimal.

The second colonization occurred in the Oligocene and consisted of primates and the Caviomorpha (a rodent suborder including guinea pigs and porcupines) (Hoffstetter, 1974; Lavocat, 1974). As reviewed earlier, there is geological evidence hinting at a discontinuous land bridge in the middle Eocene between northern Central America and the slowly rising Andes of Colombia (Sullivan, 1974; McKenna, 1980). No other major group of mammals seems to have entered South America, however, before the third colonization, which occurred after the Panamanian isthmus arose sometime in the later Pliocene (Marshall et al., 1979).

OLIGOCENE PRIMATES IN THE OLD WORLD

As described previously, the only major site for Oligocene primates in the Old World is in the Fayum Depression of Egypt. This discussion of

Table 4.1 Classification of Oligocene Primates

Taxa	*Epoch (and Area)*	*Dentition*
Suborder ANTHROPOIDEA		
Infraorder PARAPITHECOIDEA		
Family PARAPITHECIDAE		
Qatrania	E. Olig. (Egypt)	
Apidium	E. Olig. (Egypt)	2.1.3.3
Parapithecus	E. Olig. (Egypt)	0–1.1.3.3
Infraorder CATARRHINI		
Family PROPLIOPITHECIDAE		
Propliopithecus	E. Olig. (Egypt)	2.1.2.3
Aegyptopithecus	E. Olig. (Egypt)	2.1.2.3
Infraorder PLATYRRHINI		
Family CEBIDAE		
Tremacebus	L. Olig. (Argentina)	?.1.3.3
Family *incertae sedis*		
Branisella	L. Olig. (Bolivia)	?.1.3.3
Dolichocebus	L. Olig. (Argentina)	?.1.3.3
Infraorder *incertae sedis*		
Amphipithecus	L. Eoc. (Burma)	?.1.3.3
Pondaungia	L. Eoc. (Burma)	
Oligopithecus	E. Olig. (Egypt)	?.1.2.3
Suborder PROSIMII		
Infraorder TARSIIFORMES		
Family TARSIIDAE		
Afrotarsius	E. Olig. (Egypt)	?
Family OMOMYIDAE	E. Olig. (Egypt)	?
Infraorder LEMURIFORMES		
Family LORISIDAE	E. Olig. (Egypt)	?

Old World Oligocene primates will concentrate on the fossils taken from that site.

Currently, seven genera and 12 species of primates are named from these deposits. The primates are generally divided into at least three families: two anthropoid families, the **Propliopithecidae** (sometimes called the **Pliopithecidae**) and the **Parapithecidae,** and a prosimian family, the **Tarsiidae.** Propliopithecid species include *Aegyptopithecus zeuxis, Propliopithecus haeckeli, P. markgrafi, P. ankeli,* and *P.* (= *Aeolopithecus*) *chirobates.* Parapithecid species include *Parapithecus fraasi, P. grangeri, Apidium moustafai, A. phiomense,* and *Qatrania wingi.* The tarsiid species is *Afrotarsius chatrathi.* The affinities of the 12th species, *Oligopithecus savagei,* are as yet uncertain. In addition, three recently discovered teeth have been referred to the Omomyidae and the **Lorisidae** (lorises and galagos) and, if confirmed by further finds, would represent the first occurrence of these two families in Africa (Simons et al., 1986). Table 4.1 lists the Old and New World Oligocene primates.

History and Stratigraphy of the Fayum Site

The Fayum Depression of Egypt is about 150 km southwest of Cairo and covers an area of about 1,700 km^2. The history of paleontological discoveries at the Fayum site has been published in considerable detail (Simons and Wood, 1968; Conroy, 1974, 1976*a;* Simons et al., 1978). Only the major episodes of this history are reviewed here.

The first detailed geological study of the Eastern Desert of Egypt was undertaken by the German geologist and explorer Georg Schweinfurth in 1877. In the course of his study, he recovered several invertebrate and vertebrate remains from an island near the center of Lake Qarun, a brackish body of water in the lowest part of the Fayum Depression. More-detailed geological surveys were begun in 1898 by the Egyptian Geological Survey, and additional fossil localities were discovered northwest of Lake Qarun in the course of the survey.

Paleontological expeditions to the Fayum badlands were initiated by the American Museum of Natural History in 1906. The expedition was under the overall direction of Henry Fairfield Osborn with Walter Granger in charge of the field operations. Two of the earlier excavations were enlarged by the American Museum party and subsequently became known as American Museum quarries A and B.

At this time the American Museum employed the German geologist and private collector Richard Markgraf for surface prospecting. It was Markgraf who discovered the type specimens of the first Fayum primates: *Parapithecus fraasi, Apidium phiomense, Propliopithecus haeckeli,* and *Propliopithecus (= Moeripithecus) markgrafi.*

In 1961 a group from Yale University under the direction of Elwyn Simons (now of Duke University) began further geological and paleontological explorations in the upper Eocene and Oligocene badlands of the Fayum. Increasing numbers of small vertebrate remains, including primates, have been recovered since 1961 due in large part to the surface collecting techniques originally developed by Markgraf. For example, only five primate lower jaws were known from the Fayum before 1960, but since that time hundreds of jaw fragments and isolated teeth have been recovered. In addition, numerous primate skeletal remains were discovered and described for the first time (Conroy 1974, 1976*a,* 1976*b*). Probably the most spectacular find was the 1966 discovery of a virtually complete skull of *Aegyptopithecus zeuxis. Aegyptopithecus* and *Apidium phiomense* are the only Oligocene anthropoid genera known from reasonably complete cranial material.

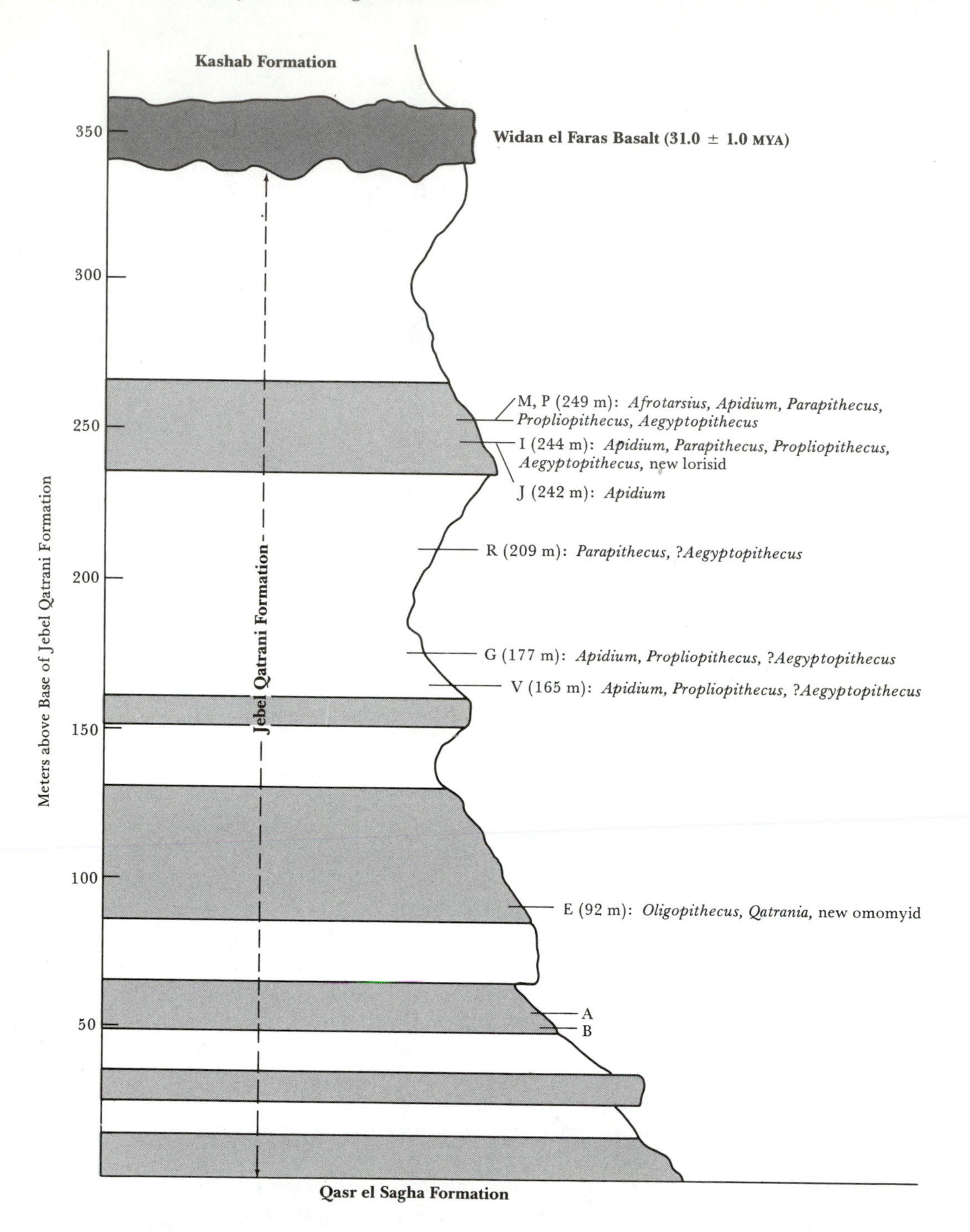

Fig. 4.3 Stratigraphic Section of the Oligocene Jebel Qatrani Formation of the Fayum, Egypt Shown is the distribution of fossil primates discovered to date. Capital letters identify fossil vertebrate localities. Alternate layers are stippled for clarity. (From Fleagle and Kay, 1987)

The sedimentary deposits from which the Fayum primates come are known as the **Jebel Qatrani Formation.** It consists of sandstone, claystone, siltstone, and carbonate units. The cross-bedded sandstones have yielded large petrified logs and disarticulated vertebrate remains; invertebrates and leaf impressions are more common in the clay and siltstones. Fossil sites are designated as quarries, and each is given a separate letter designation (Fig. 4.3).

The geologically oldest Fayum primates discovered to date are *Oligopithecus savagei* and *Qatrania wingi* from quarry E. They are probably close to 40 million years old and thus may actually date from the late Eocene. Some 170 m above the base of the Jebel Qatrani Formation, a small channel stream was discovered to be quite rich in vertebrate remains. This site, now known as quarry G, has yielded several dozen jaw fragments of the early anthropoid *Apidium moustafai* and several isolated anthropoid teeth, probably those of *Propliopithecus haeckeli* and *Propliopithecus markgrafi*.

About 80% of the primate specimens come from nine quarries (I, K, L, M, N, O, P, Q, and R) situated some 50 m up-section from quarry G. Of particular importance are quarries I and M. Four primate genera have been recovered from quarry I: *Parapithecus grangeri, Apidium phiomense, Aegyptopithecus zeuxis,* and *Propliopithecus chirobates.* Most of the primate postcranial remains also come from this quarry. The skull of *A. zeuxis* and an ulna of the same species come from quarry M (Fleagle et al, 1975; Conroy, 1976*a*). Recently three more facial fragments of *Aegyptopithecus* have been recovered from this site. They reveal the marked variation in size and robustness in this genus (Simons, 1985*a*). The tarsierlike *Afrotarsius* is also from this locality.

Unfortunately, the Fayum primates discovered early in the century (*Parapithecus fraasi, Propliopithecus haeckeli, Propliopithecus markgrafi*) are represented only by single specimens of uncertain geologic age. In addition, the type specimen of *Apidium phiomense* is from some uncertain location high in the formation.

Many of the Fayum primate bones, including skulls, jaws, and limb bones, show evidence of having been scavenged by carnivores (Gebo and Simons, 1984).

Overlying the Jebel Qatrani Formation, some 100 m above the level of quarry I, is the Widan el Faras Basalt. A new potassium–argon analysis of the basalt gives a date of approximately 31 million years (Fleagle et al., 1986). This is obviously the minimum age for all the Fayum primates. Because of the time involved for the extensive erosion evidenced by the disconformity between the basalt and the underlying Jebel Qatrani Formation, the primates could well be between 35 and 40 million years old (that is, late Eocene to early Oligocene). Thus all the Fayum primates definitely predate the Miocene by about 15 million years. In addition, they all predate the earliest known primates from South America by about the same margin. The implications of this will be examined later in the chapter.

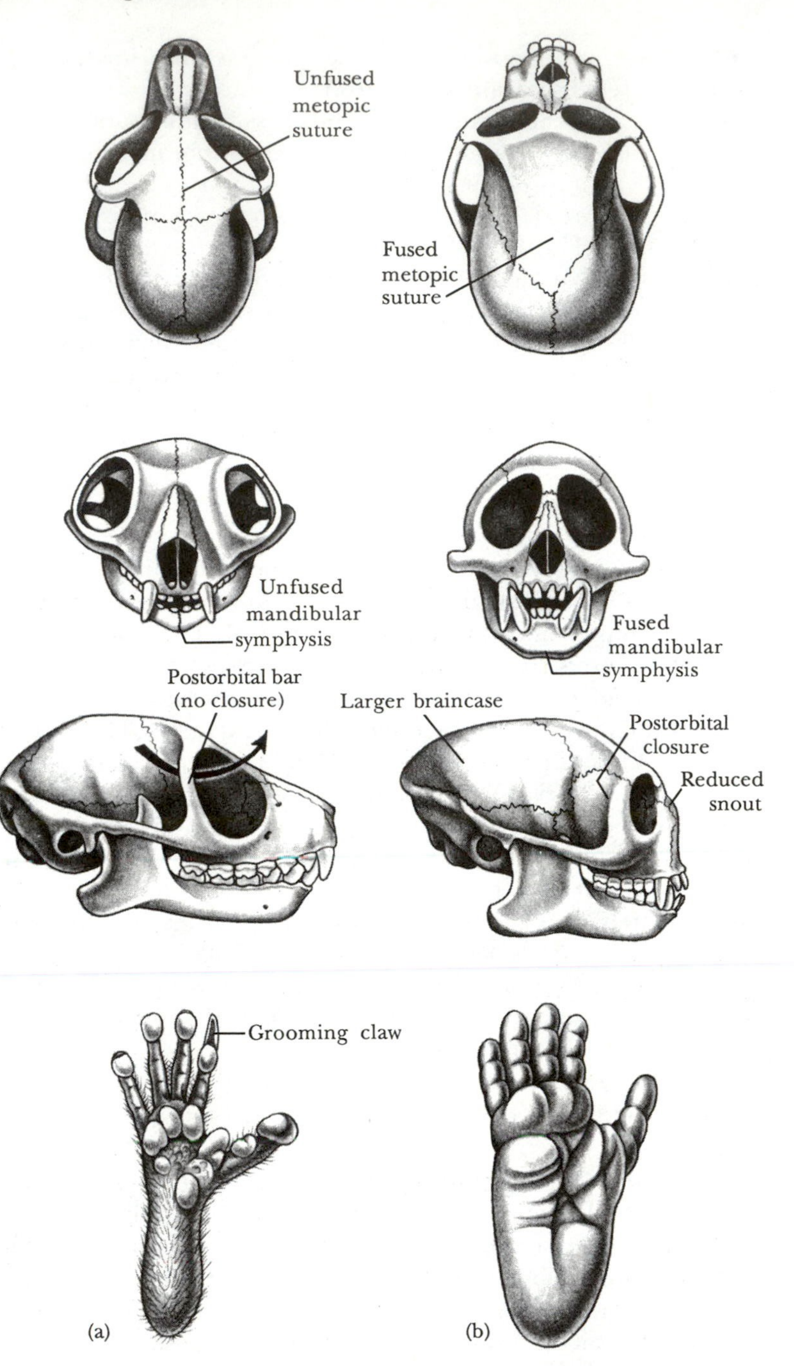

Fig. 4.4 Several Distinctions between (a) Prosimian and (b) Anthropoid Grades of Evolution (Redrawn from Rosenberger, 1986; Jolly and Plog, *Anthropology and Archaeology,* McGraw-Hill, 1986, reproduced by permission; Hershkovitz, 1977; and Swindler and Wood, 1973)

General Morphology of the Fayum Primates

Many of the cranial features distinguishing anthropoids from prosimians (Fig. 4.4) are found in some or all of the Fayum primates:

1. A **metopic suture** (the midline suture between the frontal bones) that fuses early in life
2. Almost complete postorbital closure
3. A fused mandibular symphysis
4. An absence of the stapedial artery and canal
5. The lacrimal bone lying within the orbit
6. The presence of a **central sulcus** (transverse groove separating the primary motor and sensory cortices) instead of a coronolateral sulcus in the brain (see Fig. 3.19).

However, some ancestral cranial features remain. For example, *Aegyptopithecus* retains certain aspects of the cranial venous drainage that are identical to those found in many prosimians, and both parapithecids and propliopithecids retain nontubular ectotympanics and relatively small brains. In Table 4.2 many of the cranial characters of the Fayum anthropoids can be contrasted with the prosimian condition.

Anthropoid dental features also typify many of the Fayum primates, particularly the propliopithecids. General propliopithecid tendencies include (1) lower incisors that are somewhat broad and spatulate; (2) large and sexually dimorphic canines; (3) anterior lower premolars with honing facets for sharpening the posterior edge of the upper canines; (4) absent paraconids; (5) lower molars with broad talonid basins surrounded by low, rounded cusps; and (6) a dental formula of 2.1.2.3 as in all modern Old World anthropoids. Parapithecids, on the other hand, retain three premolars on each side of the upper and lower jaws. They also have more prosimianlike premolars, a paraconid on some specimens, and a most peculiar anterior dentition with the absence of permanent central incisors (*Parapithecus grangeri* at least).

The postcranial skeletons of both propliopithecids and parapithecids indicate that all the Fayum primates were arboreal quadrupeds with varying amounts of leaping abilities. There is no indication that any of them were adapted for terrestrial locomotion, nor that they practiced vertical clinging and leaping or arm-swinging locomotion to any significant extent.

These cranial and postcranial features are discussed in more detail in the following sections.

Parapithecidae

Parapithecids are the most common primates from the Fayum and are currently assigned to five species: *Qatrania wingi, Apidium moustafai, Apidium phiomense, Parapithecus fraasi,* and *Parapithecus (= Simonsius) grangeri.*

Table 4.2 Morphological Characters of the Fayum Primates Compared with Other Primates

Character	*Nonanthropoid Condition*	*Parapithecids*[a]	*Platyrrhines*	*Aegyptopithecus/ Propliopithecus*
Cranial				
Ectotympanic bone	Ring-shaped	Ring-shaped *(A)*	Ring-shaped	Ring-shaped
Internal carotid and stapedial arteries	Both present	Carotid only *(A)*	Carotid only	Carotid only
Nasal bones	Broad posteriorly	Narrow *(A)*	Narrow	Narrow
Metopic suture	Unfused	Fused *(A)*	Fused	Fused
Postorbital closure	No	Yes *(A, Pa)*	Yes	Yes
Olfactory lobes	Very large	Large *(A)*	Reduced	Large
Zygomatic-parietal contact	No closure	Yes *(A)*	Variable	No
Fused mandibular symphysis	No	Yes	Yes	Yes
Prominence of inferior transverse torus	Absent	Weak *(A, Pg)*	Weak	Moderate
Ramus shape	Anterior margin slopes posteriorly	Anterior margin slopes posteriorly *(A, Pg, Pf)*	Variable	Vertical
Ramus anteroposterior length	Very short	Very short *(Pf, A)*	Short to long	Long
Jaw depth	Extremely shallow	Extremely shallow	Variable	Deep
Mandibular shape	V-shaped	V-shaped	Variable	V-shaped
Dental				
Lower incisors	2 ?spatulate	2 spatulate *(A)*, 0 *(Pg)*, ? *(Pf, Q)*	2 spatulate	2 spatulate
Canine sexual dimorphism	?	Strong *(A)*	Weak to strong	Strong
Upper and lower P^2	Present	Present *(Pg, A, Pf)*	Present	Absent
P^4 shape	Waisted	Oval *(Pq, A)*	Oval	Oval
P^{3-4} paraconule	Absent	Present *(Pg, A)*	Absent	Absent
P_4 trigonid	Lingually open	Lingually open	Variable	Lingually closed
P_4 metaconid size	Very small	Very small *(Q)* to small *(A, Pg)*	Subequal[b] with protoconid	Subequal[b] with protoconid
Relative crown height of molar trigonid/talonid	Variable	Little height disparity (all)	Variable	Little height disparity
Lower molar trigonids	Lingually open	Lingually closed	Lingually closed	Lingually closed
Molar paraconids	Present	Present on M_{1-3} *(Q)*; Reduced or absent *(Pg, Pf, A)*	Variable on M_1	Absent
M_{1-2} hypoconulid size	Absent	Large *(A, Q)* (Smaller *Pg, Pf*)	Small or absent	Large
M_{1-2} hypoconulid position	Cusp absent or variable	Central *(Q, A, Pg, Pf)*	Lingual if present	Central
Lower molar crown height	Low	Low *(A, Pf, Q)* High *(Pq)*	High/low	Low
Lower molars waisted	No	Yes *(Pf, A, Pg)* No *(Q)*	No	No

Table 4.2 (continued)

Character	*Cercopithecoids*	*Hominoids*	*Oligopithecus*	*Afrotarsius*
Cranial				
Ectotympanic bone	Tubular	Tubular	?	?
Internal carotid and stapedial arteries	Carotid only	Carotid only	?	?
Nasal bones	Narrow	Narrow	?	?
Metopic suture	Fused	Fused	?	?
Postorbital closure	Yes	Yes	?	?
Olfactory lobes	Reduced	Reduced	?	?
Zygomatic-parietal contact	No	No	?	?
Fused mandibular symphysis	Yes	Yes	?	?
Prominence of inferior transverse torus	Strong	Moderate	?	?
Ramus shape	Vertical	Vertical	?	?
Ramus anteroposterior length	Long	Long	?	?
Jaw depth	Deep	Deep	?	?
Mandibular shape	U-shaped	U-shaped	?	?
Dental				
Lower incisors	2 spatulate	2 spatulate	?	?
Canine sexual dimorphism	Strong	Weak to strong	?	?
Upper and lower P^2	Absent	Absent	Absent	?
P^4 shape	Oval	Oval	?	?
P^{3-4} paraconule	Absent	Absent	?	?
P_4 trigonid	Lingually closed	Lingually closed	Lingually open	Lingually open
P_4 metaconid size	Subequal[b] with protoconid	Subequal[b] with protoconid	Subequal[b] with protoconid	?
Relative crown height of molar trigonid/talonid	Little height disparity	Little height disparity	Moderate disparity	Trigonid much higher than talonid
Lower molar trigonids	Lingually closed	Lingually closed	Lingually open	Lingually open
Molar paraconids	Absent	Absent	Present on M_1	Present on M_{1-3}
M_{1-2} hypoconulid size	Absent	Large	Large	Small
M_{1-2} hypoconulid position	Cusp absent	Buccal	Lingual	Central
Lower molar crown height	High	Low	Low	Low
Lower molars waisted	Yes	Variable	No	No

Table 4.2 continued next page

Table 4.2 (continued)

Character	Nonanthropoid Condition	Parapithecids[a]	Platyrrhines	Aegyptopithecus/ Propliopithecus
M_3 size	Smaller than M_2	Smaller than M_2 *(Pq, Pf, Q)* Equal to or larger than M_2 *(A)*	Small	Equal to or larger than M_2
M_3 hypoconulid	Present	Present (all)	Absent	Present
Paraconule and metaconule on M^1 or M^2	Present	Present *(A, Pq)*	Variable	Variable
Hypocone	Variable	Present	Present	Present
Nannopithex fold	Present	Absent *(A, Pg)*	Absent	Absent
Postcranial				
Entepicondylar foramen	Present	Present	Present	Present
Medial epicondyle	Prominent	Prominent	Prominent	Prominent
Dorsal epitrochlear fossa	Present	Present	Present	Present
Olecranon fossa	Shallow	Shallow	Shallow	Shallow
Capitulum shape	Elongate	Elongate	Elongate	Elongate
Medial trochlea lip	Absent	Moderate	Moderate	Moderate
Olecranon Length	Long	Long	Long	Long
Iliac gluteal blade	Narrow	Broad	Broad	?
Iliac plane	Moderate	Moderate	Moderate	?
Ischial tuberosity	Narrow	Narrow	Narrow	?
Lesser trochanter	Large	Large	Small	?
Distal condyles	Deep	Deep	Shallow	?
Tibia shaft	Narrow	Narrow	Moderate	Moderate
Distal joint	Variable	Fibrous long, posterior	Fibrous/synovial long, posterior	Synovial short, anterior

Parapithecids, while demonstrating considerable dental diversity, share specializations that support their status as a monophyletic group:

1. Enlarged lower second premolars
2. Upper premolars with well-developed paraconules
3. "Waisted" lower molars
4. Upper molars with enlarged paraconules and metaconules
5. Greatly reduced cingula on the cheek teeth
6. Mandibles that increase in depth posteriorly
7. Retention of three premolars in the upper and lower jaws (Harrison, 1987) (Fig. 4.5)

In addition, several species of parapithecids also have unique dental specializations. In particular, *Parapithecus grangeri* lacks lower permanent incisors (making it the only known primate, living or fossil, with this condition), and *Parapithecus fraasi* may only have a single pair of lower incisors (Kay and Simons, 1983*a;* Simons, 1986).

The parapithecid skull is not well known, but enough is preserved to indicate that the anthropoid features of a fused frontal suture, postorbital closure, and a large promontory artery are present. The ectotym-

Table 4.2 (continued)

Character	*Nonanthropoid Condition*	*Parapithecids*[a]	*Platyrrhines*	*Aegyptopithecus/ Propliopithecus*
M_3 size	Variable	Variable	?	Smaller than M_3
M_3 hypoconulid	Variable	Present	?	Present
Paraconule and metaconule on M^1 or M^2	Absent	Variable	?	?
Hypocone	Present	Present	?	?
Nannopithex fold	Absent	Absent	?	?
Postcranial				
Entepicondylar foramen	Absent	Absent		
Medial epicondyle	Reduced	Prominent		
Dorsal epitrochlear fossa	Absent	Absent		
Olecranon fossa	Deep	Deep		
Capitulum shape	High	Round		
Medial trochlea lip	Prominent	Moderate		
Olecranon Length	Long	Short		
Iliac gluteal blade	Broad	Very broad		
Iliac plane	Moderate	Very broad		
Ischial tuberosity	Expanded	Expanded/ narrow		
Lesser trochanter	Small	Small		
Distal condyles	Shallow	Shallow		
Tibia shaft	Moderate	Moderate		
Distal joint	Synovial short, anterior	Synovial variable		

[a] *A = Apidium; Pf = Parapithecus fraasi; Pg = Parapithecus grangeri; Q = Qatrania*

[b] Subequal with = same size as

Source: Adapted from Fleagle and Kay (1987).

panic is ringlike and is attached to the lateral bulla wall. One can observe from the cranium that a relatively large olfactory bulb was present, indicating that smell was an important sensory modality for these primates.

In overall postcranial morphology, the limbs more closely resemble those of platyrrhines and some Eocene primates than those of apes or Old World monkeys. The limbs indicate that parapithecids were arboreal quadrupeds with well-developed leaping abilities. For example, the anatomy of the bony pelvis is now known in some detail for parapithecids. While it is more similar in overall morphology to that of higher primates, it has a mosaic of features that cannot be clearly allied with any particular extant group. The shape of the ilium is similar to that in the **Cercopithecoidea** (Old World monkeys). However, cercopithecoids have **ischial tuberosities** (bony expansions of the ischium that support **ischial callosities,** or toughened pads of skin for prolonged sitting). Because the parapithecid pelvis lacks ischial tuberosities, it resembles more closely

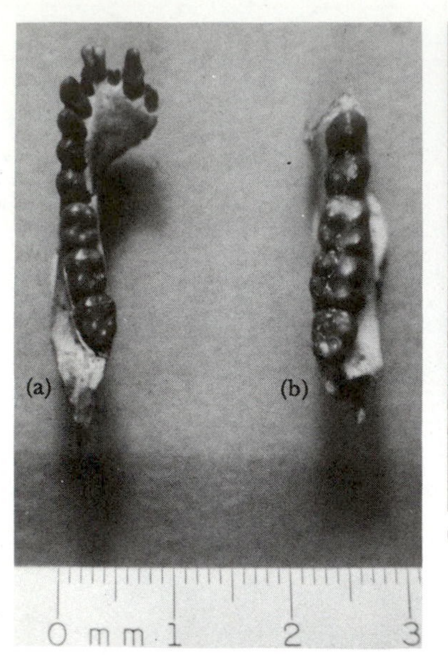

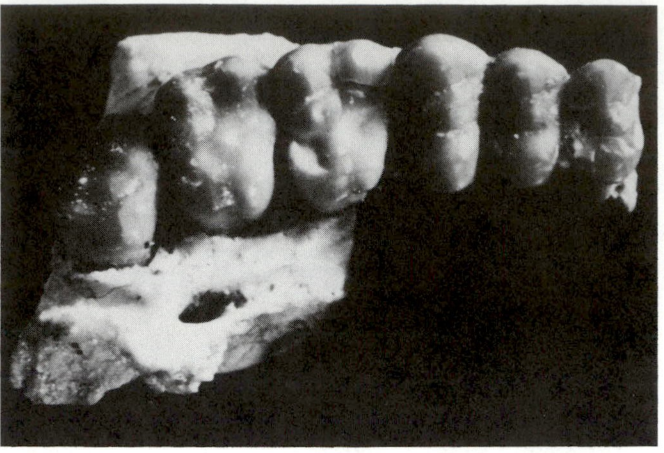

Fig. 4.5 Parapithecid Dentition (a) Lower left jaw of *Apidium;* (b) lower right jaw of *Parapithecus;* (c) upper right jaw of *Apidium.* Note the retention of three upper and lower premolars (seen here in *Apidium*), enlarged conules on the upper teeth of *Apidium,* and reduced cingula on the cheek teeth in both genera. Scale of (c) about 3.6 × natural size. (Courtesy of E. Delson)

that of the **Cebidae** (New World monkeys excluding marmosets and tamarins), which also lack this feature (Fleagle and Simons, 1979).

Apidium *Apidium* was the first Fayum primate to be discovered and described (Osborn, 1908). Initially its affinities were uncertain; it was named after Apis, the sacred bull of Egypt. The type specimen was found by Markgraf in the upper portion of the Jebel Qatrani Formation.

Two species are recognized: *Apidium moustafai* and *Apidium phiomense.* The geologically older species, *A. moustafai,* is known from several dozen jaws. *Apidium phiomense,* presumed to be the direct descendant of *A. moustafai,* was removed from higher up in the stratigraphic section; the material includes parts of the skull, jaws, teeth, and other skeletal remains.

The two species are very similar in dental morphology, although *Apidium moustafai* is about 80% the size of *A. phiomense* in most linear dental dimensions and has fewer cuspules on the molars. The dental formula for *Apidium* is 2.1.3.3, identical to that found in cebids.

Apidium had clearly reached an anthropoid grade of evolution as far as its cranial features are concerned: (1) the frontal bones of its skull fused early in life, thus obliterating the metopic suture; (2) there is evidence of postorbital closure; and (3) mandibular symphyseal fusion occurred at an early age. Although the orbits are large, they are not large enough to suggest nocturnal habits. The frontal bones show pronounced temporal lines (for the origin of the temporalis muscles), which converge toward the back of the skull to form a low **sagittal crest** (a bony ridge along the midline of the skull), an unusual feature in so small an anthropoid. This finding would suggest that the brain was small compared with the size of the chewing muscles. The olfactory bulbs were small and the face is foreshortened (Fig. 4.6).

Several fossil fragments preserving the ear region have been described (Gingerich 1973*b;* Cartmill et al., 1981). Like all extant adult anthropoids, these early anthropoids were characterized by a large promontory artery and a reduced stapedial artery. Most importantly, the ectotympanic bone was ringlike and attached to the bulla wall as in cebids. The tubular ectotympanic bone common to all later **Catarrhini** (Old World anthropoids) had not yet evolved. In every anatomical feature the ear region of the Fayum primates more closely resembles that of the recent **Platyrrhini** (New World monkeys) than it does of any prosimian including adapids (Conroy, 1980).

The postcranial specimens assigned to *Apidium* indicate arboreal adaptations, with many features similar to those found in living cebids like *Saimiri* or *Cebus* (Conroy, 1976*a*). Several partial **scapulae** (shoulder blades) and upper-limb bones (humeri, ulnae, and radii) are known. Certain features of the scapula fall within the range of extant primates

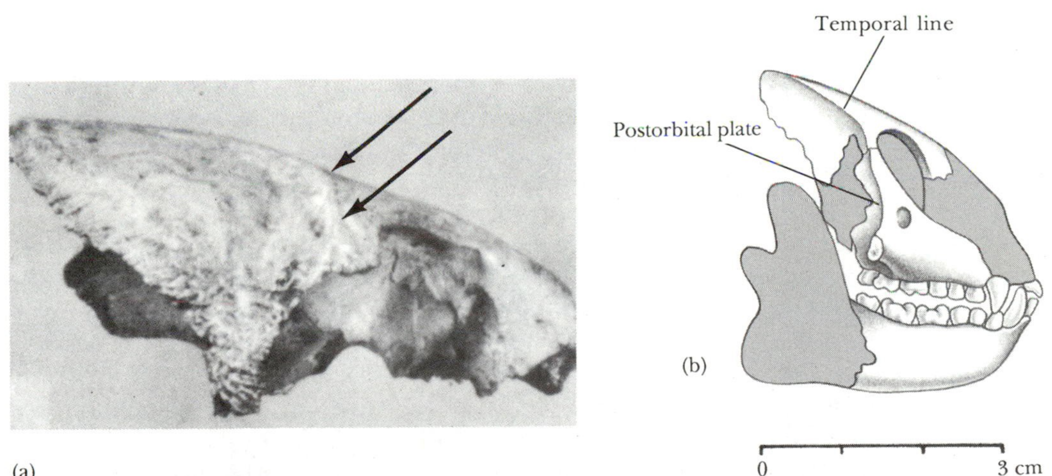

Fig. 4.6 Skull of *Apidium* (a) Right lateral view of a small frontal bone attributed to *Apidium phiomense.* Arrows point to strongly developed temporal lines that meet in the midline to form a low sagittal crest. (b) Reconstruction of the face and mandible of *Apidium phiomense.* Note the enclosed posterior orbital wall. (a, From Szalay and Delson, 1979; b, redrawn from Fleagle and Kay, 1987)

whose locomotor repertoire includes a considerable amount of leaping in addition to quadrupedal locomotion (Anapol, 1983).

At the elbow joint the size and configuration of the **medial epicondyle** (a bony projection on the medial side of the distal humerus) is a useful indicator of the relative importance of the **digital** and **carpal flexors** (muscles that flex the fingers and wrist) since they mostly arise from this bony projection. Arboreal primates have a greater need for powerful flexion of the fingers around branches and so have larger medial epicondyles than their more terrestrial relatives. As a general rule, small platyrrhines tend to have more projecting medial epicondyles than do their catarrhine counterparts. This probably reflects the fact that small New World monkeys have strongly compressed and curved nails that necessitate powerful digital flexors. On some New World monkeys, such as marmosets, all the digits of the hand are armed with sharp, recurved claws. *Apidium* displays a more platyrrhinelike medial epicondyle, suggesting that *Apidium* too may have had compressed, curved nails and been primarily arboreal.

It has recently been discovered that in this genus the distal tibia and fibula are united by a **syndesmosis** (fibrous connection) (Fleagle and Simons, 1983). This is an unusual condition in higher primates and further suggests a leaping adaptation. In contrast to tarsiers, however, the bones are not completely fused **(synostosis)** but remain distinct. The type of syndesmosis found in *Apidium* also characterizes *Microcebus* (mouse lemurs) and the small platyrrhines *Pithecia* (sakis), *Saimiri* (squirrel monkeys), *Callithrix* (marmosets), and *Cebuella* (the pygmy marmoset).

Other leaping adaptations include the following:

1. A long ischium
2. A stout femoral neck oriented perpendicularly to the femoral shaft
3. Posteriorly directed femoral condyles with pronounced elevation of the lateral patellar groove
4. A long, laterally compressed tibia
5. A low intermembral index of about 70

Many of these features are illustrated in Fig. 4.7.

The small tarsal bones associated with *Apidium* are distinct in several ways from those of plesiadapiforms and omomyids. The calcaneus is not greatly elongated anteriorly as it is in omomyids (for example, *Hemiacodon, Teilhardina, Tetonius, Necrolemur,* and *Nannopithex*). In this aspect of its morphology, *Apidium* is not specialized for vertical clinging and leaping as the evidence from the united tibia and fibula might at first suggest. The morphology of various joint surfaces indicates that **inversion** (turning the ankle so that the bottom of the foot faces medially) was the predominant position of the foot in arboreal locomotion. It is clear from this and numerous other morphological characteristics of the Fayum tarsal bones that *Apidium* was not a terrestrially adapted primate (Conroy, 1974, 1976*a*). In fact, assessing all the postcranial evidence for *Apidium* leads one to conclude that it was a generalized arboreal quadruped.

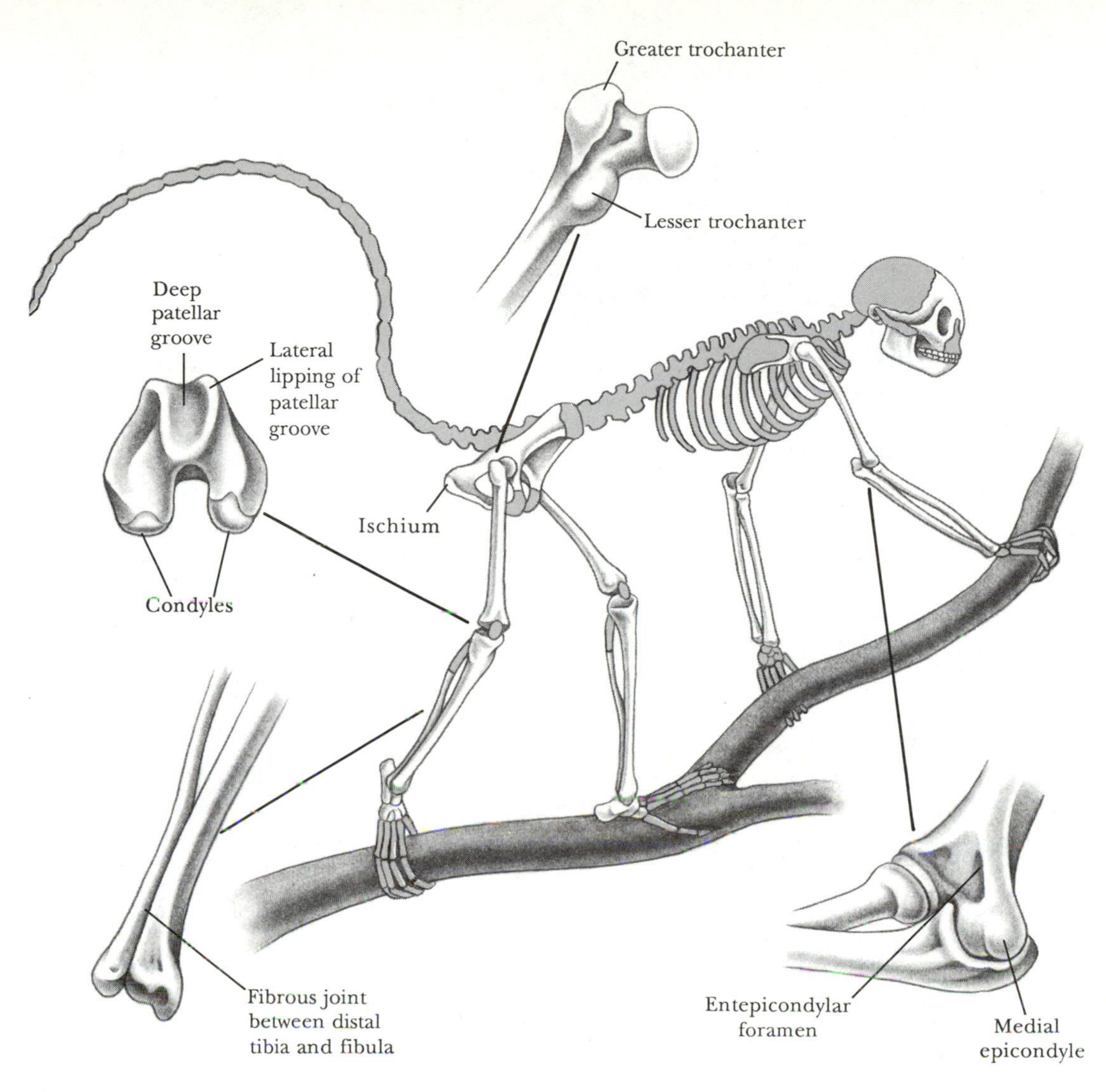

Fig. 4.7 Distinctive Skeletal Features of *Apidium phiomense* Reconstructed portions in gray. Note in the whole skeleton the low intermembral index (about 70) and the absence of ischial tuberosities. The elbow joint shows the prominent medial epicondyle and entepicondylar foramen. The proximal end of the femur shows the greater trochanter projecting over the femoral shaft ventrally and the large lesser trochanter. The knee joint has deep posteriorly directed femoral condyles and accentuated lateral lipping of the patellar groove. Note the fibrous joint (syndesmosis) between the distal tibia and fibula. (Redrawn from Fleagle and Kay, 1987)

Parapithecus Two species of the genus *Parapithecus* are recognized: *P. fraasi* and *P. grangeri.* The type species, *P. fraasi,* which was first described in 1910, consists of a mandible with canines through third molars present on each side (Schlosser, 1910). Its precise geological age is unknown. *Parapithecus grangeri* is known mainly from a number of jaws and isolated teeth. A few postcranial bones may belong to this species, but none of them is definitely associated with the dentition of *P. grangeri.*

The type specimen of *P. fraasi* was broken in the midline sometime after its discovery, and for many years debate raged about whether or not this animal had a fused mandibular symphysis. The number of preserved teeth originally suggested a lower dental formula of 1.1.3.3, as in extant tarsiers. Simons (1972) has suggested that the central incisors may have been lost when the specimen was broken and that the lower dental formula is probably 2.1.3.3 as in *Apidium.* If the recent evaluation of new specimens is correct, however, then it is possible that *P. fraasi* does have a lower dental formula of 1.1.3.3 and that *P. grangeri* has a lower dental formula of 0.1.3.3, a most peculiar dental formula for a primate (Kay and Simons, 1983*a;* Simons, 1986). For this reason some authors have placed *P. grangeri* in a distinct genus, *Simonsius.*

Qatrania *Qatrania* is a primitive member of the Parapithecidae (Simons and Kay, 1983). *Qatrania* and *Oligopithecus* (to be described later) are the oldest known primates from the Fayum. Both are known only from quarry E.

Qatrania probably weighed less than 300 g and is the smallest anthropoid known. The only specimens are two lower jaws and some isolated teeth. Even though *Qatrania* was very small, it was probably not insectivorous; its cheek teeth lack the shearing crests characteristic of that type of diet. Fruits and gums probably made up the bulk of its diet.

Propliopithecidae

Propliopithecus Four species are currently recognized in the genus *Propliopithecus: P. haeckeli, P. markgrafi, P. chirobates,* and *P. ankeli* (Simons et al., 1987). As with *Apidium,* taxonomic views about this genus have been diverse and contradictory. Some workers, such as Schlosser (1911), considered it ancestral to the **Hylobatidae** (gibbons and siamangs), whereas others, such as Kurten (1972) and Pilbeam (1967), once considered it ancestral to the **Hominidae** (humans and their immediate ancestors). This latter interpretation was based on the type and only specimen of *P. haeckeli,* which had (1) small vertical incisors as inferred from the tooth sockets **(alveoli);** (2) small canines; (3) premolars that were **homomorphic** (of similar size and shape); and (4) equal-sized, low-crowned molars.

Propliopithecus haeckeli was the first member of the genus to be found (1908). It consists of two lower jaw fragments of a single individual: on the left side is preserved the third premolar through third molar, and on the right the canine through third molar. A few isolated teeth may also belong to this species. *Propliopithecus markgrafi* is known from only a single specimen with a few teeth, and *P. ankeli* is known from one maxillary fragment with the third premolar through third molar intact and two mandibular fragments preserving most of the postcanine teeth

(Simons et al., 1987). The best-known species is *P. chirobates*, consisting of a number of jaw fragments with teeth. No cranial remains have been described for the genus. The few postcranial remains are similar to those of arboreal quadrupeds having strong grasping feet (Fleagle and Simons, 1982*a*).

The type specimen of *P. chirobates* was originally considered by Simons (1965) to be a hylobatid because of its (1) large, gibbonlike canines; (2) reduced third molars; (3) shallow mandible; and (4) gibbonlike genioglossal pit morphology. (The **genioglossal pit** is a notch on the inside of the mandibular symphysis for the origin of the genioglossus muscle, which forms the bulk of the tongue.) When *P. chirobates* was initially discovered, it was placed in a separate genus named *Aeolopithecus*. Now that many more specimens have been found, the differences originally thought to exist between *Aeolopithecus* and *Propliopithecus* can be better explained by patterns of sexual dimorphism and normal intrageneric variation (Fleagle et al., 1980). For this reason *Aeolopithecus* is no longer considered to be a valid genus name (Simons, 1985*a*).

Like all later catarrhines, the dental formula in *P. chirobates* is 2.1.2.3. The species is recognized by a number of dental features including (1) comparatively broad, low-crowned incisors; (2) sexually dimorphic canines and lower third premolars; (3) small and narrow lower fourth premolars; (4) steep-sided molar crowns with well-developed buccal cingula on the lower molars; (5) marginally placed molar cusps; and (6) reduced lower molar lengths (Fig. 4.8).

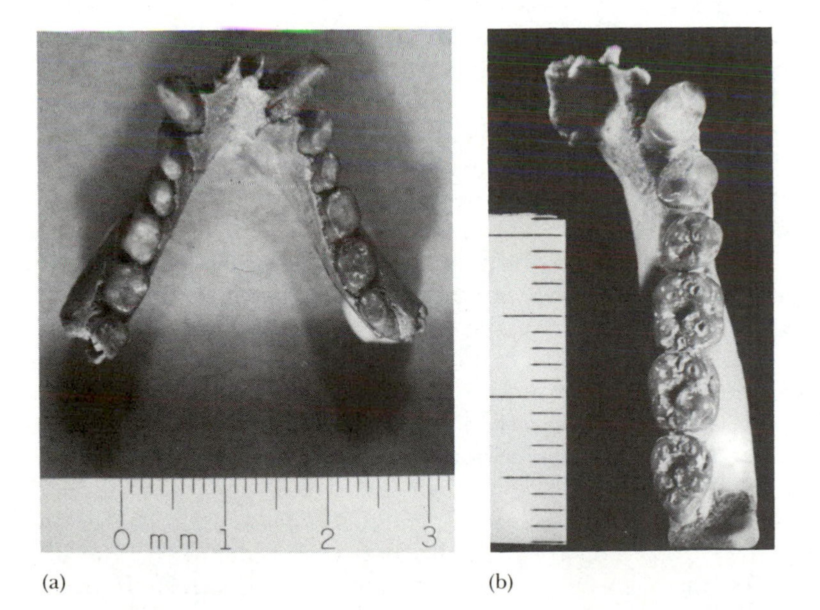

Fig. 4.8 Propliopithecid Dentition (a) Lower jaw of *Propliopithecus chirobates*. This jaw was originally considered to be a distinct proto-gibbonlike genus, *Aeolopithecus*. (b) Lower jaw of *Propliopithecus haeckeli*. Note the enlarged canine, reduced third molars and the presence of only two premolars. (Courtesy of E. Delson)

Several postcranial fragments are known for *P. chirobates*. The specimens suggest arboreal quadrupedal locomotion. For example, the tibia is rather short, and the condyles show the slight degree of asymmetry more typical of arboreal quadrupeds. In addition, the more distal insertions of the tendons of the **semitendinosus** (one of the hamstring muscles) and **gracilis** (a muscle of the medial thigh) are also characteristic of arboreal quadrupeds rather than the more proximal insertions found in leaping forms. As we shall see in Chapter 5, all the skeletal fragments are more similar to the same bones in the Miocene catarrhine *Pliopithecus* than they are to any extant primate species. *Propliopithecus* exhibits numerous primitive skeletal features more reminiscent of living **Ceboidea** (New World platyrrhines) than of either the **Hominoidea** (apes and humans) or the Cercopithecoidea (Old World monkeys) (Fleagle and Simons, 1982*a*).

Aegyptopithecus *Aegyptopithecus* was the largest of the Fayum anthropoids. Only one species, *Aegyptopithecus zeuxis,* is recognized. The first primate ever discovered in the Fayum (1906) was a wind-eroded jaw of this species, but it was not described until a half century later. In 1966 a virtually complete skull was found, and recently several more skulls of *Aegyptopithecus* have been recovered. These skulls provide the best information on the cranial anatomy of early anthropoids. The skull is similar in overall size to that of a female howler monkey *(Alouatta)* or a small guenon *(Cercopithecus mona)*. From the skull and teeth one can deduce that male body size must have averaged about 6 kg. The skulls and canines exhibit a high degree of sexual dimorphism. The muzzle is long and the brain case comparatively small (Radinsky, 1974, 1982).

The small brain case combined with large temporalis and neck muscles necessitated the development of prominent bony crests on the skull for the attachment of these muscles. The large temporalis muscle originated from the sagittal crest on the top of the skull, and the large neck muscles originated from a prominent **nuchal** (neck) **crest** on the back of the skull. Somewhat surprisingly perhaps, *Aegyptopithecus* shares certain cranial features with Eocene adapids: (1) the large protruding face; (2) a small brain case; (3) dorsally oriented orbits; (4) sharp angulation of the nuchal region with the cranial vault; and (5) **postorbital constriction** (narrowing of the skull behind the orbits) (Fig. 4.9). Like anthropoids, however, *Aegyptopithecus* has fused mandibular and frontal symphyses, postorbital closure, and **superior** and **inferior transverse tori** (Kay et al., 1981; Fleagle and Kay, 1983). (The transverse tori are thickened bony supports on the posterior side of the mandibular symphysis that serve to resist stresses passing through this part of the mandible; the inferior transverse torus is sometimes referred to as the **simian shelf.**)

The species was clearly diurnal judging from the comparatively small orbits. As in most catarrhines the lacrimal bone lies entirely within the orbital rim and does not extend onto the face (as it does in lemurs, for

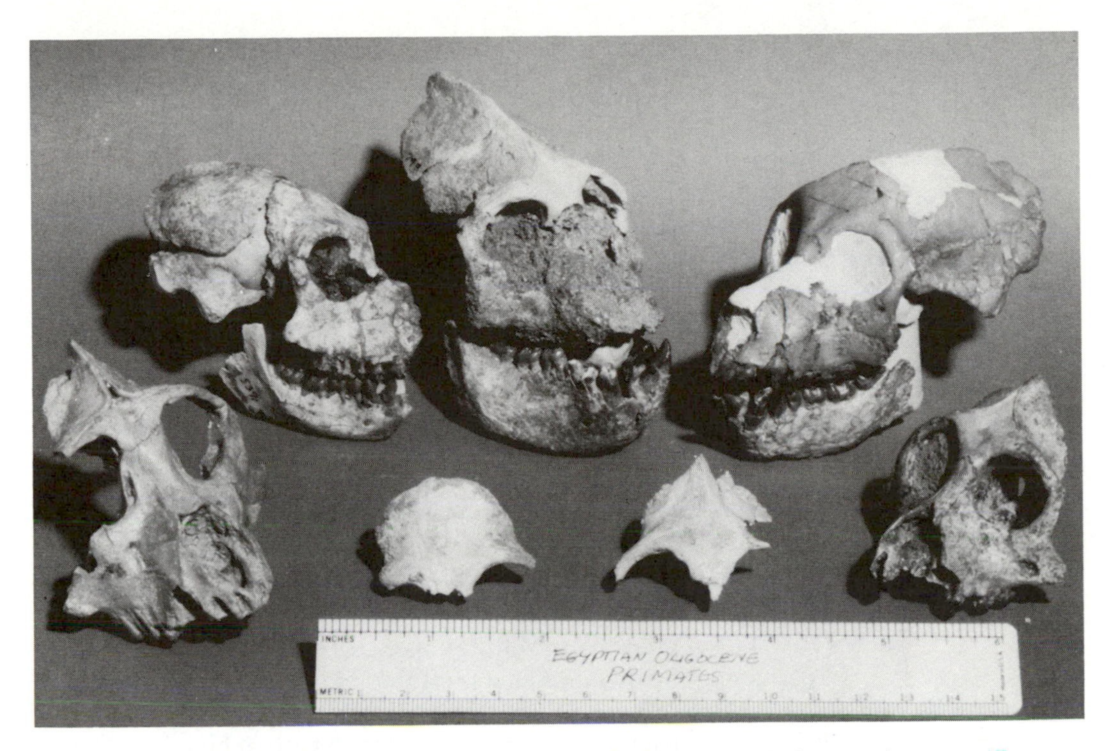

Fig. 4.9 Cranial Variation in *Aegyptopithecus zeuxis* Note the relatively small brain case and the variability in size of the nuchal and sagittal crests and elongated snout. (Courtesy of E. Simons)

example). The interorbital distance is large, comparable to that found generally in the **Colobinae** (one of the two extant subfamilies of Old world monkeys, typified by *Colobus* and *Presbytis*) and in hominoids; the interorbital distance is larger than that usually seen in the **Cercopithecinae** (the other Old World monkey subfamily, typified by *Papio* and *Macaca*). *Aegyptopithecus* appears to show a catarrhinelike arrangement of sutures on the side of the skull at **pterion** (the region corresponding to the temple) with the **alisphenoid** (the portion of the sphenoid bone that forms part of the lateral skull wall) and frontal bones interposed between the parietal and zygomatic bones (Fleagle and Rosenberger, 1983) (Fig. 4.10). As in other Fayum primates and in ceboids, the ectotympanic bone is ringlike and is attached to the lateral wall of the bulla. The infraorbital aperture is small as in all anthropoids.

The brain is more advanced than that of most prosimians in having a relatively expanded visual cortex, comparatively small olfactory bulbs, and a central sulcus. This sulcus is found in all modern anthropoids except the smallest ceboids.

The brain volume of *Aegyptopithecus* is approximately 30 cc. Depending on body-size evaluations, the encephalization quotient ranges from

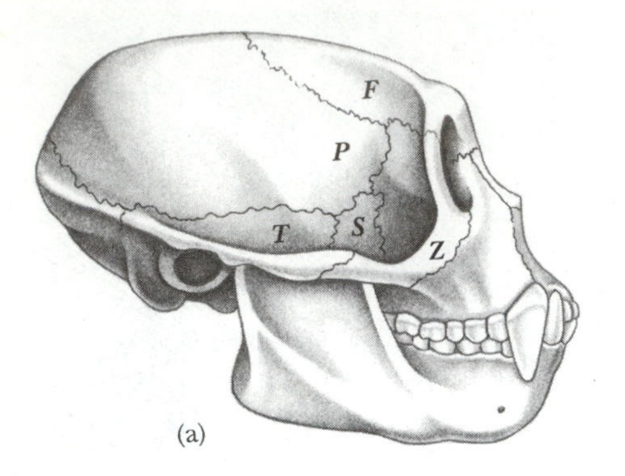

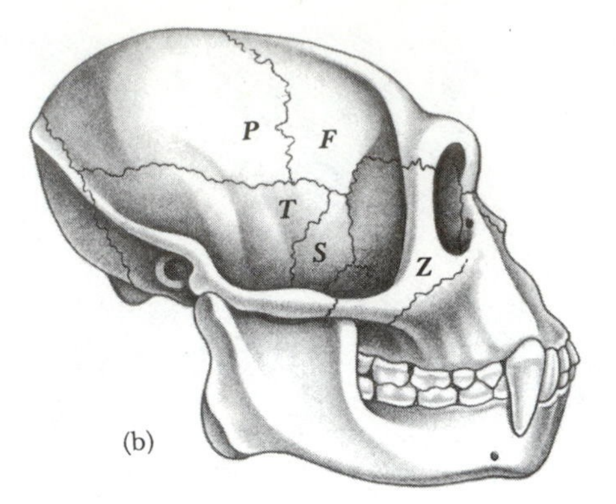

Fig. 4.10 Arrangement of Bones on the Lateral Side of the Skull (a) Platyrrhines; (b) catarrhines. *Aegyptopithecus* resembles the catarrhine condition and *Apidium* resembles the platyrrhine condition in this feature. Bones: *F*, frontal; *P*, parietal; *S*, sphenoid (alisphenoid portion); *T*, temporal; *Z*, zygomatic.

0.36 to 1.53 (Gingerich, 1977*a;* Conroy, 1987). By comparison, the EQs of living prosimians range between 0.60 and 1.73. Most extant anthropoids have EQs higher than 1.43 (Radinsky, 1974, 1982; Gurche, 1982).

In examining the dentition one finds that the incisors tend to be small and narrow, the premolars are **heteromorphic** (of different size and shape), and the molars are very broad at their base. The molars differ from those of *Propliopithecus* in having narrow molar crowns with sloping sides, an increase in length from first to third molars, and more bulbous cusps. The ascending ramus of the mandible is broad, implying large medial pterygoid and masseter muscles for powerful chewing. (The **medial pterygoid** muscle runs from the base of the skull to the inner surface of the angle of the mandible, and the **masseter** muscle runs from the zygomatic arch to the outer surface of the angle of the mandible.) The species is sexually dimorphic in terms of canine size, premolar size, and jaw depth. See Table 4.2 for a comparison of the craniodental features of parapithecids and propliopithecids.

Postcranial bones include the hallux, ulna, humerus, phalanges, and caudal vertebrae. Two complete humeri are known for *Aegyptopithecus.* They are about the size of the red howler monkey *(Alouatta seniculus)* or the dusky leaf monkey *(Presbytis obscura).* The distal articular surface is similar to that of many extant ceboids (particularly *Alouatta*) and to the Miocene catarrhine *Pliopithecus vindobonensis* (see Chapter 5). The head of the humerus faces posteriorly and is narrower than that in apes or monkeys that engage in extensive suspensory posture. In primates that frequently engage in suspensory postures the humeral head faces more medially in order to articulate with the scapula, which is situated more on the dorsal than on the lateral side of the thorax. This increases the range of movements permitted at the shoulder joint. In most cursorial mammals and in terrestrial quadrupedal primates the **greater tuberosity** is large and projects above the level of the articular head. In *Aegyptopithecus* the greater tuberosity reaches only to the level of the humeral head as in arboreal quadrupeds generally. The humerus shares several (possibly primitive) features with extant hominoids, such as a large medial epicondyle and a comparatively wide trochlea (the spool-shaped surface on the distal humerus that articulates with the ulna); the humerus shows none of the (probably derived) features that typify extant cercopithecoids (Fleagle, 1980). The brachial index is about 94, virtually identical to the same measurement in *Alouatta seniculus.* In most features the humerus of *Aegyptopithecus* is more primitive than that of the hypothetical last common ancestor of extant cercopithecoids and hominoids and supports other lines of evidence indicating that the Fayum propliopithecids were ancestral to both Old World monkeys and apes (Fleagle and Simons, 1982*b;* Harrison, 1987).

The ulna in *Aegyptopithecus* also resembles the ulna in *Alouatta* in many respects. This similarity provides strong evidence that this bone sustained the same compressive, tensile, and shearing stresses as those occurring in the ulna of living adult howler monkeys. This evidence reaffirms the conviction that the best models for depicting early anthropoid locomotor habits are to be found in the extant New World monkeys, not living Old World monkeys (Conroy, 1976*a;* Schon-Ybarra and Conroy, 1978; Fleagle et al., 1975).

Only two parts of the foot skeleton have been described, a **hallucal metatarsal** bone (the long bone in the hallux, or big toe) and a slightly damaged left **talus** (**astragalus,** or ankle bone). The Fayum metatarsal consists of the proximal two-thirds of the bone. Its overall size suggests an animal whose body weight approximated that of the South American long-haired spider monkey *(Ateles belzebuth).* Of particular interest is a well-defined oval facet at the base of the metatarsal for a small accessory bone (the **prehallux**) in the **hallucal tarsometatarsal joint.** In primitive fossil anthropoids this joint consists of three bones: the hallucal metatarsal, the medial cuneiform, and the prehallux. Among extant primates this condition is normally found only in platyrrhines and hylobatids. It is interesting to note that *Proconsul africanus,* a putative descendant of *Aegyptopithecus zeuxis,* also apparently had a prehallux incorporated into its hallucal tarsometatarsal joint (Lewis, 1972*b*). (*Proconsul* will be discussed in greater detail in Chapter 5.) Thus there is nothing in the anatomy of the hallucal tarsometatarsal joint that would preclude *Aegyptopithecus* from being broadly ancestral to all later higher primates (Wikander et al., 1986).

The hallux in *Aegyptopithecus* was clearly adapted for powerful grasping. This can be deduced from the morphology of the **peroneal tubercle** (a large, blunt projection for insertion of the peroneus longus tendon). The peroneus longus is an important muscle in arboreal primates that have powerfully grasping feet since it is a major adductor of the hallux. By way of contrast, more terrestrially adapted primates have relatively smaller peroneal longus muscles since hallucal grasping is less important for them (Jolly, 1972). Thus all the morphological evidence of the first metatarsal, including the shape of its articular surfaces, points to an animal that was arboreal (Conroy, 1976*a*, 1976*b*).

The talus is slightly smaller than expected for a primate the size of the male *A. zeuxis* and presumably represents the female of this highly dimorphic species. The bone is similar to those described for later Miocene hominoids (for example, *Dendropithecus, Limnopithecus,* and *Proconsul,* discussed in Chapter 5) and shows many of the characteristics of arboreal quadrupeds.

Other Fayum Primates

Oligopithecus *Oligopithecus savagei* and *Qatrania wingi* are the only species that have been recovered from the lower layers of the Jebel Qatrani Formation. *Oligopithecus savagei* was named for its presumed Oligocene age and for Donald Savage of the University of California, Berkeley, who found the type, and thus far only, specimen. The available material consists of a left mandibular ramus with the canine through second molar preserved. The overall size of the lower jaw matches that of the South American squirrel monkey *(Saimiri)*.

The dentition is characterized by an unusual mosaic of primitive and derived characters. *Oligopithecus* has a sectorial lower third premolar. The large honing facet on the lower third premolar is indicative of a large upper canine in this genus.

The lower dental formula of ?.1.2.3 and the presence of a sectorial lower third premolar are both derived features in catarrhines, but the molar morphology and associated patterns of dental wear in *Oligopithecus* are very reminiscent of certain Eocene adapids (Gingerich, 1977*a*, 1977*b*). Some of these more primitive dental features include (1) the primitive separation of trigonid and talonid crushing basins, (2) the elevated and sharply defined occlusal crests and cusps, (3) a trigonid that is more elevated than the talonid, (4) a small but distinct paraconid, and (5) a long and obliquely directed **cristid obliqua** (an enamel ridge on the lower molars running obliquely from the hypoconid to the back of the trigonid) (Harrison, 1987) (Fig. 4.11).

Because of this particular mixture of lower and higher primate dental features, the taxonomic position of *Oligopithecus* has not been secure. It is clearly distinguishable from all later Fayum primates, however. At one time or another it has been considered a primitive monkey, a primitive hominoid, and an Oligocene prosimian (Simons, 1962; Szalay, 1970; Cachel, 1975; Kay, 1977). Szalay and Delson (1979) are no more precise than labeling it a primitive catarrhine, and this is probably the most prudent taxonomic assessment at the present time.

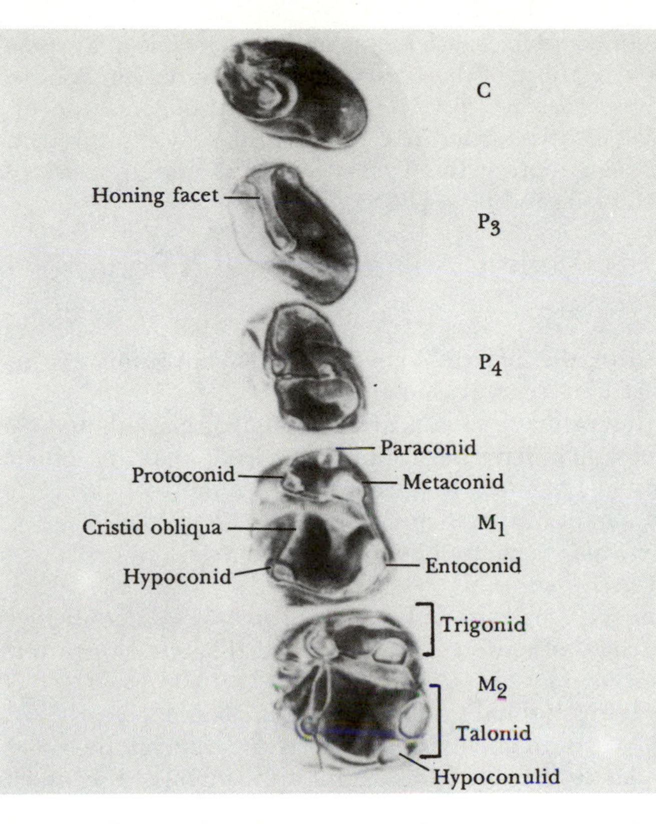

Fig. 4.11 Lower Jaw of *Oligopithecus* Note the honing facet on the lower third premolar and the cristid obliqua on the lower first and second molars. The diastema between the canine and third lower premolar is an artifact of preservation. Scale about 6 × natural size. (From Simons, 1962)

Afrotarsius The first, and so far only, fossil record of a tarsierlike primate in Africa is *Afrotarsius.* (Simons and Bown, 1985). Molar morphology of this diminutive species is most similar to that of the Eocene microchoerine *Pseudoloris;* however, among living primates the closest similarity is to *Tarsius* (tarsiers). Thus *Afrotarsius* is best placed in the family Tarsiidae. The type specimen is from quarry M and consists of a right mandibular fragment that includes the first through third lower molars and the lower parts of the crowns of the lower third and fourth premolars.

Although the precise geographical origins of *Afrotarsius* cannot be known for sure, some circumstantial evidence supports an Asian origin: (1) there are no known suitable morphological candidates for the ancestry of *Afrotarsius* in the relatively well-sampled fossil record of the Eocene Omomyidae of Europe and North America; (2) the dentition of *Afrotarsius* is most similar to that of living Southeast Asian tarsiers; and (3) at least some floral and faunal elements of the Oligocene in Egypt are more closely linked to those of the Eocene and present day in Southeast Asia than to those of the early Tertiary of Europe. If the ancestors of *Afrotarsius* did arrive in Africa from Asia, this migration most probably occurred during the regression of the Tethys Sea in the late Eocene and early Oligocene (Simons and Bown, 1985).

Body Size, Feeding Adaptations, and Behavior in the Fayum Primates

In summarizing the inferred habits of the Fayum primates, the following generalizations seem reasonable:

1. The Fayum primates were primarily frugivorous, although *Parapithecus grangeri* may have been more folivorous than the other species and *Qatrania wingi* may have been more gummivorous.
2. A predominantly arboreal milieu is indicated for most species. Again, *P. grangeri* may be an exception; its high molar crowns suggest some semiterrestrial habits.
3. The relatively small size of the orbits for *Apidium* and *Aegyptopithecus* suggests that these two genera had diurnal habits, similar to those of extant anthropoids.
4. The small infraorbital foramina of *Apidium* and *Aegyptopithecus* indicate a poorly developed tactile sensory apparatus in the snout, a feature similar to that found in extant anthropoids. (Kay and Simons, 1980)

Inferences can also be drawn about the mating behavior of at least some of the Fayum primates. For example, both propliopithecid genera, *Propliopithecus* and *Aegyptopithecus*, are sexually dimorphic in tooth size, mandibular size, and body weight. This allows some generalizations to be made about probable social behaviors in these early catarrhines. Among nonhuman higher primates, monogamous social groups consisting of a permanently mated male and female plus their offspring are found in several smaller species of New World monkeys (for example, marmosets, *Aotus, Callicebus*, and *Pithecea*), in all species of gibbons, and in a few island-dwelling colobines. Such species are characterized by very low levels of sexual dimorphism. By contrast, polygynous higher-primate species in which there is intense male–male competition for mating access to females invariably show considerable sexual dimorphism in canine and/or body size. The amount of canine dimorphism in *Aegyptopithecus* and *Propliopithecus* is greater than that found in any living monogamous higher primate suggesting a complex polygynous social structure for these earliest anthropoids (Kay et al., 1981; Sussman and Kinzey, 1984).

The body-size distribution of modern fruit-eating arboreal primates has three peaks that are probably explained by the limitations of fruit as a protein source (Kay and Simons, 1980) (Fig. 4.12). Although fruits

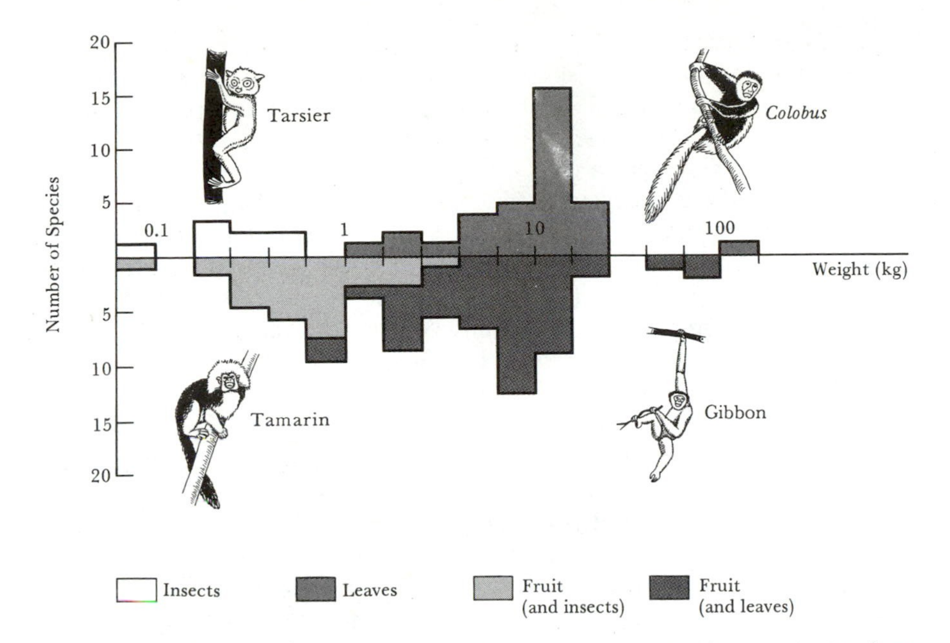

Fig. 4.12 Relationship between Body Size and Dietary Preference in Modern Primates Insectivores and folivores are above the line, frugivores below. Predicted weight (in kilograms) of Fayum primates: *Qatrania* = 0.3; *Apidium phiomense* = 1.6, *A. moustafai* = 0.85; *Parapithecus fraasi* = 1.7, *P. grangeri* = 3.0; *Propliopithecus haeckeli* = 4.0, *P. chirobates* = 4.2, *P. markgrafi* = 4.0, *P. ankeli* = 5.7; *Aegyptopithecus zeuxis* = 6.7; *Oligopithecus* = 1.5; *Afrotarsius* = 0.1 (Graph adapted from Kay and Simons, 1984)

eaten by most primates contain large amounts of readily available carbohydrates, they are deficient in protein (Hladik, 1977). Protein must be obtained by supplementing the diet with leaves (which contain about 20% protein by dry weight), insects, or both. Those frugivores that supplement their diet with insects, gums, or nectar tend to be smaller than those that supplement their diet with leaves.

Body weights of the Fayum primates ranged from about 0.3 kg in *Qatrania wingi* to about 6.0 kg in *Aegyptopithecus zeuxis* (Gingerich et al., 1982; Simons and Kay, 1983; Conroy, 1987). This is roughly comparable to the size range found in modern ceboids. All the Fayum primates except *Qatrania* and *Afrotarsius* were larger than any extant insectivorous primate and were primarily frugivorous. Since the teeth of *Afrotarsius* resemble those of extant tarsiers and the Eocene microchoerine *Pseudoloris,* it is reasonable to suggest an insectivorous diet for this primate (Simons and Bown, 1985). In *Qatrania,* however, molar morphology suggests a more frugivorous or gummivorous diet than a primarily insectivorous one. More specifically, molar crests are weakly developed and are distinctly different from the sharp, high crests of more insectivorous species. The larger species such as *Parapithecus grangeri, Propliopithecus,* and *Aegyptopithecus* probably ate leaves as a supplementary source of protein, whereas smaller species such as *Apidium moustafai* and *Parapithecus fraasi* may have resorted to insects for this purpose.

As discussed earlier, exant fruit-eating primates can be distinguished from leaf-eating ones on the basis of molar morphology. Leaf-eating hominoids have more prominent shearing crests on their molars than fruit-eating hominoids do. The shortened shearing blades in the Fayum anthropoids are suggestive of a mainly frugivorous diet. Only *Parapithecus grangeri* has well-developed molar shearing crests, which may indicate that it had more fiber in its diet (Kay, 1977).

It has also been noted that terrestrial species of extant Old World monkeys tend to have significantly higher molar crowns than do closely related arboreal species. This tendency undoubtedly reflects a dental adaptation for dealing with the larger amounts of grit in the diets of terrestrial species. Fayum primates, again with the exception of *P. grangeri,* have molar crown heights that reflect an arboreal milieu. Possibly *P. grangeri* foraged more often on the ground than did the other Fayum anthropoids (Kay, 1977).

In summary, the closest living ecological and morphological analogues of the Fayum primates are found among living New World monkeys. The two groups are similar in range of body size, diurnal habits, and predominantly frugivorous diets.

OLIGOCENE PRIMATES IN THE NEW WORLD

Primates first appear in the South American fossil record in the late Oligocene, some 25 to 30 MYA: *Branisella* from the late Oligocene of Bolivia, and *Dolichocebus* and *Tremacebus* from the late Oligocene (perhaps early Miocene) of Patagonia (southern Argentina) (Fig. 4.13). All have reached an anthropoid grade of evolution; no fossil prosimians have been uncovered in South America, whether during the Oligocene or later.

Unfortunately, the three Oligocene genera are known from only a handful of specimens from separate sites in Bolivia and Argentina. For this reason, few overall generalizations can yet be made about the morphology, diet, or behavior of these early anthropoids; therefore, each genus will be discussed individually. At the end of the chapter we will examine some of the hypotheses put forward to explain this sudden appearance of higher primates in the late Oligocene of South America.

An interesting issue we shall briefly explore in this section is the question of whether the callitrichids (marmosets and tamarins) represent primitive or specialized platyrrhine morphotypes. The few fossil mon-

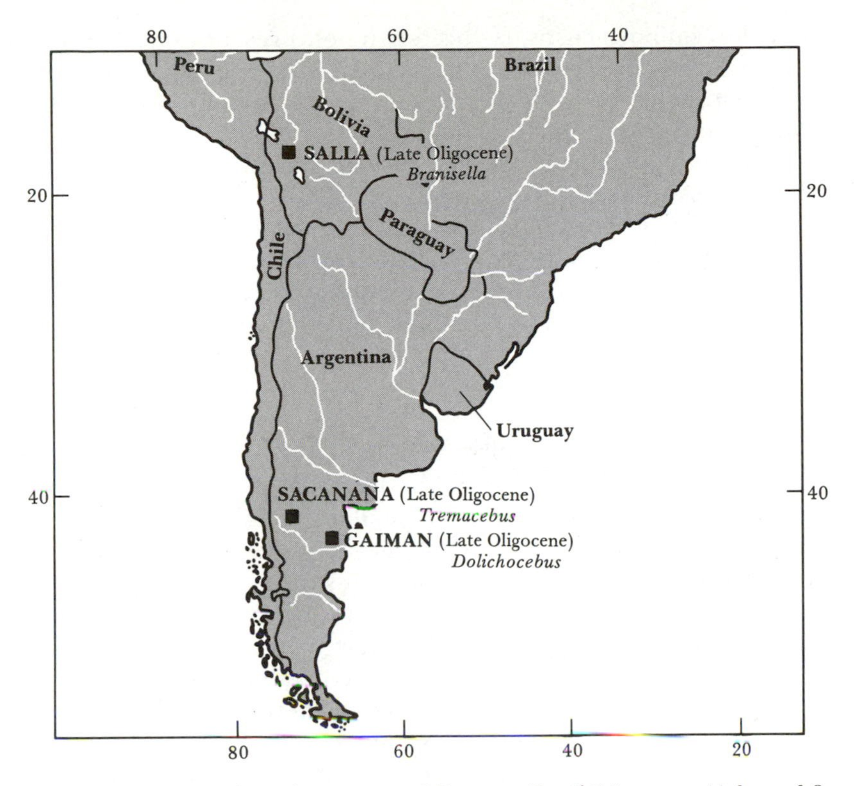

Fig. 4.13 Localities of South American Oligocene Fossil Primates (Adapted from Hershkovitz, 1974*b*)

keys from the Oligocene of South America provide our only clues as to what the earliest South American primates actually looked like.

Branisella

The geologically oldest platyrrhine fossil is *Branisella boliviana* from late Oligocene sediments in Bolivia; it is about 27 million years old (Hoffstetter, 1969; Rosenberger, 1981; MacFadden et al., 1985). Besides *Branisella*, this fauna has yielded remains of marsupials, edentates, rodents, various notoungulates, birds, turtles, and frogs.

The type specimen of *Branisella* is part of an upper jaw preserving the fourth premolar through the second molar, the roots of the second and third premolars, and part of an upper jaw indicating the presence of an upper third molar. A mandibular specimen preserving the lower second molar has also been described. This jaw fragment resembles parapithecids in having a comparatively shallow mandible and (probably) a fused mandibular symphysis (Rosenberger, 1981).

In its low bulbous cusps, reduced conules, presence of a hypocone, and general morphological configuration (including dental formula), *Branisella* most closely resembles extant forms like the squirrel monkey *(Saimiri)*. These dental traits also indicate a primarily frugivorous diet. Like the modern squirrel monkey, *Branisella* was most probably a short-faced animal. And like all modern cebids, *Branisella* has three premolars and molars. The upper molars are quadritubercular; this is somewhat unexpected in such an ancient platyrrhine since tritubercular molars are often thought to be the primitive molar configuration by those who view the callitrichid dentition as the primitive morphotype.

Tremacebus

Tremacebus is known from a fairly complete skull and a lower-jaw fragment (Hershkovitz, 1974*b;* Fleagle and Bown, 1983) (Fig. 4.14). Th overall skull size is similar to that of smaller living cebids like the titis *(Callicebus)*, the night monkey *(Aotus)*, and the squirrel monkey *(Saimiri)* and thus probably represents an animal that weighed about 1 kg (Fleagle and Rosenberger, 1983). The upper dental formula of 2.1.3.3 is similar to that of all other extant noncallitrichid New World monkeys. The cheek teeth are similar to those of *Callicebus* in size, proportions, and crown wear. The relative orbital size is intermediate between the unexpanded orbit of *Callicebus* and the greatly expanded orbit of *Aotus*, suggesting perhaps that *Tremacebus* was more crepuscular in its habits.

According to one reconstruction, *Tremacebus* had a combination of lower and higher primate features: it attained a higher-primate grade with respect to most dental and cranial characters while retaining some lower-primate features with respect to (1) its more laterally directed orbits, (2) its (inferred) posteriorly directed foramen magnum, and most importantly, (3) its (apparent) incomplete postorbital closure (Hershkovitz, 1974*b*). In this view, the complex of such characters as a callitrichidlike skull, complex four-cusped molars, and well-defined temporal lines all appear to suggest herbivorous adaptations, although the dentally similar *Callicebus* is predominantly frugivorous.

More recent evaluations of this important fossil have led to significantly different interpretations of the postorbital region (Rose and Fleagle, 1981; Fleagle and Rosenberger, 1983). These more recent studies suggest that the postorbital opening was the result of postmortem breakage and thus does not illustrate the true amount of postorbital closure present. Apparently, postorbital closure is as advanced in *Tremacebus* as it is in extant New World monkeys. It has been suggested that *Tremacebus* may have phylogenetic affinities to *Aotus* (Rosenberger, 1980).

The lower premolars and molars of *Tremacebus* are very primitive and do not show any clear-cut features linking them with any particular extant platyrrhine. They are most comparable to the Miocene platyrrhine *Homunculus* (Fleagle and Bown, 1983).

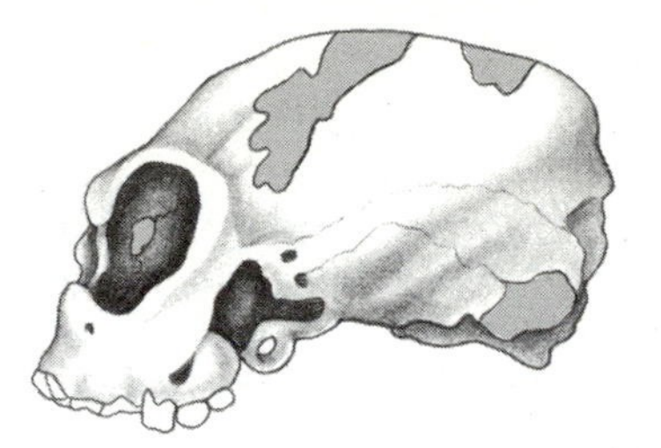

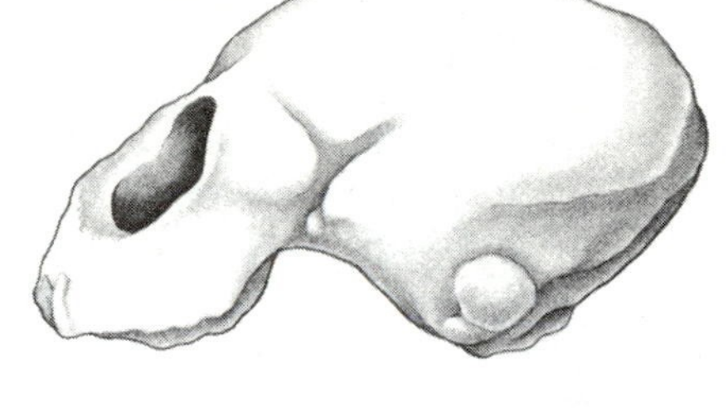

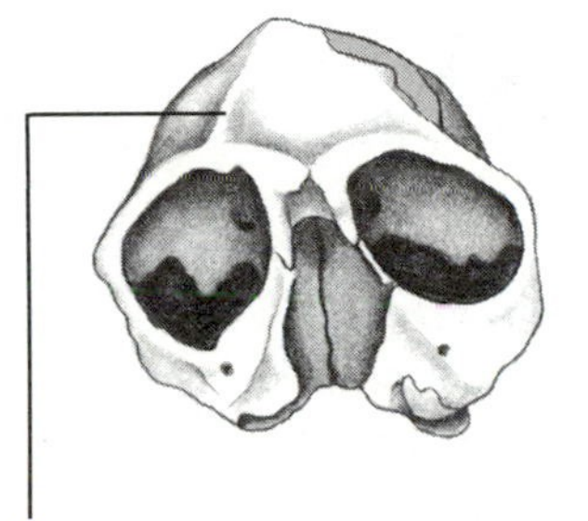

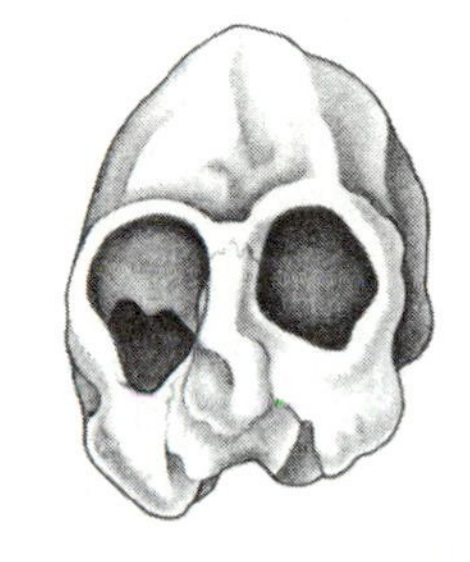

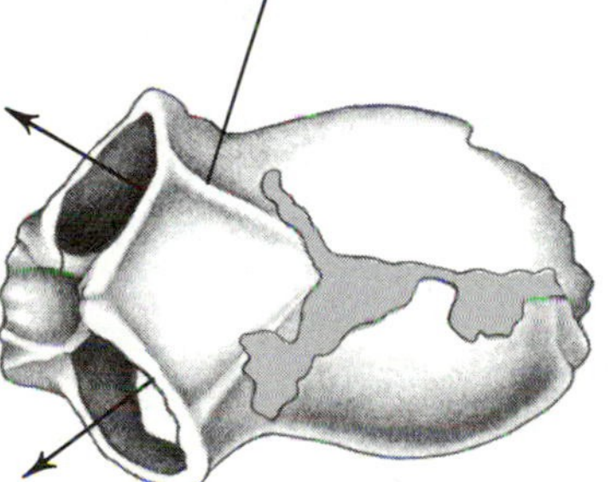

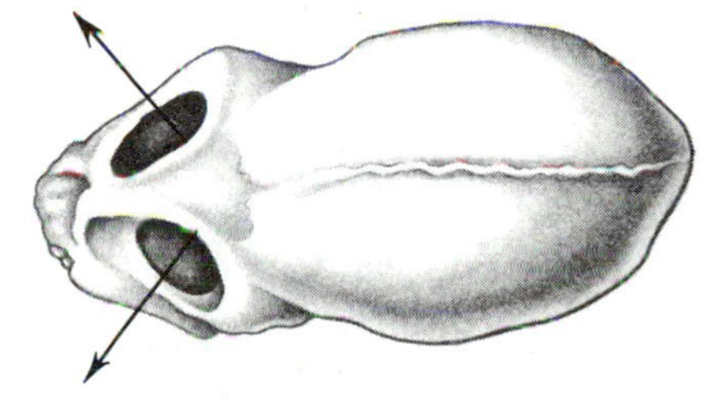

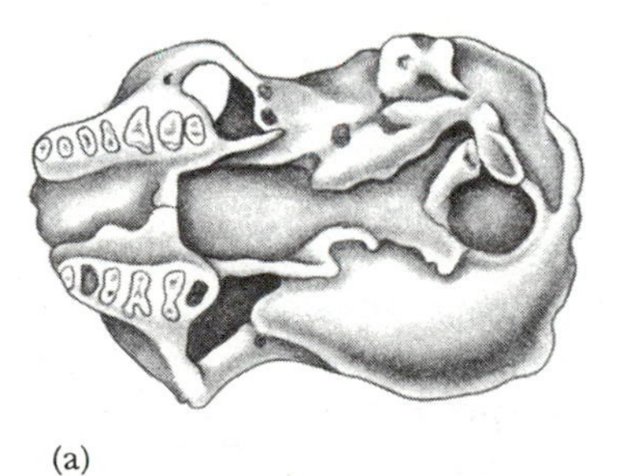

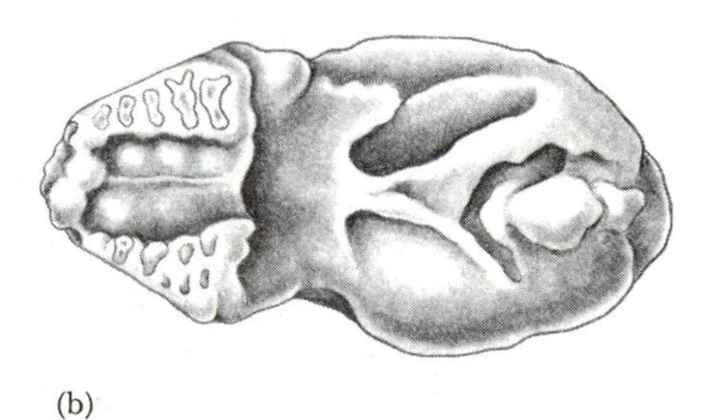

Fig. 4.14 Cranial Comparisons of Two Oligocene South American Primates (a) *Tremacebus,* and (b) *Dolichocebus.* Note the well-defined temporal lines in *Tremacebus;* the long, narrow skull shape in *Dolichocebus;* and the fact that in both fossils the orbits are more laterally directed (arrows show orientation) than in most anthropoids. (Redrawn from Hershkovitz, 1974*b*)

Dolichocebus

The third Oligocene genus, *Dolichocebus,* is known from a distorted skull and several isolated teeth from late Oligocene deposits in Patagonia; it is the oldest anthropoid cranium from the New World. The upper dental formula is presumably 2.1.3.3, but the last molar along with the crowns of all remaining teeth were broken away and lost during removal of the specimen from its stone matrix (Hershkovitz, 1970). Its overall skull size is similar to *Tremacebus* but is longer and narrower in shape. There is no evidence for enlarged orbits in this genus. Postorbital closure, configuration of the ectotympanic (fused to the lateral bulla wall), and sutural patterns at pterion are all typically platyrrhinelike (Rose and Fleagle, 1981) (Fig. 4.14).

The medial orbital walls are perforated by a large interorbital opening similar to that found in *Saimiri* and no other mammal. This has been taken to indicate possible phyletic affinities between this genus and *Saimiri,* thus signifying a distinct lineage for this group over the past 25 million years (Rosenberger, 1979).

The upper molars of this Oligocene primate are quadritubercular, lending further support to the view that the tritubercular upper molars of living marmosets and tamarins are secondarily derived, not primitive.

Having examined the morphological evidence of Old and New World Oligocene primates, we now consider their origins, phylogeny, and classification.

PHYLOGENY AND CLASSIFICATION OF OLIGOCENE PRIMATES

Origins of the Anthropoidea

Anthropoids (higher primates, which include monkeys, apes, and humans) appear suddenly in the late Eocene–early Oligocene sediments of Africa and possibly Southeast Asia without any obvious antecedents. In the New World primate fossils do not show up until the late Oligocene (about 25 MYA) (MacFadden et al., 1985; Fleagle et al., 1986). The scarcity of Oligocene primate fossils has led to much speculation and controversy about anthropoid origins.

Over the years several poorly known fossil candidates have been put forward as the antecedents of these earliest anthropoids, but there is no unanimity concerning the validity of such claims. None of the candidates deserves the sweeping claims made on its behalf. In fact, any dogmatic interpretation of the prosimian–anthropoid transition goes far beyond what the fossil evidence warrants at this stage (Conroy, 1978).

This somewhat pessimistic assessment stems from two basic concerns, both of which were discussed in Chapter 1. First, there is the doubt cast on the validity of using strictly paleontological criteria to determine taxonomic groupings, because the criteria do not always correlate well with the biology of living primates; nor are the criteria universally applied.

Second, there is the difficulty of ascertaining direct fossil lineages. One can only determine with any confidence whether a primate group has achieved a particular grade of evolution and can thus be considered ancestral in a general way to subsequent groups of the same grade. For example, one can speak of the Oligocene primates from the Fayum as having attained an anthropoid grade of evolution, which entitles them to be considered broadly ancestral to later undisputed anthropoids without necessarily implying that the Fayum primates are directly ancestral to them.

With these two points in mind, let us consider in some detail two poorly known genera, *Pondaungia* and *Amphipithecus*, both from the Eocene Pondaung Formation of Burma, and which have been considered early anthropoids by many workers. Extended correlations with radiometrically dated rocks in China and Mongolia indicate that this fauna existed some 40 to 44 MYA. Thus these primates may predate the earliest-known African anthropoids from the Fayum of Egypt by as much as 5 million years (Ciochon et al., 1985).

Pondaungia *Pondaungia* was discovered near Pangan, Burma, in the early 1900s. The type specimen, which was not described until more than a decade after it was found (Pilgrim, 1927), consists of a fragment of the upper jaw with first and second molars, a left lower-jaw fragment with second and third molars, and a right lower-jaw fragment with the third molar. A second specimen of this genus, a fragment of lower jaw containing the second and third molars, has recently been described (Maw et al., 1979). Although many workers believe that these fossils reflect an anthropoid grade of evolution, this view is not unanimous. It has been considered everything from a lemur (Madden, 1975) to a condylarth (Von Koenigswald, 1965).

The lemuroid similarities are found in the morphology of the upper molars. Like Eocene notharctines, their hypocones (actually pseudohypocones) evolved through cleavage of the protocone (see Fig. 3.10b). This would seem to be a clear-cut similarity between *Pondaungia* and Eocene lemuroids if it were not for the fact that some South American primates also exhibit a rather similar morphology. It was noted long ago that in ceboids (New World monkeys) the size of the hypocone and its connection to the trigone could be arranged in a size-graded series from a tritubercular to a quadritubercular tooth with no abrupt transitions

(Thomas, 1913). At one extreme (in callitrichids, or marmosets and tamarins) the hypocones are essentially absent, but they become progressively larger in the larger cebids. Ceboids like *Saimiri* and *Callimico* (Goeldi's marmoset) occupy the middle part of this range (Rosenberger, 1977). Some workers consider a morphology like that of the titis *(Callicebus)* to be primitive (Gregory, 1920), whereas others regard the *Saimiri* model as primitive and all other morphotypes as derived (Orlosky and Swindler, 1975; Rosenberger, 1977). The correct interpretation of this morphocline polarity is elusive, and the debate illustrates one of the main uncertainties underlying any analysis of evolutionary relationships, cladistic or otherwise. In any event, the taxonomic significance of the pseudohypocone has yet to be demonstrated in modern primate populations.

The lower molars are larger and **bunodont** (having low, rounded cusps). The trigonid is slightly higher than the talonid, and a small paraconid is present. Guy Pilgrim (1927) pointed out the overall similarity of the lower molars to those of the notharctine *Pelycodus* but noted that the crests on the trigonid and talonid characteristic of that genus were absent in *Pondaungia.* In spite of these general resemblances to Eocene notharctines, Pilgrim considered *Pondaungia* to be an anthropoid because of the low-crowned cusps and the large basin-shaped area of the lower third molar.

Amphipithecus The first specimen of *Amphipithecus* was discovered in 1923 by Barnum Brown of the American Museum of Natural History in upper Eocene sandstones near Mogaung, Burma. The specimen was subsequently overlooked until Edwin Colbert described it in 1937. The fossil consists of a left lower-jaw fragment with the fourth premolar and first molar preserved. It is distinguished by the great depth of the lower jaw and the short, vertical symphyseal region. The dental formula was inferred to be ?.1.3.3.

As with *Pondaungia,* opinions have differed widely as to the proper taxonomic placement of this genus. It has been considered a lemuroid by some (Von Koenigswald, 1965; Szalay, 1970, 1972*b*) and an anthropoid by others (Colbert, 1937; Simons, 1971, 1972). Gingerich (1980*b*) has argued that *Amphipithecus* and *Pondaungia* may be transitional primates linking higher primates to an adapid origin; Szalay and Delson (1979) have also concluded that *Amphipithecus* "does not seem far removed from a notharctine, or some other, unknown, primitive adapid."

In comparing *Amphipithecus* with various anthropoids, Colbert (1937) emphasized certain similarities with New World primates, particularly the presence of three premolars, the deep mandibular ramus, and the abbreviated symphysis. He remarked specifically on the strong morphological resemblance between *Amphipithecus* and *Alouatta.* Ultimately, however, he dismissed the significance of the similarities with New World monkeys, concluding that the "resemblance is not close enough to indicate any true affinity."

Colbert's primary contention was that Old World monkeys have very large second premolars, whereas in *Amphipithecus* the second premolar is obviously small. He therefore considered the small premolar in *Amphipithecus* to be a retention of a lemuroid or tarsioid characteristic. However, one should note that some New World monkeys (for example, the woolly spider monkey, *Brachyteles*) do not have enlarged second premolars. The size of this tooth in New World monkeys is functionally related to the size of the upper canine blade against which it occludes (Zingeser, 1973).

To account for the dental similarity between *Amphipithecus* and certain New World monkeys, Colbert (1937) suggested that "it may be rather a parallelism in the development of these teeth." Resorting to parallel evolution to explain morphological similarity should be done judiciously, however; otherwise, the paleontological tenet that genetic similarity underlies morphological similarity becomes practically inoperable. Parallel evolution should be invoked only when one has evidence from sources beyond the realm of gross morphology alone that make it the only viable explanation.

Colbert (1937) provisionally placed *Amphipithecus* within the Simiidae (his name for Anthropoidea) for the following reasons:

1. It has a deep lower jaw.
2. It has a well-developed inferior transverse torus (simian shelf).
3. It has a deep genioglossal pit.
4. It has an abbreviated, fused symphysis with crowding of the canine and premolar teeth (Fig. 4.15).

A recently discovered specimen confirms these features and also shows that the lower first molar retains a small paraconid but has no hypoconulid (Ciochon et al., 1985).

Amphipithecus exhibits an interesting mixture of prosimian and anthropoid traits, and for this reason some workers have questioned its anthropoid status. Subtle differences between *Amphipithecus* and the Fayum anthropoids in lower first molar morphology have been cited in support of the view that this genus is an adapid (Szalay and Delson, 1979; Gingerich, 1980*b*). For example, the presence of a small paraconid on the lower first molar and the absence of molar hypoconulids are sometimes touted as adapid features. However, paraconids are sometimes found in Fayum anthropoids and in living New World monkeys, and the absence of hypoconulids typifies some of the latter anthropoids as well.

Some workers have questioned the anthropoid status of *Pondaungia* (and of *Amphipithecus*) on paleozoogeographic grounds, arguing that these fossils were part of an endemic Eurasian fauna that could not possibly have reached North Africa by the early Oligocene (Von Koenigswald, 1965; Madden, 1975). Suffice it to say, however, that some other endemic Eurasian mammals did reach Africa by the early Oligocene, notably creodonts and anthracotheres (Simons and Wood, 1968; Fleagle et al., 1986).

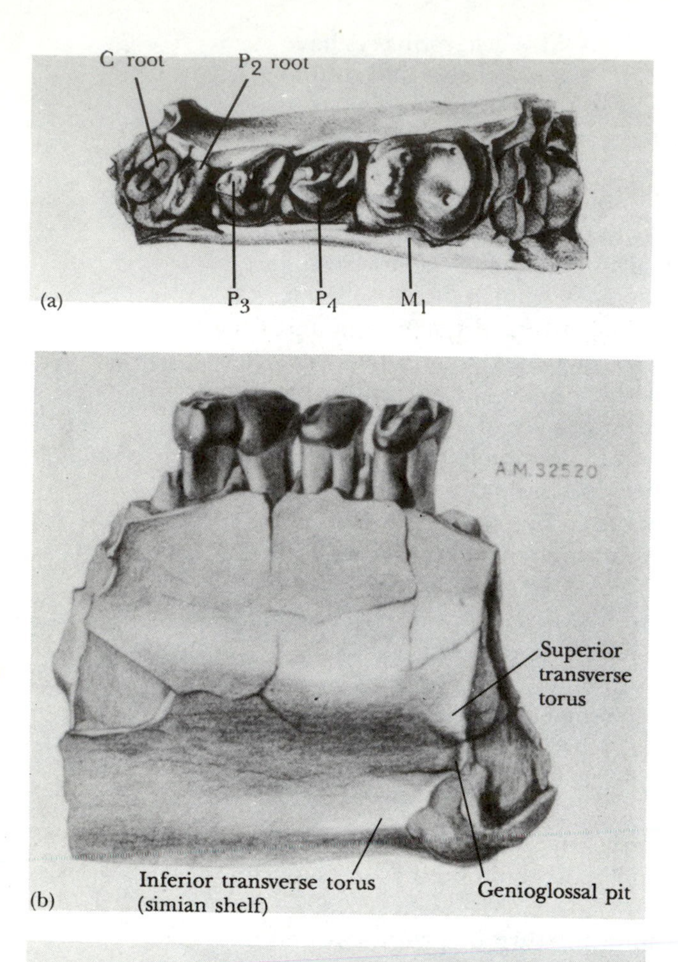

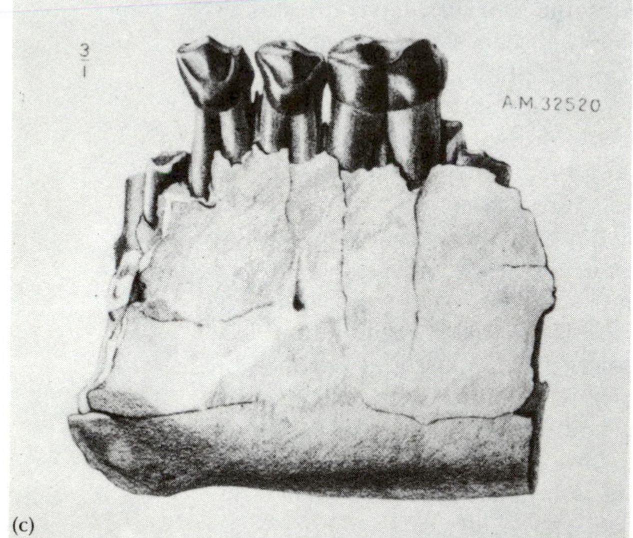

Fig. 4.15 Type Specimen of *Amphipithecus mogaungensis* Preserved is a left mandible with roots of the canine and second premolar, the third and fourth premolars, and the first molar. (a) Occlusal, (b) medial, and (c) lateral views. The specimen is about 2.4 × natural size. Some of its alleged anthropoid features include a deep jaw, a well-developed inferior transverse torus (simian shelf), a deep genioglossal pit between the superior and inferior transverse tori, a shortened and fused mandibular symphysis, and crowding of canine and premolar teeth. (From Colbert, 1937)

Thus the two Burmese genera present a mosaic of lower- and higher-primate characters with the latter predominant, indicating that by later Eocene times in southern Asia primates may have entered an anthropoid adaptive grade. This finding raises the interesting possibility that the earliest anthropoids could have originated in southern Asia rather than in Africa (Ciochon et al., 1985).

Phylogeny of the Fayum Primates

The Fayum parapithecids and proliopithecids are a diverse lot, yet all have attained an anthropoid grade of evolution. The presence of *Afrotarsius* indicates that prosimian primates inhabited this region as well. The cladistic relationship of these primates to other fossil and modern anthropoids is still problematic, however. The view adopted here is diagrammed in Fig. 4.16d.

Parapithecid Affinities Where do the parapithecids fit in? The phylogenetic relationship of these primates to recent higher primates has been the subject of much speculation over the years. They have been variously considered as

1. A **paraphyletic group** (one that does not include all of the descendants of the common ancestor of that group), with *Parapithecus* related to the cercopithecoids and *Apidium* to the Miocene hominoid *Oreopithecus* (Simons, 1960)
2. Early representatives of Old World monkeys (Simons, 1972)
3. A group ancestral to the playtrrhines (Hoffstetter, 1980)
4. The sister group of all other catarrhines (Szalay and Delson, 1979)
5. The sister group of all other anthropoids (Fleagle and Kay, 1987; Harrison, 1987)

The view that the parapithecids have some special relationship to Old World monkeys has very few adherents today. Simons (1972) originally proposed this hypothesis on the basis that the parapithecids are short-faced anthropoids with molar morphology similar to the living talapoin monkey *(Cercopithecus talapoin)*. However, if the parapithecids are the sister group of the Old World monkeys, then many of the specialized characters shared by Old World monkeys and apes, such as the loss of the second premolar and the presence of a tubular ectotympanic, must have evolved in parallel since they are not present in the parapithecids. In addition, if the recently suggested lower dental formulas for *P. fraasi* (1.1.3.3) and *P. grangeri* (0.1.3.3) are valid, then *Parapithecus* cannot be ancestral to later cercopithecoids: once teeth have been lost in an ancestral form, they can never be regained in a descendant. Thus *Parapithecus* could not be ancestral to the Old World monkeys and apes, which have a dental formula of 2.1.2.3.

Other workers minimized resemblances between parapithecids and cercopithecoids by stressing that most of these features are just primitive

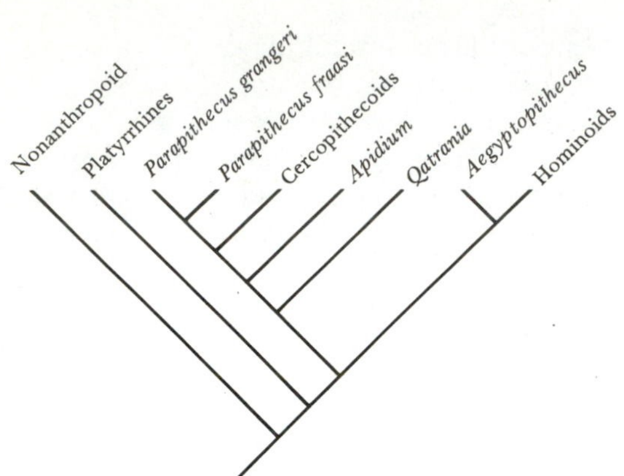

(a) Simons (e.g., 1972, 1986; Gebo and Simons 1987)

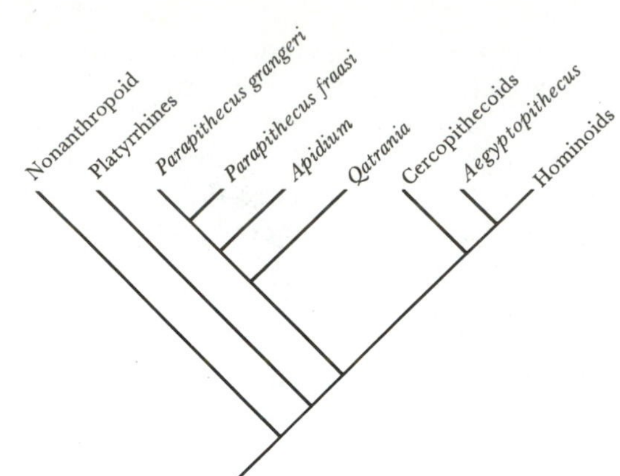

(b) Szalay and Delson (1979)

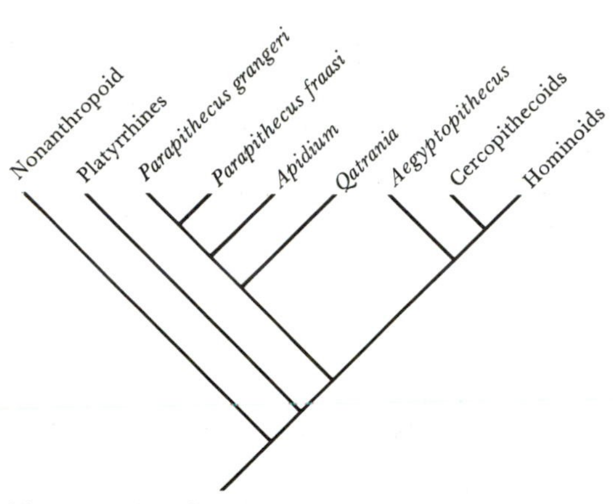

(c) Andrews (1985)

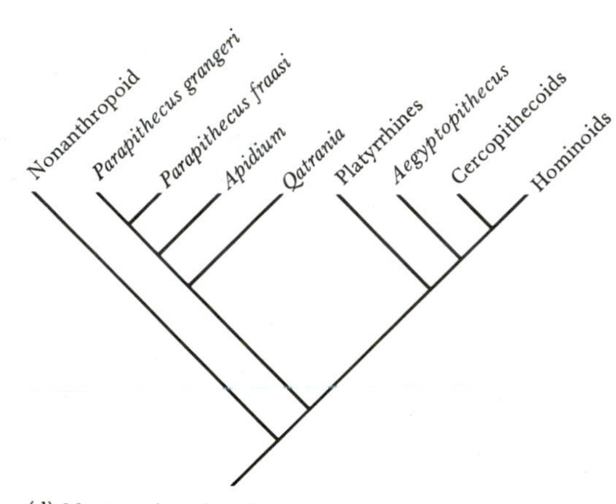

(d) Most parsimonious hypothesis

Fig. 4.16 Alternate Views of Fayum Anthropoids Note that cladogram (a) places parapithecids as the sister group of Old World monkeys (cercopithecoids); cladograms (b) and (c) place parapithecids within the catarrhines (Old World monkeys and apes), whereas cladogram (d) places parapithecids before the platyrrhine–catarrhine split. Propliopithecids *(Aegyptopithecus)* are viewed as either the sister group of apes and humans (hominoids) (a and b) or as the sister group of monkeys and apes (catarrhines) (c and d). (From Fleagle and Kay, 1987)

catarrhine characters. They noted that *C. talapoin* is a highly derived modern species and thus not a proper model for the ancestral cercopithecoid morphotype (Delson, 1975*a,* 1975*b,* 1977; Delson and Andrews, 1975). Accordingly, parapithecids were considered cercopithecoid **vicars** (filling the same ecological niche as monkeys) but not necessarily their sister group. According to Delson, one derived feature in *Parapithecus* that excludes it from the ancestry of later cercopithecoids is the sectorial lower second premolar. Delson believes that such a specialization for honing the upper canine is unlikely to have been lost and redeveloped again on the third premolar as it was in later cercopithecoids. Gingerich (1975*d*) has nicely demonstrated, however, how this transfer of honing function from one premolar to the next may have taken place through time. In a specimen of the adapid *Leptadapis magnus,* the upper canine hones against both the small first premolar and the larger second premolar. Thus, as the first premolar atrophied during adapid evolution, the honing between the upper canine and second premolar could have continued without interruption. A similar morphology may be seen in the fossil adapid, *Mahgarita stevensi* (Wilson and Szalay, 1976).

Based on his deductions of what the ancestral cercopithecoid dental morphotype should be, Delson (1975*a,* 1975*b*) considered three features found in *Parapithecus* to rule out its inclusion in the Cercopithecoidea: (1) small fourth premolar metaconids; (2) short third molars with little or no hypoconulid development; and (3) the honing mechanism mentioned above. In addition, the configuration of the ectotympanic bone found in all living Old World monkeys suggests that they are descended from a common ancestor already possessing a tubular ectotympanic, not a ringlike one as is found in parapithecids. Thus, considerations of craniodental anatomy strengthen the view that the parapithecids are not the sister group of cercopithecoid monkeys but are best regarded as the sister group of all the Anthropoidea.

The parapithecids have clearly not reached the catarrhine stage since they retain three premolars, lack a tubular ectotympanic, have an **entepicondylar foramen** on the distal humerus, and lack expanded ischial tuberosities. However, parapithecids certainly share derived features with the Anthropoidea that indicate a close phyletic relationship:

1. Early fusion of the metopic suture
2. Extensive postorbital closure
3. An enlarged promontory artery and a reduced stapedial artery
4. A fused mandibular symphysis
5. Late eruption of the lower third molar after all of the deciduous premolars have been replaced
6. A large dorsal **epitrochlear fossa** on the posterior surface of the medial epicondyle of the humerus for a strong ligament uniting the humerus and the ulna

The fact that parapithecids retain a few nonanthropoid features, such as a single-cusped lower fourth premolar and large **coronoid** and **angular processes** of the mandible (for insertion of chewing muscles), suggests that they are probably best considered as the sister group of all the Anthropoidea (Harrison, 1987) (Fig. 4.16d).

The situation with *Oligopithecus* is a little more uncertain since some dental features are very primitive; these include elevated and sharply defined occlusal crests, an elevated trigonid, a distinct paraconid, and a long, oblique cristid obliqua. Other traits are much more advanced, such as a dental formula of ?.1.2.3 and a sectorial third premolar.

Propliopithecid Affinities Propliopithecids have clearly reached an anthropoid grade of evolution. Many of their morphological features are listed in Table 4.3. One might argue that they should be considered catarrhines since they share such catarrhine features as a reduced post-glenoid foramen, loss of the second premolars, sectorial third premolars, the same molar structure, and a catarrhine configuration of the bones in the region of pterion (see Fig. 4.10). However, propliopithecids have not yet evolved the catarrhinelike tubular ectotympanic, and the morphology of the humerus is more primitive than that of the hypothetical last common ancestor of extant cercopithecoids and hominoids; for these reasons it is best to consider the Propliopithecidae the sister group of the Catarrhini.

We can summarize recent views on the phyletic position of parapithecids and propliopithecids as follow. Parapithecids (*Apidium* and

Table 4.3 Ancestral Anthropoid and Catarrhine Craniodental Features

Ancestral Anthropoid Condition
1. Face is small in relation to the size of the **neurocranium** (portion of the skull enclosing the brain).
2. Premaxilla is relatively small.
3. Postorbital closure: the orbits are completely separated by a bony septum from the **temporal fossa** (space behind the orbits that is enclosed by the skull wall and zygomatic arch and filled by the temporalis muscle).
4. Ethmoid makes up part of the medial orbital wall.
5. Lacrimal bone is contained within the orbital margin.
6. Stapedial artery is small or absent; promontory artery is enlarged.
7. Ectotympanic is ringlike and attached to the lateral bulla wall.
8. Metopic suture of the frontal bone fuses early in life.
9. Mandibular symphysis fuses early in life.
10. Dental formula is 2.1.3.3 with loss of P1.
11. Upper and lower canines are sexually dimorphic.
12. Paraconid is vestigial.
Ancestral Catarrhine Condition Includes Features Cited above Except That:
1. Ectotympanic is tubular, not ringlike.
2. Mandibular symphysis has a superior transverse torus larger than the inferior transverse torus (simian shelf).
3. Dental formula is 2.1.2.3 with loss of P2.
4. Hypocone is large and well developed.
5. P_3 is single-cusped, moderately high-crowned, long and narrow, with a honing facet for contact with the upper canine.
6. Paraconid is absent.

Parapithecus) are most similar to small platyrrhines among extant anthropoids, but they also retain a number of primitive morphological features found today only in prosimians. Since many of these primitive features are lost in later anthropoids (both platyrrhines and catarrhines), it appears likely that parapithecids are more primitive than any extant anthropoid and thus predate the playtyrrhine–catarrhine split (Fig. 4.16d). This assessment is of course very different from the earlier view (Simons, 1972) that parapithecids are related specifically to Old World monkeys (Fig. 4.16a).

On the other hand, propliopithecids *(Aegyptopithecus* and *Propliopithecus)* share many dental similarities to later catarrhines, particularly *Pliopithecus* from the Miocene of Europe and *Proconsul, Limnopithecus,* and *Dendropithecus* from the Miocene of East Africa. In terms of postcranial anatomy, propliopithecids share many primitive features with *Pliopithecus;* however, their cranial and skeletal anatomy is more primitive than that of any later catarrhine, and they share no unequivocal features with any living catarrhine. Thus they are suitable phyletic ancestors for all later catarrhines (Fleagle and Kay, 1983) (Fig. 4.16d).

Origin of New World Primates

New World anthropoids show up suddenly in the late Oligocene, with no antecedents in South America. Because that continent was relatively isolated by water, the issue of geographical origin figures prominently in hypotheses for the phyletic origin of New World anthropoids.

The numerous anatomical similarities between the Oligocene anthropoids of the Fayum and extant New World monkeys have prompted several workers to suggest that the New World monkeys originated in Africa and rafted across the Atlantic sometime in the early Tertiary (Lavocat, 1974; Hoffstetter, 1974, 1980) (see Fig. 4.2). The competing scenario is that higher primates evolved in North America and Eurasia and subsequently migrated to warmer climates in South America and Africa respectively in response to the cooling trends at higher latitudes in the early Oligocene (see Fig. 4.1). If the common ancestor of both New and Old World anthropoids had itself reached an anthropoid grade of evolution, then the Anthropoidea would be a monophyletic taxon. However, if the New and Old World higher primates evolved independently from some North American and/or Eurasian lower primate group like adapids or omomyids, then the Anthropoidea would be a **diphyletic taxon** (arising from two separate groups).

Two arguments have been used to bolster the rafting hypothesis from Africa to South America: first, Africa is the only continent known to have fossil anthropoids at the time these animals reached South America (that is, in the Oligocene). Second, many of the Fayum primates share morphological features with the Platyrrhini, for example, three premolars (parapithecids), a nontubelike ectotympanic (parapithecids and propliopithecids), and detailed postcranial similarities. It is possible, however, that these similarities are primitive rather than derived traits.

Could primates have rafted across the Atlantic when it was anywhere from 500 to 3,000 km wide by the end of the Eocene? The odds against it are great. For example, the presence of the Recent fossil monkey *Xenothrix* on Jamaica, some 600 km from South America, is the longest inferred rafting journey known for a primate. Moreover, neither rodents nor primates have ever been rafted great distances to isolated oceanic islands (Simons, 1976).

The presence of caviomorph rodents in the Oligocene of South America has been used as evidence both for and against the African migration hypothesis. Some have suggested that the ancestors of the caviomorph rodents are to be found among the Oligocene Phiomorpha of Africa (an extinct group of rodents similar to modern African cane rats) (Hoffstetter, 1974; Lavocat, 1974, 1980). Others have strenuously denied this possibility (Wood, 1980). Support cited for the former thesis was the apparent lack of ancestral caviomorphlike rodents from the Eocene of North or South America. However, recent discoveries of phiomorphs have been made in Eocene sites of North America (Wood, 1972, 1973, 1980; Cachel, 1981). Thus the geographical origins of platyrrhine monkeys based on caviomorph origins remain unresolved.

Malcolm McKenna (1980) of the American Museum of Natural History prefers a model in which primates arrived in northern South America in the early Cenozoic via the Caribbean; exposed oceanic ridges or volcanic arcs may have increased the chances of migration along this route: "The Caribbean has long been an area of complex motions, including various examples of plate convergence, and probably never presented the kind of open oceanic barrier created by the Atlantic, whose primary tectonic mode was separation and subsidence." He argues that *Branisella* is very similar in molar morphology to the omomyid *Rooneyia* from the Oligocene of Texas and suggests that New World monkeys may be derived from some omomyid group.

Taxonomy of New World Oligocene Anthropoids

The fossil evidence from the New World Oligocene is frustratingly meager, and the taxonomic relationships of these few fossils—to each other, to Old World primates, and to later New World primates—are inconclusive. Nevertheless, a few comments can be made about their taxonomy.

The fact that the upper molars of the early South American primate, *Branisella,* are quadritubercular lends further support to the view that the tritubercular upper molars of living marmosets are secondarily derived, not primitive for that group. One opposing view is that of Hershkovitz (1974), who argues that the tritubercular molars of marmosets are the primitive condition and that the presence of quadritubercular molars in *Branisella* effectively excludes it from the Platyrrhini.

The fact that the upper molars of *Branisella* are very similar in overall morphology to those of *Aegyptopithecus zeuxis* from the Oligocene of Egypt provides some support for the argument that Africa, not North America, was the geographical source of the platyrrhine ancestor (Fleagle and Bown, 1983). In spite of the fact that most workers would consider *Branisella* to be an early platyrrhine (New World anthropoid), its precise relationship to later platyrrhines is totally obscure.

The taxonomic relationships of the other two Oligocene New World primates, *Tremacebus* and *Dolichocebus,* are also uncertain. The molar morphology of *Tremacebus* is most similar to that of the titi *(Callicebus)* or night monkey *(Aotus),* and Rosenberger (1984) has actually suggested that it is the ancestor of the living night monkey. Rosenberger (1979) has also argued that *Dolichocebus* is the ancestor of the living squirrel monkey *(Saimiri)* because both have an interorbital foramen linking right and left orbits, a derived feature. Hershkovitz (1982) has countered this argument by suggesting that the supposed foramen is an artifact of preservation.

Despite the paucity of our knowledge of the earliest New World primates, they are important data points in the study of primate evolution for two major reasons: First, they provide the only evidence for possibly linking Oligocene higher primates from Africa to the New World. Second, they provide corroborating evidence for the view that the small-bodied, claw-bearing callitrichids are secondarily derived platyrrhines adapted for eating exudate and for vertical clinging and are not primitive platyrrhine prototypes (Sussman and Kinzey, 1984; Ford, 1986*a,* 1986*b;* Fleagle et al., 1987).

At this point it is important to reemphasize the two most significant events of primate evolution in the Oligocene: (1) an anthropoid grade of evolution was reached in the Old World which spread soon thereafter to the New World; and (2) fossil primates virtually disappear from the Northern Hemisphere. By the end of the Oligocene and into the early Miocene we are on the brink of another major radiation of higher primates, that of the Hominoidea. This will be the subject of Chapter 5.

CHAPTER 5
Miocene Primates

Paleoclimates and Biogeography
Summary of the Miocene Fossil Record
Rise of the Hominoidea
Miocene Monkeys
Dietary Preferences, Behavior, and Habitats of Miocene Primates
Phylogeny and Classification of Miocene Primates

The Miocene epoch lasted for nearly 20 million years, from about 24 to 5 MYA (Berggren and Van Couvering, 1974; LaBrecque et al., 1977). In the early Miocene fossil record (approximately 24 to 16 MYA) we witness the first appearance of primitive Old World apes and monkeys in East Africa. Moreover, with the exception of a few monkey specimens in South America, higher primates are absent from the fossil record of all other continents during this period. The primitive apes are very diverse and abundant in these early Miocene African faunas, whereas the monkeys are quite rare. These proportions change dramatically, however, beginning in the middle Miocene (approximately 16 to 10 MYA). Monkeys become more abundantly represented in fossil faunas and apes less so, beginning a trend that continues to the present day. A second major event in the middle Miocene was the migration of African primates into Eurasia. We will discuss the ecological and geographical factors involved in these trends and migration patterns.

By the late Miocene (approximately 10 to 5 MYA) primitive apes are almost nonexistent in the fossil record of Africa, although they survived in some abundance in southern Asia and China until about 7 MYA. Monkeys clearly dominate the primate faunas of Europe. By the end of the Miocene a very different apelike creature makes its first appearance in the fossil record—the hominids. Their story will be told in Chapter 6.

PALEOCLIMATES AND BIOGEOGRAPHY

It is difficult to generalize about world climates in the Miocene other than to say that climates overall were probably warmer than in the preceding Oligocene. By the end of the Miocene the general trend of continental climate was toward drier and cooler conditions. The transformation of the circum-Mediterranean region from subtropical environments to progressively cooler and more seasonal climates is documented in the Tertiary climatic and floral records. Seasonal **sclerophyllous vegetation** (leaves with thickened cell walls resistant to water loss), adapted to summer droughts, progressively evolved and spread geographically, so that by the late Miocene it had replaced forests previously adapted to moister, more tropical conditions (Axelrod, 1975; Axelrod and Raven, 1978).

Continental configuration, with several important differences, were roughly comparable to those of today. In the early Miocene, sea levels were probably higher than they are today. Much of Europe was still inundated by small seaways and inlets, as were parts of eastern Asia. The western and eastern coastal belts of southern North America were largely covered by seas. Water barriers restricted mammals from migrating between North America and Asia via the Bering Strait and between North America and South America via the isthmus of Panama. By the end of the Miocene many of the seas were receding. Perhaps the most dramatic illustration of this process was the drying up of the Mediterranean Sea into a land-locked basin about 6 MYA.

It is difficult to generalize about Miocene climates in Asia because the data indicate a constantly changing environment from area to area, in different latitudes, and at different elevations. However, one can say that plate tectonics played a major role in determining Asian paleoclimates. The floating island continent of India, which had collided with the Asian landmass prior to the Miocene, continued to slide underneath Asia resulting in the formation of the Tibetan Plateau and the Himalayan Mountains. Data from the major Miocene primate-bearing locality in China, the site of Lufeng in Yunnan Province, indicate that Miocene climates were greatly affected by the uplift of the Himalayas and the Tibetan Plateau. Early Miocene deposits suggest drier and cooler climates that gradually warmed up towards the end of the epoch. By the mid-Miocene the climate was warmer and wetter; lake and swamp deposits were common. The latter indicate hot, wet climates typical of the southern Asian tropical zone. A more tropical monsoon climate prevailed towards the late Miocene (Wanyong et al., 1986).

By way of contrast, the paleobotanical evidence from northern China suggests a somewhat different scenario. The early Miocene paleoflora in northern China is indicative of warm and moist conditions. However, western China was more arid. This arid climate ultimately spread over much of northern China during the Miocene (Gengwu, 1987).

Since so many of the Miocene higher primates to be reviewed in this chapter come from sites in East Africa, the climate and ecology of this particular region will be discussed in more detail.

East African Rift System

Since all the African Miocene hominoids and many of the African Miocene monkeys come from the equatorial region of East Africa, the following comments on geology and paleoecology are necessarily restricted to that region of the continent. Today East Africa is ecologically isolated from Eurasia even though it has been connected to the Arabian Peninsula since the Miocene (Thomas, 1985). The landscape is broken by the great East African Rift System stretching 3,000 km from the Red Sea in the north to the southern end of Lake Malawi (bordering Mozambique, Tanzania, and Malawi) in the south (Fig. 5.1).

The first geological observations of the East African Rift System were made by the naturalist and explorer Joseph Thomson during his journey through Masai Land in 1883. He penetrated inland as far as Lake Victoria and Lake Baringo. The earliest geological reports concerning the region of Lake Turkana (formerly Lake Rudolf) were in the narratives of Count Samuel Teleki's exploration and hunting trip to northern Kenya in 1887–1888. It was the publication of J. W. Gregory's work eight years later, however, that marked the beginning of real geological understanding of the rift valley (Gregory, 1896). He was the first to realize that the rift was a true **graben** (fault trough) and observed that much of the faulting was of relatively recent origin. In fact, it was Gregory who introduced the term **rift valley (rift system),** which he defined as "a long strip of country let down between normal faults—or a parallel series of step faults—as if a fractured arch had been pulled apart by tension so that the keystone dropped in *en bloc* or in strips" (Gregory, 1896). Rifts, are a response not only to tension but also to regional uplift, which tends to stretch the surface rocks causing gaps to develop (Holmes, 1965).

The East African Rift System has some of the characteristics of a midocean ridge, and it is laterally continuous with the Red Sea and Gulf of Aden spreading centers. However, the rate of crustal extension (about 0.5 mm per year), is much less than that at midocean ridges (1–10 cm per year). All three spreading centers (Kenya, Red Sea, and Gulf of Aden) began to form in pre-Miocene times. In the early Miocene the Kenyan Rift passed through a period of broad crustal flexing and local faulting accompanied by volcanism: large domes of great volcanoes rose to 1,000 m or more above the surrounding countryside. This stage was succeeded in the late Miocene by voluminous floodlike eruptions. Most of the Miocene hominoid fossil sites are associated with these ancient volcanoes, the remnants of which now stand in various stages of erosion along the Kenya–Uganda border. During subsequent epochs, the East African landscape changed fairly rapidly toward its present configuration, so that now equatorial Africa is dominated by the low-lying Congo Basin in the west and by the rift system in the east.

Paleoecological deductions have been made by comparing the pattern of community structure in East African fossil faunas to that of various modern African ecological communities (Andrews et al., 1979; Evans et al., 1981; Van Couvering and Van Couvering, 1976). Several modern ecological community types have been distinguished: lowland forest, montane forest, woodland, and bushland (Fig. 5.2).

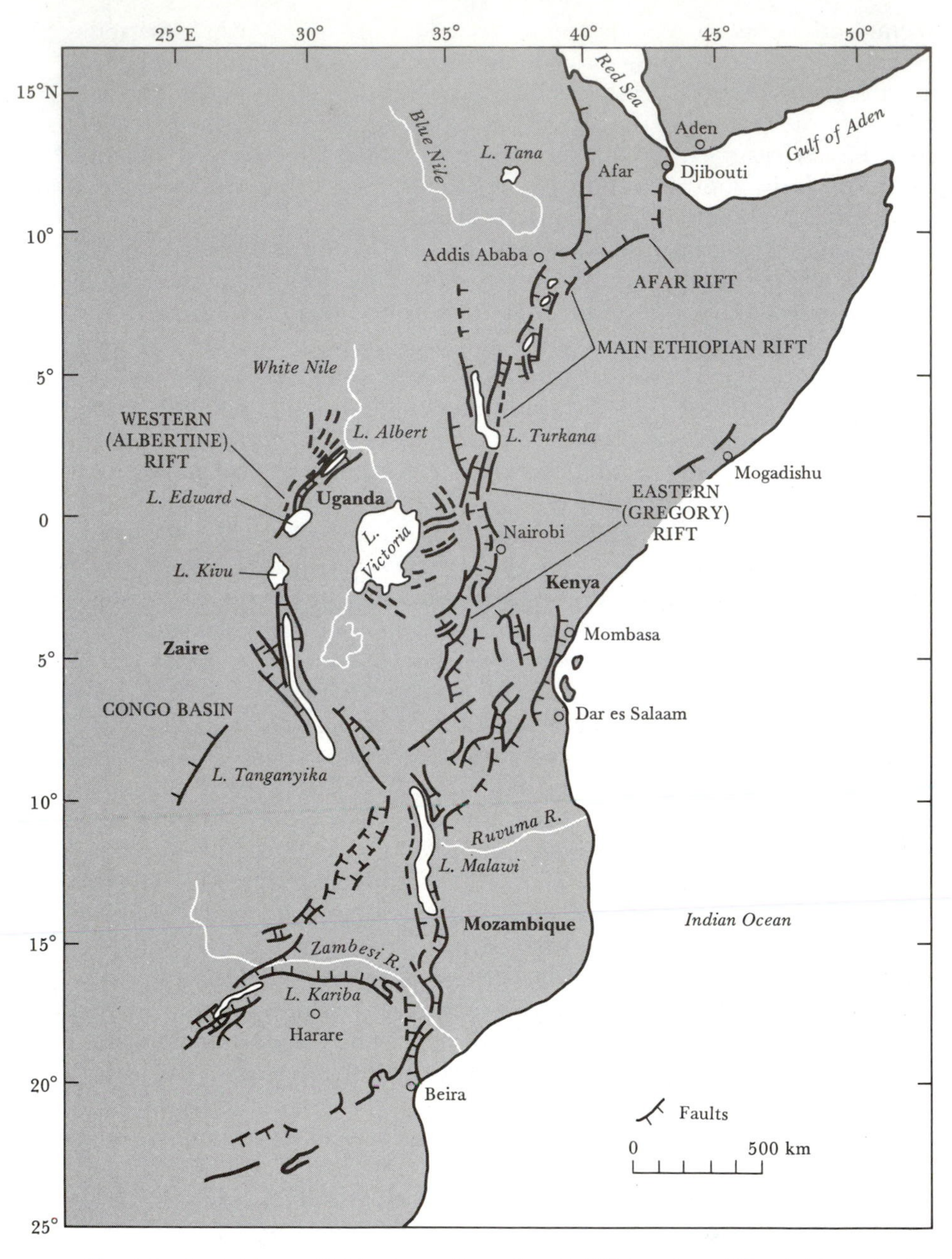

Fig. 5.1 East African Rift System Stretches 3,000 km from Mozambique in the South to the Red Sea in the North. Most of the major Miocene primate localities are in the interrift zone in Kenya and Uganda between the Eastern and Western rifts. Highlands from the shoulders of both major rifts. (Adapted from Bishop, 1978)

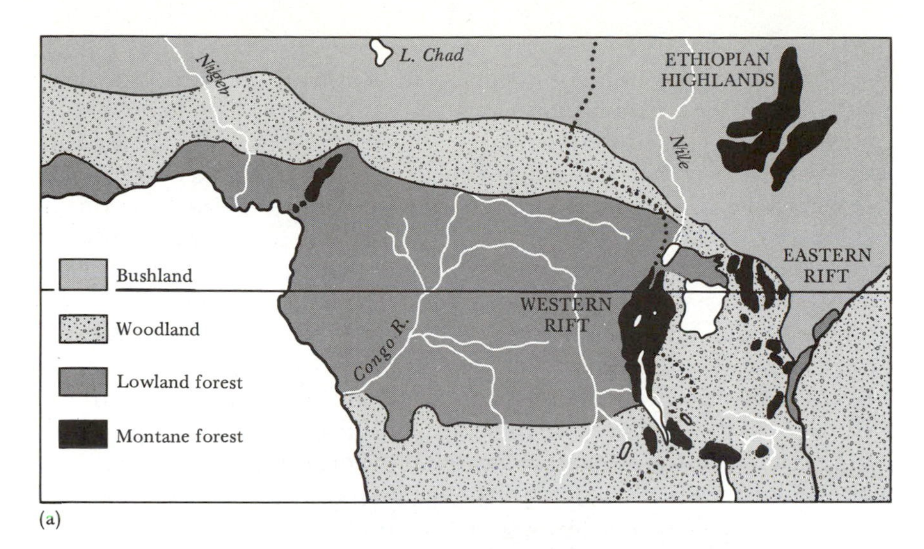

(a)

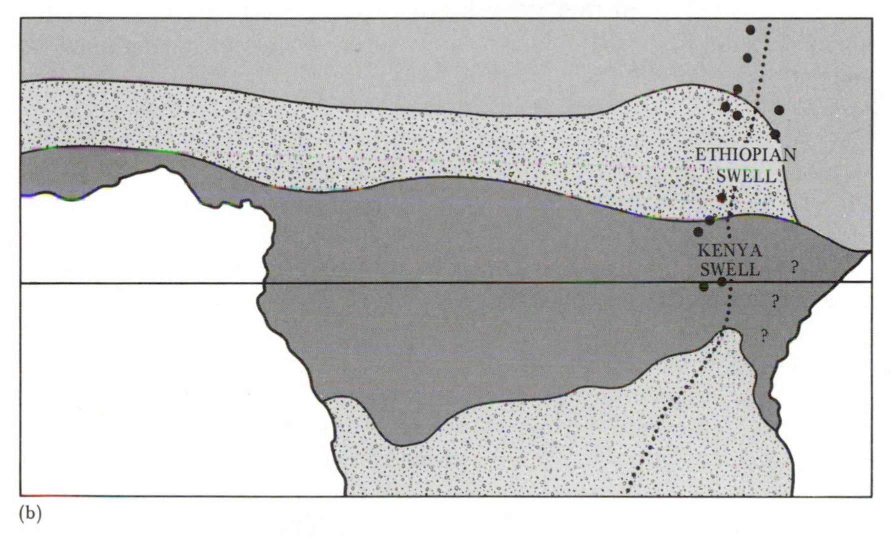

(b)

Fig. 5.2 Vegetation Map of Equatorial Africa (a) Present-day pattern; (b) probable early Miocene distribution. Dotted line indicates continental divide, and black circles represent volcanic mountains. Note that the rift system of present-day East Africa was not evident in the early Miocene; it began to form in the middle to late Miocene. (From Andrews and Van Couvering, 1975)

Before the 1960s East African early Miocene environments were usually interpreted as having semiarid woodland–bushland and grassland ecologies, much like those of the same region today (Clark and Leakey, 1951; Chesters, 1957). This view is now changing. Peter Andrews and John Van Couvering (1975) have reinterpreted the paleoenvironment of the East African early Miocene by utilizing two meteorological facts: (1) rainfall is heavy where moisture-bearing winds from the sea are forced up over mountain ranges, and (2) land on the leeward (or rainshadow) side of mountain ranges is usually dry. The present East African climate is influenced by the highlands bordering the Eastern and Western rifts that prevent both Atlantic and Indian Ocean rainfall from reaching the interior. Today the moisture precipitated in equatorial West Africa and the Congo Basin is brought to it by warm prevailing winds of the South Atlantic, and this moisture-laden air penetrates inland to the western shoulder of the Western Rift valley. These highlands capture the moisture remaining after the passage of the air mass over the Congo Basin. The northeast monsoon in the November to April season and the year-round southeast trade winds also bring moisture to East Africa from the Indian Ocean. This moisture precipitates predominately at the coast and over the highlands bordering the Eastern Rift. The result of these rainfall patterns is that today much of the interrift area is in a rainshadow causing semiarid conditions.

It is now apparent that these highlands did not begin to form until the early to middle Miocene and thus would not have had the same climatic influence in the early Miocene (Andrews and Van Couvering, 1975; Kortlandt, 1983). During the early Tertiary the African land surface in the equatorial belt probably rose gently from the Atlantic coast in the west to a continental divide near the line of the present-day rift system and down again to the Indian Ocean in the east. The early Miocene climate, vegetation, and geology probably resembled those of the well-forested volcanic mountains and lowlands of the present-day Virunga region between Lakes Kivu and Edward (Fig. 5.1). Based on this scenario, the climate of eastern equatorial Africa would have been wetter in the Oligocene and early Miocene than it is at present.

These patterns undoubtedly changed with the major regional uplift beginning in the middle to late Miocene. Before the uplift the eastward extension of the high year-round rainfall zone characteristic of the Congo Basin would have covered most of East Africa. Moreover, the great volcanoes that formed in the early Miocene probably provided the earliest East African montane forest habitats on their upper slopes. The absence of the Western Rift highlands in the early Miocene would have promoted climatic conditions favoring lowland forest. The abundance of mahogany among the Miocene paleofloras at several sites near Lake Victoria suggests the presence of extensive tracts of evergreen forest, and the occurrence of *Celtis* (a member of the elm family) is indicative of lowland forest (Chesters, 1957).

Most of the early Miocene hominoids of Africa are thought to have inhabited forested environments. Only one species *(Proconsul nyanzae)* has been found in faunal associations even remotely suggesting nonforested conditions (Andrews, 1978*a*).

Middle Miocene faunas from East Africa show closer affinities with today's floodplain and woodland–bushland communities in terms of overall species diversity. On the other hand, the presence of species regarded as forest indicators points to the existence of nearby forests (Andrews and Walker, 1976; Shipman et al., 1981). Middle Miocene faunas clearly indicate a major ecological shift from the more forested habitats of the early Miocene to more open-country habitats of the middle Miocene (Andrews and Evans, 1979; Evans et al., 1981; Andrews et al., 1981; Pickford, 1981, 1983).

Unfortunately, late Miocene faunas are very rare in East Africa, making it difficult to infer paleoecological information for that period. However, we know that the climatic trend toward cooling and drying, seen in the middle Miocene, continued in the late Miocene.

Land Movements and Dispersal Routes

The shape of Africa has essentially remained unchanged since the Mesozoic era. Nevertheless, several important geologic events have taken place since that time:

1. Africa's connection with the Northern Hemisphere was broken in the early Cenozoic by the Tethys Sea.
2. Arabia separated from the main African continental mass at the Red Sea Rift in the Miocene.
3. The Afro-Arabian tectonic plate rejoined Eurasia in the region of the Zagros and Caucasus mountains and at Gibraltar in the middle Miocene, between 18 and 14 MYA.
4. The Mediterranean Sea dried up in the late Miocene, about 6 MYA.

Eurasia, like Africa, changed dramatically from an earlier Cenozoic configuration of a relatively uniform and low topography to a continent with a more highly variable landscape (Bernor, 1983). As discussed earlier, this change in African topography resulted from the development of the East African Rift System. In Eurasia, however, this topographic change resulted from the mountain-building forces of the Alpine and Himalayan systems. Following the docking of the Afro-Arabian plate with Eurasia, about 18 to 15 MYA (Fig. 5.3a), a land corridor was established between Africa and Eurasia via Arabia, permitting a substantial immigration of African mammals into Eurasia. It was during this period that hominoids made their first appearance in Eurasia. This period also records the development of more seasonal environments and the early evolution of more open-country woodland habitats. About 15 to 8 MYA the Tethys Sea began to regress to form the Mediterranean, Black, and Caspian sea basins, and open-country habitats expanded (Fig. 5.3b,c). As environments became increasingly seasonal, the geographic ranges of hominoid primates began to shrink. By 7 to 8 MYA Eurasian hominoids had disappeared except for **relict populations** (residual populations persisting long after most of the group has become extinct) in the subtropical and tropical environments of Southeast Asia (Bernor, 1983).

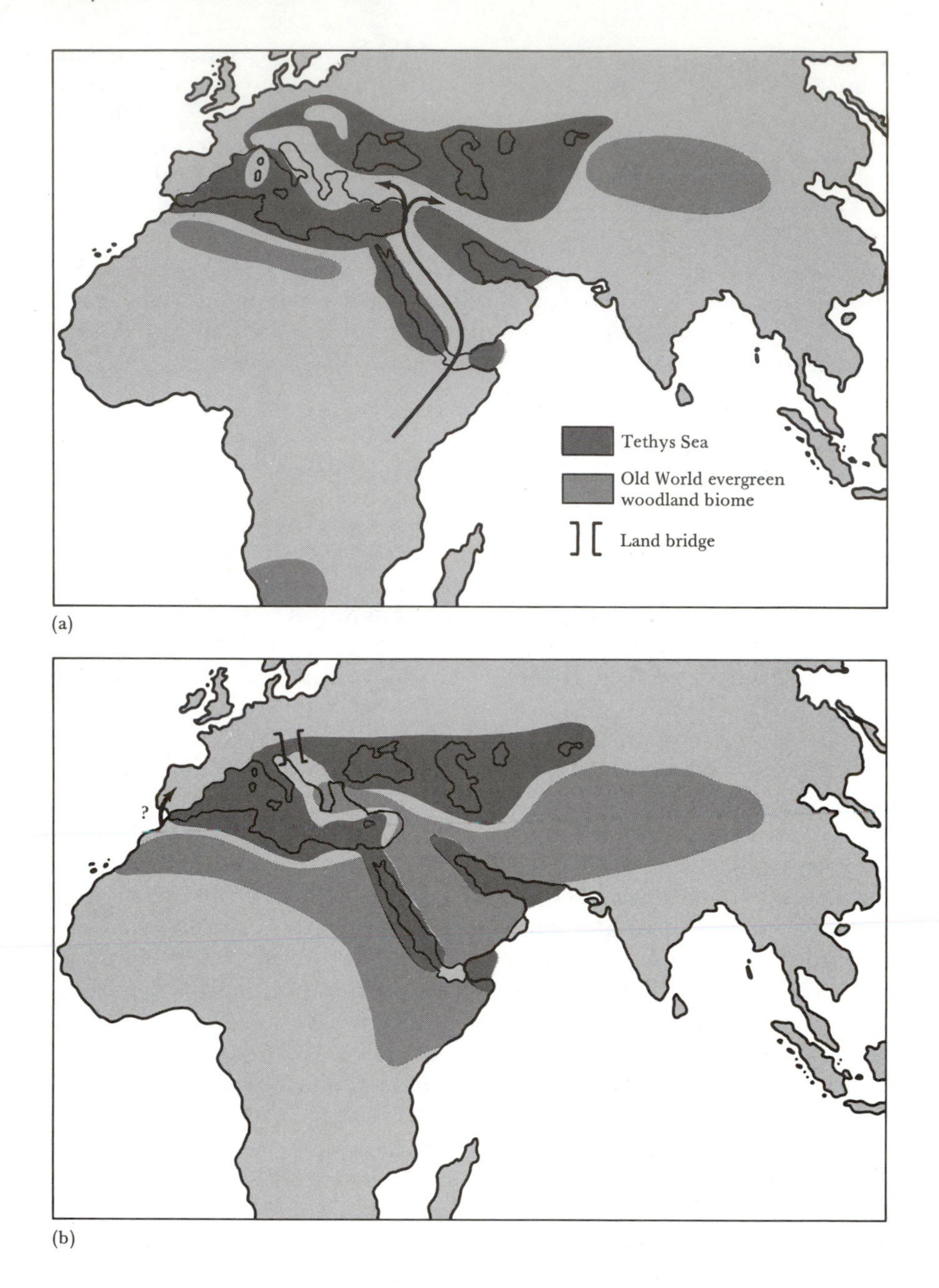

(a)

(b)

During the Tertiary in southern Asia, the uplift and subsequent erosion of the Himalayas resulted in a thick sequence of terrestrial sediments that were deposited at the base of the emerging mountain range. These sediments, known as the **Siwalik Group,** are the most thoroughly studied Miocene localities in Eurasia (Pilgrim, 1908, 1913, 1934; Col-

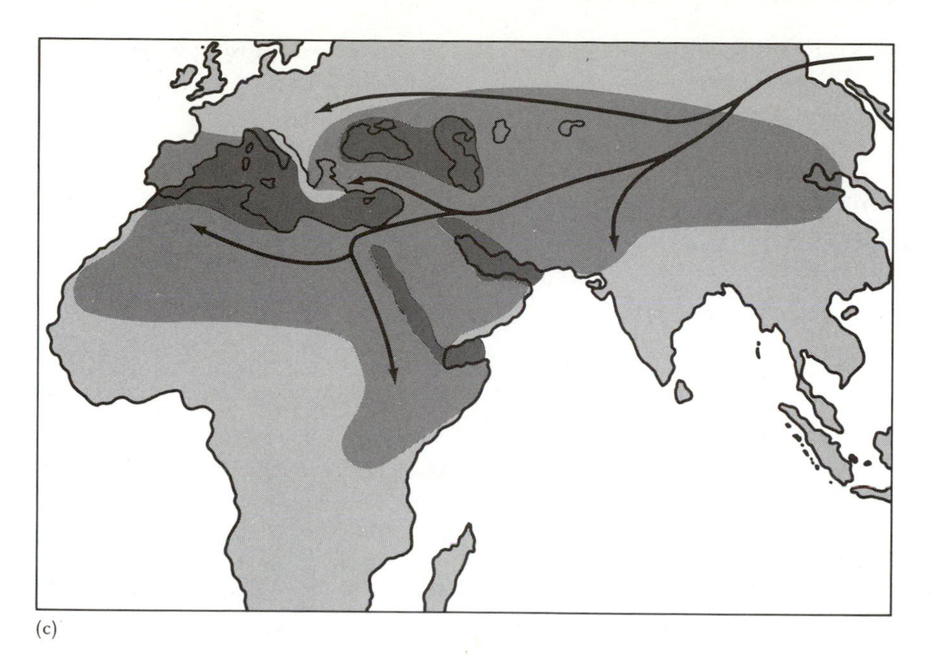

Fig. 5.3 Regression of the Tethys Sea during the Miocene and Establishment of Mammal Migration Routes between Africa and Eurasia Note the expansion of the woodland biome. (a) 18 to 15 MYA; (b) 15 to 12 MYA; (c) 12 to 8 MYA. (Adapted from Bernor, 1983)

bert, 1935; Lewis, 1937; Barry et al., 1980, 1982, 1985). These Neogene Siwalik rocks can be traced continuously along the base of the mountains all the way from western Pakistan to Assam, a state in northeastern India.

The sedimentary rocks of the Siwaliks were laid down from the rising Himalayas as a consequence of the collision between the Indian and Asian plates. This collision began some 40 to 50 MYA, and the crust has been compressed continuously since then. India continues to slide beneath Asia, producing the Tibetan Plateau and the Himalayas. There has been a significant topographical difference between the mountain front and the Siwalik Basin (at least 500 to 1,000 m) since at least the late Oligocene or early Miocene. It seems, however, that the Himalayas themselves became a significant ecological barrier only in the Pliocene or Pleistocene since they were at the same elevation as Tibet until the later Tertiary (Wang et al., 1982).

The spread of African mammals, including primates, into Asia during the early Miocene depended to a large degree on regional tectonic and sea-level events. Several episodes of lowered sea-levels during the Miocene increased the likelihood of mammal migrations into Asia. The potential land corridor linking Africa, Saudi Arabia, and Iran / Iraq allowed primates to reach southwestern Asia during one of these periods of lower sea levels in the early to middle Miocene. Subsequent dispersal into China occurred across southwestern Asia through Indo-Pakistan (Bernor et al., 1988).

SUMMARY OF THE MIOCENE FOSSIL RECORD

African Fossil Record

Early Miocene The East African early Miocene faunas, including the primates, were extremely varied. These faunas remained relatively stable between about 24 and 15 MYA. The transition to the East African middle Miocene faunas occurred approximately 16 to 15 MYA, and the transition to the late Miocene faunas about 12 MYA.

The early Miocene primate sites fall into two geographical groups, one in North Africa and the other in the rift valleys of East Africa (Fig. 5.4a). The sites in North Africa yield early Miocene monkeys. These will be discussed later in the chapter. Virtually all of the Miocene primitive apes, and many of the monkeys as well, come from the Eastern Rift and interrift zones of East Africa (Fig. 5.4b). These sites are associated with the eroded remnants of the great Miocene volcanoes of East Africa.

The volcanoes of Tinderet and Kisingiri are associated with the fossil sites of Songhor and Rusinga Island respectively, the two most prolific sources of early Miocene hominoids. Tinderet is also associated with the fossil site of Koru, where the first Miocene hominoids were found south of the Sahara, and Fort Ternan, which has provided some of the earliest dated specimens of *Ramapithecus* (Fig. 5.4c). Because of their association with lavas and tuffs from these volcanic sources, the interrift sites have been radiometrically dated (Bishop et al., 1969; Van Couvering and Miller, 1969; Van Couvering and Van Couvering, 1976). The Western Rift valley deposits are not volcanic and have not been radiometrically dated.

All of the interrift early Miocene fossil localities would have been within the forest zone. At least two of the fossil sites, Songhor and Koru, may have been at or near the montane forest level (Andrews and Van Couvering, 1975; Pickford, 1981, 1983).

The main source of early Miocene fossil vertebrates in Africa is the tuffaceous-sedimentary sequence below the lavas of the Miocene Kisingiri volcano. Rusinga and Mfwangano islands are erosional remnants of the upper flank of the volcano, and their peaks rise from Lake Victoria some 15 km north of the central vent area. The sites of Karungu to the south and Chianda to the north lie beneath the outermost extension of the Kisingiri lavas. Volcanic activity at Kisingiri probably began in the early Miocene. Most of the hominoid fossils at Rusinga associated with this volcano are 17 to 18 million years old (Van Couvering and Miller, 1969). The hominoid fossils associated with the Tinderet volcano (Son-

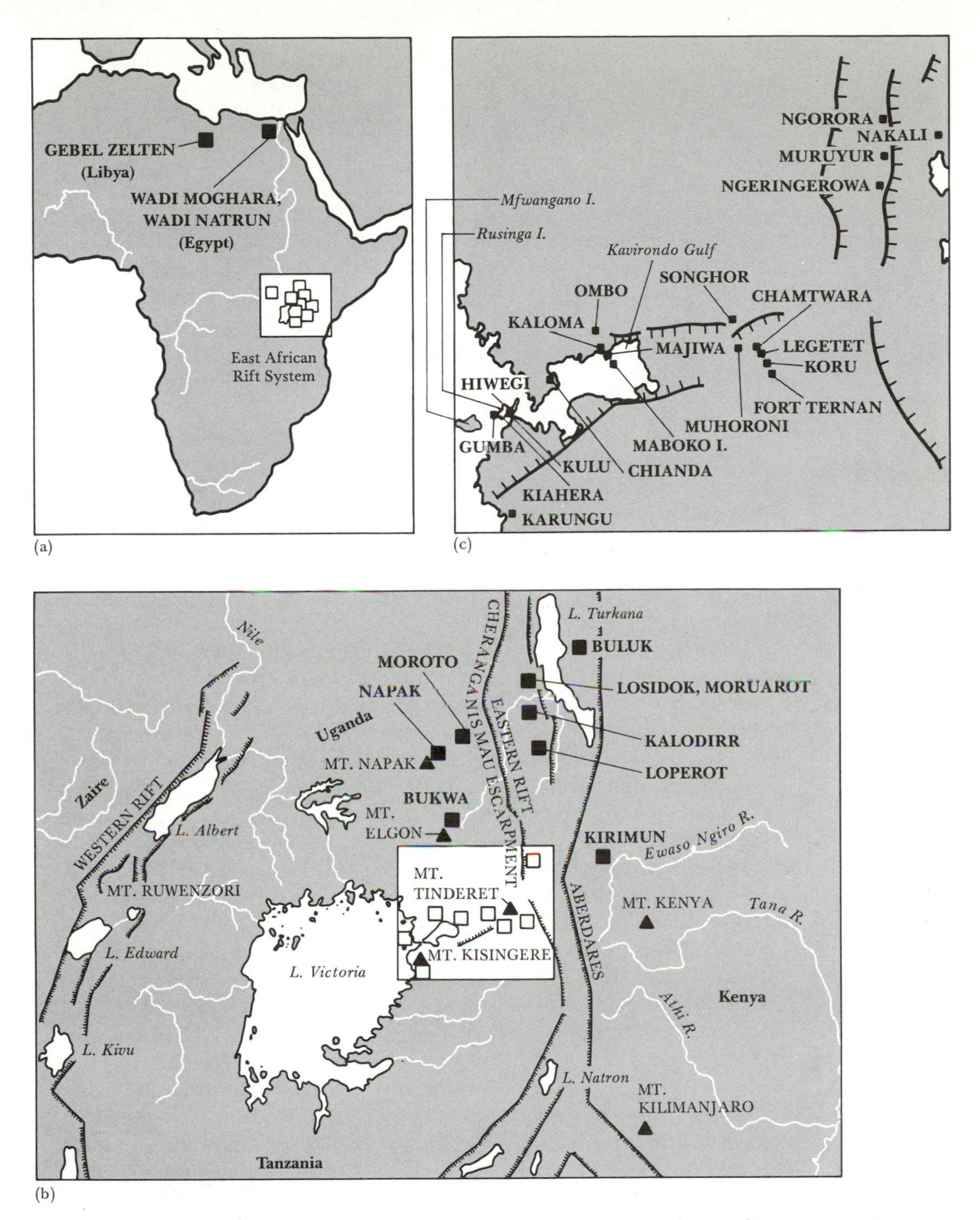

Fig. 5.4 Fossil Sites in Africa Yielding Miocene Apes and Monkeys (a) Sites in North and East Africa; area in the square is expanded to give (b), sites in East Africa; area in square is enlarged to give (c), fossil sites in western Kenya. Triangles represent volcanoes. (a, Adapted from Savage and Russell, 1983; b, adapted from Andrews and Van Couvering, 1975; c, adapted from Pickford, 1983)

ghor and Koru) are about 19 million years old. Other sites of the Kavirondo Gulf such as Maboko Island lie beneath the 12-million-year old Kisumu lava that flooded the region from the northwest. The Maboko Island fauna is probably 15 million years old and the Fort Ternan fauna close to 14 million years old (Andrews et al., 1981; Shipman et al., 1981). Several ancient volcanic regions in Uganda have also yielded important early Miocene hominoid fossils (Bishop 1964, 1967; Walker, 1969). For example, the Bukwa site at Mt. Elgon may be as much as 23 million years old.

It is apparent from this series of dated local faunas that a distinctive East African mammalian assemblage persisted with relatively little evolutionary change from approximately 23 to 16 MYA. This is undoubtedly due to the fact that sub-Saharan Africa was relatively isolated in the Oligocene and the earliest Miocene and that a high proportion of endemic taxa had developed in the African fauna (Andrews and Van Couvering, 1975; Van Couvering and Van Couvering, 1976; Pickford, 1983).

A detailed analysis of the East African Miocene biostratigraphy has been developed (Pickford, 1981). At Rusinga Island, Songhor, and Napak the large mammals dominating the early Miocene faunas are primitive elephants (Deinotherioidea) and rhinoceroses; medium-sized mammals are represented by rock hyraxes (Hyracoidea) and mouse deer (Tragulidae); and the smaller mammals include cane rats (Thryonomyidae) and flying squirrels (Anomaluridae) and rabbitlike picas (Ochotonidae). Insectivores, especially elephant shrews (Macroscelidae), are also locally common. There is also a notable proliferation of primates.

The early Miocene fauna from Songhor shows similarities to present-day African lowland forest communities both in terms of overall faunal diversity and in patterns of locomotor and feeding diversity. The Songhor fauna also has some excellent forest indicator species like flying squirrels, elephant shrews, and prosimians. Nine primate species are present; in modern faunal communities this number of primate species is found only in fully forested conditions. The fauna is unusual in having only eight artiodactyl species. The Koru fauna also suggests forest habitats, but the evidence is not as clear-cut as at Songhor (Evans et al., 1981).

Monkeys are rare in the early Miocene faunas of East Africa. It seems that in the early Miocene the hominoids occupied the ecological niches now more fully exploited by cercopithecoids. In some instances up to five or six species of fossil hominoids have been discovered in the same deposits. This number of different hominoid species is never found together in any one locality today, although it is not uncommon to find such a diversity of modern African monkeys sharing the same patch of forest (Andrews, 1981*a*).

Middle and Late Miocene In the middle Miocene faunas of Fort Ternan and Ngorora, there are instead hyracoids and very few tragulids. Comprising by far the greatest proportion of the fauna at Fort Ternan and

Ngorora are the Bovidae (bushbucks, buffaloes, gazelles, etc.) and the Giraffidae (giraffes and okapis). Small mammals and primates are rare. There are fewer rodents, and they belong to the Cricetidae, a family new in Africa at that time (Andrews and Van Couvering, 1975; Shipman et al., 1981). Middle Miocene sites such as Fort Ternan and Maboko Island have faunas that are strikingly different from and more modern looking than those in the early Miocene sites. This undoubtedly reflects changes in local ecology (Churcher, 1970; Gentry, 1970; Benefit and Pickford, 1986).

Middle Miocene faunas clearly reflect the evolution of more open woodland habitats and a reduction of forested environments. This ecological pattern continues into the late Miocene, but there is little in the way of fossil primate evidence in Africa at this time. A few isolated primate teeth from deposits near Lake Baringo in Kenya are the only indications that hominoids were present. These teeth tell very little, however, about the biology of late Miocene apes.

Eurasian Fossil Record

As in Africa, the Miocene in Eurasia was a time of fundamental changes in mammalian evolution. The following summary is drawn chiefly from Raymond Bernor's recent work (1983).

The continental Miocene of Europe is divided into five land-mammal ages: two for the early Miocene (Agenian and Orleanian), one for the middle Miocene (Astaracian), and two for the late Miocene (Vallesian and Turolian). Strictly speaking, the names for these land-mammal ages apply only to European biostratigraphy, but they have sometimes been used to characterize mammalian faunas in Asia as well. (African, South American, and North American sediments have also been characterized by separate sets of land-mammal ages. In this text only occasional reference to North American and European land-mammal ages will be made; see Appendix A for a complete delineation of these.)

The Agenian age (25 to 20 MYA) was characterized by an extensive Tethys Sea, which was an effective geographical barrier between the Eurasian and African landmasses. No primates are known from Eurasia during this time period.

The Orleanian age (20 to 15 MYA) records the first major emigration of African land mammals into Eurasia. Among the earliest emigrants were members of the Proboscidea (elephants and their relatives), which reached Eurasia about 18 MYA (Berggren and Van Couvering, 1974; Barry et al., 1985). Hominoid primates appeared in Eurasia (in France) for the first time about 16 MYA.

The Astaracian age (15 to 12 MYA) was an important episode in large-mammal evolution. The geographic expansion of open-country woodland faunas in both Eurasia and Africa occurred during this time. A major faunal immigration from Africa into western Europe occurred around 15 MYA and included proboscideans, bovids, and pigs.

During the Vallesian age (12 to 10 MYA) primitive apes disappeared from Europe.

The succeeding Turolian age (10 to 5 MYA) was a time of major faunal turnover in Eurasia. It was during this period that many hominoids became extinct in Europe and parts of Asia. The Turolian expanse of more open-country faunas and sclerophyllous woodlands across Eurasia is coincident with these hominoid extinctions. Monkeys first appeared in Eurasia during this time and may also have contributed to the demise of hominoid primates.

Miocene primate fossils are known from numerous sites in Europe and Asia. Some of the more important sites in Europe are St. Gaudens, La Grive, and Sansan (France), Can Llobateres (Spain), Rudabanya (Hungary), Neudorf an der March (Czechoslovakia), Eppelsheim (Germany), Monte Bamboli (Italy), Pikermi and Salonika (Greece), and Goriach (Austria). Important sites in Asia include Candir, Pasalar, and Mt. Sinap in Turkey; the Siwaliks of Pakistan and India; and Lufeng in Yunnan Province, China (Fig. 5.5).

As mentioned earlier, the Siwalik Group of sediments is among the most thoroughly studied localities of the Miocene of Eurasia. The Siwaliks were originally divided on the basis of mammalian fossils into lower, middle, and upper units (Pilgrim, 1910). These units were subsequently formalized improperly into formations and thus given lithologic and chronostratigraphic significance. The lower Siwaliks conventionally include the Chinji Formation, the middle Siwaliks include the Nagri and Dhok Pathan formations, and the upper Siwaliks include the Soan Formation in Pakistan and the Tatrot and Pinjor formations and the Boulder Conglomerate in India. The entire sequence in Pakistan and India averages about 5 km in thickness.

Very few of the Siwalik sediments can be radiometrically dated. For this reason their ages have been determined on the basis of paleomagnetic stratigraphy and faunal comparisons (see Chapter 1). Recent geological work has established a reasonable paleomagnetic column for the Siwaliks. The Chinji Formation is dated at 13.1 to 10.1 MYA, the Nagri Formation at 10.1 to 7.9 MYA, and the Dhok Pathan Formation at 7.9 to 5.1 MYA. The Pinjor–Tatrot boundary is dated at 2.4 MYA (Opdyke et al., 1979; Johnson et al., 1982). The bulk of the hominoid fossils come from the Nagri Formation, although the hominoids first appeared in the Chinji and apparently disappeared after the Dhok Pathan (Pilbeam et al., 1977*a*, 1977*b;* Raza et al., 1983). Monkeys *(Presbytis sivalensis)* made their first appearance between about 7.4 and 5.3 MYA (Barry et al., 1980, 1982).

The interval of approximately 8 million years spanned by the middle and upper Siwalik formations was a period of major change in southern Asian mammalian faunas and records a major replacement of many lower Siwalik faunas by an immigrant suite of herbivorous mammals including

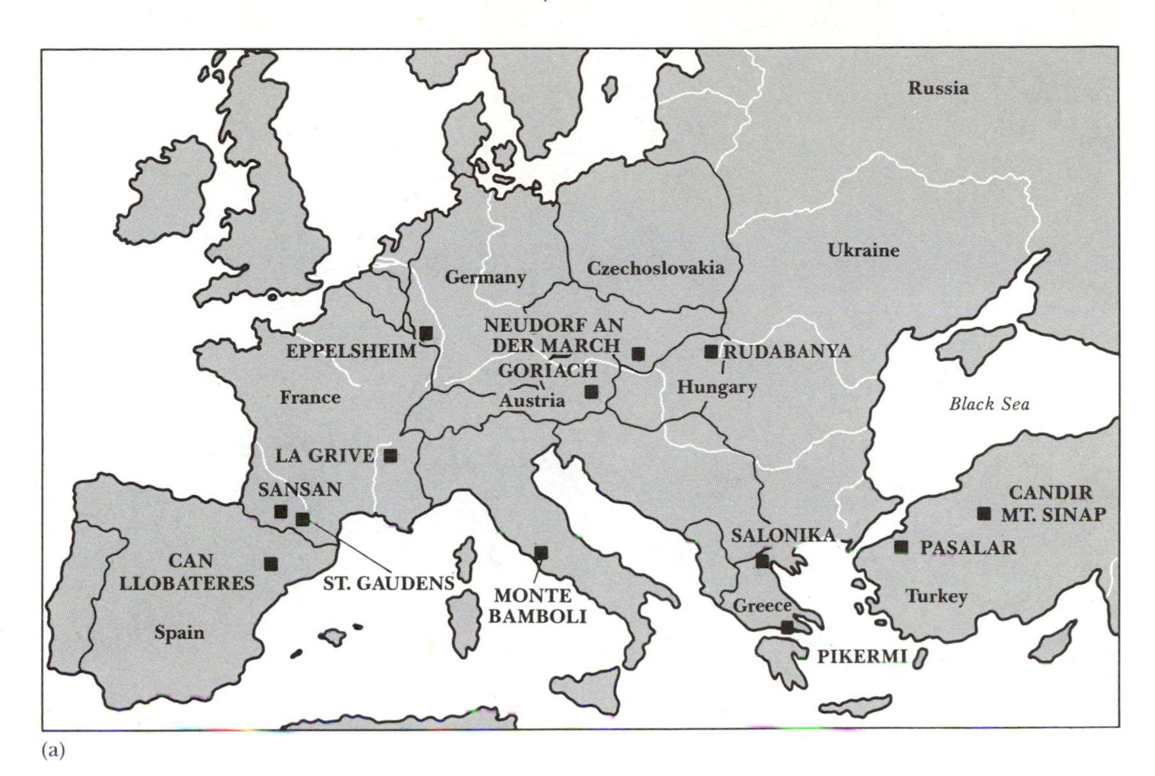

(a)

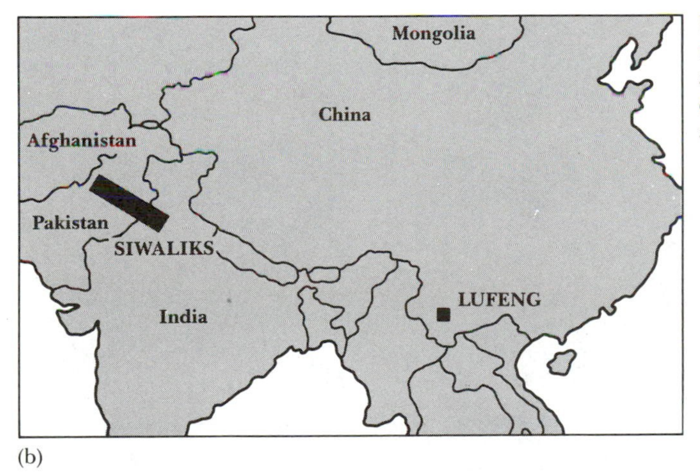

(b)

Fig. 5.5 Some Miocene Fossil Primate Localities in Eurasia (a) Europe and Asia Minor; (b) southern and eastern Asia. (Adapted from Savage and Russell, 1983)

the Equidae (horses), bovids, Suidae (pigs), and giraffids. A relatively high species richness of small to medium-sized ungulate herbivores suggests the presence of diverse vegetation types that probably included forest, woodland, and grassland (Badgley and Behrensmeyer, 1980).

Miocene primates are rare from the New World. Higher primates are absent altogether from North America, and only Argentina and Col-

Fig. 5.6 Miocene Fossil Primate Localities in South America

ombia in South America have produced Miocene monkeys (Fig. 5.6). These will be discussed later in the chapter.

RISE OF THE HOMINOIDEA

Miocene hominoids have been known from Eurasia for more than a century (Lartet, 1837, 1856; Lydekker, 1879) and from Africa for just over half a century (Hopwood, 1933). Dozens of genera and species have been named in the years since the first Miocene hominoids were discovered. At first glance this might appear to document an explosive radiation of primitive hominoids in the Miocene. Although it is true that primitive hominoids flourished in the Miocene, the plethora of names reflects an explosive "radiation" of careless taxonomy as much as it does that of fossil hominoids. For many years paleontologists described almost every newly discovered specimen—even a single tooth—as a new primate taxon. This practice was particularly widespread among early paleontologists working in Europe and southern Asia. One can accept some excuse for the practice, however, since much less was then known about anatomy and variability in living or extinct primates. After all, the French paleontologist Edouard Lartet first described the fossil hominoid *Dryopithecus* in 1856, three years before Darwin's *Origin of Species* was published and more than 50 years before fossil hominoids were discovered in Africa.

It is difficult to present a coherent overview of Miocene hominoids within some generally accepted taxonomic and phylogenetic framework. Major taxonomic revisions of these fossils occurred in the 1960s and 1970s, and there have been fundamental changes in our concepts of hominoid phylogeny in the years since then (Sarich and Cronin, 1976; Andrews and Cronin, 1982; Sibley and Ahlquist, 1984; Templeton, 1984; Andrews, 1986). The last section of the chapter will discuss historical taxonomies of hominoids, conflicting evidence from recent biomolecular and morphological data, and the changing definitions of classical terminology.

In the meantime the task at hand is to define some nomenclature for these Miocene hominoids in order to discuss them in a meaningful and unambiguous way. For this purpose the terminology expressed in several recent publications (Pilbeam, 1979; Fleagle and Kay, 1983; Ward and Pilbeam, 1983) will be amalgamated in order to recognize three distinct clusters of Miocene hominoids: **dryomorphs** (primitive early and middle Miocene hominoids of East Africa and Eurasia characterized by thin enamel on the molar teeth); **ramamorphs** (middle Miocene hominoids of Eurasia and East Africa characterized by thickened enamel on the molar teeth); and **pliomorphs** (early and middle Miocene hominoids of Eurasia sharing many primitive catarrhine features).

Dryomorph genera include *Proconsul, Rangwapithecus, Limnopithecus, Dendropithecus, Micropithecus,* and *Dryopithecus;* ramamorph genera include *Sivapithecus, Ramapithecus* (often considered part of *Sivapithecus*), and *Gigantopithecus;* and pliomorph genera include *Pliopithecus* and *Laccopithecus* (Table 5.1).

It is important to note from Table 5.1 that these three "morph" categories are not taxonomic terms; for example, dryomorphs include genera classified in the families **Proconsulidae** *(Proconsul)* and **Pongidae** *(Dryopithecus).*

Dryomorphs make their first undisputed appearance in the early Miocene of East Africa approximately 20 MYA. The dryomorphs present during this period are mostly species of *Proconsul, Limnopithecus, Micropithecus, Dendropithecus,* and *Rangwapithecus* (Table 5.2). The genus *Dryopithecus* is known only from Eurasia.

Table 5.1 Classification of Miocene Hominoids

Taxa	*Epoch (and Area)*
Infraorder Catarrhini	
Superfamily Hominoidea	
Family Proconsulidae	E. Mio. (Africa and Asia)
Dendropithecus	E. Mio. (Africa)
Dionysopithecus	?E. Mio. (Asia)
Limnopithecus	E. Mio. (Africa)
Micropithecus	E. Mio. (Africa)
Proconsul	E. Mio. (Africa)
Rangwapithecus	E. Mio. (Africa)
Family Oreopithecidae	E.–L. Mio. (Africa, Eur.)
Nyanzapithecus	E.–M. Mio. (Africa)
Oreopithecus	L. Mio. (Eur.)
Family Pongidae	M. Mio.–Pleist. (Africa, Eur., Asia)
Dryopithecus	M.–L. Mio. (Eur.)
Gigantopithecus	L. Mio.–Pleist. (Asia)
Lufengpithecus	L. Mio. (Asia)
Sivapithecus	M.–L. Mio. (Africa, Eur., Asia)
Family Pliopithecidae	M.–L. Mio. (Eur., Asia)
Laccopithecus	L. Mio. (Asia)
Pliopithecus	M.–L. Mio. (Eur.)
Family *incertae sedis*	
Afropithecus	E.–?M. Mio. (Africa, Saudi Arabia)
Turkanapithecus	E. Mio. (Africa)

As one approaches the middle Miocene in East Africa, approximately 16 to 12 MYA, ramamorphs become the dominant hominoid group with the dryomorphs clearly on the wane. The major ramamorph fossils during this period have been placed in several rather similar genera, which include *Ramapithecus, Kenyapithecus,* and *Sivapithecus.* (Some authorities include both *Ramapithecus* and *Kenyapithecus* within the genus *Sivapithecus,* a point of view we shall adopt here.) Monkeys also become a common primate at several middle Miocene sites.

As previously mentioned, higher primates are virtually absent from the African fossil record of the late Miocene.

In most of the early Miocene hominoid localities, forested-to-woodland environments have been sampled. These localities are characterized by hominoid primates covering a wide spectrum of body sizes and adaptations. By the middle Miocene, the indications are that woodland and bushland had become established in the same general areas. This ecological change was accompanied by a great reduction in the diversity of hominoid taxa, as well as a reduction in their population density (as fossils) relative to other mammals. Toward the late Miocene, savanna-

Table 5.2 Temporal Distribution of Miocene Primates in Western Kenya

Age (MYA)	*Stratigraphic Units*	*Dominant Hominoids*	*Common Hominoids*	*Rare Hominoids*
10.5–8	Ngeringerowa	—	—	Monkeys
12–10.5	Ngorora	—	—	Ramapithecine, small-bodied apes, monkeys
14.5–12	Fort Ternan[a]	*Pliopithecus, Ramapithecus wickeri*[a]	—	?*Proconsul* sp.
	Muruyur	—	—	?*Proconsul* sp.
16–14.5	Maboko Formation[a]	Monkeys	Ramapithecines (*Sivapithecus africanus*[a], *Ramapithecus* sp.)	?*Rangwapithecus vancouveringi, Limnopithecus legetet,* ?*Proconsul nyanzae*
18.5–16	Rusinga Island: Kulu Formation	—	—	*Dendropithecus macinnesi, P. nyanzae*
	Hiwegi Formation[a]	*Dendropithecus macinnesi, Proconsul africanus, P. nyanzae*[a]	—	*L. legetet,* ?*R gordoni*
	Kiahera Formation	*D. macinnesi, P. africanus, P. nyanzae*	—	?*R. gordoni, R. vancouveringi*
20–18.5	Songhor[a]	Songhor-type assemblages, *Rangwapithecus gordoni*[a]	*Proconsul major*[a], *Dendropithecus songhorensis*[b]	*R. vancouveringi, L. legetet, D. macinnesi*
	Chamtwara[a]	—	—	—
	Legetet Formation	*Limnopithecus legetet*[a], *Micropithecus clarki*	*D. macinnesi, D. songhorensis*[b]	*P. africanus*[a]
	Koru Formation	—	*P. major*	
?23	Muhoroni agglomerates	—	?*P. major* or *P. nyanzae*	—

[a] Type locality for the taxon
[b] May be synonymous with *D. macinnesi*
Source: Pickford (1983).

oriented fossil assemblages become more common and cercopithecoids become more diverse (Pickford, 1983).

It is interesting to note that most of the early Miocene hominoids (dryomorphs) had thin-enameled teeth and most of the middle Miocene hominoids (ramamorphs) had thick-enameled teeth. This suggests that enamel thickness increased coincidentally with major ecological changes that were taking place in East Africa during this period, namely the change from wetter, more forested conditions to drier, more woodland–bushland environments.

Hominoids first appeared in Eurasia during the Astaracian land-mammal age, 16 MYA, with the occurrence of the pliomorph *Pliopithecus,* which flourished in Europe until 11 or 12 MYA. The dryomorph *Dryopithecus* first appeared in Europe during the Astaracian age as well.

Astaracian hominoids appear to have segregated into distinct regional provinces. *Dryopithecus* and *Pliopithecus* were limited mainly to western, southern, and central Europe; *Proconsul, Dendropithecus, Rangwapithecus,* and *Limnopithecus* (all dryomorphs) were limited to East Africa; and *Sivapithecus, Ramapithecus,* and *Gigantopithecus* (all ramamorphs) were limited mainly to Indo-Pakistan and China (but were also found in East Africa, Central Europe, Turkey, and Greece).

Sivapithecus made its first appearance in Eurasia about 15 MYA and disappeared about 7 MYA in the Siwaliks of South Asia. *Dryopithecus* first appeared about a little over 12 MYA and lasted until about 9.3 MYA in Europe. It was associated with more forest-adapted faunas.

During the Vallesian age (12 to 10 MYA), *Pliopithecus* disappeared from western and southern Europe, as did most species of *Dryopithecus.* However, one pliomorph, *Laccopithecus,* was still present in China (Lufeng) about 8 MYA.

During the Turolian age (10 to 5 MYA) hominoids became extinct from Eurasia, except for the ramamorph *Gigantopithecus,* which lasted until the Pleistocene in China. Ramamorphs such as *Sivapithecus* and *Ramapithecus* flourished until their disappearance in Asia about 7 MYA.

To summarize these major events as they relate to hominoid evolution in Eurasia: (1) hominoids were exclusive to Africa until about 17 MYA. In Eurasia *Pliopithecus* first appeared about 16 MYA and flourished until approximately 11 MYA. *Sivapithecus* first appeared about 15 MYA and became extinct in Europe by 10 to 9 MYA and in the Siwaliks of Indo-Pakistan by 8 to 7 MYA. *Dryopithecus* first appeared about 13 to 12 MYA and died out about 3 million years later (Fig. 5.7).

Dryomorphs

Dryomorphs are found predominantly in Africa. Of the various dryomorph genera, only one *(Dryopithecus)* occurs in Europe; the others occur in East Africa (although *Dionysopithecus* may represent an Asian version of *Micropithecus;* see Table 5.1). The East African dryomorphs were a very successful group. Several of these species lasted from the early Miocene (about 22 MYA) to the middle Miocene (about 14 MYA) with little apparent morphological change, thus making them some of the longest-lived species in the fossil primate record. The Eurasian dryomorph, *Dryopithecus,* first appeared in Europe about 12 to 13 MYA and died out about 9 to 10 MYA.

All African Miocene sites bearing hominoid fossils are found in just two countries, Kenya and Uganda. The great majority of fossil remains are dryomorphs, although a few ramamorphs have been found; pliomorphs to date have been restricted to Eurasian sites.

The first fossil hominoids from East Africa were collected at Koru in Kenya by H. L. Gordon in 1926. Two of the richest Miocene sites in Kenya—Songhor and Rusinga Island—were discovered by L. S. B. Leakey

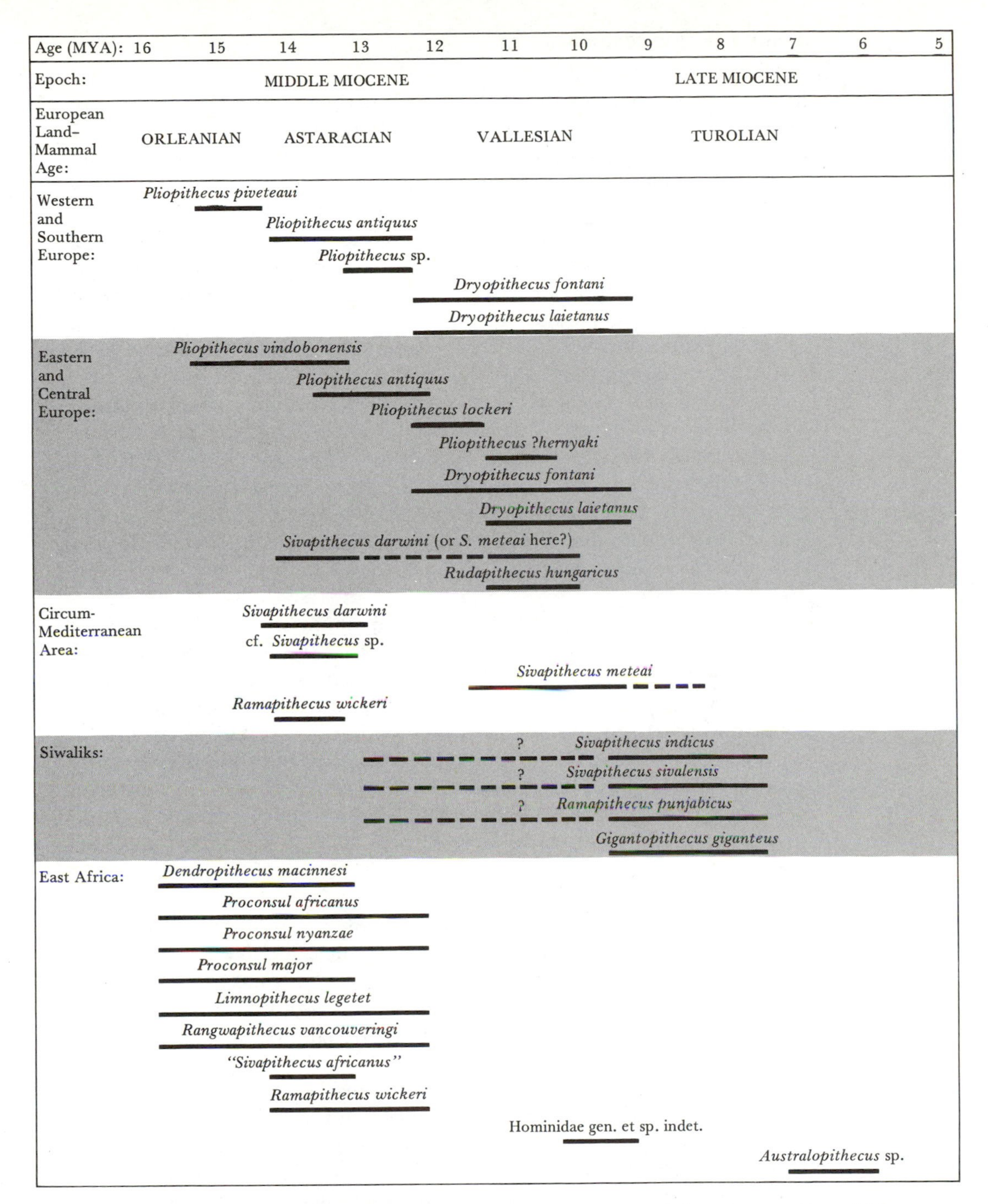

Fig. 5.7 Age Ranges of Middle and Late Miocene Hominoids Solid horizontal bars represent chronological ranges of fossil species; dashed lines give more general estimates. Many of the East African species extend back into the early Miocene, not shown on this timeline; moreover, dates of extension for some of the Middle Miocene East African hominoids are uncertain (compare with Table 5.2). Note also that a number of species discussed in this chapter, particularly the Chinese specimens, are not indicated. (Adapted from Bernor, 1983)

and D. G. MacInnes in 1931 and 1932. Further collecting at these sites was carried out by the British–Kenya Miocene Expeditions of 1947 through 1951, organized by Le Gros Clark and Leakey. This pioneering work culminated in the publication of several monographs describing the fossil hominoids discovered during these expeditions (Le Gros Clark and Leakey, 1951; Le Gros Clark and Thomas, 1951; Napier and Davis, 1959). These works became classics and have exerted a powerful influence in the field of paleoprimatology ever since. Work has continued at these sites off and on to the present day.

In 1978, P. Andrews made a comprehensive review of the East African Miocene hominoids. He recognized seven species within two families, the Pongidae (great apes) and the Hylobatidae (gibbons and siamangs). He included two genera, *Proconsul* and *Limnopithecus,* in the Pongidae and one genus, *Dendropithecus,* in the Hylobatidae. More recently, a third pongid genus, *Rangwapithecus,* has been recognized. Most workers recognize the following species in these five genera: *Proconsul africanus, Proconsul nyanzae, Proconsul major, Limnopithecus legetet, Rangwapithecus vancouveringi, Rangwapithecus gordoni, Dendropithecus macinnesi,* and *Micropithecus clarki.*

More than 1,500 specimens of fossil hominoids have now been recovered from the Miocene of East Africa. Although most of these specimens are teeth and jaw bones, one species at least *(Proconsul africanus)* is well represented by both cranial and postcranial material. Fragmentary cranial remains are known for other *Proconsul* species, but as yet very little is known about skull morphology in these other fossil hominoids (Teaford et al., 1988). Some postcranial material of *P. nyanzae, P. major,* and *Dendropithecus macinnesi* has also been described (Le Gros Clark and Thomas, 1951; Walker and Pickford, 1983; Rose, 1983; Langdon, 1985).

Proconsul The first comprehensive study of *Proconsul* was undertaken by Le Gros Clark and Leakey (1951). They concluded that *Proconsul* species were members of the Pongidae (which at that time was comprised of the great apes: gorillas, chimpanzees, orangutans, and their ancestors). They also concluded that many of the characters in which they differed from the living great apes were merely primitive ones that should not be used to exclude them from the Pongidae. Leakey (1936) later changed his mind and created the new family **Proconsulidae** for the African fossil hominoids in order to further distinguish them from the Eurasian dryopithecines. Several specimens were later found that were considered distinct from *Proconsul* and were placed in a new species, *Sivapithecus africanus,* in the belief that they resembled the Asian fossil hominoids more closely than the African ones (Le Gros Clark and Leakey, 1951). (See the discussion later in the chapter on Eurasian ramamorphs.)

There are three generally accepted species of the genus *Proconsul: P. africanus, P. nyanzae,* and *P. major.* Andrews (1978*a*) included two more:

P. gordoni and *P. vancouveringi.* The latter two species are now usually placed in a closely related but separate genus, *Rangwapithecus.* Both genera are known only from the early to middle Miocene of East Africa. All these species share many primitive catarrhine features:

1. A narrow nose
2. A broad interorbital region
3. Small frontoethmoidal sinuses (the primate sinus system will be discussed and illustrated later in the chapter)
4. **Gracile** (slender) mandibles
5. Slender canines
6. Single-cusped and semisectorial third premolars
7. Molars with thin enamel
8. Prominent molar cingula
9. Well-developed molar occlusal ridges with pointed cusps

P. africanus was first described by A. T. Hopwood in 1933. He was the first to suggest an ancestor–descendant relationship between it and *Pan* (chimpanzees), a view later expanded upon by Pilbeam (1969). As we shall see, however, recent studies in both paleontology and molecular biology make it highly unlikely that any modern great ape lineage was distinct by the early Miocene.

The geographical distribution of *P. africanus* is limited to western Kenya. Fossils of this species are most common in early Miocene sediments of Rusinga Island, Mfwangano Island, and Koru. They are also found at Songhor and the middle Miocene site of Fort Ternan (Andrews, 1978*a;* Bosler, 1981). The early Miocene occurrences of *P. africanus* probably range between 20 and 18 MYA. The youngest record of this species is at Fort Ternan, where the fossil-bearing deposits are bracketed by dates of 14 and 12.5 MYA (Bishop et al., 1969; Andrews and Walker, 1976). The known time range for this species is therefore at least 5.5 million years.

Proconsul africanus is the smallest of the *Proconsul* species. It is intermediate in dental size between gibbons (*Hylobates)* and the pygmy chimpanzee *(Pan paniscus).* Body-weight estimates for one female specimen known from much of the skull and postcranial skeleton range between 10 and 12 kg (Walker et al., 1983; Conroy, 1987). The length of the upper premolar–molar series is less than 40 mm, and the length of the lower dental series is less than 45 mm. The skull is lightly built and somewhat **prognathous** (with the premaxilla projecting forward), and it lacks **supraorbital tori** (brow ridges) or strong muscular markings (Le Gros Clark and Leakey, 1951; Walker et al., 1983).

A distorted skull of a female *Proconsul africanus* was found by Mary Leakey in 1948 in early Miocene sediments on Rusinga Island (Le Gros Clark and Leakey, 1951). Unfortunately, the occipital portion of the skull did not articulate with the rest of the specimen because large parts of the intervening skull cap were missing. Thirty-five years later two of the missing pieces of the skull were found, making it possible to attach the

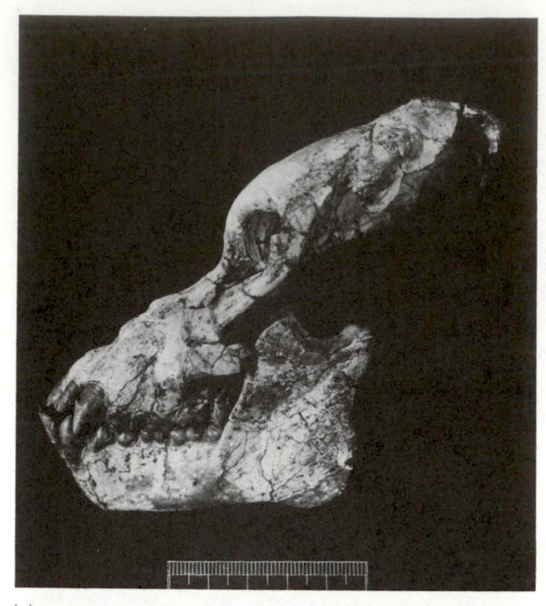

(a)

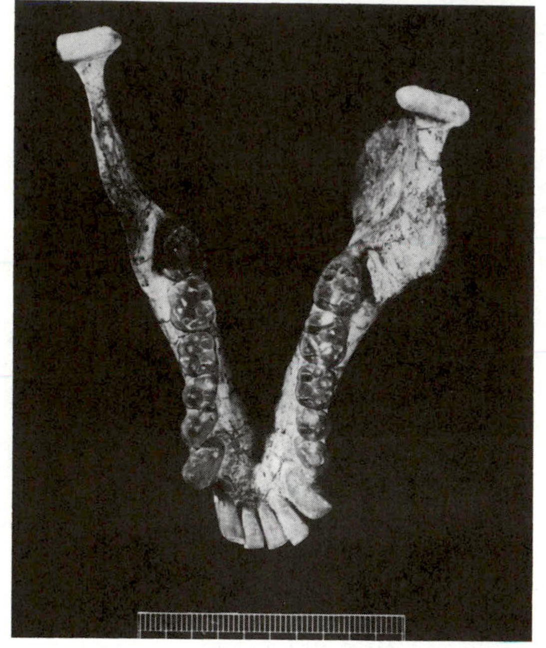

(b)

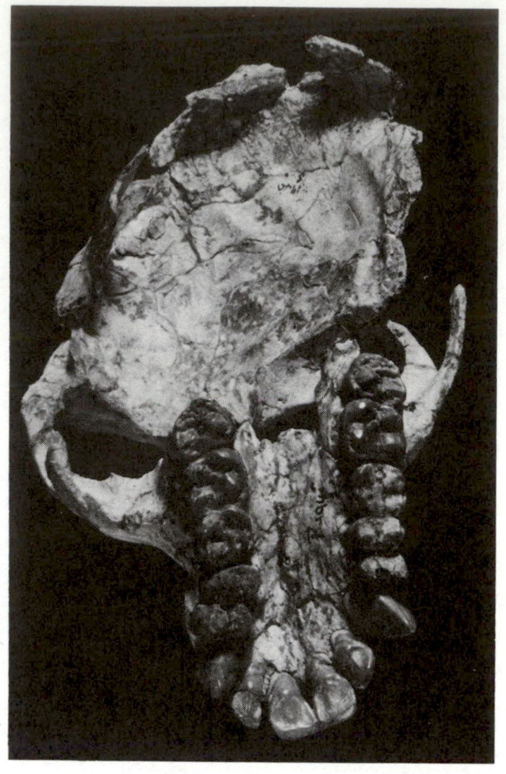

(c)

Fig. 5.8 Skull of *Proconsul africanus* from the Early Miocene of Rusinga Island, Kenya (a) Lateral view; (b) occlusal view of the upper jaw; (c) lower jaw (about 0.9 × natural size). (Courtesy of P. Andrews)

main skull fragment to the occipital portion. The skull is now complete in the dorsal midline from the face to the foramen magnum (Walker et al., 1983) (Fig. 5.8).

An accurate endocranial volume determination of 167 cc has been made for this Miocene hominoid (Walker et al., 1983). Body weight, based on postcranial material, is estimated at 10 to 12 kg. Thus the encephalization quotient (EQ) for this species is approximately 1.5. Walker and colleagues divided this value by 2.87, the EQ for *Homo sapiens,* in order to express relative brain size as a percentage of human relative brain size. Values of this new quotient in *Pan troglodytes, Pongo pygmaeus,* and *Gorilla gorilla* are 41%, 32%, and 17% respectively. Similar values for New and Old World monkeys range between 23% and 82%. Thus on this scale there is no difference in relative EQ between monkeys and apes, and by these calculations the gorilla has the smallest relative brain size of all the catarrhine primates. The relative EQ for *P. africanus* is about 49%. Importantly, however, relative EQs in monkeys of about the same body size as *P. africanus* range between 23% and 41%. It appears, then, that *P. africanus* had a relatively bigger brain than modern monkeys of comparable body size (Walker et al., 1983).

The postcranial skeleton of *P. africanus* shares several features with *Pan:* (1) the general configuration of the shoulder and elbow joints; (2) the value for the brachial index; and (3) the morphology of the fibula and hallucal metatarsal (Napier and Davis, 1959; Conroy and Fleagle, 1972; Zwell and Conroy, 1973; Walker and Pickford, 1983). Many other postcranial features, particularly in the wrist joint, are more monkeylike (Schon and Ziemer, 1973; Corruccini et al., 1975, 1976; Morbeck, 1975; McHenry and Corruccini, 1983; Beard et al., 1986).

Proconsul nyanzae was larger than *P. africanus.* Estimates of body size based on newly discovered lower-leg and foot bones suggest a weight similar to that of female chimpanzees (approximately 40 kg) (Walker and Pickford, 1983). Overall dental size, including canine size, is most similar to the chimpanzee as well. This species was strongly sexually dimorphic, a feature that is particularly well reflected in canine size. Distinctive dental features of the species include the following: (1) on the upper molars the lingual cingulum is beaded and the posterior cingulum is well developed; (2) the lower first molars are very small relative to the lower second molars, and the lower third molars are slightly smaller than the lower second molars; and (3) the length of the upper postcanine tooth row is 40 to 50 mm, and the length of the lower tooth row is 45 to 55 mm. The maximum depth of the mandibular body is greater than that in *P. africanus,* but this feature is variable (Andrews, 1978*a*) (Fig. 5.9).

Proconsul nyanzae is known from lower Miocene deposits of Rusinga Island, Mfwangano Island, Songhor, Koru, and Karungu, all in western Kenya (Andrews and Walker, 1976; Bosler, 1981). The complete dentition, mandible, maxilla, and parts of the face are known for this species. Only a few limb-bone fragments have so far been described (Andrews and Walker, 1976; Andrews, 1978a; Beard et al., 1986). In general, the known postcranial parts of *P. nyanzae* seem to be larger, isometrically scaled versions of *P. africanus.*

Proconsul major is known mainly from the lower Miocene deposits of Songhor and Koru in western Kenya and possibly from the lower Mio-

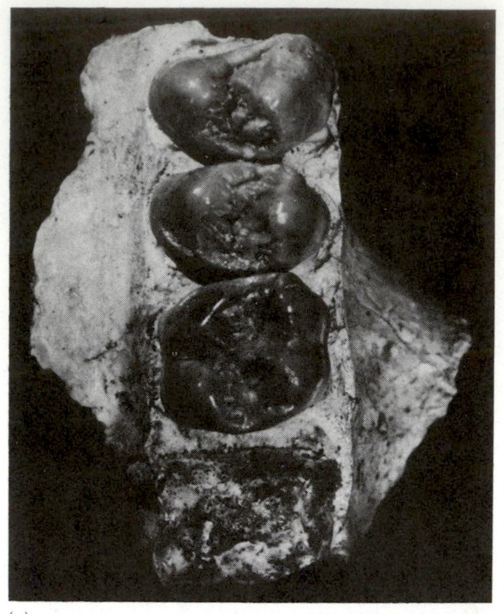

(a)

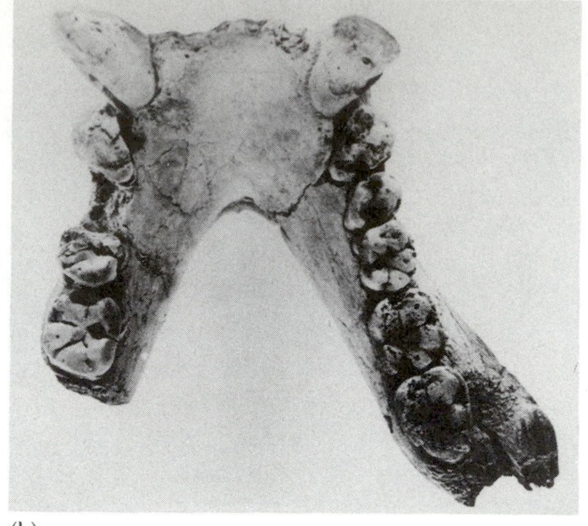

(b)

Fig. 5.9 Dentition of *Pronconsul nyanzae* (a) Upper left jaw, third premolar through first molar (mesial end at top); (b) lower jaw. Scales about 2 × natural size in (a) and 1 × in (b). (Courtesy of P. Andrews)

cene deposits of Losidok, Moruarot, and Kirimun in northern Kenya. (Specimens from the northern Kenyan sites may belong to the same genus as fossils recently described from Buluk and Kalodirr named *Afropithecus,* to be discussed later in the chapter.) *Proconsul major* is also known from the lower Miocene of Napak and the middle Miocene of Moroto in Uganda (Pilbeam, 1969), although it has recently been suggested that the latter specimen should no longer be considered *P. major* (Martin, 1981; Pickford, 1982). All of the mandible and lower dentition except for the central incisors is known, as is the maxilla and most of the upper dentition (Fig. 5.10). Parts of the lumbar vertebrae have been described from Moroto, the same site yielding the palate of *P. major* (Walker and Rose, 1968).

Proconsul major is the largest species of the genus. In size its dentition approaches that of a female gorilla. The length of the lower postcanine tooth row may exceed 65 mm. The mandibular body is larger than that in *P. nyanzae,* and the mandibular symphysis is even more massive than that in ramamorphs (such as *Sivapithecus indicus*) that otherwise approach *P. major* in size. As in all species of *Proconsul,* the simian shelf (inferior transverse torus) is absent.

The main difference between *P. nyanzae* and *P. major* is one of size. The incisors and canines are morphologically identical in the two species. However, there are a few dental features that distinguish the two species. For example, the lower third molar is more elongated in *P. major,* and the distal cusps are characteristically atrophied. In *P. nyanzae* the third molar is a broad rectangular tooth and is not reduced at all (Andrews, 1978*a;* Pilbeam, 1969).

(a)

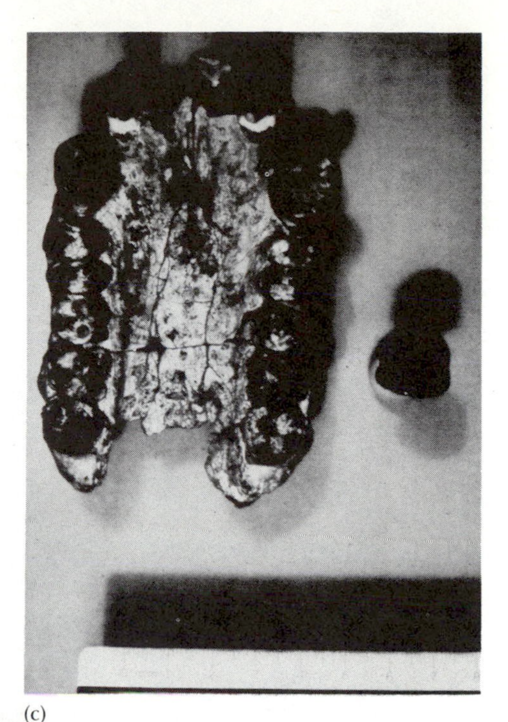

(c)

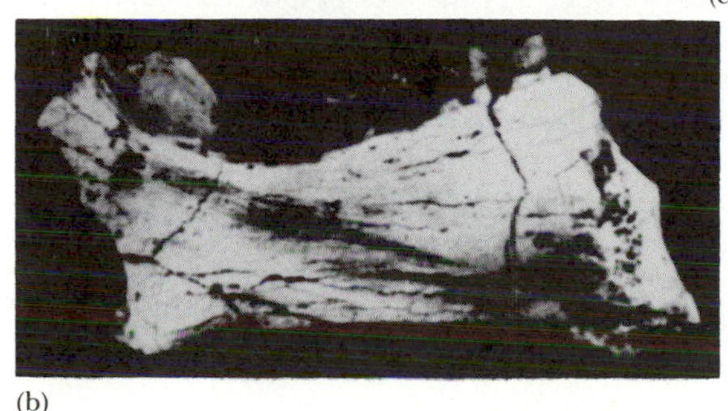

(b)

Fig. 5.10 Dentition of *Proconsul major* (a) Occlusal and (b) medial views of lower jaw from the early Miocene of Kenya; scale about 0.8 × natural size. (c) Occlusal view of the palate of a Miocene specimen from Moroto, Uganda, that may belong to *P. major*. (a,b, Courtesy of P. Andrews; c, courtesy of E. Delson)

The geographical distribution of these three *Proconsul* species suggests that there were at least two population centers of these large hominoids in the East African early Miocene, one centered on the southern sites around the Kisingiri volcano (Rusinga Island, Mfwangano Island, and Karungu) and the other on the northern sites around Tinderet (Songhor and Koru), Napak, and Moroto (Andrews, 1978*a;* Bosler, 1981).

Rangwapithecus As mentioned earlier, this genus now consists of two species, *Rangwapithecus gordoni* and *R. vancouveringi,* which were originally considered members of *Proconsul.* The two *Rangwapithecus* species overlap the bottom half of the *Proconsul* size range in dental dimensions. They differ from *Proconsul* in having (1) elongated upper premolars and molars; (2) massive cingular development on the upper postcanine dentition; (3) gracile maxillary bodies with extensive development of the

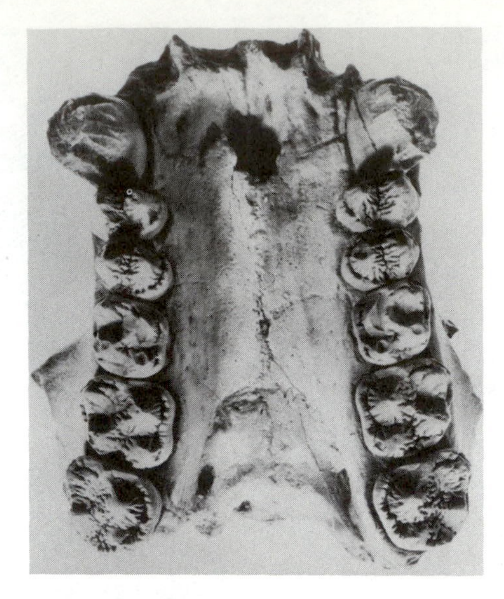

Fig. 5.11 Occlusal View of the upper jaw of *Rangwapithecus gordoni* Scale about 1.3×natural size. (Courtesy of P. Andrews)

maxillary sinus; and (4) gracile mandibles that are very deep relative to transverse thickness (Andrews, 1978*a*, 1978*b*).

Rangwapithecus gordoni is similar in size to *P. africanus*. Its lower dentition has the following characteristics: (1) the incisors are high-crowned and narrow; (2) the canines are high-crowned and bilaterally compressed; (3) the lower third premolars are very compressed bilaterally with nearly parallel buccal and lingual sides; (4) the fourth premolars are elongated; (5) the molars are elongated with low cusps; and (6) occlusal ridges and buccal cingula are poorly defined. The upper and lower postcanine tooth rows are about 40 and 44 mm long respectively. The species is known from the lower Miocene of Songhor and a few specimens from Rusinga Island, Koru, and Mfwangano Island in Kenya (Bosler, 1981). The complete mandibular and maxillary dentition is known (Andrews, 1978*a*) (Fig. 5.11).

Rangwapithecus vancouveringi is a small species with a dentition the size of siamangs *(Symphalangus)*. The species is known from the lower Miocene of Rusinga Island, Mfwangano Island, and Songhor (Bosler, 1981).

Rangwapithecus gordoni and *R. vancouveringi* are usually considered to be so similar morphologically that there seems little doubt about their close relationship. Recent studies on more complete material, however, have cast some doubt on this view. Harrison (1986) has observed dental characteristics in *R. vancouveringi* that he believes link it to an unusual, difficult-to-classify hominoid called *Oreopithecus bambolii.*

Known from late Miocene deposits of Italy, *Oreopithecus* is one of the most enigmatic catarrhine primates in the fossil record. It was originally described by Gervais (1872) on the basis of a nearly complete lower subadult dentition from the area around Monte Bamboli in the province of Grossetto, Tuscany (see Fig. 5.5a). Up until the 1950s, most discussions of the genus revolved around the question of whether it was a distinctive hominoid or cercopithecoid, or in some way intermediate between the two groups. In the 1950s, it was even claimed to have some special phyletic relationship to *Homo*. Most researchers since then have accepted *Oreopithecus* as an aberrant hominoid, although some authors still persist in believing it to be a sister group of the Cercopithecidae based on shared derived dental features.

Harrison (1987), who has reviewed the morphology of *Oreopithecus*, argues that dental similarities between it and cercopithecids are probably the result of functional convergence and that the derived postcranial characters shared by *Oreopithecus* and hominoids are too numerous and detailed to be anything but homologies implying close relationship. For this reason he considers it a hominoid but has placed it in a separate family, the **Oreopithecidae.** In overall body proportions, *Oreopithecus* is most similar to female orangutans; presumably its mode of locomotion and substrate use (how it uses vegetation and arboreal supports) were similar as well (Jungers, 1987).

Harrison (1986) has found similarities between *R. vancouveringi* and several medium-sized anthropoids from the middle Miocene site of Maboko Island that have a very distinctive dental morphology reminiscent of *Oreopithecus*. For this reason he has assigned *R. vancouveringi* and the Maboko specimens to a new genus, *Nyanzapithecus*. By referring this genus to the Oreopithecidae as well, he suggests that the origins of this family may be traced back to at least the middle Miocene, if not the early Miocene, of East Africa.

Limnopithecus One of several small-bodied hominoids from the Miocene of East Africa is *Limnopithecus*. The single species, *L. legetet*, is known from the lower Miocene sites of Bukwa, Napak, Songhor, Koru, Ombo, and Rusinga Island and from the middle Miocene of Fort Ternan and Maboko Island. The complete dentition is known, as well as the premaxilla and the mandible.

Dental characteristics of the genus include the following:

1. Relatively large and broad central incisors
2. Well-developed canines
3. Low, rounded molar cusps with distinct buccal cingula on the lower molars and lingual cingula on the upper molars
4. Poorly developed occlusal ridges
5. A distinct size increase from lower first to third molars
6. Slight protoconule development

The lower postcanine dental row measures about 27 mm (Andrews, 1978*a*) (Fig. 5.12). The dental size of this species is slightly smaller than that of *Hylobates*.

Limnopithecus was first described from very inadequate material from Koru (Hopwood, 1933). It was considered at that time to be a gibbonlike primate in contrast to the chimpanzeelike *Proconsul africanus*. Simons (1972), among others, included *Limnopithecus* in the Hylobatidae, regarding

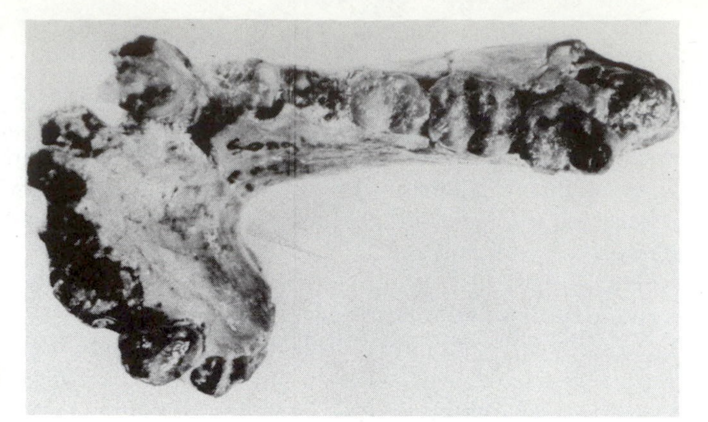

Fig. 5.12 Lower Jaw of *Limnopithecus legetet* Scale about 2 × natural size. (Courtesy of P. Andrews)

it as an ancestor of the gibbons. At one time two species were attributed to *Limnopithecus*. Andrews (1974), however, recognized that *L. legetet,* the type species of *Limnopithecus,* was more closely related to *Proconsul* than it was to *L. macinnesi,* the more gibbonlike species of *Limnopithecus*. Consequently, *L. legetet* was removed from the Hylobatidae and transferred to the pongid subfamily Dryopithecinae. By the rules of taxonomic nomenclature, the generic name *Limnopithecus* could no longer be used for the fossil hylobatid *L. macinnesi*. As a result, a new generic name was chosen for this species, *Dendropithecus* (Andrews, 1974; Andrews and Simons, 1977).

Among the many morphological features *L. legetet* shares with *Proconsul africanus* are the following:

1. A long, narrow maxillary sinus
2. A narrow nasal aperture
3. A mandibular symphysis with a superior transverse torus
4. Broad, low-crowned incisors
5. Spatulate upper central incisors
6. Rounded upper canines that are not bilaterally compressed and that lack double mesial grooves (longitudinal grooves running along the crown of the tooth and onto the roots)
7. Triangular lower third premolars whose function is grinding rather than honing
8. Square upper first molars and upper third molars that are very reduced relative to the upper second molars
9. Lower first molars with strong buccal cingula
10. Elongated lower third molars with large heellike hypoconulids (Andrews, 1978*a*)

Dendropithecus Another of the small-bodied Miocene hominoids is *Dendropithecus*. The genus has a single species, *D. macinnesi,* which, as mentioned earlier, once was considered a member of the genus *Limnopithecus*.

Dendropithecus macinnesi is known from the lower Miocene of Rusinga Island, Mfwangano Island, Karungu, Songhor, and Koru. Among the specimens are the maxilla and mandible and all the teeth. Postcranial material in direct association with dental remains is also known and is most comparable in preserved parts to the South American spider monkey, *Ateles* (Le Gros Clark and Thomas, 1951).

In dental size, *D. macinnesi* is similar to the siamang. The incisors are high-crowned and strongly compressed mesiodistally. The canines are bladelike in males (with double mesial grooves) and show a marked degree of sexual dimorphism. In fact, the populations from Rusinga and Mfwangano islands show greater degrees of sexual dimorphism in canine size than are found in extant baboons or gorillas. Because of this, it is possible that more than one species is being sampled (Andrews, 1978*a*).

The lower third premolar is sectorial as in gibbons, and the upper third premolar has a strongly projecting buccal cusp. The five main cusps of the lower molars are arranged around the periphery of the crowns and are connected by well-defined ridges that enclose large trigonid and talonid basins. The buccal cingula of the lower molars are variably developed, but lingual cingula on the upper molars are prominent. The upper molars are of simple construction with well-defined trigones and relatively small hypocones. The upper third molar is usually reduced.

The palate is long and narrow, and the maxillary sinus is well developed. The body and symphysis of the mandible are **robust** (thick-boned) with a well-developed superior transverse torus. The inferior transverse torus usually projects back at least to the extent of the superior torus (Fig. 5.13).

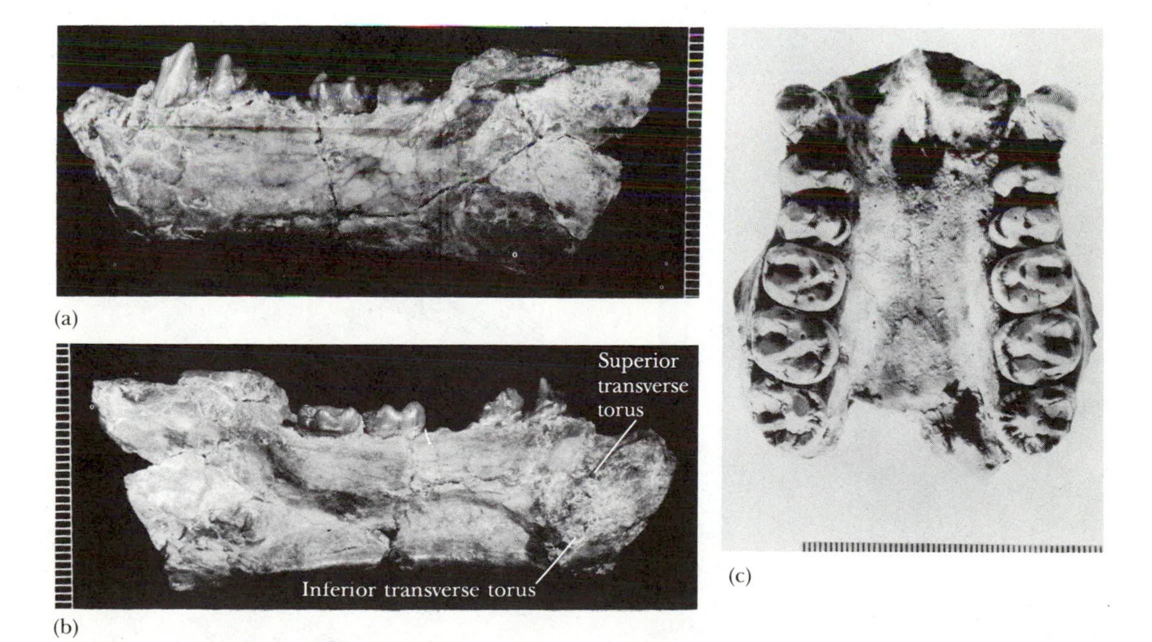

Fig. 5.13 Dentition of *Dendropithecus macinnesi* (a) Lateral and (b) medial views of the lower jaw; (c) occlusal view of the palate. Note the superior and inferior transverse tori in (b). Scale bars show millimeters. (Courtesy of P. Andrews)

The genus shares many dental similarities with *Pliopithecus* from the Miocene of Europe (to be discussed later in the chapter) and with *Limnopithecus,* and all three have traditionally been considered hylobatid ancestors (Le Gros Clark and Thomas, 1951; Simons, 1972). *Dendropithecus* and *Pliopithecus* are now often placed in the family **Pliopithecidae,** although some authors have suggested that *Dendropithecus* is more closely related to other dryomorphs than it is to *Pliopithecus* (Fleagle and Simons, 1978; Fleagle and Kay, 1983).

Micropithecus The fifth genus of Miocene dryomorph from the early Miocene of East Africa is *Micropithecus clarki* (Fleagle, 1975; Fleagle and Simons, 1978) (Fig. 5.14). The genus is known from sites in Uganda (Napak) and Kenya (Koru). *Micropithecus clarki* is the smallest known hominoid primate, living or fossil, and is about the size of *Cebus albifrons* (2 to 4 kg). It is slightly smaller than *Propliopithecus chirobates* or *P. haeckeli* from the Fayum (see Chapter 4). It is similar to extant gibbons in facial morphology. Dentally it most resembles the small hominoids *Dendropithecus* and *Limnopithecus* from the early Miocene of Kenya. *Micropithecus* has large canines and incisors relative to cheek-tooth size. The canines are daggerlike, and the anterior lower premolar is a laterally compressed honing tooth. The upper molars are unique among early hominoids in terms of their triangular shape and abbreviated cingula. A frontal fragment from Napak, originally assigned to a colobine monkey (Pilbeam and Walker, 1968), is probably that of *Micropithecus.*

Recently a specimen from Kiangsu Province, China, has been described that consists of a maxillary fragment preserving the first through third

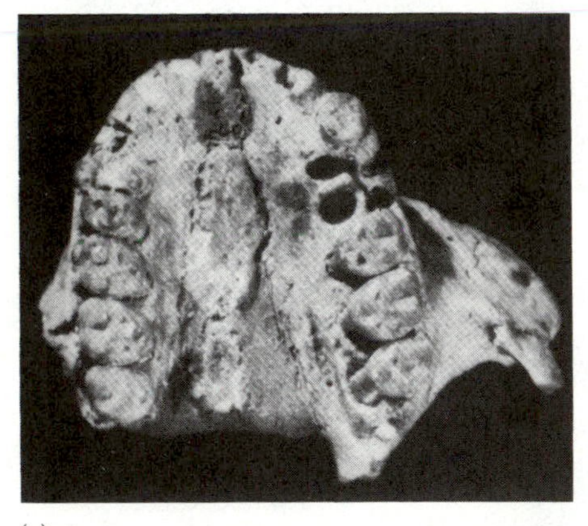

(a)

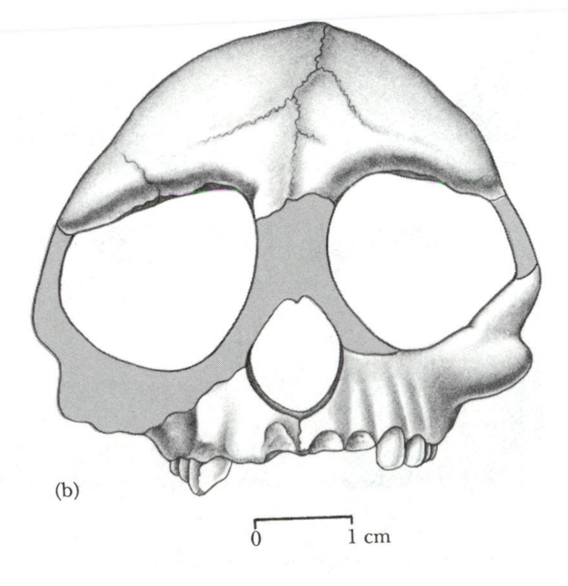

Fig. 5.14 *Micropithecus clarki* (a) Palate 1.5 × natural size); (b) facial reconstruction. (From Fleagle and Simons, 1978)

molars. It has been given the name *Dionysopithecus shuangouensis* (Li, 1978); however, it may be synonymous with *Micropithecus* since the morphology of the two fossils is very similar.

Craniodental Comparison of African Dryomorphs and Modern African Hominoids As we shall see in the later section "Phylogeny and Classification of Miocene Hominoids," many authors originally tried to establish direct ancestor–descendant relationships between various African Miocene hominoids and modern apes (Hopwood, 1933; Pilbeam, 1969; Simons, 1972). In order to properly assess such hypotheses, it is necessary to compare some craniodental features of modern apes with these East African fossil hominoids.

In the region of the snout, the premaxilla is relatively longer in extant great apes than in Miocene dryomorphs. The morphology of the **alveolar processes** (the portion of the upper jaw housing the teeth) also differs. They are long in modern apes, often projecting as bony tuberosities beyond the third molar, but they are less strongly developed in Miocene dryomorphs.

The overall shape of the palate is quite distinctive in modern great apes. The distance between the canines is as great or greater than the distance between the molars, thus giving the palate a somewhat rectangular appearance. In dryomorphs (except for *Proconsul major*) the palate narrows anteriorly to form a more V-shaped configuration (Leakey, 1968; Zwell, 1972).

All modern hominoids preserve the primitive primate heritage of **maxillary** and **sphenoidal sinuses** (large air-filled spaces in the skull); the former penetrate down between the roots of the molars so that the floor of the sinus is divided up into many air chambers. The structure of these sinuses varies within the group. In orangutans the maxillary sinus has expanded at the expense of the sphenoidal sinus and occupies an enormous area within both the maxilla and the sphenoid. It even extends into other areas of the skull including the pterygoid plates and palatine processes and the temporal and frontal bones. Chimpanzees and gorillas share with humans the extensive development of the **ethmoidal sinuses,** outgrowths of which are the twin **frontal sinuses** (Cave and Haines, 1940; Ward and Pilbeam, 1983; Ward and Kimbel, 1983) (Fig. 5.15).

Rangwapithecus gordoni has the most extensive maxillary sinus system of all the East African Miocene hominoids. It is followed closely in this feature by *R. vancouveringi.* The floor of the sinus penetrates deeply between the roots of the molars and extends laterally into the zygomatic process of the maxilla as well. *Dendropithecus macinnesi* also has a fairly extensive maxillary sinus. By contrast, *Proconsul africanus* seems to have a more restricted maxillary sinus. A frontoethmoidal sinus similar to that of the modern African apes and humans is found in *P. africanus* and *P. major.* The presence of a true frontal sinus in *Proconsul* has been taken

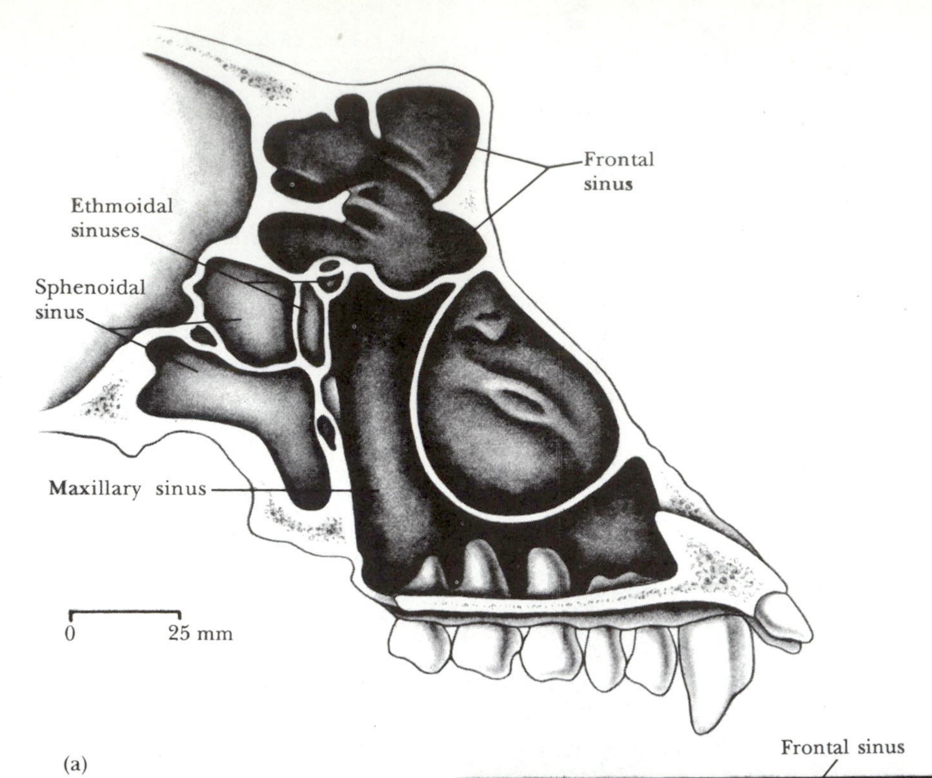

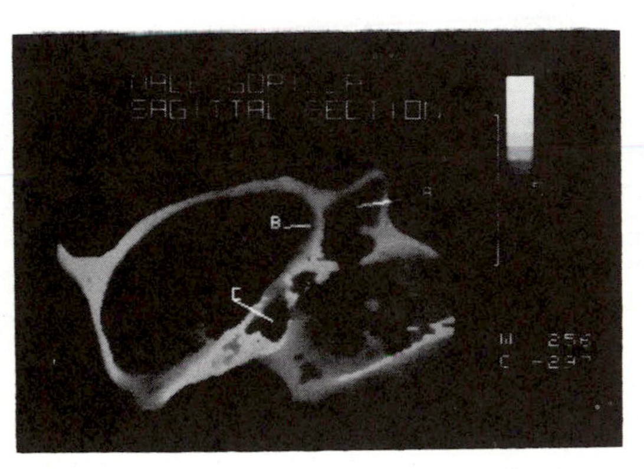

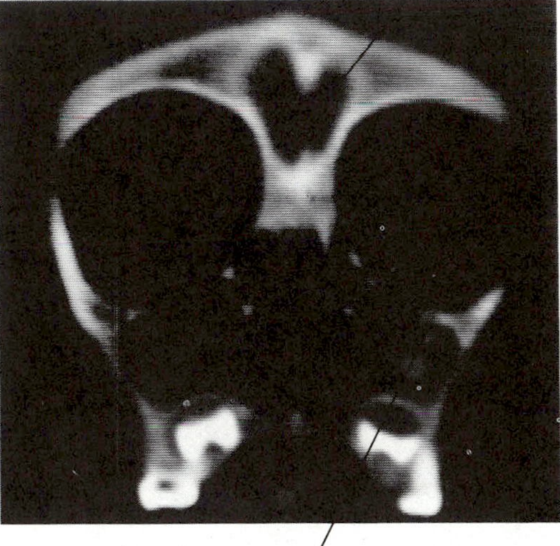

Fig. 5.15 Sinus System in Modern African Apes (a) Parasagittal section of an adult male gorilla showing position of sinuses mentioned in the text. (b) Three-dimensional computer-generated sagittal section of male gorilla skull produced from contiguous 2-mm-thick CT scans: A, frontal sinus; B, anterior limit of the frontal cortex; C, sphenoidal sinus; scale bar indicates 5 cm (c) Coronal section (side to side through the crown of the skull) from a 2-mm-thick CT scan of a chimpanzee showing the paranasal sinuses. In both species note the extensive pneumatization (system of air space). (a, Redrawn from Cave, 1961)

by some to be a taxonomically significant feature, since among modern hominoids it is found only in humans and African apes; it is absent in orangutans, gibbons, and all other primates (Ward and Pilbeam 1983). It is one piece of evidence that *Proconsul* may be the sister group of modern African apes.

The inferior transverse torus (simian shelf) of the mandible extends back as far as the fourth premolar or first molar in the orangutan and to the first molar in the gorilla and (sometimes) in the chimpanzee. The mandibular symphysis in most African dryomorphs differs strikingly from that of modern apes in that none of them have a well-developed simian shelf (although the inferior transverse torus may be developed in some Eurasian species, for example *Sivapithecus indicus* and *Dryopithecus fontani*). In the African species the superior transverse torus is always well developed and relatively large, reaching its maximum development in *Rangwapithecus gordoni* and *Proconsul major*. *Dendropithecus* differs from other dryomorphs but resembles the Oligocene primate *Aegyptopithecus* in having an inferior torus that sometimes projects further back than the superior torus.

Modern hominoids have relatively large incisors compared to many other mammals. The central incisors are broad and spatulate, and the upper incisors usually have a central pillar or tubercle. Chimpanzees and orangutans have relatively larger incisors than those of other apes, which reflects the increased amount of incisor use needed in their frugivorous diets (Hylander, 1975). In general, dryomorphs have relatively smaller incisors than modern apes. As a rule, hylobatids have upper central incisors that are narrower, higher-crowned, and less spatulate than those of the other extant apes. *Dendropithecus* and *Pliopithecus* resemble hylobatids in these features.

In modern apes the canines are variable in size but are always pointed and tusklike. The upper canines tend to be blunted over time by wearing against the lower third premolar. In gibbons and siamangs, however, the canine is sharpened by honing against the lower third premolar. The canines in dryomorphs tend to be less tusklike and robust than in modern apes but dryomorphs do possess the sectorial canine–lower third premolar complex.

The upper molars of apes are usually broader than long. They are four-cusped with a well-developed hypocone in addition to the three main cusps of the trigone. Accessory cusps are uncommon. Wear is usually much heavier on the lingual side, mirroring the heavier buccal wear of the lower molars.

The lower molars in modern apes are invariably longer than they are broad. In the East African dryomorphs the lower molars are relatively much more elongated than in modern apes. They generally have a pattern of five cusps and rarely have additional cusps. The occlusal ridges are usually less accentuated (except in hylobatids) compared with the Miocene hominoids. Cusp morphology of the lower molars in Miocene hominoids is basically similar to that of the modern apes. Many of the East African fossil hominoids have a strong buccal cingulum on the lower molars in contrast to the absence of such a structure in most Eurasian fossil hominoids and living apes.

From this brief survey it seems that the *Proconsul* species were unlike modern apes in palatofacial morphology but shared a primitive catarrhine pattern more reminiscent of extant cercopithecoids. By middle Miocene times, however, some fossil hominoids, such as the ramamorph *Sivapithecus,* had developed the palatofacial morphology typical of hominoids, a phenomenon that appears to be true for the postcranial material as well (McHenry et al., 1980).

Postcranial Adaptations of African Miocene Dryomorphs The elucidation of locomotor adaptations in *Proconsul* and *Dendropithecus* has important implications for primate paleobiology. If the early Miocene fossils already had many of the unique postcranial specializations found in extant hominoids, then an early Miocene divergence of lineages leading to chimpanzees and gorillas would not have to imply extensive parallel evolution of the hominoid skeleton. On the other hand, if the fossils do not have these specializations, then either there occurred extensive parallel evolution after an early Miocene divergence, or hominoids must have shared a late common ancestor that had these unique hominoid locomotor adaptations. The latter alternative is certainly more likely if one accepts the molecular-clock evidence (discussed later in the chapter) of a late separation between great apes and humans (Sarich and Cronin, 1976, 1977; Greenfield, 1979, 1980, 1983; Cronin, 1983).

Our understanding of early Miocene hominoid postcranial material is based on a fossil record that consists mainly of unassociated fragments of different individuals from various sites. Although many skeletal parts are known from various *Proconsul* and *Dendropithecus* species, major uncertainties have arisen due to lack of knowledge of limb proportions and to difficulties in assigning isolated postcranial remains to species based mainly on teeth and jaws. Most of our knowledge concerning the postcranial adaptations of Miocene hominoids has been based on the partial skeleton of *Proconsul africanus* originally described by Napier and Davis (1959). Postcranial adaptations of *Dendropithecus* (formerly *Limnopithecus*) were originally described by Le Gros Clark and Thomas (1951).

When only the forelimb of *P. africanus* was known, hypotheses varied considerably as to the possible locomotor adaptations of this species (Napier and Davis, 1959; Conroy and Fleagle, 1972; Zwell and Conroy, 1973; Schon and Ziemer, 1973; Morbeck, 1975; O'Connor, 1975; Corruccini et al., 1975, 1976; McHenry and Corruccini, 1983). These hypotheses rested largely on the evidence of the elbow joint, distal ulna, and carpal bones (Fig. 5.16). John Napier and Peter Davis (1959) originally described *P. africanus* as a medium-sized quadrupedal, semibrachiating animal that lacked special adaptations for living on the ground.

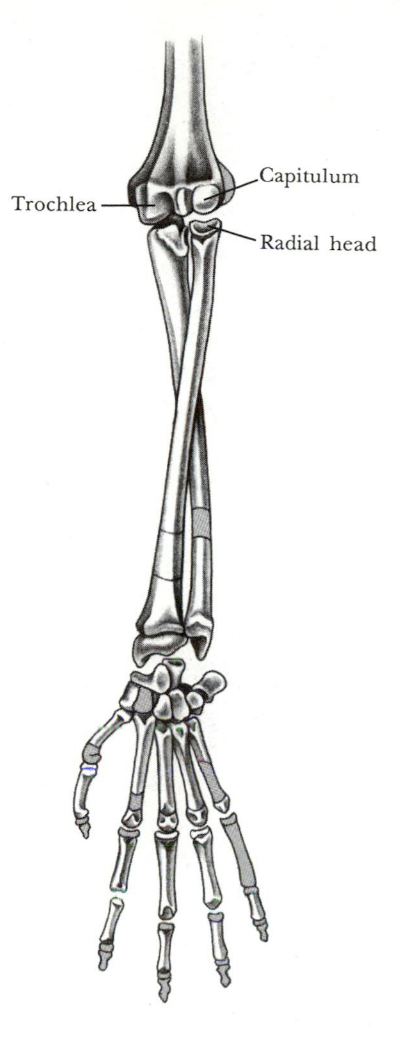

Fig. 5.16 Reconstruction of the Left Forearm and Hand Skeleton of *Proconsul africanus* Note the wide trochlea and spherical capitulum. The latter, combined with the rounded radial head, permits increased range of pronation and supination. Scale about 0.4 × natural size. (Redrawn from Walker and Pickford, 1983)

They considered *Presbytis* (leaf monkeys and langurs of southern Asia) to be the closest living analogy in terms of postcranial morphology and inferred locomotor behavior. This interpretation was challenged by O. J. Lewis (1969, 1971, 1972*a*) on the basis of a series of studies on wrist-joint mechanisms of living hominoids, cercopithecoids, and cebids. Lewis held that the wrist of *P. africanus* showed some clear specializations for suspensory habits and brachiation. He related these features to the increased capacity in brachiators for **pronation** (rotation of the hand so that the surface of the palm faces downward or posteriorly) and **supination** (rotation of the hand so that the surface of the palm faces upward or anteriorly). In addition, he pointed to a locking mechanism at the midcarpal joint that resisted tensile forces across the carpus during brachiation. It was pointed out, however, that the most accomplished brachiator, the gibbon, has a wrist joint lacking many of these supposedly

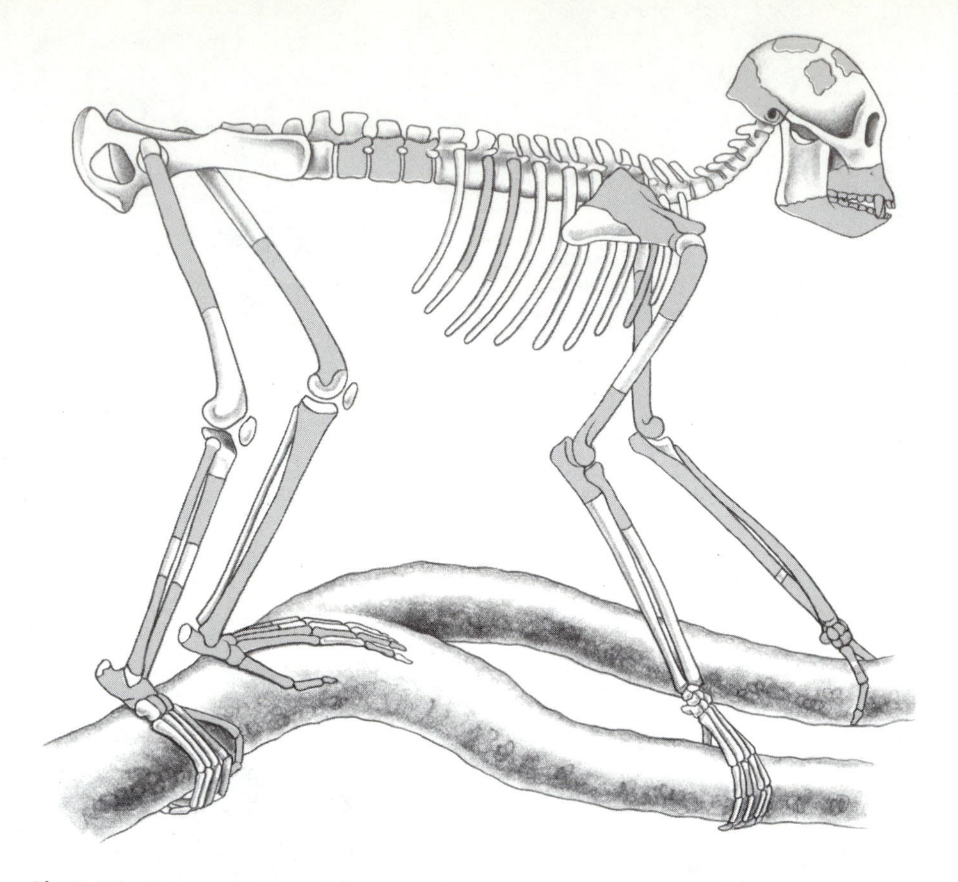

Fig. 5.17 Reconstruction of the Complete Skeleton of *Proconsul africanus* About ⅕ × natural size. (Redrawn from Walker and Pickford, 1983)

brachiating features. Ironically, the primates that do show such features are the knuckle-walking great apes (Conroy and Fleagle, 1972). (**Knuckle walking** is a type of quadrupedal gait used by chimpanzees and gorillas in which the upper body is supported by the dorsal surfaces of the middle phalanges of the hands.)

Overall, the anatomy of the wrist of *Proconsul* is still very monkeylike. Comparative anatomical and statistical analyses of the *Proconsul* carpal bones suggest that the wrist attained its close-packed and habitual position when the pronated hand was **palmigrade** (placed flat on the ground), as occurs in the quadrupedal stance and locomotion of arboreal monkeys; terrestrial monkeys tend to be more **digitigrade** (supported on the digits of the fingers and toes, rather than on the palms and soles) (Schon and Ziemer, 1973; Corruccini et al., 1975, 1976; Beard et al., 1986).

Recent fieldwork on Rusinga Island has resulted in the discovery of parts of five more *Proconsul africanus* individuals. There are two adults, a subadult, and two infants. Nearly all parts of the skeleton are now represented in one or another of the individuals (Walker et al., 1985) (Fig. 5.17).

Part of the new postcranial material includes a right proximal humerus and partial scapula. The fragmentary scapula looks like a small version of a chimpanzee scapula (Walker and Pickford, 1983). The articular surface of the humeral head faces posteromedially (as in hominoids) rather than directly posteriorly (as in quadrupedal monkeys). The difference in these two conditions is due to the fact that in hominoids the shoulder region is more flexible and better adapted for movements in all planes, whereas the cercopithecoid shoulder is basically specialized for **parasagittal** limb movement (that is, in a plane parallel to the long axis of the body).

The elbow region of *P. africanus* has received considerable attention. There are many osteological features in the elbow that reflect functional differences between modern hominoids and cercopithecoids (Fig. 5.18). Henry McHenry and Robert Corruccini (1975*a*) describe these differences as follows.

In hominoids the distal humerus is wide with an especially wide trochlea. The **capitulum** (the small, rounded head on the distal humerus that articulates with the radius) is spherical and smooth, and there is a lateral trochlear ridge separating the radial and ulnar articulations. This braces the ulna and reduces the contribution of the radial articulation to joint stability. The wide proximal ulna—in particular, the relatively wide **sigmoid notch** (the concave depression in the proximal end of the ulna that articulates with the trochlea)—indicates a broad fit between the ulna and the trochlea. This wide ulna–trochlea contact is the main determinant of joint stability, thus freeing the radius for pronation and supination. The head of the radius is round, and the **radial notch** (the depression near the proximal end of the ulna that articulates with the head of the radius) is shallow, reflecting the increased potential for independent rotation.

Other characteristics of the hominoid elbow joint to note are the short **olecranon process** (the projection at the proximal end of the ulna for attachment of the **triceps,** the major antigravity muscle of the upper extremity in quadrupeds), and the shallow **olecranon fossa** (the depression at the posterior distal end of the humerus that receives the olecranon process during extension of the forearm). The medial epicondyle is large, projecting medially from the trochlea, and the lateral epicondyle is not prominent, reflecting the fact that forearm and wrist extensors are reduced relative to those in cercopithecoids.

The cercopithecoid elbow joint contrasts with the hominoid condition in a number of ways. Elbow stability in monkeys depends on both radial and ulnar articulations. The radial head is elliptical rather than spherical. The trochlea is narrow mediolaterally and thick in the anteroposterior plane. The medial edge of the trochlea is a concave, downward-projecting rim, which allows the cercopithecoid elbow to lock when the forearm is extended and prone. The medial epicondyle is smaller and directed more posteriorly, the lateral epicondyle is more prominent,

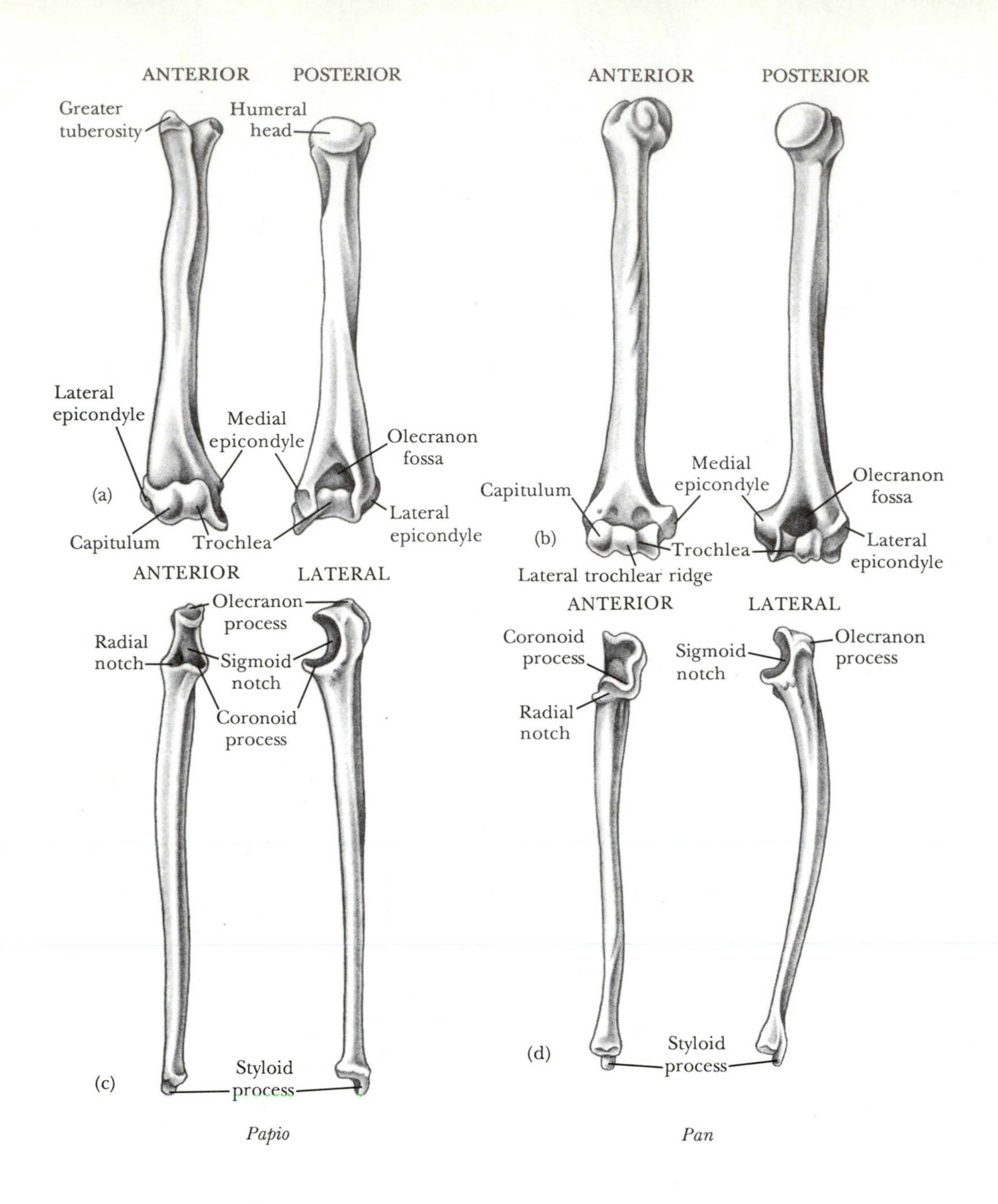

Fig. 5.18 Comparisons of Cercopithecoid *(Papio)* and Hominoid *(Pan)* Forelimb Bones (a,b) Front and back views of the baboon humerus compared with that of the chimpanzee. Note differences in position of the humeral head and size and shape of the trochlea, capitulum, and medial epicondyle. (c,d) Front and side views of the baboon ulna compared with that of a chimpanzee. Note differences in size and orientation of the olecranon process. *P. africanus* resembles the hominoid condition in forelimb morphology. (Redrawn from Swindler and Wood, 1973.)

and the olecranon fossa is deeper than in hominoids. The cercopithecoid ulna has an elongated olecranon process, which is associated with its large triceps. The large olecranon also prevents elbow hyperextension. The sigmoid notch is narrow.

Based on a multivariate analysis of the distal humerus in *Proconsul africanus,* McHenry and Corruccini (1975*a*) concluded that it is more similar to the cercopithecoid humerus than to that in extant hominoids, although in some ways it is intermediate between the two. In the ulna the olecranon and styloid processes are relatively longer than comparable parts in the living great apes.

The **talocrural** (ankle) **joint** in *P. africanus* is quite distinctive in that the trochlear surface of the talus is highly curved and quite deeply grooved. In many features this joint is similar to that of a number of arboreal anthropoids but lacks specialized features evident in extant groups, especially the emphasized development of lateral trochlear margins best developed in some cercopithecines, or the more squared-off and flattened trochlear surface seen in great apes. In functional terms the morphological variations in extant anthropoids relate to the degree to which the foot is "preset" into an inverted position due to the shape of the joint surfaces, the degree to which **dorsiflexion** and **plantarflexion** (movements of ankle flexion and extension) occur, and the means by which the joint becomes **close packed** in dorsiflexion. In *Proconsul* the inversion set of the joint was probably intermediate between that seen in many monkeys and great apes. In the talus the cup-shaped articulation for the **tibial malleolus** and the presence of a dorsiflexion "stop" are features mostly seen in cercopithecoids. They enable the joint to take up its close-packed position in dorsiflexion when the propulsive thrust is being generated toward the end of the stance phase (Conroy and Rose, 1983) (Fig. 5.19).

Features of the **subtalar joint** (between the talus and the calcaneum) complex allowing inversion and **eversion** (turning outward) closely approximate the condition seen in many monkeys. As a general impression, the *Proconsul* tarsus is quite long compared with overall foot length but narrow for its length, as in monkeys. This comparative gracility is evident in the proportions of all the tarsal bones (Conroy and Rose, 1983).

It is evident that the *Proconsul* foot has features characteristic of both monkeys and apes. In its major proportions the *Proconsul* foot is a gracile version of an African ape foot. Quadrupedal walking and running, as seen in generalized (nonbrachiating or leaping) arboreal New and Old World monkeys, was probably an important part of the locomotor repertoire of these primitive hominoids. This is suggested by some features of the talocrural, subtalar, talonavicular, and digital joints, together with the presence of a **transverse arch** (the side-to-side concavity of the sole of the foot formed by the configuration of the tarsometatarsal joints) (Conroy and Rose, 1983).

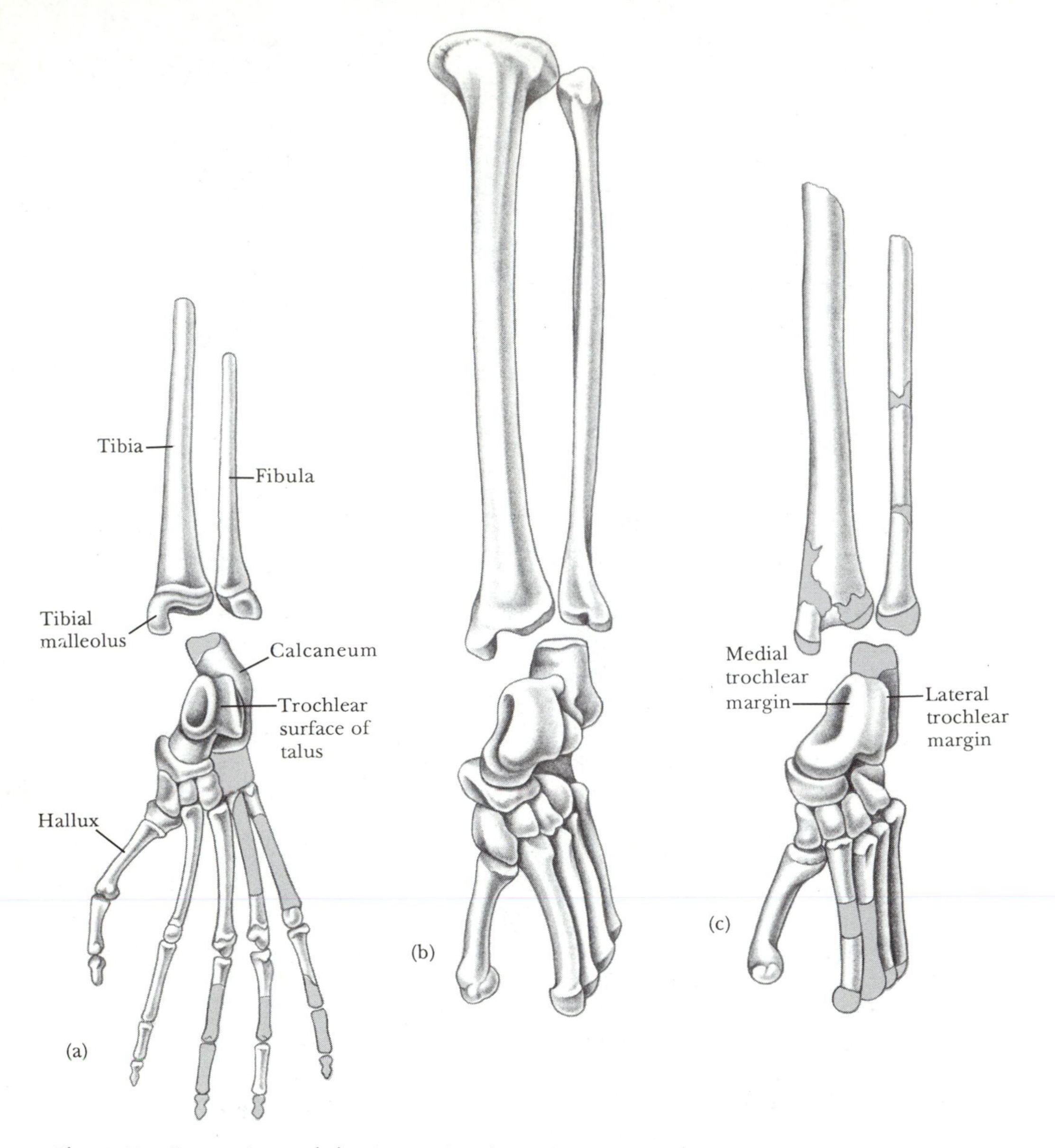

Fig. 5.19 Comparison of the Lower Leg in *Proconsul* and Chimpanzee (a) *Proconsul africanus;* (b) *Pan troglodytes;* (c) *Proconsul nyanzae.* Note the opposable, grasping hallux in all three. About ⅓ × natural size. (Redrawn from Walker and Pickford, 1983)

From the recent discoveries on Rusinga Island, more of the important hindlimb skeleton of *P. africanus* and *P. nyanzae* has been recovered (Walker and Pickford, 1983; Rose, 1983). These new data permit limb proportions of *P. africanus* to be accurately calculated for the first time. Importantly, the body weight for this individual can now be reliably estimated, and hypotheses concerning the locomotor adaptations of this species based on the previously known forelimb anatomy can be checked against the new hindlimb evidence. In addition, the body weight of *P. nyanzae* can be estimated based on its hindlimb skeleton, and meaningful comparisons can be made between the two species.

For the first time an accurate brachial index can be calculated for an African Miocene hominoid. The brachial index of 96 in *P. africanus* is close to that of many other extant primates. It is particularly close to the values recorded for *Pan*. Although the brachial index is well within the range recorded for *Pan*, the crural and intermembral indices are outside *Pan*'s range. The crural index of 92 lies within the range of values for quadrupedal primates of diverse locomotor types, as does the intermembral index of 87. In fact, the overall limb proportions are not like those of any extant primate (Walker and Pickford, 1983). The skeleton has the general size and robustness of *Colobus guereza*, suggesting a body weight of approximately 11 kg.

The hindlimb skeleton of *Proconsul nyanzae* suggests an animal similar in size to female *Pan troglodytes*—possibly around 40 kg. The weight of *P. nyanzae* seems to be at least three times that of *P. africanus*, clearly signifying that they represent two distinct species (Walker and Pickford, 1983).

In general, the robustness of the skeleton of *P. africanus* is surprising. The limb bones of this individual seem quite strongly built, and the articular surfaces are relatively large for such a small primate. The scapula and proximal humerus suggest that ranges of motion at the shoulder joint were large. Similarly, ranges of elbow-joint motion seem to have been extensive, and the range of supination and pronation of the forearm were fairly large in all elbow-joint positions. The radius is stout and strutlike, whereas the ulna is relatively straight. There are no overt signs in the hand of any special knuckle-walking postures; it is safest to assume that the animal was palmigrade. The fibula and the hallux are stout, both indicators of strong grasping and climbing capabilities. It appears that in the forelimb there is a hominoid–cercopithecoid gradient of features along the limb such that the more hominoidlike features are found more proximally, at the shoulder. In the hindlimb the gradient is reversed with the more hominoidlike features toward the ankle (Walker and Pickford, 1983).

Dryopithecus *Dryopithecus* is the only dryomorph present in the Miocene of Europe; however, the dryomorph *Micropithecus* (=*Dionysopithecus*) is present in the Miocene of Asia. In fact, the first fossil anthropoid jaw ever described was that of the dryomorph *Dryopithecus fontani* (Lartet, 1856) (Fig. 5.20). It was discovered in Miocene sediments near the village of St. Gaudens, France, by the naturalist M. Fontan. A hominoid humerus was also described from the same site. Since oak leaves had been found at other sites in France with a fauna similar to that at St.

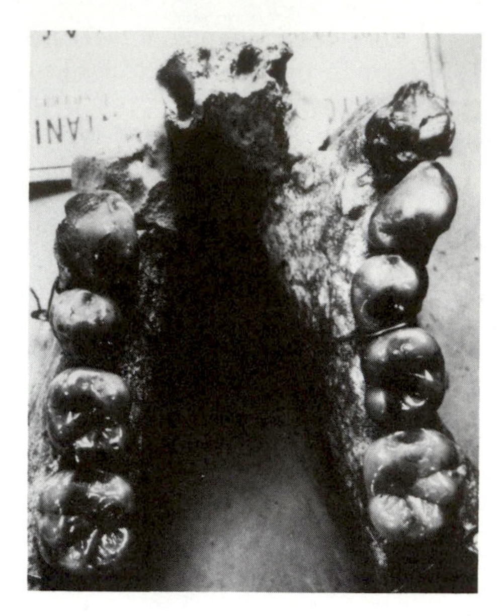

Fig. 5.20 Type Specimen of *Dryopithecus fontani* Discovered in the mid-19th century from Miocene deposits in France. (Courtesy of E. Delson).

Gaudens, Lartet named the fossil ape after the dryads, or oak nymphs, of Greek mythology (Simons, 1972). Two other lower jaws from the same site were later discovered and assigned to the same species (Gaudry, 1890; Harle, 1898).

Over the past century, a number of other specimens also referred to *Dryopithecus* have been recovered from other parts of Europe including Spain, Germany, Czechoslovakia, Hungary, and Greece (Branco, 1898; Schlosser, 1901; Woodward, 1914; Villalta and Crusafont, 1944). As was the typical practice of the day, many of these specimens, even isolated teeth, were given separate species names. Much of this taxonomic confusion has been clarified by Simons and Pilbeam (1965) and more recently by Kay and Simons (1983*b*). These authors essentially agree that all the European specimens assigned to *Dryopithecus* can be accommodated within two species, *D. fontani* and *D. laietanus*.

Dryopithecus can be distinguished from African dryomorphlike *Proconsul* species chiefly by dental features.

1. The incisors tend to be narrower and less spatulate, and the canines are more compressed laterally.
2. The buccal cusps of the upper third premolars are not as high or projecting as the lingual cusps.
3. The upper molars do not have the well-marked occlusal ridges or the development of protoconules seen in *Proconsul*.
4. There are no prominent lingual cingula on the upper molars or buccal cingula on the lower molars.

5. Cusp projection and wrinkling of the occlusal enamel is generally not as great as in *Proconsul.*
6. The lower third molars are not as large, nor do they show prominent development of hypoconulids.
7. Large mandibular inferior transverse tori are present but not superior ones. (Andrews, 1978*a*)

Ramamorphs

While dryomorphs are found primarily in the Miocene of East Africa and Europe, ramamorphs are found predominantly in Eurasia and are relatively rare in East Africa. The number of ramamorph genera and species is a matter of some dispute, since workers at different sites have tended to invent new generic names to describe their finds. For example, ramamorph "genera" have at one time or another included *Sivapithecus, Ramapithecus, Kenyapithecus, Gigantopithecus, Ouranopithecus, Graecopithecus,* to name just a few. This discussion will follow the view that all these Miocene ramamorphs can be reduced to two genera, *Sivapithecus* and *Gigantopithecus.* A number of species are included in *Sivapithecus (S. indicus, S. sivalensis, S. simonsi, S. africanus, S. darwini* and *S. meteai),* and one species is included in *Gigantopithecus (G. giganteus)* (Kay, 1982). One other *Gigantopithecus* species from the Plio-Pleistocene of China, *G. blacki,* has also been described.

Gigantopithecus was probably the largest primate that ever lived. Based on dental dimensions it could have weighed anywhere from about 150 to 230 kg (Conroy, 1987). The dentition of these animals is characterized by small, vertical incisors and short, stubby canines. The cheek teeth are large with flattened crowns and thickened enamel. The jaws housing these teeth are thick and deep (Fig. 5.21). The total morphological pattern of jaws and teeth indicates a diet consisting of hard, fibrous material, similar perhaps to the diet of the giant panda of today. Because of their great size, *Gigantopithecus* must have been mainly terrestrial.

The various ramamorph genera are listed in Table 5.1.

Assigning ramamorph genera to higher taxa has also been difficult (Greenfield, 1980; Kelley and Pilbeam, 1986). The ramamorphs *Sivapithecus* and *Gigantopithecus* are now usually considered to be in the same subfamily, Ramapithecinae, or the same family, Ramapithecidae (= Sivapithecidae) (but note that in Table 5.1 we consider them part of the Pongidae). We will consider the taxonomy of ramamorphs in greater detail later in the chapter.

A number of ramamorph fossils have been found at the same sites and even in the same strata, as dryomorphs and pliomorphs. Although a few fossils bear morphological affinities to both dryomorphs and ramamorphs, these two groups can usually be distinguished from one another,

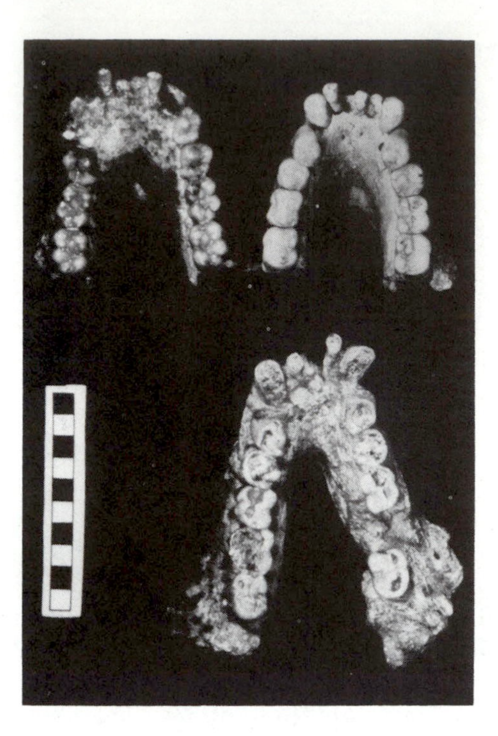

Fig. 5.21 Dentition of *Gigantopithecus* Three mandibles of *G. blacki* from the early Pleistocene of China. Judging from the size of the jaws and teeth, *Gigantopithecus* was probably the largest primate that ever lived. Note the robust jaws and large molars and premolars. Scale in centimeters. (Courtesy of E. Delson)

as we shall see shortly. Separating ramamorphs from pliomorphs, on the other hand, has never been an issue.

Since there is still so much disagreement about the taxonomy of the ramamorphs, we will cover the fossil evidence for this group by geographical region, rather than by taxon.

Ramamorphs from Europe and Asia Minor Ramamorphs have been found at several sites in Czechoslovakia, Greece, Hungary, and Turkey.

From Neudorf an der March in Czechoslovakia, several isolated hominoid teeth have been described (Abel, 1902; Steininger et al., 1976). These teeth probably belong to *Sivapithecus darwini*. Specimens of *S. darwini* taken from this site (and from a site in Turkey to be discussed shortly) are the oldest-dated *Sivapithecus* uncovered so far (see Fig. 5.7).

Andrews (1978*a*, 1978*b*) has suggested that *S. darwini* from Neudorf an der March could be derived from a form similar to *Proconsul major*. In fact, he considered the series *P. major–S. darwini–S. meteai* (from Turkey)–*G. giganteus* (from India)–*G. blacki* (from China) to be one of the best-documented lineages in the fossil record.

Since 1973 a team of Greek and French paleontologists have recovered numerous jaws and teeth of hominoids from a site near Salonika, Greece. The fauna suggests an age of about 10 or 11 MYA. L. de Bonis and J. Melentis take the position that a single gorilla-sized species called *Ouranopithecus macedoniensis* is represented (de Bonis et al., 1974; de Bonis and Melentis 1977, 1978, 1980). They consider this species to be close to *Sivapithecus* and possibly near the ancestry of *Gigantopithecus.* As mentioned earlier, most workers would now assign this material to *Sivapithecus* because the molars are low-crowned and thick-enameled. Other characteristic ramamorph features include the broad, shallow mandibular body, the short and stout upper canines, the prominent maxillary canine fossa, and the size disparity between the upper central and lateral incisors. It is also possible that more than one species is being sampled at this site since the teeth show greater metric variability than seen in the most sexually dimorphic of extant apes. The smaller species falls within the range of variation of *S. indicus* in most dental dimensions, whereas the larger species is considerably larger than *S. indicus* but smaller than *G. giganteus* (found in the Siwaliks) (Kay, 1982).

Elsewhere in Greece a lower jaw of a ramamorph has also been found at Pyrgos on the outskirts of Athens. This specimen was described by G. H. R. von Koenigswald (1972) as *Graecopithecus freybergi.* The specimen has recently been restudied by Martin and Andrews (1984), who include it in the taxon *Sivapithecus meteai,* a species otherwise known only from the middle to late Miocene of Turkey.

Several species of Miocene hominoids have been recovered from deposits in northeastern Hungary at a site called Rudabanya (Kretzoi, 1975). This site contains an extensive Miocene fauna with a suggested age of approximately 12 MYA. The flora indicates a Mediterranean to subtropical climate with traces of wet forest to grassland and higher-mountain elements.

Kretzoi (1975) recognized three higher primate species at Rudabanya: a large species of *Pliopithecus* and two larger hominoids that he named *Rudapithecus hungaricus* and *Bodvapithecus altipalatus.* The latter is known from a nearly adult mandible that preserves all the lower teeth on one side or the other except for the third molar. The type specimen is a maxilla with the third premolar through first molar preserved. These specimens resemble the large *Sivapithecus* species from India, *S. indicus,* in size and dental proportions and thus most probably represent this latter species (Kay, 1982). It should be noted, however, that part of a frontal bone reveals that this specimen had a rather broad interorbital region quite unlike that known from *S. indicus* specimens from Pakistan (see the following subsection, "Ramamorphs from the Siwaliks").

Many authors have argued that *Rudapithecus* is very similar to the small species of *Sivapithecus, S. sivalensis* (previously known as *Ramapithecus*) (Andrews, 1978*a;* Greenfield, 1979; Pilbeam, 1979; Wolpoff, 1980). However, others have argued that *Rudapithecus* may be **conspecific** (of the same species) with *Dryopithecus fontani* since both share the dryomorph pattern of very thin enamel coupled with gracile jaws (Kay and Simons, 1983*b*).

Ramamorphs are known from three Turkish localities. Pasalar, in northwestern Anatolia, has yielded about 100 isolated teeth of at least

20 individuals, which can be grouped into perhaps two species. Pasalar deposits have been dated by faunal correlation at 15 to 16.5 million years old (Andrews and Tobien, 1977), but they are possibly as young as 13 to 15 million years old. The Candir Formation, northeast of Ankara, has yielded the type specimen of *Sivapithecus alpani* (now referred to *S. sivalensis*), a mandible about 12 million years old. The middle Sinap series exposed on Mt. Sinap, northwest of Ankara in central Anatolia, has produced *Sivapithecus (= Ankarapithecus) meteai,* which consists of mandibular fragments with teeth and a palate and lower face (Ozansoy, 1957; Andrews and Tekkaya, 1980).

The molars from the larger species at Pasalar are about the size of those in *S. indicus.* Unlike the teeth of any Siwalik hominoid, the lower molars have a prominent buccal cingulum reminiscent of the condition in *Proconsul* and *Sivapithecus darwini.* The molar enamel is quite thick. The smaller species is similar in having (1) well-developed molar cingula; (2) low, rounded molar cusps; and (3) thick molar enamel. Andrews and H. Tobien (1977) assigned the large species to *S. darwini,* claiming the Pasalar specimens were similar to specimens of *S. darwini* from Neudorf an der March in the size of the third molar and in the molar combination of low, rounded cusps, prominent cingula, and thick enamel. The smaller specimens Andrews and Tobien assigned to *Ramapithecus wickeri* (= *S. africanus)* on the basis of third-molar size but they recognized that the presence of cingula makes them unlike any other *Ramapithecus* specimens (Kay and Simons, 1983*b*).

The morphology of the mandible from Candir, Turkey, is similar to that of other small *Sivapithecus* specimens. It is very shallow and broad in cross section across the molar region, and its symphysis has both superior and inferior transverse tori. In overall size of the cheek teeth, *S. alpani* is quite a bit smaller than the size typically seen in Siwalik or African ramamorphs.

Two hominoid specimens come from the middle Sinap series. The type specimen was described under the name *Ankarapithecus meteai* (Ozansoy, 1957, 1965) (but was later classified as *Sivapithecus*). It is a lower jaw with the left fourth premolar through third molar associated with a symphysis with the crowns of the left canine through third molar and right second incisor through canine. A recently described specimen consisting of the complete palate with all the teeth intact comes from a higher stratigraphic level, but the two specimens presumably belong to the same species (Andrews and Tekkaya, 1980). The teeth are in the size range of *S. indicus.* The proportions of the palate and face closely resemble comparable parts of *S. indicus* recently recovered from Pakistan. The Mt. Sinap face is important for the details it reveals about the structure of the palate and face of *S. indicus.* Andrews and Tekkaya (1980) have concluded that the face is similar to that in modern great apes and is unlike *Proconsul* in overall proportions. In addition to the many pongidlike features, the deep and widely flaring zygomatic process, marked prognathism, short upper face, and narrow interorbital distance are features shared only with the orangutan among extant hominoids. In addition, Andrews and Tekkaya have argued that the relatively large first incisor relative to the second incisor and the large squared molars are also orangutanlike features.

Ramamorphs from the Siwaliks (Pakistan and India) By far the most prolific site for ramamorphs has been the Siwalik Hills of southern Asia. Only two fossil anthropoid specimens were recovered from the Siwaliks before 1910, although numerous other fossil mammals had been collected in the previous 80 years (Falconer, 1868; Lydekker, 1879).

Guy Pilgrim (1910, 1915) described several additional species of Siwalik fossil anthropoids: *Dryopithecus punjabicus* from parts of a small lower and upper jaw; *D. giganteus* from a very large lower molar; *Paleosimia* from an upper molar; and *Sivapithecus* from a lower jaw. *Dryopithecus punjabicus* was later reclassified as *Ramapithecus punjabicus* and is often now considered to be *Sivapithecus sivalensis; D. giganteus* became *Gigantopithecus giganteus. Sivapithecus* and *Paleosimia* (later referred to *Sivapithecus*) were both considered by Remane (1921) to be closer to the orangutan than to the African apes.

Further collecting in the 1920s by teams from the American Museum of Natural History produced three more hominoid mandibles, which again were all given separate species names: *Dryopithecus frickae, D. cautleyi,* and *D. pilgrimi* (Brown et al., 1924). These are all probably synonymous with *S. sivalensis* (Simons and Pilbeam, 1965).

Further important discoveries were made by G. E. Lewis in the early 1930s. He described several new hominoid genera from the Siwaliks, including *Ramapithecus.* In fact, he was the first to suggest the possible hominid affinities of this genus by noting that it had a parabolic dental arch (since shown to be incorrect), a slightly prognathous face, no gaps in the dental series, and a relatively small canine.

Fossil collecting in the Siwaliks continued into the 1950s and 1960s, and more anthropoid remains were discovered (Prasad 1962, 1964). Since 1973, a cooperative field program between Yale (later Harvard) and the Geological Survey of Pakistan has uncovered hundreds of new anthropoid specimens (Pilbeam et al., 1977a, 1977b). There are at least 150 hominoid specimens, representing about 100 individuals of Miocene hominoids, now known from the Siwaliks.

During this century the Siwalik primates have been categorized into as many as 14 different genera and 31 species (one genus and one species of Lorisidae, four genera and five species of Cercopithecidae, and nine genera and 25 species of Pongidae). The task of devising synonyms for this plethora of names commenced with Lewis (1937) and continues to this day, a process that will be discussed more fully later in the chapter. The most recent taxonomic review has been that of Kay and Simons (1983*b*). They have concluded that dental variability within *Sivapithecus*

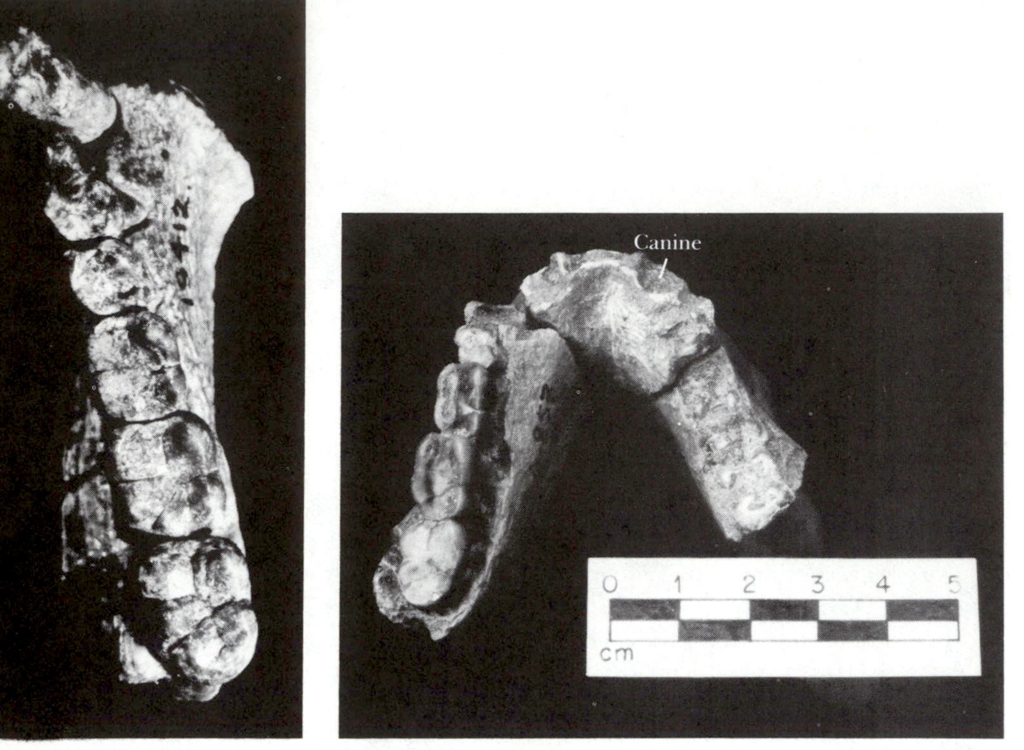

Fig. 5.22 Comparison of the Lower Jaws of Siwalik Hominoids (a) Left lower jaw fragment of a specimen originally described as *Dryopithecus cautleyi* by Brown et al. (1924) but now generally classified as *Sivapithecus sivalensis*. Scale about 1.2 × natural size. (b) Lower jaw of a specimen often classified as *Ramapithecus punjabicus* but also attributed by some workers to *Sivapithecus sivalensis*. Note that the *Ramapithecus* specimen is foreshortened and has a smaller canine (judged by root socket size) than does *S. sivalensis*. Whether these differences represent sexual dimorphism in a single species or morphological distinctions warranting taxonomic separation is an area of current debate. (a, Courtesy of E. Delson; b, courtesy of D. Pilbeam)

sivalensis is sufficient to include all the specimens formerly attributed by Simons and Pilbeam (1956) to *Ramapithecus punjabicus* (Fig. 5.22).

It is well known that intraspecific dental variability in primates is lowest in the upper and lower first and second molars and that sexual dimorphism is greatest in the canines and anterior lower premolars. More than one species is likely to be present in a fossil sample if the **coefficient of variation** (a statistical measure of sample variability independent of size) of upper and lower first and second molars exceeds about 8.5 in teeth with similar morphology. High coefficients of variation or bimodal distributions (curves with two peaks) in the canines and third premolars are indicative of sexual dimorphism. Such dental comparisons indicate that *S. indicus* and *S. sivalensis* are distinct species even though they are found in the same deposits (Fig. 5.23). Neither species appears to be strongly sexually dimorphic in canine or premolar dimensions (Kay and Simons, 1983*b*).

Two Siwalik species are quite different from *S. sivalensis* and *S. indicus* in terms of dental dimensions: *Gigantopithecus giganteus* (formerly *G. bilaspurensis*) and *S. simonsi* (which consists of several mandibles formerly assigned to *Dryopithecus laietanus* by Simons and Pilbeam, 1965) (Kay, 1982). Thus, according to Kay and Simons (1983*b*), only four *Sivapithecus* species are present in Eurasia: *S. indicus, S. sivalensis* (including *Ramapithecus*), *S. simonsi*—all three from the Siwaliks—and *S. darwini* from Czechoslovakia and Turkey. *Sivapithecus africanus* is known only from East Africa (and possibly also from Pasalar, Turkey). I would also add *S. meteai* from Turkey to this list.

Very few postcranial remains of ramamorphs have been positively identified from the Siwaliks, and none is found in direct association with dental remains. Several foot bones, some finger bones, and parts of the humerus, radius, and femur have recently been described (Pilbeam et al., 1977*a;* Pilbeam et al., 1980; Raza et al., 1983; Rose, 1986). The bones range in size from those of a female gorilla to those of a pygmy chimpanzee. It is possible that the larger bones belong to *Gigantopithecus.* If these postcranial bones have been correctly identified, it would indicate that *Gigantopithecus* and/or *Sivapithecus* was truly **megadont** (having cheek teeth very large with respect to body size). The morphology of the foot bones suggests that rotation around the long axis of the foot was considerable at the **transverse tarsal joint** (consisting of parallel articulations

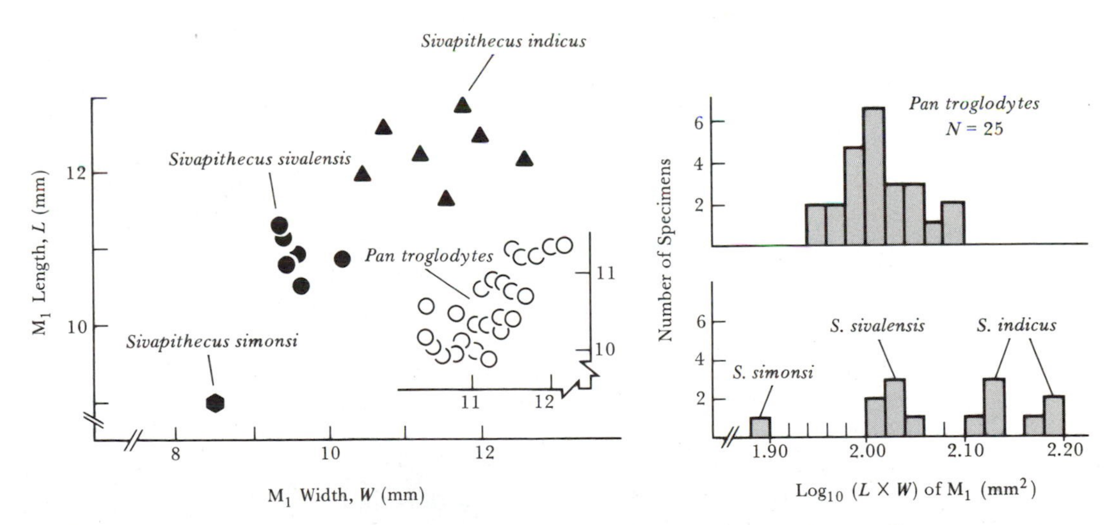

Fig. 5.23 Comparison of Intraspecific Dental Variability in Modern Chimpanzees and Siwalik Ramamorph Sample. Note that lower-first-molar dental variability in the overall Siwalik sample is far greater than that seen in modern chimpanzees, thus confirming that more than one fossil species is being sampled. (From Kay, 1982)

between the calcaneus and cuboid and between the talus and navicular bones) and at the subtalar joint. The articular surfaces of the finger bones are similar to those of palmigrade quadrupeds, although the degrees of shaft curvature and robustness imply that the fingers were also subjected to tensile stresses. This evidence is compatible with the view that *Sivapithecus* engaged in arboreal quadrupedalism and climbing behavior (Rose, 1986).

Ramamorphs from China In the last three decades ramamorphs have been recovered from lignite beds in Lufeng, Yunnan Province, in south-central China. In the mid-1950s, five lower cheek teeth were described as a new species, *Dryopithecus keiyuannensis* (Woo, 1957). Another series of larger teeth was later recovered from the same site and attributed to the same species. Simons and Pilbeam (1965) attributed the first set of specimens to *Dryopithecus (Sivapithecus) sivalensis* and the second to *D. (S.) indicus.* (Note that the taxa given in parentheses identify subgenera, not equivalent genus names.)

Extensive excavations at the same locality since the 1970s have produced a wealth of new hominoid material that includes jaws, skulls, teeth, and a few postcranial bones (Xu and Lu, 1979; Wu et al., 1981, 1983, 1984, 1986). The hominoid fossils were found in deposits representing lake and swamp environments (Wanyong et al., 1986).

The Chinese paleoanthropologists have divided the ramamorph fossils into two genera, *Sivapithecus* and *Ramapithecus.* As of 1986 the ramamorph fossil material consisted of over 350 skeletal parts for each genus. Most of the material consists of cranial and dental remains, the bulk of it isolated teeth; the only postcranial material is a scapula, a clavicle, and two phalanges. As we shall see shortly, there is considerable doubt whether these fossils belong to one, both, or neither of these two genera. English translations of many of the original Chinese articles dealing with these fossils can be found in Etler (1984).

The Lufeng hominoids bear some close examination, since there is doubt that the fossils can truly be said to separate out into separate genera (Fig. 5.24). The Lufeng crania (three assigned by the Chinese to *Ramapithecus* and two to *Sivapithecus*) share the following features:

1. The supraorbital tori are poorly developed and discontinuous over the wide and concave *glabella* (region on the frontal bone between the brow ridges).
2. A **supratoral sulcus** (bony groove between the brow ridges and the rest of the frontal bone) is absent.
3. The interorbital distance is very wide.
4. The orbital contours are square with rounded corners.
5. The nasal aperture is narrow and pear-shaped.
6. The face is short.

The main morphological features of the 10 mandibles (five assigned to each genus) include the following:

(a)

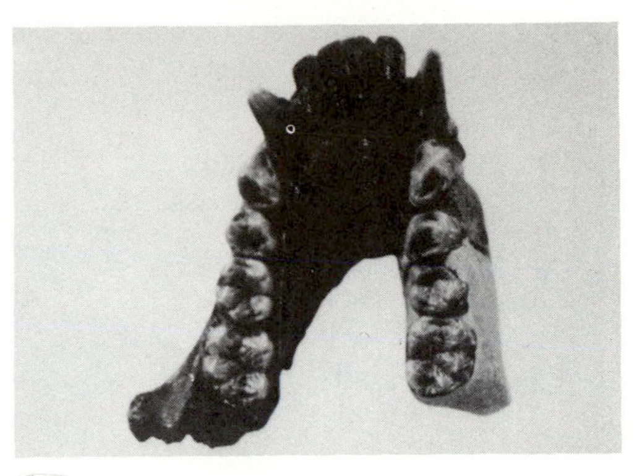

(b)

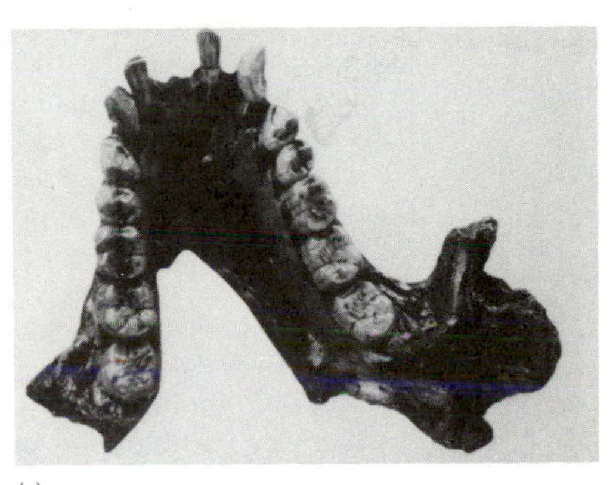

(c)

Fig. 5.24 Miocene Hominoids from Lufeng, China (a,b) Skull and lower jaw of a specimen originally attributed to *Sivapithecus;* (c) lower jaw originally attributed to *Ramapithecus*. It now appears likely that these hominoid specimens are members of a single dimorphic species, tentatively identified as *S. lufengensis;* however, there is evidence that it is generically distinct from other Eurasian *Sivapithecus* specimens, in which case it should be given a new generic name. (Courtesy of R. Wu)

1. The frontal part of the symphysis and the incisors are oriented almost vertically.
2. The depth of the symphyseal region is large compared with the posterior portion of the mandible.
3. The tooth rows diverge slightly.
4. The superior transverse tori are not pronounced.
5. The ascending rami are broad and perpendicular to the mandibular body.

Many of these features are similar to those found in the modern orangutan (Wu et al., 1984).

The differences in dentition and cranial shape are summarized in Table 5.3.

Table 5.3 Craniodental Differences in the Ramamorph Fossils from Lufeng, China

Character Trait	*Fossils Attributed to Sivapithecus*	*Fossils Attributed to Ramapithecus*
Temporal crest	Strong and converging	Weak and separated
Sagittal crest	Strong	None or very slight
Midface	Very wide	Relatively narrow
Zygomatic arches	Laterally expansive	Less expansive
Canine juga	Strongly developed	Weakly developed
Canine fossae	Very deep	Relatively shallow
Canines	Large	Relatively small
Maxillary dental arcade	U-shaped	V-shaped

The few postcranial remains have been described by Wu et al. (1986). The size and shape of the coracoid process and glenoid cavity of the scapula are described as orangutanlike. The morphology of the clavicle is also said to be orangutanlike. In contrast, the phalanges seem to be less like those in the orangutan: for example, the articular surface for flexion and extension is less extensive in the *Ramapithecus* phalanges than in orangutans, suggesting that this movement between proximal and middle phalanges is more extensive in modern orangutans than in this Miocene hominoid. It seems clear that the phalanges were adapted to grasping and hanging since they are very long and strongly curved in a longitudinal direction.

Many of the morphological differences noted between the Chinese *Sivapithecus* and *Ramapithecus* specimens are similar to those found between male and female orangutans respectively, so it is possible that these two "genera" really represent the males and females of a single, sexually dimorphic genus (Wu et al., 1986). Another suggestion recently put forward is that the Chinese *Sivapithecus* and *Ramapithecus* are indeed two separate genera but with very different patterns of dental sexual dimorphism, with *Sivapithecus* more similar to *Pongo* and *Ramapithecus* more similar to *Homo* (Lieberman et al., 1985).

Through the courtesy of Professor Wu Rukang, I have had a chance to study the original specimens in Beijing. I would agree with his interpretation that the *Sivapithecus* and *Ramapithecus* skulls are male and female of a single, sexually dimorphic species. If this is so, then the original names (*Ramapithecus lufengensis* and *Sivapithecus yunnanesis*) should be changed. According to the rules of nomenclature, the correct name for this sexually dimorphic hominoid species should now be *Sivapithecus lufengensis*. This raises one other problem, however: the crania of *S. lufengensis* bear no significant resemblance to other *Sivapithecus* crania such as *S. indicus* from the Siwaliks in Pakistan (Table 5.4) and *S. meteai*

Table 5.4 Cranial Differences between *Sivapithecus lufengensis* (China) and *Sivapithecus indicus* (Pakistan)

Character	*S. lufengensis*	*S. indicus*
Orbits	Oval-shaped	Rectangular
Interorbital space	Wide and concave	Narrow and convex
Lateral orbital wall	Robust	Gracile
Biorbital breadth (maximum width across the orbits)	Wide	Narrow
Midfacial region	Short	Long
Zygomatic arches	Gracile	Robust

from Turkey. Thus I would favor placing *S. lufengensis* in a new genus of Miocene hominoid, *Lufengpithecus*.

Comparison of Asian Ramamorphs and Dryomorphs Although the phylogenetic affinities of the ramamorphs are still subject to debate, there seems little doubt that as a group they are distinct from the Miocene dryomorphs.

The Asian ramamorphs share many morphological features distinguishing them from dryomorphs:

1. Thick-enameled molars with relatively bunodont cusps and simple occlusal patterns
2. Stoutly built central incisors
3. Low-crowned and robust canines
4. Relatively deep mandibular bodies and symphyses
5. Laterally flaring zygomatic arches
6. Anteriorly abbreviated mandibles and premaxillae suggesting a relatively **orthognathous** (nonprojecting) face
7. A lower first molar enlarged relative to the third molar.

Most of these features foreshadow the australopithecine condition (to be discussed in Chapter 6) and probably reflect increased lateral chewing stresses in these animals. In spite of these presumed heavy chewing stresses, however, dental wear striations in *Sivapithecus* cannot be distinguished from those in the frugivorous chimpanzee (Teaford and Walker, 1984).

The anatomy of the lower face also distinguishes ramamorphs and dryomorphs. These two groups differ in premaxillary morphology in the same way that African apes and orangutans do. The differences are based primarily on the disposition of the incisive canal and its relationship to the hard palate. (The **incisive canal** is a passage for small arteries and nerves running between the **nasal incisive fossa,** or opening into the nasal cavity, and the **oral incisive fossa,** or opening into the oral cavity.) In dryomorphs like *Proconsul* and *Dendropithecus* the oral incisive fossa is a transversely broad basin that opens directly into the oral cavity. The **subnasal alveolar process** (housing the roots of the upper incisors) appears as a flattened oval in sagittal section. No true incisive canal is present (Ward and Kimbel, 1983; Ward and Pilbeam, 1983) (Fig. 5.25).

The pattern found in Asian ramamorphs differs from the dryomorph condition. The *nasoalveolar clivus* (the portion of the premaxilla from the nasal cavity to the incisor root sockets) arcs back into the nasal cavity without terminating at a broad and deep nasal incisive fossa, as it does in *Proconsul.* The incisive canal opens into the floor of the fossa as a narrow cleft. This pattern is found today in the orangutan. As for some of the other Miocene Eurasian specimens, the *Rudapithecus* palate from Hungary has the dryomorph configuration, and palates of *Sivapithecus meteai* and the Chinese *Sivapithecus* are similar to the orangutan configuration.

Other features of the face and palate also distinguish dryomorphs from ramamorphs. For example, the combination of anteriorly diminished maxillary sinuses, zygomatic flare, and laterally rotated canine roots produces a well-developed **canine fossa** (a slight depression on the lower face just behind the ridge formed by the canine root) in *Sivapithecus* and *Ramapithecus* (Ward and Pilbeam, 1983).

Ramamorphs from Africa Only a few specimens from East Africa document the existence of ramamorphs in the African Miocene. These specimens all come from the middle Miocene Kenyan sites of Fort Ternan, Maboko Island, Majiwa, and Kaloma (Pickford, 1982).

The fauna from Fort Ternan is 12.5 to 14 million years old and has yielded some of the earliest dated specimens of *Ramapithecus* (Bishop et al., 1969). Four ramamorph specimens from this site were originally described as *Kenyapithecus wickeri* (Leakey, 1962). These specimens are now referred to *S. africanus* (Kay, 1982). Additional fragmentary specimens have been reported from this site in recent years (Andrews and Walker, 1976). Only one ramamorph species (among the various fossil hominoids present at the site) is known from Fort Ternan. In most morphological details this animal closely resembles the Asian specimens of *Sivapithecus* (including *Ramapithecus*):

1. It has a highly arched palate and a well-developed canine fossa.
2. The zygomatic process projects laterally from the maxilla at the level of the first molar.
3. The mandible is shallow and broad, and the mandibular symphysis has a well-developed inferior transverse torus extending back to the level of the first molar.
4. The molars lack cingula and are low-crowned with thick enamel.
5. The third premolar is a narrow, compressed tooth with its longest axis set at an angle to the mesiodistal plane of the tooth row.
6. A variably developed third premolar hone for the upper canine is present.

7. Upper and lower canines are small, and the lower incisors must have been small as well.

The overall size of the postcanine dentition is similar to that of *S. sivalensis* from the Siwaliks of southern Asia.

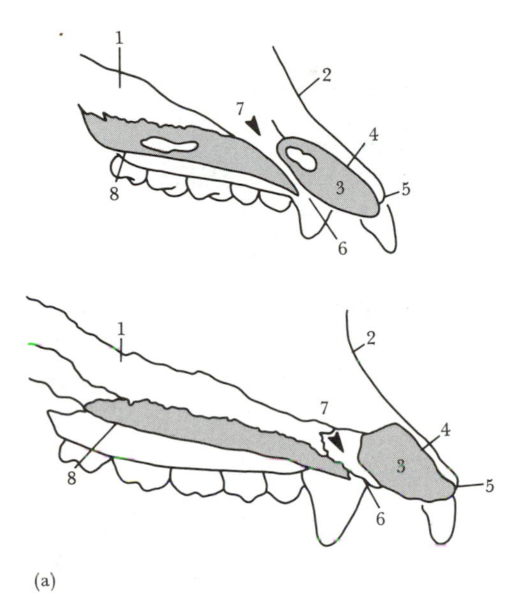

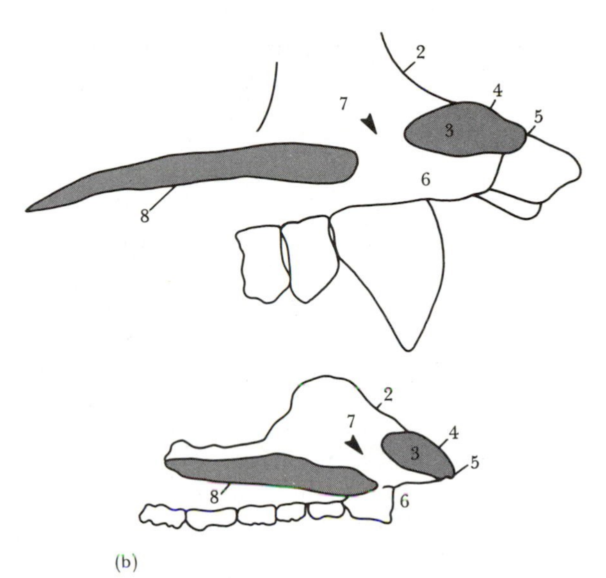

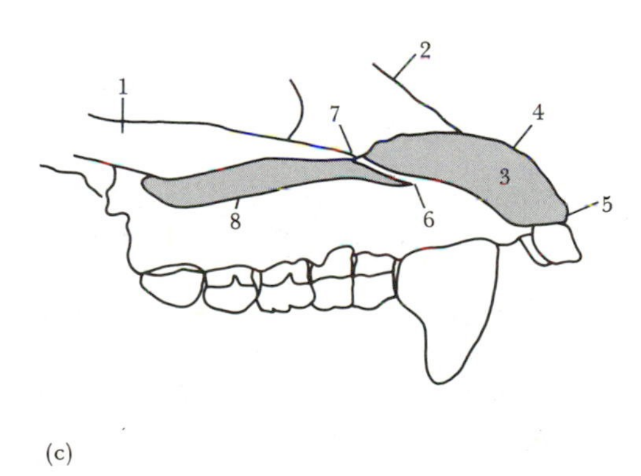

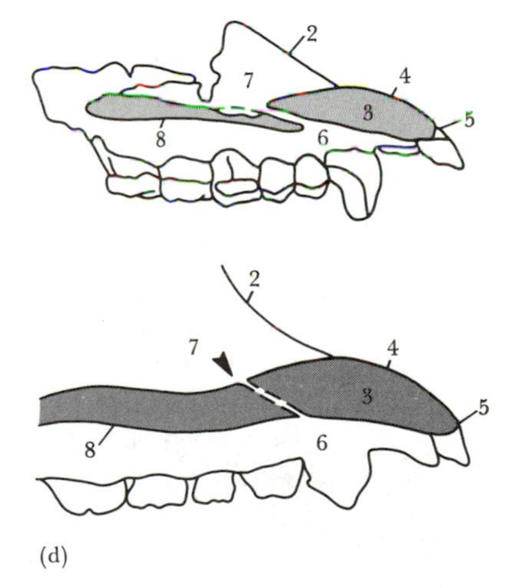

Fig. 5.25 Anatomy of the Lower Face in Miocene and Extant Hominoids Sagittal sections through the premaxilla and palate (shaded areas): (a) *Pan* (top) and *Gorilla* (bottom); (b) *Proconsul major* (top) and *Rangwapithecus vancouveringi* (bottom); (c) *Pongo;* (d) *Sivapithecus indicus* (top) and *S. meteai* (bottom). Anatomical parts: *1,* vomer; *2,* lateral margin of nasal aperture; *3,* subnasal alveolar process; *4,* nasoalveolar clivus; *5,* prosthion; *6,* oral incisive fossa; *7* and arrowhead, nasal incisive fossa; *8,* hard palate. Note in both specimens in (b) that the premaxilla is broadly separated from the palate. (Adapted from Ward and Kimbel, 1983)

The ramamorph specimen from Kaloma is a mandible that is similar in size to *Proconsul africanus,* yet much more robust in the mandibular body. No superior transverse torus is present. Features of molar morphology (such as increased enamel thickness and reduced cingula), mandibular robustness, and differential molar-wear gradients on these ramamorph specimens are markedly different from the complex of morphological features that characterize the early and middle Miocene dryomorphs (such as thinner molar enamel, prominent molar cingula, more slender mandibular bodies, and a less pronounced molar-wear gradient). In these features the specimens from Kaloma, Majiwa, Maboko Island, and Fort Ternan resemble ramamorphs (as well as *Australopithecus* and *Homo;* see Chapter 6) (Pickford, 1982). These morphological differences between dryomorphs and ramamorphs signify a change in diet from relatively soft and easily chewed foods in the former group to more resistant foodstuffs in the latter. This is consistent with the paleoenvironmental evidence indicating that these middle Miocene sites were more open woodland environments compared with the earlier Miocene sites.

The fossil remains attributable to ramamorphs in East Africa seem to fall into two distinct size groups, the larger of which includes the type material of *Sivapithecus africanus* and *Kenyapithecus wickeri.* The smaller specimens are closer in size to *Ramapithecus punjabicus* from Asia. Pickford (1982) has suggested that all these African ramamorphs may belong to one highly sexually dimorphic species, which he refers to as *Kenyapithecus africanus.* As noted earlier, Kay (1982) refers these specimens to *S. africanus.*

Pliomorphs

Members of the family **Pliopithecidae** are the earliest hominoids to appear in Eurasia. They first appear in European deposits dated at approximately 16 MYA *(Pliopithecus)* and continue to about 8 MYA in China *(Laccopithecus).* Over the years a number of different European and Asian genera and species have been described, some on very scanty material (Bernor et al., 1988). However, all the European species can probably be included within the genus *Pliopithecus* and the Asian species within the genus *Laccopithecus.*

Miocene pliomorphs have often been considered gibbonlike and were formerly included in the Hylobatidae by many authors. Some authors include these primitive catarrhines, along with the Oligocene primates *Propliopithecus* and *Aeygptopithecus,* in the family Propliopithecidae (= Pliopithecidae). It is difficult to see the logic of such a classification, since Miocene pliopithecids (at least to this author) look nothing like these Oligocene primates. The only features they seem to have in common are probably best regarded as primitive retentions.

Pliopithecus *Pliopithecus* consists of several species and occurs throughout Europe between about 16 and 12 MYA (Szalay and Delson, 1979).

As mentioned, *Pliopithecus* has traditionally been regarded as an ancestral gibbon. This view was first promulgated as long ago as 1837 by the French paleontologist Edouard Lartet. The type species, a mandible having all its teeth and lacking only the ascending rami, is from Sansan, France. In spite of the fact that *Pliopithecus* is one of the best-known Miocene hominoids, being represented by three partial skeletons and hundreds of isolated cranial and dental remains, its phylogenetic affinities are still the subject of considerable controversy (Groves, 1972, 1974; Tuttle, 1972; Frisch, 1973; Simons and Fleagle, 1973; Delson and Andrews, 1975; Ciochon and Corruccini, 1977) (Fig. 5.26). Since Gervais first named the genus in 1845, many different species of *Pliopithecus* have been described (*P. vindobonensis, P. lockeri, P. antiquus,* and *P. piveteaui,* to name just a few). Specimens of *Pliopithecus* and *Dryopithecus* have rarely been discovered in the same European sites. Possibly the two genera were adapted to different ecological niches (Simons and Fleagle, 1973).

The skull and dentition of *Pliopithecus* are more primitive than those of extant gibbons. Several morphological features of the skull and jaw (for example, the robust mandible, the deep and fused mandibular symphysis, and the high mandibular condyles) indicate that *Pliopithecus* fed mainly on tough vegetation. The large area for the origin of the **temporalis** muscle (a major mastication muscle that inserts into the coronoid process of the mandible) combined with the anterior location of the masseter muscle clearly indicate a more powerful chewing apparatus than is typical of modern gibbons. The incisors are much narrower than those seen among most extant frugivorous primates. Thus food preferences were more similar to those of leaf-eating colobine monkeys *(Colobus)* and howler monkeys *(Alouatta)* than they were to modern gibbons.

As mentioned earlier, Pliopithecidae exhibit many ancestral catarrhine features, particularly in the maxilla, mandible, and dentition. What are some of these ancestral catarrhine features?

The ancestral condition in the catarrhine mandible probably included the moderate degree of tooth-row divergence seen in dryomorphs, coupled with a relatively deep mandibular body having a symphysis with an inferior transverse torus. These traits generally characterize monkeys, gibbons, and Oligocene anthropoids. This scenario implies that some early Miocene hominoids, including *Proconsul* and *Limnopithecus,* lost the primitive inferior transverse torus and developed instead a superior transverse torus (Delson and Andrews, 1975; Andrews and Cronin, 1982).

The ancestral condition of the catarrhine maxilla probably included projecting snouts with high, narrow noses; narrow, shallow palates; small maxillary sinuses; gracile zygomatic arches; gracile alveolar processes; and comparatively large, wide-set orbits.

The ancestral condition of the dentition probably included relatively small incisors and canines, sectorial third premolars that were laterally

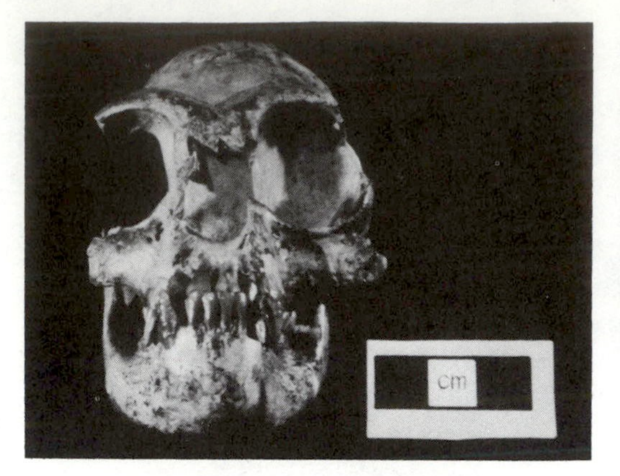

(a)

(b)

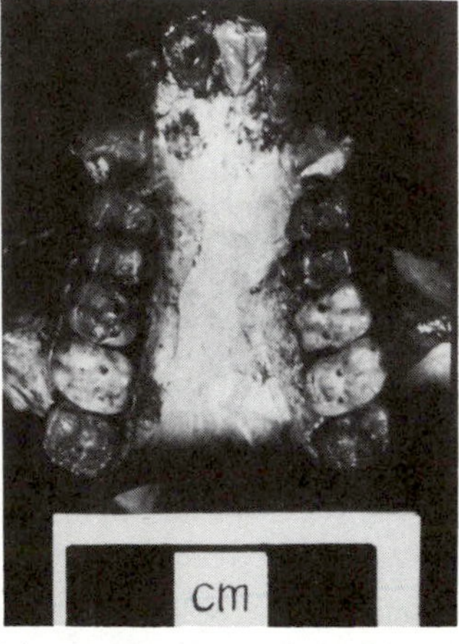

(c)

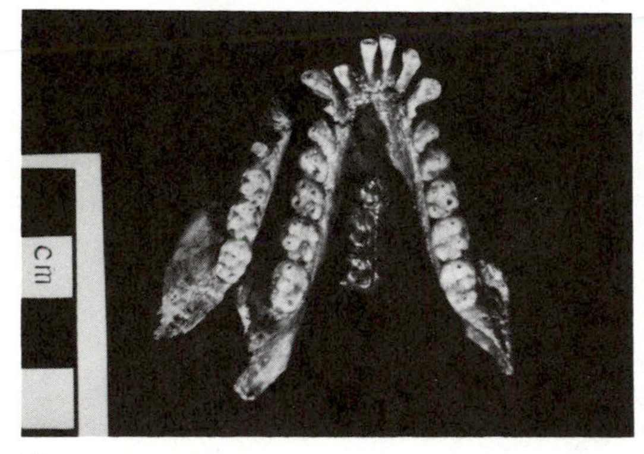

(d)

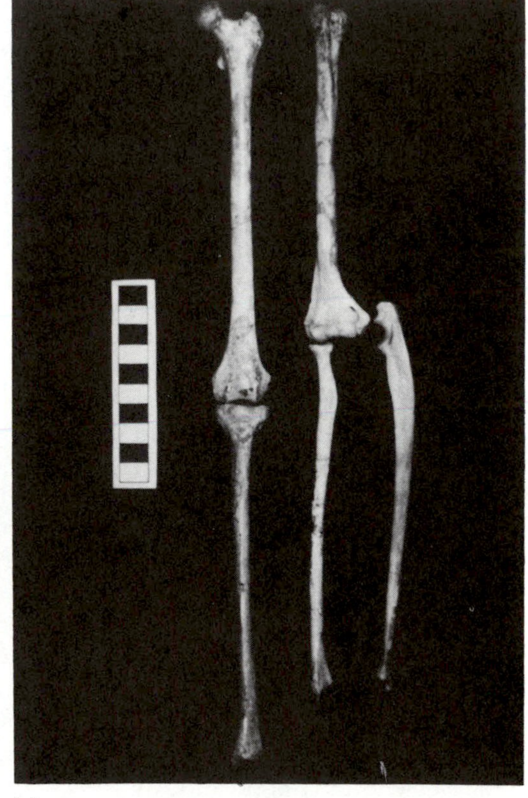

(e)

Fig. 5.26 Cranial and Postcranial Remains of *Pliopithecus* from the Miocene of Europe (a) Frontal view of the skull; (b) lateral view of the skull; (c) occlusal view of the palate; (d) several lower jaw specimens; (e) hindlimb *(left)* and forelimbs *(right)* bones. Note particularly the rather gibbonlike appearance of the face and the elongated and slender postcranial bones. (Courtesy of E. Delson)

compressed, relatively broad upper premolars and molars, small lower first molars and elongated third molars, molar cingula, paraconids occurring frequently on first molars, hypoconid–protoconid ridges, and a tendency for talonids to be broader than trigonids.

Dryomorphs retain most of these ancestral features except for the sectorial third premolars, which are not as well developed as in some modern apes. In contrast, the middle Miocene ramamorphs have evolved more-derived features such as the development of thick enamel, progressive size increase of lower fourth premolars and lower first molars relative to third premolars and third molars, greater wear gradient from first molars through third molars, and loss of cingula. It is interesting to point out that recent studies have shown that thick molar enamel may be the ancestral condition for great apes rather than the derived one (Martin, 1985), illustrating again the difficulties involved in assessing morphocline polarities.

Studies of the postcranial skeleton of *Pliopithecus* were made possible by the discovery of three partial skeletons from the middle Miocene of Czechoslovakia. In many features the postcranial skeleton is reminiscent of platyrrhines such as *Alouatta* (howler monkey) and *Lagothrix* (woolly monkey). The forelimb does not show any obvious adaptations to terrestrial quadrupedalism nor to gibbonlike brachiation. Certain ancestral catarrhine postcranial features are present in pliomorphs:

1. A low intermembral index (that is, the forelimbs are shorter than the hindlimbs)
2. Unspecialized hand phalanges
3. A large olecranon process for increased leverage of the triceps muscle
4. The presence of an entepicondylar foramen on the distal humerus for passage of the brachial artery and the median nerve
5. Six or seven caudal vertebrae and only three sacral vertebrae (and the possible existence of a long tail)
6. A direct articulation between the distal ulna and the wrist bones
7. The lack of a long auditory tube formed by the ectotympanic (Ankel, 1965; Zapfe, 1958, 1960; Groves, 1972)

In a multivariate statistical analysis of the forelimb, Ciochon and Corruccini (1977) concluded that *Pliopithecus* did not display the hominoid morphotype and thus denied any specific phylogenetic relationship between this genus and *Hylobates*. If *Pliopithecus* is considered an ancestral hylobatid, then it would be necessary to assume that several features present in both gibbons and great apes evolved independently, such as elongation of the external auditory meatus, development of a **meniscus** (a cartilaginous pad) in the wrist joint, and loss of the entepicondylar foramen, among others. Because of this concern, several authors have more recently concluded that it might be better to regard *Pliopithecus* as a primitive catarrhine phyletically closer to *Aegyptopithecus* than to other Miocene hominoids (Remane, 1965; Groves, 1972; Delson and Andrews,

1975). According to this view, *Pliopithecus* should be considered a Miocene survivor of the Oligocene catarrhines. Thus the Pliopithecidae (or Propliopithecidae) would include *Pliopithecus, Aegyptopithecus, Propliopithecus,* and perhaps *Oligopithecus.* Some investigators would even include *Dendropithecus* in the *Pliopithecidae* (Delson and Andrews, 1975; Ciochon and Corruccini, 1977). This family would then contain the common ancestors of both the Cercopithecoidea and the Hominoidea. However, I do not have much confidence in linking *Pliopithecus* with *Aegyptopithecus,* since the skulls of each look completely different.

Other Pliomorphs Pliomorphs have recently been described from several middle and late Miocene sites in China (Li, 1978; Wu and Yuerong, 1984, 1985). The most prolific of these sites, Lufeng, has yielded a total of a dozen maxillae and mandibles, nearly 60 isolated teeth, and one incomplete skull. These pliopithecids have been assigned to the genus *Laccopithecus.* The genus is characterized by the following dental characteristics: (1) the premolars are relatively large, and the lower fourth premolars are highly molarized; (2) the canines show marked sexual dimorphism; (3) the mandibular bodies are deep; and (4) the mandibular symphyses extend back to the level of third premolars. Both superior and inferior mandibular tori are well developed. The lower incisors are high crowned and narrow, resembling *Dendropithecus* and *Micropithecus* in these features. The dentition is similar to European *Pliopithecus* in overall size. The Chinese authors consider *Laccopithecus* to be part of or near the ancestry of modern gibbons.

Recently Discovered Miocene Hominoids

Recent collecting in the Kavirondo Gulf region of Lake Victoria in western Kenya and the Lake Turkana region of northern Kenya has produced additional samples of various Miocene hominoids (Andrews et al., 1981; Harrison, 1981, 1985*a,* 1985*b,* 1986; Pickford, 1982; Leakey and Walker, 1985; McDougall and Watkins, 1985; Leakey and Leakey, 1986*a,* 1986*b*). For example, deposits on Maboko Island, dating to approximately 15 to 16 MYA, have yielded at least five species of hominoids.

The most common hominoid species recently discovered at Maboko Island probably represents a new taxon that has its closest affinities with *Rangwapithecus vancouveringi* from the early Miocene of Rusinga Island. As mentioned earlier, Harrison (1986) has described this new genus as *Nyanzapithecus* and has also assigned *R. vancouveringi* to it as well. *Nyanzapithecus* has a highly distinctive suite of derived characters in molar and premolar morphology that it shares with *Oreopithecus bambolii* from the late Miocene of Europe. For this reason, Harrison (1986) includes both the middle Miocene African *Nyanzapithecus* and the late Miocene European *Oreopithecus* in the single family **Oreopithecidae,** implying an early Miocene African origin for this catarrhine family.

A new early Miocene site on the western side of Lake Turkana, Kalodirr, has also yielded hominoid fossils (Leakey and Leakey, 1986*a*, 1986*b*). Based on faunal comparisons, the hominoids are 18 to 16 MYA. One new taxon, *Afropithecus turkanensis* (which includes a partial cranium, several mandibles, isolated teeth, and associated postcranial bones), appears to be distinct from other known African and Asian hominoids. It differs from other Miocene genera in the basal flare of the upper molars and premolars, particularly the upper fourth premolar. This has the effect of making the cusp tips appear more closely packed together than the cusp bases. It is about the size of *Proconsul major* but differs in mandibular morphology, particularly its undeveloped superior transverse torus.

Afropithecus resembles Eurasian *Sivapithecus* in a number of ways: both have (1) large central incisors that are **procumbent** (projecting more horizontally than vertically), coupled with relatively small, asymmetrical lateral incisors; (2) a long superior aspect of the premaxilla; and (3) small supraorbital tori with high hafting of the cranial vault onto the face. The lateral facial profile differs greatly, however. In *Sivapithecus* the lower part of the face is very prognathous while the midface is rather concave, whereas in *Afropithecus* this lateral profile is much more linear (Fig. 5.27). This difference also pertains when comparison is made with the known large hominoids from Lufeng, China. The Lufeng specimens resemble *Afropithecus* only in the conformation of the temporal lines and the wide interorbital distance (Leakey and Leakey, 1986*a*).

A second new specimen from the same site, *Turkanapithecus,* has also recently been described (Leakey and Leakey, 1986*b*). Its skull is slightly smaller than that of female *Proconsul africanus* but differs from the latter in many important respects:

1. The maxilla has a distinctive squarish shape.
2. The nasal aperture is larger.
3. The nasal bones are longer, and they widen at either end.
4. The zygomatic process of the maxilla has greater depth beneath the orbit.
5. The root of the zygomatic process is above the first and not the second molar.
6. The interorbital distance is wider.
7. The supraorbital tori are distinct.
8. The **supraglabellar region** (forehead) is rather flat.

Miocene material from the northern Kenya site of Buluk includes three different hominoid species and at least one cercopithecoid species. The fauna is 16 to 18 million years old (Leakey and Walker, 1985; McDougall and Watkins, 1985). The small hominoid is about the size of *Micropithecus clarki,* the middle-sized hominoid is about the size of *Sivapithecus (= Kenyapithecus) africanus,* and the large hominoid is the size of *Sivapithecus indicus.* Leakey and Walker (1985) originally included the larger species in *Sivapithecus,* but Leakey now includes it within his new genus *Afropithecus.*

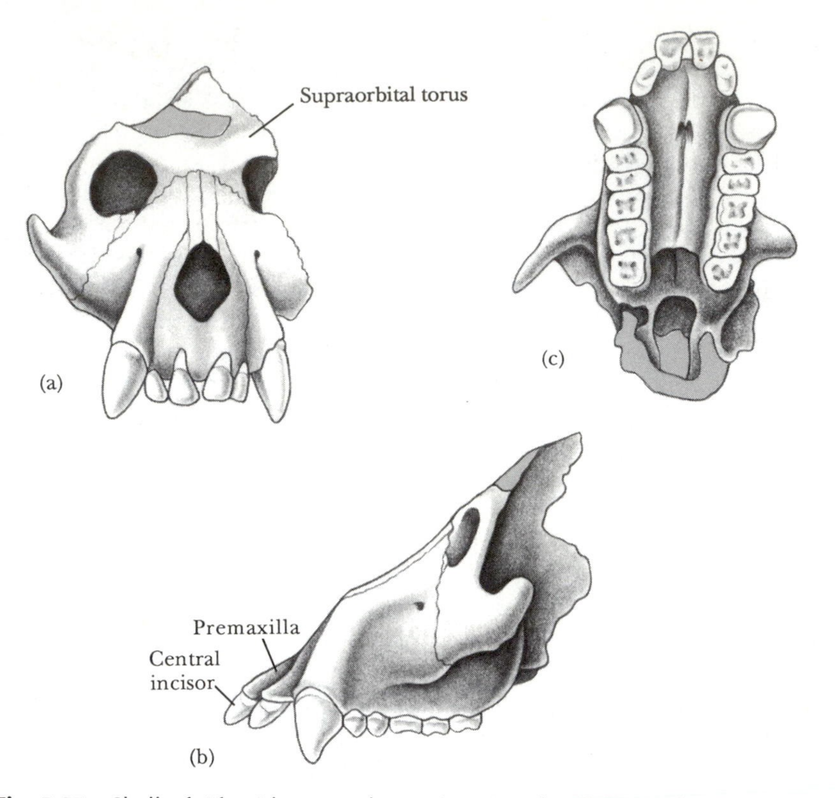

Fig. 5.27 Skull of *Afropithecus turkanensis,* a Newly Discovered Miocene Hominoid from Kalodirr, Kenya (a) Front, (b) side, and (c) basal views of the upper jaw and face. Note the small supraorbital tori, the long premaxilla, and the procumbent incisors, all traits shared with *Sivapithecus.* It differs from *Sivapithecus* in its more linear midface profile. Scale about ⅓ × natural size. (Redrawn from Leakey et al., 1988)

This new material from Kalodirr, Maboko, and Buluk provides convincing evidence that early to middle Miocene hominoids in Africa were more diverse than previously suspected.

MIOCENE MONKEYS

Although fossil monkeys appear in the early Miocene fossil record of Africa along with the primitive apes, they are extremely rare. By the middle Miocene, however, their abundance and diversity are well established in East Africa. This adaptive radiation of Old World monkeys in the middle to late Miocene parallels the apparent evolutionary decline of the apes in these same faunas. There is probably a causal relationship between these two evolutionary trends.

Miocene monkeys from the New World are known from only a few sites in Colombia and Argentina (see Fig. 5.6). As we shall see, most of these Miocene monkeys share morphological similarities with modern primate genera of South America. There have, of course, never been any fossil apes in the New World.

Table 5.5 lists the fossil monkeys of the Old and New Worlds from the Miocene through the Recent epochs.

Old World Miocene Monkeys (Cercopithecoidea)

Old World monkeys are all placed within the superfamily Cercopithecoidea. Two families are recognized. The **Victoriapithecidae** (subfamily **Victoriapithecinae**) includes the earliest Old World fossil monkey genera: *Prohylobates* from the early Miocene of North Africa (Egypt and Libya) and *Victoriapithecus* from the middle Miocene of East Africa (Kenya) (see Fig. 5.4). The **Cercopithecidae** are comprised of the "modern" genera (some of which reach back to the late Miocene) and are divided into the two subfamilies **Colobinae** and **Cercopithecinae** (see Table 5.5). Colobines and cercopithecines are clearly distinct by the middle to late Miocene in Africa.

In craniodental morphology, ancestral cercopithecoids (Miocene victoriapithecids) show a combination of cercopithecine and colobine traits: they follow the cercopithecine pattern in their dental morphology, while their facial structure resembles that of colobines.

Modern macaques, which are cercopithecines, have the least-specialized dentition of any cercopithecoid monkey. For this reason, their dental morphology is considered to reflect the ancestral Old World monkey condition (Delson, 1975*a*, 1975*b*). Such ancestral dental features include the following:

1. High-crowned bilophodont cheek teeth
2. Cusps separated by shallow notches
3. Long, narrow lower third molars with distinct hypoconulids
4. A well-developed honing mechanism between the upper canine and the lower third premolar, especially in males
5. Large upper canine teeth with a mesial groove from apex to root base
6. An absence of cingula
7. Fairly similar proportions in the four incisor teeth

In terms of facial morphology, ancestral cercopithecoids were probably more similar to colobines and gibbons than to cercopithecines. This ancestral pattern would have included the following features:

1. A relatively short and broad face
2. A wide interorbital region with short, broad nasal bones
3. A lacrimal fossa that extended beyond the orbit onto the maxilla

Postcranial elements of victoriapithecines are relatively few in number. The limb bones in general, and the elbow joint in particular, indicate that these early Old World monkeys were generalized arboreal

Table 5.5 Classification of Old and New World Fossil Monkeys (Miocene through Recent)

Taxa	*Epoch (and Area)*
Infraorder CATARRHINI	
Superfamily CERCOPITHECOIDEA	
Family VICTORIAPITHECIDAE	E.–M. Mio. (Africa)
Subfamily VICTORIAPITHECINAE	
Prohylobates	E. Mio. (N. and E. Africa)
Victoriapithecus	?E.–M. Mio. (Kenya)
Family CERCOPITHECIDAE	L. Mio–Rec. (Africa, Eur., Asia)
Subfamily CERCOPITHECINAE	
Cercocebus[a]	Plio-Pleist.–Rec. (Africa)
Cercopithecus[a]	Plio.–Rec. (Africa)
Dinopithecus[a]	Plio.–Rec. (S. Africa)
Gorgopithecus	Pleist. (S. Africa)
Macaca[a]	L. Mio.–Rec. (N. Africa, Eur., Asia)
Papio[a]	Plio.–Rec. (Africa)
Paradolichopithecus	Plio. (Eur.)
Parapapio	L. Mio.–E. Pleist. (Africa)
Procynocephalus	Plio. (Asia)
Theropithecus[a]	Plio-Pleist.–Rec. (Africa, ?Asia)
Subfamily COLOBINAE	
Cercopithecoides	Plio. (Africa)
Colobus[a]	L. Mio.–Rec. (Africa)
Dolichopithecus	Plio. (Eur.)
Libypithecus	L. Mio.–Plio. (N. Africa)
Mesopithecus	L. Mio.–Plio. (Eur., W. Asia)
Microcolobus	L. Mio. (Africa)
Paracolobus	Plio-Pleist. (Africa)
Presbytis[a]	L. Mio.–Rec. (Asia)
Rhinocolobus	Plio-Pleist. (Africa)
Rhinopithecus[a]	E. Pleist.–Rec. (Asia)
Infraorder PLATYRRHINI	
Superfamily CEBOIDEA	
Family CALLITRICHIDAE	M. Mio.–Rec. (S. America)
Subfamily CALLITRICHINAE	
Micodon	M. Mio. (Colombia)
Family CEBIDAE	L. Olig.–Rec. (S. and C. America)
Subfamily AOTINAE	
Aotus[a]	M. Mio.–Rec. (Colombia)
Homunculus	E. Mio. (Argentina)
Subfamily ATELINAE	
Stirtonia	M. Mio. (Colombia)
Subfamily CEBINAE	
Neosaimiri	M. Mio. (Colombia)
Subfamily PITHECIINAE	
Cebupithecia	M. Mio. (Colombia)
Mohanamico	M. Mio. (Colombia)
Subfamily *incertae sedis*	
Soriacebus	E. Mio. (Argentina)
Xenothrix	Rec. (Jamaica)

[a] Extant genera

quadrupeds that were not particularly specialized for exclusive arboreal or terrestrial substrates. A modern equivalent might be the vervet monkey *(Cercopithecus aethiops).*

With this general picture of the ancestral cercopithecoid morphology in mind, let us now turn to the actual Miocene fossil evidence.

Some of the oldest undisputed monkeys from the Old World come from the early Miocene deposits of Wadi Moghara in Egypt and Gebel Zelten in Libya (approximately 18 to 20 MYA) (Simons, 1969, 1970; Delson, 1979). Three fragmentary and worn mandibles of *Prohylobates tandyi* are known from Wadi Moghara, and one specimen of the same genus is known from Gebel Zelten. The reason a cercopithecoid has the inappropriate generic name of *Prohylobates* (literally "early gibbon") is that its initial describer thought it to be directly ancestral to modern gibbons (Fourtau, 1918). Other earlier workers thought it to be a direct descendant of the Oligocene *Propliopithecus.* Molar crowns are relatively high, and the intercusp notches are shallow. The lower third molar is short but has a strong hypoconulid. There are no molar cingula. These are all cercopithecinelike features. The only colobinelike feature is the relatively deep mandibular corpus.

Most of the remaining Miocene monkeys from the Old World come from the Lake Victoria region of East Africa. A colobinelike frontal bone and a cercopithecinelike upper molar have been described from the early Miocene site of Napak, Uganda (Pilbeam and Walker, 1968).

A large sample of early Miocene cercopithecoids (referred to *Prohylobates*) has been reported from Buluk in northern Kenya (Leakey, 1985). The specimens are particularly important because they have features in common with both *Prohylobates* and *Victoriapithecus.* This latter genus is best known from Maboko Island, Kenya, and is dated to about 15 MYA (von Koenigswald, 1969). The upper and lower molars of the Buluk and Maboko Island cercopithecoids are all high-crowned with low occlusal-surface relief. Cusp tips are close together with marked flaring toward the base. Most significantly, the upper molars are not fully bilophodont: the anterior loph is always present, whereas the posterior loph is not fully developed. Upper molars display a crista obliqua in varying degrees of development. The lower cheek teeth are fully bilophodont and have no buccal cingula. These features characterize the subfamily Victoriapithecinae, separating these early cercopithecoids from the later Colobinae and Cercopithecinae.

Recently, middle Miocene cercopithecoids have been described from the Kenyan sites of Ngeringerowa, Ngorora, and Nakali. These have been dated at 10.5 to 8.5 MYA (Benefit and Pickford, 1986). These specimens are the only cercopithecoid remains definitely dated at 15 to 6 MYA from sub-Saharan Africa (with the possible exception of a single cercopithecine tooth from Ongoliba, Zaire, described by D. Hooijer in 1963). The mandible of the small colobine from Ngeringerowa has been assigned to a new genus and species, *Microcolobus tugenensis.* It is a pecu-

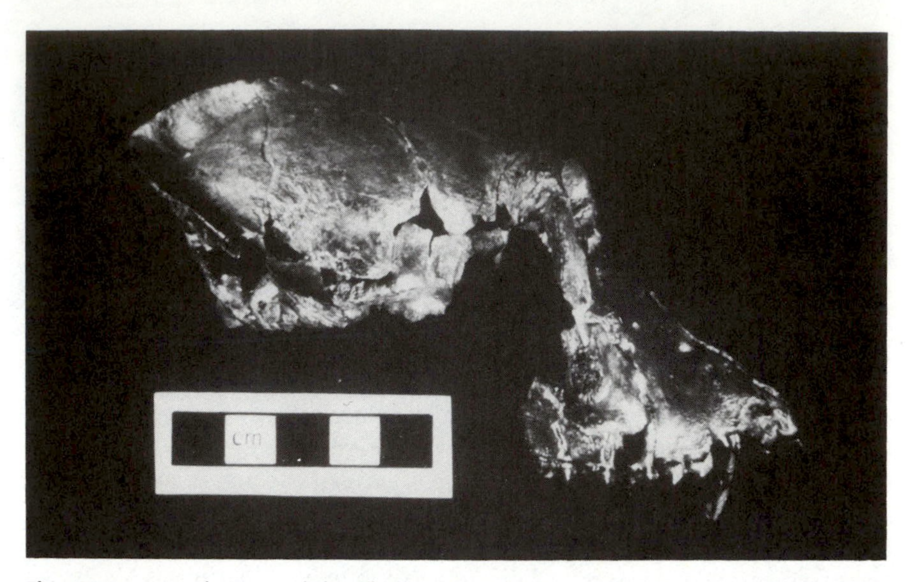

Fig. 5.28 Lateral View of the Skull of the Colobine *Libypithecus markgrafi* from Northern Egypt. The large, long canines suggest the specimen is a male. (Courtesy of E. Delson)

liar colobine in that it lacks an inferior transverse torus. These finds are the oldest undisputed colobines from Africa.

The presence of both cercopithecine and colobine fossils from Algeria (*Colobus flandrini, Macaca sylvanus*) and Egypt (*Libypithecus markgrafi*) prove that both groups had reached North Africa by the late Miocene (Fig. 5.28). By the Plio-Pleistocene some colobine genera (for example, *Cercopithecoides*) had reached South Africa as well.

In Europe the oldest known colobine is *Mesopithecus pentelici* from 11 to 6 million-year-old deposits in southern and central Europe and Afghanistan (Heintz et al., 1981) (Fig. 5.29). *Mesopithecus* is clearly colobine in dental and cranial features, although unlike most colobines it had a fairly long face. Interestingly, however, the postcranial skeleton is very macaquelike (hence cercopithecinelike) and seems terrestrially adapted (Jolly, 1967; Simons, 1970; Delson, 1975*b*).

Later circum-Mediterranean colobines include *Dolichopithecus* (dated at 4 to 2 MYA) from southern Europe. This large colobine was more terrestrially adapted than any other colobine, living or extinct. No colobines have survived in Europe up to the present.

A few colobine specimens are known from the Dhok Pathan Formation of the Siwaliks of southern Asia. These have been given various generic names in the past, but all probably represent a single small colobine species called *Presbytis sivalensis*. Monkeys do not appear in China until the latest Pliocene or earliest Pleistocene (Yuerong and Jablonski, 1987).

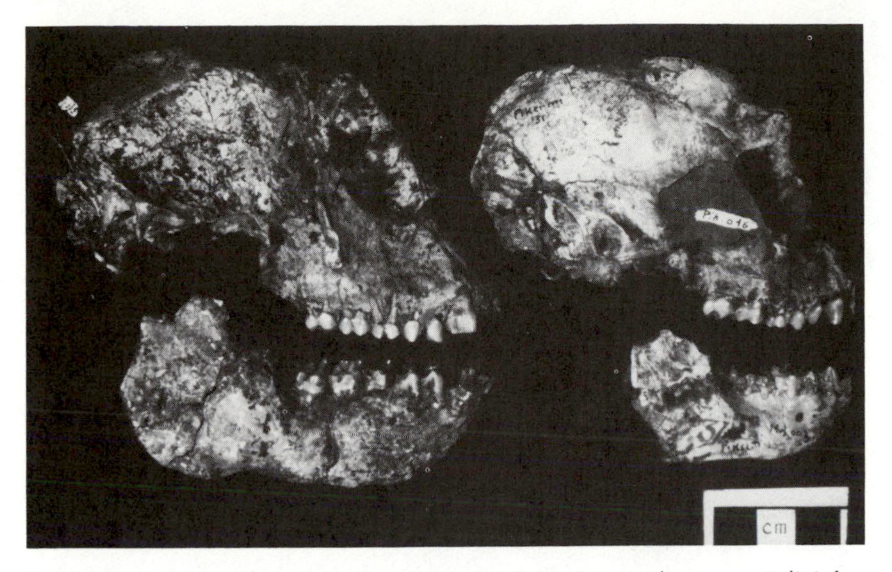

Fig. 5.29 Lateral View of Skulls of the Colobine *Mesopithecus pentelici* from Europe (Courtesy of E. Delson)

The earliest undoubted cercopithecines are the late Miocene populations from Algeria mentioned earlier and macaquelike fossils from Wadi Natrun in Egypt that are associated with the colobine *Libypithecus.* Although colobines were the dominant monkeys in the late Miocene to early Pliocene of Europe, macaques became the commonest and later the only cercopithecid in Europe throughout the later Pliocene and into the Pleistocene. The relationships between fossil and living Old World monkeys will be discussed in the last section of the chapter.

New World Miocene Monkeys (Ceboidea)

Miocene monkeys from the New World are known from only two South American countries, Argentina *(Homunculus* and *Soriacebus)* and Colombia *(Stirtonia, Cebupithecia,* and *Neosaimiri)* (see Table 5.5).

Homunculus is known from early Miocene sediments near the mouth of the Rio Gallegos in Argentina. The genus includes several mandibular specimens with teeth, a toothless cranial fragment preserving much of the left side of the face, and several limb bones. The dental formula is 2.1.3.3 / 2.1.3.3 as in all cebids, and the mandible is V-shaped. Judging from the size of its teeth, *Homunculus* was about the size of extant species of *Cebus* (capuchin monkeys). Some authors have stressed a dental similarity between *Homunculus* and *Aotus* (the night monkey); others have emphasized similarities with *Alouatta* (howler monkeys) (Stirton, 1951); and still others have pointed to a similarity with the subfamily Pitheciinae (sakis and uakaris) (Rosenberger, 1980).

The cranial fragment shows moderate-sized orbits, suggesting that *Homunculus* was diurnal. As in all higher primates, postorbital closure seems complete. The postcranial evidence is perplexing. While the radius and femur suggest an animal with the body size and limb proportions of *Cebus, Aotus* or *Callicebus* (all modern cebids), morphometric analyses of the femur indicate affinities with callitrichids (marmosets and tamarins). Ciochon and Corruccini (1975) do not regard this as having any phylogenetic significance, however, since they consider the callitrichid condition to be the ancestral platyrrhine morphotype. As already mentioned in Chapter 4, however, callitrichids are probably not primitive morphotypes but rather secondarily derived forms. They have several uniquely derived features:

1. The combination of twinning and a simple uterus
2. Extended family monogamy
3. Secondarily derived claws that are not homologous with primitive eutherian claws
4. A clawed thumb
5. Reduction or loss of the third molar and reduction of the second molar

Some members of the group have become adapted to a highly gummivorous diet, an unusual strategem for anthropoids (Rosenberger, 1980; Sussman and Kinzey, 1984).

A second Miocene genus, *Soriacebus,* has recently been described from Pinturas, Argentina (Fleagle et al., 1987). Its combination of extremely large, procumbent incisors and canines associated with small, narrow molars cannot be easily matched in any modern platyrrhine. It probably weighed 1,500 to 2,000 g, and its dental specializations suggest it had a diet of insects and plant exudates. Although *Soriacebus* is more primitive than living marmosets and tamarins in its dental formula and more specialized in its anterior dentition, it resembles the Callitrichinae (true marmosets) in many aspects of its premolar and molar morphology and may be close to the origin of that group (Fleagle et al., 1987).

Stirtonia, from the Miocene of La Venta, Colombia, is the largest fossil ceboid (Hershkovitz, 1970). The type specimen is a nearly complete mandible preserving both canines and most of the cheek teeth. The dental formula is 2.1.3.3. The incisors were apparently very small but were aligned side by side instead of in the more staggered position seen in *Homunculus.* R. Stirton (1951) believed this species to show clear morphological similarities with *Alouatta.* Kay et al. (1986) have recently discovered several new specimens of *Stirtonia* in Colombia. These specimens possess well-developed molar shearing crests and molar stylar shelves. The combination of these molar features and evidence of very small incisors clearly supports the view that the diet was at least as folivorous as in living *Alouatta* and *Brachyteles* (the woolly spider monkey).

Cebupithecia is known from a single specimen from the same La Venta fauna of Colombia (Stirton, 1951). The specimen preserves maxillary and mandibular portions of the face with the canines and the third premolar through second molar intact. The articular ends of some postcranial bones are also known. Stirton (1951) argued for a close phylogenetic relationship with *Pithecia* (sakis) mainly due to its possession of procumbent incisors and laterally splayed canines. These dental specializations may be adaptations for opening hard nuts and fruits to remove the seeds (Rose and Fleagle, 1981).

The third monkey genus from the La Venta fauna is *Neosaimiri.* The type and only specimen is a nearly complete mandible preserving most of the dentition. Most authors consider the form to be closely related to the extant squirrel monkeys *(Saimiri).*

DIETARY PREFERENCES, BEHAVIOR, AND HABITATS OF MIOCENE PRIMATES

Earlier in the chapter it was noted that Old World monkeys did not undergo an adaptive radiation until the middle to late Miocene, roughly coincident with the extinction of many hominoid species from Eurasia and Africa. One explanation for this replacement lies in the relationship between the dietary and habitat preferences of each group.

Andrews (1982) has suggested that the ancestral catarrhine habitat preference was for tropical forest, more specifically lowland wet evergreen forest. He has further pointed out that frugivory in mammals is generally associated with forest habitats. Consequently the retention of frugivorous diets in most hominoids (the "primitive" dietary condition) and the associated primitive character of their dentition can be related directly to the fact that they have retained the ancestral catarrhine habitat preference for forests.

By way of contrast, Andrews considers the ancestral habitat of the Cercopithecoidea to be savanna, and he regards this condition as derived with respect to the ancestral catarrhine condition (Andrews and Aiello, 1984). His conclusion differs markedly from that of others (Napier, 1970; Delson, 1975*a*, 1975*b*) who postulate that the ancestral monkeys combined a mixed leaf-and-fruit diet in a deciduous-forest environment. To Andrews, the primitive teeth and diet in hominoids contrasts with the specialized teeth and diet in cercopithecoids and is one explanation for why monkeys replaced hominoids in the evolutionary record. The derived adaptations of the cercopithecoid digestive tract give these primates an advantage over hominoids in dietary terms. The colobine stomach is specialized to digest cellulose, a vast new food source not available to hominoids. The cercopithecine tolerance of plant secondary compounds enables them to eat unripe fruit that hominoids cannot stomach; even though cercopithecines and hominoids may eat similar fruit from the same trees, the monkeys can thus start exploiting the food resource earlier in its growth than the hominoids can.

Studies of limb bones suggest that Miocene hominoids were generalized arboreal quadrupeds, although their combination of apelike and monkeylike postcranial features is unlike those found in any extant primate (Rose, 1983; Conroy and Rose, 1983). There is no morphological evidence indicative of specialized apelike locomotor activities either in the form of knuckle walking or true brachiation.

PHYLOGENY AND CLASSIFICATION OF MIOCENE PRIMATES

Hominoidea

As mentioned earlier in the chapter, a multitude of redundant genera have been named for the many hominoid fossils unearthed throughout Europe, Asia, and East Africa over the last century. Several attempts have been made over the last two decades at reducing and clarifying this plethora of taxa. Moreover, recent biomolecular data altering our understanding of the taxonomic relationships among the living primates has in turn raised questions about the phyletic affinities of fossil hominoids.

Before outlining the most recently proposed and accepted theories, it behooves us to look briefly at the history of hominoid classification. Although many of the taxonomic issues may be confusing to the student, it is important to have some familiarity with the terminology used in the 1960s and early 1970s, since many of the major papers on these fossil hominoids were written during that period.

Simons and Pilbeam's (1965) Taxonomy In the first major taxonomic revision of Miocene hominoids, Elwyn Simons and his Yale colleague David Pilbeam (1965) were working under the assumption common by the 1960s that all living and fossil hominoids could be divided into three families: Pongidae, Hylobatidae, and Hominidae. They assigned all the Miocene hominoids from Africa, Europe, and Asia to a single subfamily, the Dryopithecinae, within the family Pongidae (Fig. 5.30). Most of these fossil pongids were assigned to the genus *Dryopithecus,* and within this genus three subgenera were recognized: *Dryopithecus,* mainly for the European dryopithecines; *Proconsul,* for the African dryopithecines; and *Sivapithecus,* mainly for the Asian dryopithecines. The European dryopithecines were divided into two species, *Dryopithecus (Dryopithecus) fontani* and *D. (D.) laietanus,* the Asian dryopithecines into two species, *Dryopithe-*

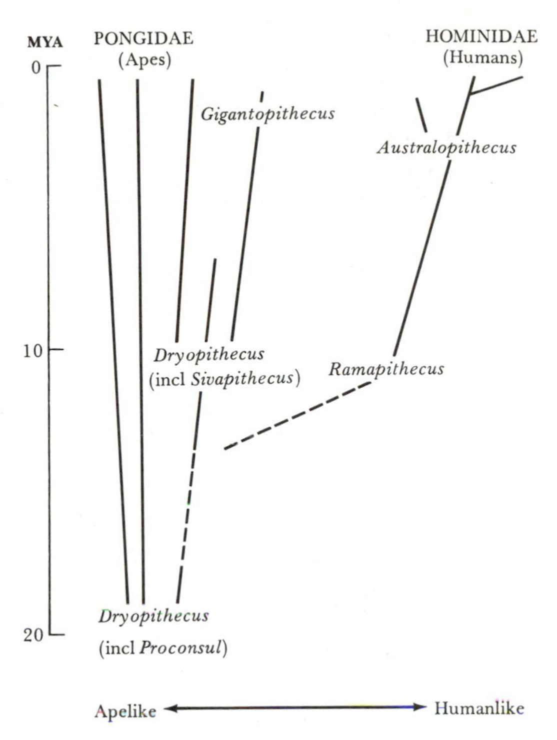

Fig. 5.30 Evolution of Miocene Hominoids as Viewed by Simons and Pilbeam (1965) Reproduced with permission, from the *Annual Review of Anthropology*, Vol. 8. © 1979 by Annual Reviews, Inc.

cus (Sivapithecus) indicus and *D. (S.) sivalensis,* and the African dryopithecines into three species, *Dryopithecus (Proconsul) africanus, D. (P.) nyanzae,* and *D. (P.) major.* Within this framework *D. africanus* was considered ancestral to modern chimpanzees and *D. major* to modern gorillas (Pilbeam, 1969). This conclusion implied that the evolutionary lineages leading to each of the African great apes had already separated from one another by the early Miocene, between 18 and 20 MYA. In addition, it was suggested that *D. sivalensis* may have been related to the ancestry of the orangutan. Ironically, the latter suggestion concerning orangutan phylogeny now seems more convincing than the former suggestion concerning African ape phylogeny (see later discussion).

Gigantopithecus was a second dryopithecine genus accepted by Simons and Pilbeam. At the time of their taxonomic revision only one species, *G. blacki* from the Pleistocene of China, was known. A second species, *G. bilaspurensis* was later described from the late Miocene of northern India (Simons and Chopra, 1969). (This specimen, along with the molar found by Pilgrim early in the 20th century, is now generally referred to *G. giganteus.*) This Indian species seems to be directly ancestral to the Chinese species.

The second Miocene hominoid family recognized by Simons and Pilbeam was the Hominidae, the family of humans and their immediate fossil ancestors. Only one Miocene genus and species, *Ramapithecus punjabicus,* was allocated to this family. Thus for many years *Ramapithecus* was considered the earliest known hominid in the fossil record, and discussions about this genus have had a profound effect on paleoanthropological thinking over the past quarter century (Conroy and Pilbeam, 1975). As we shall see, this view has shifted radically within the past few years.

The third hominoid family, the Hylobatidae, was not dealt with in Simons and Pilbeam's revision, but the prevailing view among most paleoanthropologists at the time was that several Miocene genera, such as *Limnopithecus* from East Africa and *Pliopithecus* from Europe, represented ancestral gibbons.

Thus until a few years ago many paleoanthropologists thought that the evolutionary lineages leading to chimpanzees, gorillas, orangutans, and gibbons were all distinct in the early Miocene by approximately 18 MYA, and that the lineage leading to humans was distinct from the other apes by the middle Miocene about 14 MYA (Conroy and Pilbeam, 1975).

Until the mid-1970s most paleontologists accepted the taxonomic and phylogenetic conclusions of Simons and Pilbeam and continued to work under the assumption that the great apes were more closely related to one another than any one of them was to humans or to the lesser ape, the gibbon. For that reason humans, gibbons, and great apes (and closely related ancestral forms) were traditionally classified within the families Hominidae, Hylobatidae, and Pongidae respectively, and all the Miocene anthropoids were forced into one of these three families as well. Accordingly, all apelike fossils (including *Dryopithecus, Sivapithecus, Proconsul,* and *Rangwapithecus*) were classified as pongids (subfamily Dryopithecinae); all small and generalized gibbonlike fossils (including *Pliopithecus* and *Dendropithecus*) were classified as hylobatids (subfamily Pliopithecinae), and all humanlike fossils (including *Ramapithecus* and *Australopithecus*) were classified as hominids (subfamily Homininae).

With the benefit of hindsight provided by several decades of further advances in paleontology and molecular biology (see following section), most paleontologists would now agree that the diversity of Miocene hominoids was underestimated by assigning so many of them to the single genus, *Dryopithecus.* It is now generally accepted that the subgenera *Dryopithecus, Proconsul,* and *Sivapithecus* (for the European, African, and Asian Miocene hominoids respectively) should all be elevated to generic rank. This taxonomic allocation more faithfully represents the morphological diversity seen among the three groups.

Theories in Light of Biomolecular Data and Recent Fossil Evidence A major rethinking of Miocene hominoid phylogeny emerged by the late 1970s and into the 1980s. This reevaluation was greatly influenced by new fossil discoveries, more sophisticated studies in molecular anthropology, and the wave of enthusiasm for phylogenetic systematics (cladistics) (Fig. 5.31). These factors led paleontologists to conclude that the traditional taxonomic trichotomy of anthropoid primates was incorrect. Two things were becoming clear: (1) the African apes are more closely related to humans than either group is to the orangutan; and (2) modern pongid species could not have diverged from other hominoids as long ago as the early Miocene because they share so many molecular similarities (Sibley and Ahlquist, 1984; Templeton, 1984).

Comparisons of biomolecular data in modern apes and humans have been crucial in establishing both of these points. These molecular techniques can be divided conveniently into four main categories: (1) those (such as immunodiffusion and electrophoresis) that measure differences in whole proteins without knowledge of the specific changes in the protein structure; (2) those (such as DNA hybridization) that detect differences in DNA molecules without indicating exactly which nucleotides have changed; (3) those (such as amino-acid sequencing) that detect specific amino-acid changes in proteins; and (4) those (such as DNA sequencing) that actually break DNA into recognizable nucleotide fragments (Andrews, 1986).

Data from such studies show a fairly consistent branching sequence within anthropoid evolution: (1) the separation of gibbons from the great ape–human clade; (2) the separation of the orangutan from the African great ape–human clade; and (3) the separation of chimpanzees, gorillas, and humans after a long common evolutionary pathway postdating separation from the orangutan (Cronin, 1983) (Fig. 5.32).

Molecular anthropologists have attempted to date these major cladogenic events by devising various molecular clocks. A **molecular clock** relies on the premise that molecular differences between taxa accumulate at a relatively constant rate when averaged over geologic time: a greater molecular difference implies an earlier divergence date. The clock must ultimately be calibrated against the fossil record, however. Suppose two taxa differ in some molecular parameter by X units and that they diverged Y million years ago as judged from the fossil record. Thus X number of changes have occurred over Y millions of years (assuming relatively constant rates of change). This relationship sets the clock. It is then a simple matter to calculate the divergence times of other taxa by measuring how much the taxa differ in the same molecular parameter. As an illustration, suppose two taxa that diverged 60 MYA differed by 20 units of some molecular parameter. Change in this parameter is thus assumed to occur at the rate of 1 unit per 3 million years. Consequently the divergence date of two taxa differing by 10 such units would be estimated at 30 MYA, 5 units at 15 MYA, and so on.

A number of such molecular clocks have been proposed for calculating divergence times within hominoid evolution. On average they suggest the following:

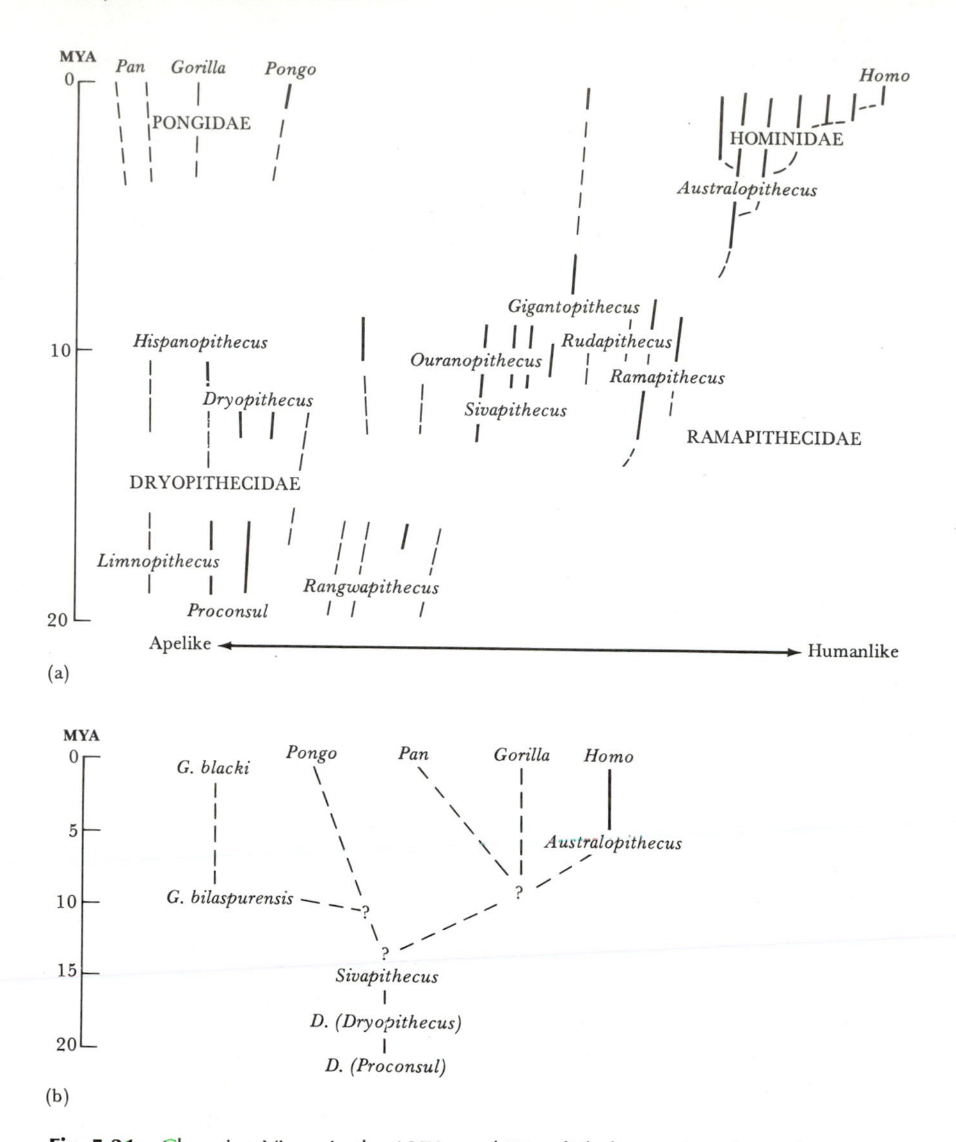

Fig. 5.31 Changing Views in the 1970s and 80s of Phylogenetic Relationships among Miocene and Living Hominoids (a) As envisioned by Pilbeam (1979). His assessment here is much more cautious than his views in the 1960s. For example, very few taxa are now connected directly to other fossil taxa or to living species. (b) As seen by Greenfield (1980). His late-divergence hypothesis accords well with both cladistic and molecular phylogenies, but certain question marks remain. (c) As conceived by Andrews (1978b). Solid lines indicate the life of the species as known in 1978 from the fossil record, and dashed lines indicate possible relationships. Where more than one phylogenetic possibility is known, the lines are separated by a question mark. Note the absence of ancestor–descendant relationships linking the Miocene hominoids with extant hominids or pongids. (Fig. 5.31a reproduced, with permission, from the *Annual Review of Anthropology*, Vol. 8. © 1979 by Annual Reviews, Inc.)

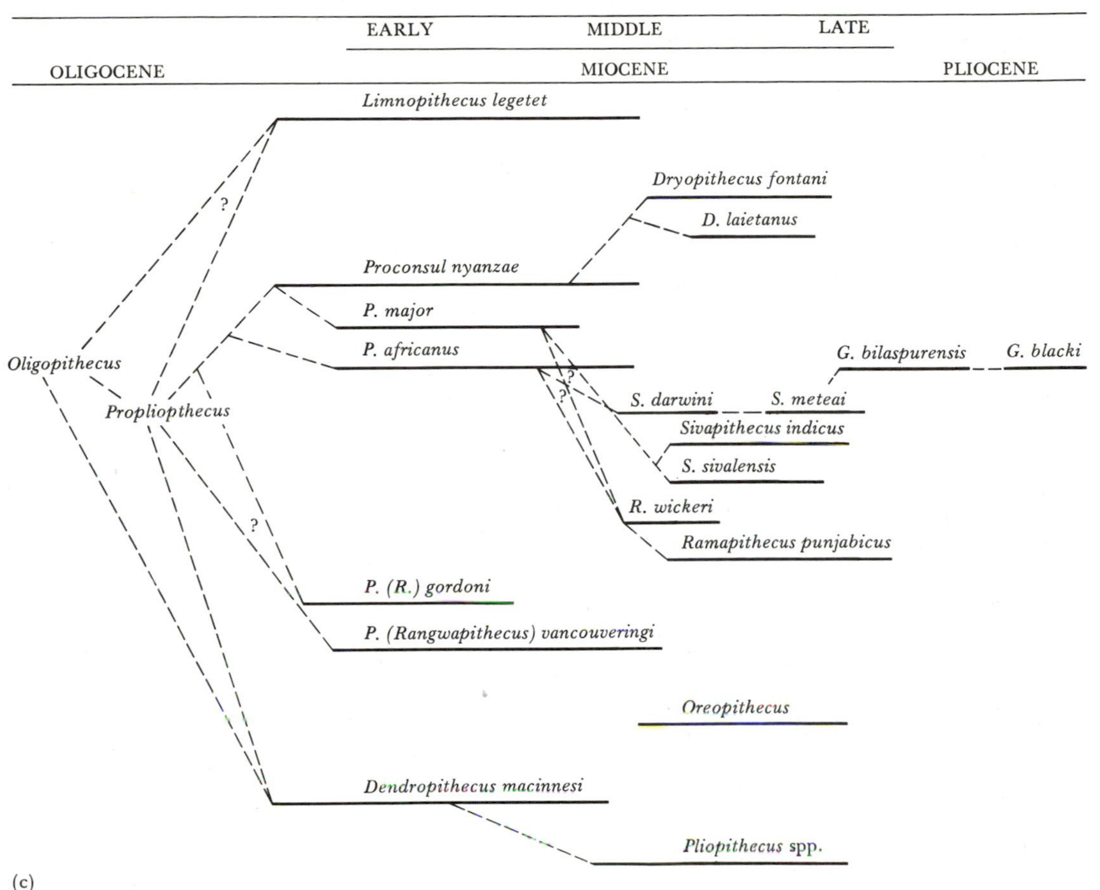

(c)

1. Separation of the gibbon from the great ape and human clade occurred about 12 MYA.
2. Separation of the orangutan from the african ape and human clade occurred about 10 MYA.
3. Separation of African apes from humans happened about 5 MYA. (Sarich and Wilson, 1967; Andrews and Cronin, 1982; Sibley and Ahlquist, 1984; Andrews, 1986)

If these dates are anywhere near to being correct, then the traditional timetable for chimpanzee, gorilla, and human divergence offered by paleoanthropologists in the 1960s and early 1970s (such as in Fig. 5.30) is clearly in error.

Molecular anthropology has infused paleontology with more objective means of assessing phylogeny and cladogenesis, but it is still by no means the panacea that some claim it to be. For example, it is sometimes difficult to ascertain which molecular configurations are primitive and which are derived, so that similarity in molecular structure may not in

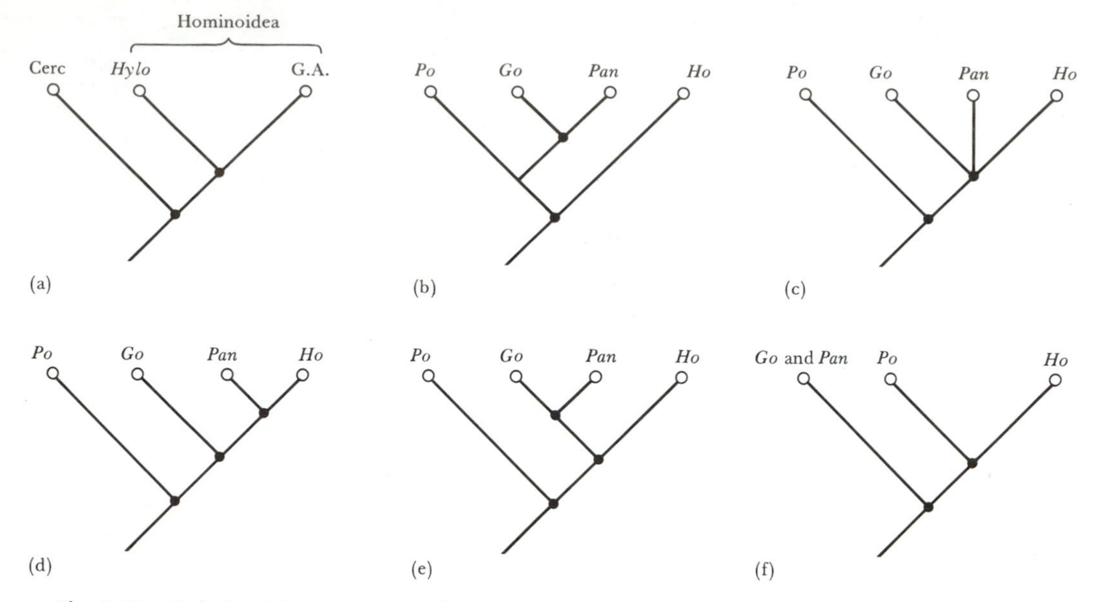

Fig. 5.32 Relationships among Modern Apes and Humans *Ho, Homo; Pan,* chimpanzee; *Go,* Gorilla; *Po,* Pongo (orangutan); *Hylo, Hylobates* (gibbons and siamangs); *Cerc, Cercopithecoidea* (Old World monkeys). (a) Great apes and humans are considered by most workers to form a clade (labeled G.A.), which is most closely related to the lesser-ape clade (gibbons and siamangs). Views (b–f) show different arrangements within the G.A. clade. (b) In the traditional view (held in the 1920s to the mid-1960s), extant great apes were considered more similar to one another than to humans. Note: this is a grade rather than a clade view. (c) Most biomolecular data indicate that the African apes (gorillas and chimpanzees) and humans form a clade in which all three lines diverged at about the same time. (d) A few biomolecular studies show chimpanzees and humans forming a sister group with gorillas. (e) Most workers relying primarily on morphological evidence agree with the molecular data for an African ape–human clade (as in views c and d) but see gorillas and chimpanzees as comprising a sister group with humans. (f) A recent view (Schwartz, 1984) has humans and orangutans forming a sister group with the African great apes. (From Spuhler, 1988)

and of itself indicate a special relationship. In addition, it seems increasingly clear that molecular change, even at the DNA level, does not occur at constant rates in all vertebrate groups (Goodman et al., 1983; Andrews, 1986; Britten, 1986).

Morphological evidence would also support the pattern of these three major cladogenic events noted above. Most students of primate evolution agree that the hominoids (gibbons, chimpanzees, gorillas, orangutans, and humans) are a monophyletic group. Within this clade numerous morphological features of the wrist, shoulder musculature, and vertebral column also attest to the linkage between the great apes and humans to the exclusion of gibbons. Within the great ape–human clade there is also strong morphological evidence (for example, the many anatomical specializations for knuckle walking) linking gorillas and chimpanzees to the exclusion of orangutans.

Many anatomical features of the craniofacial skeleton are shared by humans, chimpanzees, and gorillas:

1. The nasal aperture is broad.
2. The subnasal plane is truncated and stepped down to the floor of the nasal cavity.
3. The orbits are approximately square and often broader than they are high.
4. The interorbital distance is large.
5. The infraorbital foramina are usually three or less in number and are situated on or close to the zygomaticomaxillary suture.
6. The zygomatic bone is usually curved and has a pronounced posterior slope.
7. The one or two **zygomatic foramina** (holes in the zygomatic for transmission of nerves and blood vessels to the side of the face) are small and are situated at or below the lower rim of the orbits.
8. The glabella is thickened.
9. The palate has small incisive fossae and large, oval-shaped **greater palatine foramina** (small holes in the posterior part of the hard palate for passage of nerves and blood vessels to the back of the palate).
10. The teeth are basically similar in cusp pattern. (Andrews and Cronin, 1982)

According to the principles of cladistics these relationships should be reflected in the taxonomic label applied to these animals. In cladistic terminology the African apes and humans are sister groups, which means that using Pongidae as the family designation for all the great apes (to the exclusion of humans) is no longer valid. Many new taxonomic schemes have been proposed in the past few years to reflect these relationships. Some favor the designation Paninae for the African great apes as a subfamily division of the Hominidae (Gantt, 1983). The term *Pongidae* would then be retained solely for the orangutan and related extinct forms. A somewhat similar proposal would use the subfamily designation Gorillinae for the African apes (Andrews and Cronin, 1982). Unfortunately, almost all the new taxonomic proposals result in the expansion of the Hominidae to include at least some of the great apes. Milford Wolpoff (1983) has suggested that the term *hominid* be retained to include only "those taxa on the lineage leading to *Homo sapiens,* and any collateral sidebranches on this lineage, after the divergence of this lineage with the one leading to the African apes." Consequently, Russell Ciochon (1983) has proposed that the subfamily name Paninae be raised to family rank, resulting in a three-family division of the extant great apes and humans: Panidae, Pongidae, and Hominidae. This proposal would retain the classical usage of the term *hominid* while not distorting the true cladistic relationships of the great apes. This newer terminology creates havoc with such familiar terms as *pongid, hominid,* and *ape* and violates one of the main purposes of any taxonomic classification, namely ease of communication. The taxonomic issues involved are complex, and the reader is encouraged to review the papers cited here (and references therein) for further discussion.

Because the role of ramamorphs in hominoid evolution is especially controversial, we shall now consider ramamorph phylogeny in more detail.

Phylogeny of the Ramamorphs The Eurasian fossils, *Sivapithecus, Ramapithecus,* and *Gigantopithecus,* are usually considered to be in the same subfamily, Ramapithecinae, or the same family, Ramapithecidae (or Sivapithecidae). But what are the possible phyletic affinities of this group? Kay and Simons (1983*b*) list five alternatives for ramamorphs:

1. In or near the ancestry of humans, African apes, and orangutans, with all these forms passing through a ramamorphlike stage in their evolution
2. Specifically related to orangutans alone and not to either hominids or African great apes
3. Specifically related to both African great apes and humans but not to orangutans
4. Related to living African great apes alone
5. Related to hominids alone

Proposition 1 has gained in popularity recently, particularly among those who favor a recent time of divergence among the modern groups mainly on the basis of molecular evidence (Zihlman et al., 1978; Greenfield, 1980; Wolpoff, 1983). Proposition 2 has been championed recently by Andrews and others (Andrews and Tekkaya, 1980; Andrews and Cronin, 1982; Ward and Pilbeam, 1983). They point to certain possibly shared derived features of the palate, face, and dentition linking *Sivapithecus* and orangutans (Fig. 5.33). There are no current proponents of views 3 and 4. Kay and Simons support view 5.

There have been suggestions that *Sivapithecus* might be related to the ancestry of orangutans ever since it was first described. Molecular data and new paleontological finds from Turkey, India, and Pakistan seem to substantiate these suggestions. The most important *Sivapithecus* finds to date have been part of the facial skeleton of *S. meteai* from Turkey (Andrews and Tekkaya, 1980) and *S. indicus* from Pakistan (Pilbeam and Smith, 1981; Pilbeam, 1982; Preuss, 1982). For the first time the craniofacial architecture of these hominoids can be studied. Most important, such study reveals that the craniofacial morphology of *Sivapithecus* is very orangutanlike.

Characters shared between the orangutan and *Sivapithecus* include the following:

1. The nasal aperture is higher than broad, and oval-shaped.
2. The subnasal alveolar plane is smooth, not "stepped down" to the hard palate.
3. The orbits are higher than broad.
4. The interorbital distance is very narrow.

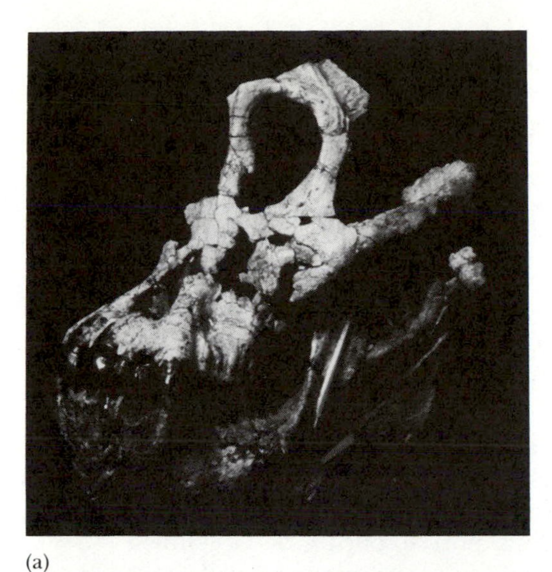

(a)

Fig. 5.33 *Sivapithecus* as the Current Candidate for the Ancestor of the Modern Orangutan (a) Three-quarter view of the skull of *Sivapithecus indicus* from Pakistan; (b) frontal view of *Sivapithecus* with chimpanzee *(left)* and orangutan *(right)*; (c) lateral view of *Sivapithecus* with chimpanzee *(left)* and orangutan *(right)*. In the shape of the eye socket, the construction of the bony ridge over the orbits, dimensions of the incisors, and the detailed anatomy of the lower face, *Sivapithecus* can be seen to resemble the orangutan more closely than the chimpanzee. Scale in (b,c) about ¼ × natural size. (Courtesy of D. Pilbeam)

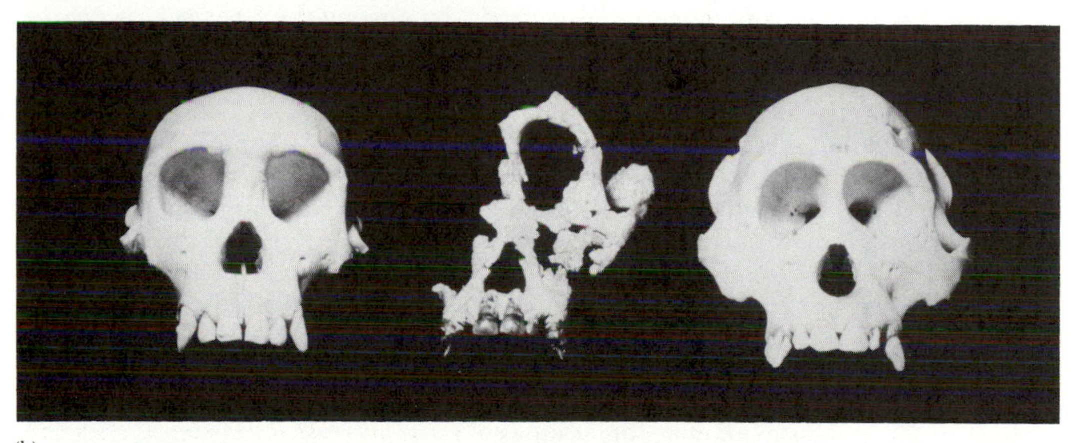

(b)

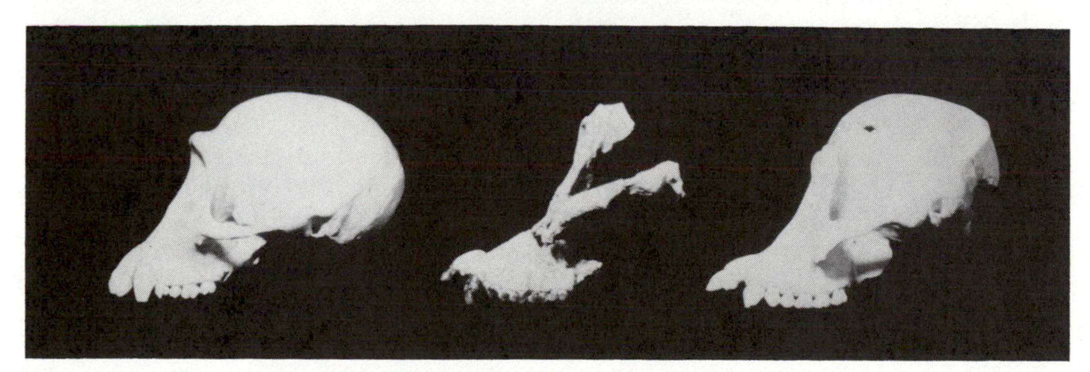

(c)

5. The infraorbital foramina are few in number and well removed from the zygomaticomaxillary suture.
6. The zygomatic bone is flattened and faces anteriorly.
7. The zygomatic foramina are relatively large.
8. There is no glabellar thickening.
9. The incisive foramina are extremely small.
10. The greater palatine foramina are slitlike.
11. There is a great size discrepancy between the first and second incisors.
12. The molars have thick enamel. (Andrews and Cronin, 1982)

It thus appears that some of the characters distinguishing the orangutan from the African great apes are already present in *Sivapithecus.*

Since the morphological features of *Sivapithecus* are for the most part shared with *Ramapithecus,* it has been suggested that the latter is part of the same orangutan clade. This has been recognized implicitly by Pilbeam et al. (1977*b*) in combining ramapithecines and sivapithecines in the family Ramapithecidae, and more explicitly by Greenfield (1979), who has incorporated *Ramapithecus* into the genus *Sivapithecus.*

Two consequences ensue from recognizing the *Sivapithecus–Ramapithecus* group as a clade. First, if one member of the clade, say *S. meteai,* is accepted as being closely related to the orangutan, then all the other species of the clade, including *Ramapithecus,* are also linked in the same association. Second, the proposed relationship between *Sivapithecus* and the orangutan provides an approximate minimum divergence date between the orangutan and its extant sister group, consisting of the African apes and humans. *Sivapithecus* is known from a time range of approximately 13 to 8 MYA. This age corresponds well with the divergence date of 10 ± 3 million years suggested by the molecular evidence and strongly indicates a minimum age of at least 11 MYA for the divergence of the orangutan clade from the clade comprised of the African apes and humans.

The taxonomic ramifications of a *Sivapithecus–Ramapithecus* clade are that the family Pongidae would have to include the orangutan and *Sivapithecus,* while the family Hominidae would have to include the subfamilies Gorillinae for *Pan* and *Gorilla* and Homininae for *Homo* and *Australopithecus.* (See Chapter 6 for a discussion of *Australopithecus* and *Homo,* two hominid genera from the Plio-Pleistocene.) This change from traditional classifications has long been advocated by molecular biologists and is one that is necessary with the recognition that the African apes are more closely related to humans than they are to the orangutan. This taxonomic scheme is still not universally accepted, however. For example, Kay and Simons (1983*b*) still argue the case that *Sivapithecus* (= *Ramapithecus*) is a hominid ancestor rather than an orangutan ancestor.

The *Sivapithecus* fossil from Buluk, Kenya, which dates to about 16 to 18 MYA, throws new light on hominoid origins and the relationships between African and Asian hominoid clades. Although Leakey has recently reclassified it as belonging to *Afropithecus*, the implications of his original assignment to a ramamorph genus are worth considering (Leakey and Walker, 1985):

> The presence of putative *Pongo* ancestors at 11–8 million years in Asia was thought to be consistent with divergence dates between *Pongo* and African hominoids calculated from molecular evidence. If there is indeed a unique ancestral relationship between the Buluk species and the later Asian ones, the fossil and molecular evidence are no longer congruent because the divergence time would be twice as great.

They discount the possibility that the Buluk species represents an extinct lineage completely independent of *Sivapithecus* in Asia. They also point out that the presence of such an old *Sivapithecus* specimen in Africa does not indicate that the African and Asian hominoid clades diverged at that time or before. "Rather than conclude that the Buluk *Sivapithecus* represents the beginning of the *Pongo* clade in Africa over 17 million years ago, we view it as an early species of a genus from which both African apes and hominids could be derived" (Leakey and Walker, 1985).

Cercopithecoidea and Ceboidea

Chapter 4 discussed the pros and cons of the claim that Oligocene parapithecids were ancestral to Old World monkeys (Cercopithecoidea). We now turn to the Miocene evidence for cercopithecoid evolution.

Earlier in the chapter it was pointed out that in their craniodental morphology Miocene monkeys represent a mixture of the two modern subfamilies of Old World monkeys: the Cercopithecinae and the Colobinae. In dentition Miocene cercopithecoids are most similar to modern cercopithecines, whereas in facial structure they most resemble colobines.

Clifford Jolly (1966) of New York University has expressed a different viewpoint. Since he considers the original dietary niche of monkeys to be leaf eating, he has suggested that the ancestral molar morphology of cercopithecoids would be more like that of the modern leaf-eating monkeys, the colobines.

The fossil material from Napak, Uganda, which consisted of a colobinelike frontal bone and a cercopithecinelike upper molar, was originally thought to prove that the Cercopithecinae and the Colobinae had already diverged by the early Miocene some 20 MYA. However, if it is correct that early cercopithecoids should be expected to have colobine-

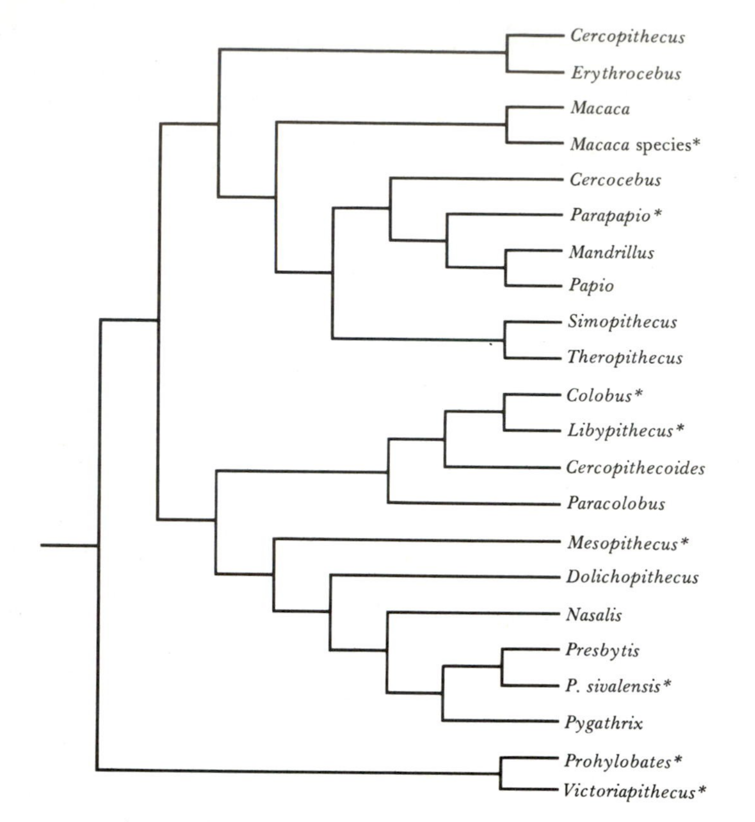

Fig. 5.34 Relationships among Fossil and Living Cercopithecoid Monkeys Starred genera are from the Miocene (Adapted from Andrews and Aiello, 1984)

like skull features and cercopithecinelike dental features, then there is no compelling fossil evidence for postulating a subfamily split in the early Miocene. As mentioned previously, it now appears that the frontal bone is actually from the hominoid *Micropithecus*.

Where does the subfamily Victoriapithecinae fit into cercopithecoid phylogeny? It is possible that victoriapithecines could be the common ancestor of both colobines and cercopithecines. However, including all the Miocene cercopithecoids from Maboko Island, Buluk, Napak, Loperot, and northern Africa in the Victoriapithecinae is not in accord with Delson's (1975*b*) suggestion that the colobines and cercopithecines were already distinguishable by the middle Miocene (Leakey, 1985).

Figure 5.34 shows the relationships between fossil and modern cercopithecoid monkeys.

As the Miocene epoch drew to a close, a major new episode in primate evolution began to unfold. While primitive apes became scarcer and monkeys more common in the fossil record, descendants of one of these Miocene ape species began their adaptive radiation throughout Africa. These more advanced apes would begin to shape the world around them in ways no other primate had even approached. The story of this last major adaptive radiation in higher primate evolution, the rise of the hominids, is the subject of our next and last chapter.

CHAPTER 6

Plio-Pleistocene Primates

Paleoclimates and Biogeography

Summary of the Plio-Pleistocene Fossil Record

Australopithecus

Homo

Morphology of Hominid Locomotion

Behavioral and Cultural Trends in Australopithecines and Archaic Humans

Phylogeny and Classification of Early Hominids

Epilog

Given the hundreds of fossil-ape specimens recovered from Miocene deposits in Africa and Eurasia, it is perhaps surprising that fossil apes are virtually unknown from the succeeding epoch, the Pliocene, which began about 5 MYA. Instead, what one sees in the Pliocene fossil record of Africa are several examples of a new kind of higher primate: it walked on two legs rather than four, used tools rather than teeth for tearing and cutting, had a relatively large brain, and had evolved behavioral and social mechanisms enabling it to survive on the African savanna.

It is common to hear *Australopithecus* referred to as small-brained, small-sized, or both. Although these terms may be apt in comparison to *Homo sapiens,* they are somewhat misleading. Australopithecines had brains roughly three times larger than those in *Proconsul africanus,* and their body size was within the range of modern chimpanzees, which are very powerful animals indeed.

Australopithecus, which appeared some 5 MYA in Kenya and 3 MYA in South Africa, lasted throughout the Pliocene and into the early part of the next-to-last epoch in earth's history, the Pleistocene, which began approximately 1.8 MYA. The combined time span covered by this discussion is sometimes called the **Plio-Pleistocene.**[1]

Also making its first appearance during the Plio-Pleistocene was an early species of our own genus, *Homo,* which has been found in East African deposits dating back to nearly 2.5 MYA and in South African deposits dating back to slightly less than 2.0 MYA (Howell et al., 1987; Brain, 1988). In fact, some species of *Australopithecus* and *Homo* coexisted in East and South Africa for more than a million years, from at least 2.4 to 1.2 MYA (White, 1988). This new genus reveals several important evolutionary distinctions from *Australopithecus,* including more gracile jaws and teeth, more frequent use of stone-tool technology, and most impor-

[1] The final epoch in the geological record is termed, appropriately enough, Recent (also called the Holocene), which commenced about 10,000 years ago. This text will not cover primate evolution during the Recent epoch.

tantly new levels of cerebral organization in terms of both absolute size and complexity (Toth, 1987; Tobias, 1987). These trends continued throughout the Plio-Pleistocene in the various *Homo* lineages, culminating in early members of our own species, *H. sapiens,* several hundred thousand years ago.

PALEOCLIMATES AND BIOGEOGRAPHY

In popular accounts the Pleistocene is often referred to as the time of the ice ages. Oxygen-isotope analyses of deep-sea sediments from the southern oceans and of fossil plant data from various parts of Europe, North America, and Asia confirm that temperatures fell fairly rapidly toward the end of the Miocene and into the Plio-Pleistocene (Flint, 1971) (Fig. 6.1). (See Chapter 2 for a discussion of how temperatures can be inferred from oxygen-isotope analyses of Foraminifera.) These falling temperatures probably initiated the development of the west Antarctic ice sheet, which built up to a size far greater than that seen today. Glaciers appear to have developed repeatedly in Antarctica, Alaska, and Iceland over the past 10 million years, although large middle-latitude ice sheets apparently did not appear until the Pleistocene.

The resulting lower sea levels near the end of the Miocene contributed to the so-called **Mediterranean crisis:** about 6 MYA the Mediterranean Sea dried up and became an evaporation basin in which vast quantities of salt and gypsum were deposited (Hsu et al., 1977; Brain, 1981*a,* 1983). The complete drying up of such a large body of water had profound biological and geological consequences. Of great interest is the fact that the expanded dry-land connections between Africa and Europe permitted the free exchange of faunas and floras (Brain, 1983).

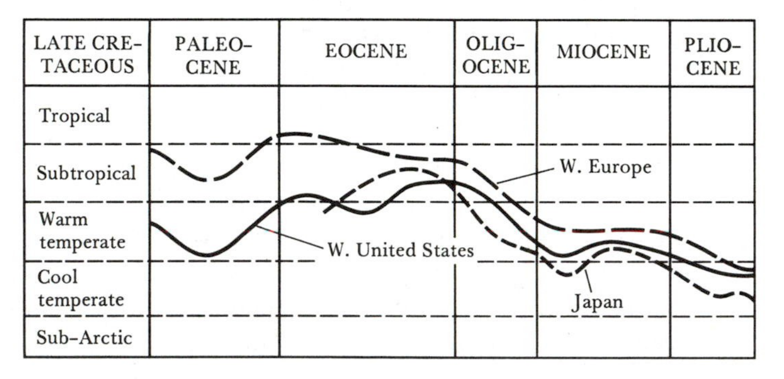

Fig. 6.1 Climatic Cooling through the Cenozoic of North America, Europe, and Asia Inferred from Fossil Plants Note the cooling trend apparent near the end of the Miocene. (From Flint, 1971)

These terminal Miocene events had far-reaching implications for hominid paleobiology. During the Antarctic ice-cap buildup, cold upwelling water flowed along the west coast of Africa and, by drawing moisture-laden air off the land, increased the aridity of the coastal areas of southern Africa (Ward et al., 1983). Any flora strongly dependent on moisture would have suffered during this period as more open, drought-resistant forms of vegetation spread. This trend must have affected arboreal hominoids of the late Miocene in Africa since the equatorial forests were undoubtedly shrinking. The continuation of this process greatly enlarged the area of the transitional ecological zone between forest and adjacent savanna. It is tempting to view this transitional ecological zone, which is neither forest nor savanna, as the area in which behavioral and anatomical changes were taking place in hominid evolution. These changes led from arboreal quadrupedalism to terrestrial bipedalism (Brain, 1981*a*).

One recently proposed ecological model of hominid origins stresses the importance of such fringe environments. Milford Wolpoff (1982) of the University of Michigan has speculated that at the time the African apes and australopithecines split from their common ancestor, the initial ecological niche available to both would most likely have been a rather broad one, including dense to fairly open savanna and some even more densely forested areas. The result of the growth of the fringe environment might have been to lessen competition between the new sibling species by partitioning the niche occupied by the parent species into two narrower and less overlapping adaptive zones. The australopithecines would have been the ones gravitating toward the more open areas and the African apes the ones moving into the more densely forested regions. Wolpoff's speculation goes on to suggest that competition would have been further reduced if the apes and australopithecines had developed their own dietary, dental, and locomotor specializations. Such specializations of early hominids would have included the evolution of a powerful masticatory apparatus, bipedalism, the use of rudimentary tools and weapons, and a set of social changes including the use of home bases, the division of labor, and food sharing. A more detailed discussion of the paleoclimates and biogeography of Plio-Pleistocene hominids is given in the following section.

SUMMARY OF THE PLIO-PLEISTOCENE FOSSIL RECORD

As mentioned at the beginning of the chapter, there is no fossil record of apes in Eurasia and Africa after the Miocene (except for *Gigantopithecus* from the Pleistocene in China). Old World monkeys, however, continue to appear throughout the Plio-Pleistocene in both Eurasia and Africa. The two modern cercopithecid subfamilies, Cercopithecinae and Colobinae, which were well differentiated by the end of the Miocene, are represented in fossil sites on both continents. New World monkeys are only rarely found in Plio-Pleistocene localities in South America, although several have been recovered from Recent sites in the Caribbean. See Table 5.5 for a listing of Old and New World fossil monkeys in the Miocene through Recent epochs and Table 1.3 for a listing of extant genera; we will not concern ourselves further with fossil monkeys in this text.

The predominant primate group of interest in the Plio-Pleistocene fossil record is unquestionably the **Hominidae,** the hominoid family that makes its first appearance in the early Pliocene (Table 6.1). Although some Miocene fossils (such as *Ramapithecus*) have been attributed by some workers to the Hominidae, ***Australopithecus*** was the earliest undisputed hominid, appearing approximately 5 MYA in Kenya and 3 MYA in South Africa. *Australopithecus* lasted throughout the Pliocene, disappearing from the fossil record about 1 MYA, in the early part of the Pleistocene.

Most paleoprimatologists recognize at least four species of *Australopithecus: A. africanus, A. afarensis, A. robustus,* and *A. boisei,* although there is as yet no unanimity of opinion about the number of distinct species or the classification of individual australopithecine fossils (Kimbel et al., 1988). These species are often further separated according to body type as either **gracile** (slender, thin-boned) or **robust** (thick-boned); in this text we shall consider the first two species gracile and the latter two robust (although there is disagreement about this distinction also, as we shall see later in

Table 6.1 African Plio-Pleistocene Hominids

Taxa	*Epoch and Area*
Family HOMINIDAE	
Subfamily AUSTRALOPITHECINAE	
Australopithecus	Plio-Pleistocene (E. and S. Africa)
A. afarensis	Pliocene (E. Africa)
A. africanus	Pliocene (S. Africa, ?E. Africa)
A. robustus	Plio-Pleistocene (S. Africa)
A. boisei	Plio-Pleistocene (E. Africa)
Subfamily HOMININAE	
Homo	Plio-Pleistocene (worldwide)
H. habilis	Plio-Pleistocene (E. and S. Africa)
H. erectus	Pleistocene (Africa, Eurasia)
H. sapiens	Pleistocene (worldwide)

the chapter). Two generalizations about early hominid evolution in Africa can be made: (1) only gracile australopithecines, either *A. africanus* or *A. afarensis,* were present prior to about 2.5 MYA; and (2) robust australopithecines, first appearing about 2.5 MYA, survived until about 1 MYA. As we shall see later in the chapter, these generalizations have had a significant impact on theories of australopithecine phylogeny.

The other hominid genus, ***Homo,*** first appeared about 2.5 to 2.0 MYA in East Africa (although some of these earliest specimens are difficult to distinguish from gracile australopithecines). The three undisputed species of *Homo* found in Plio-Pleistocene deposits are *H. habilis, H. erectus,* and later *H. sapiens.* This text will cover the *Homo* fossil record only briefly, and will emphasize taxonomic considerations and comparisons with australopithecine morphology. The subject of human evolution is, of course, complex and highly controversial and deserves an entire volume unto itself.

Australopithecines were confined to the African continent. Some investigators have suggested their presence in the Far East, but these claims are not generally accepted. There are two widely separate areas from which australopithecines are known: East African Rift Valley deposits and South African cave breccias. (A **breccia** is a rock composed of angular fragments of older rocks that have been cemented together.) The East African deposits generally consist of lake and river sediments that have embedded in them occasional layers of volcanic **tuff** (rock composed of fragments of volcanic material that have been blown into the atmosphere by volcanic activity.) Many of these tuffs have been radiometrically dated by the methods discussed in Chapter 1. The cave breccias, on the other hand, cannot be radiometrically dated by any of the currently available techniques. Figure 6.2 shows the main hominid-bearing sites of the Plio-Pleistocene in East and South Africa.

Before examining the morphology and behavior of *Australopithecus* and early *Homo,* we will examine the fossil evidence briefly by site.

South African Sites

There are five major South African cave sites: Sterkfontein, Kromdraai, Swartkrans, Makapansgat, and Taung. The first three are located within 4 km of one another just a few miles from Johannesburg; Makapansgat is in the northern Transvaal; and Taung is on the edge of the Kalahari Desert, north of Kimberley (Fig. 6.2a).

The South African caves are all formed out of Precambrian dolomitic (magnesium-containing) limestone. C. K. Brain (1981*b*) has described a sequence of stages in the formation of such caves:

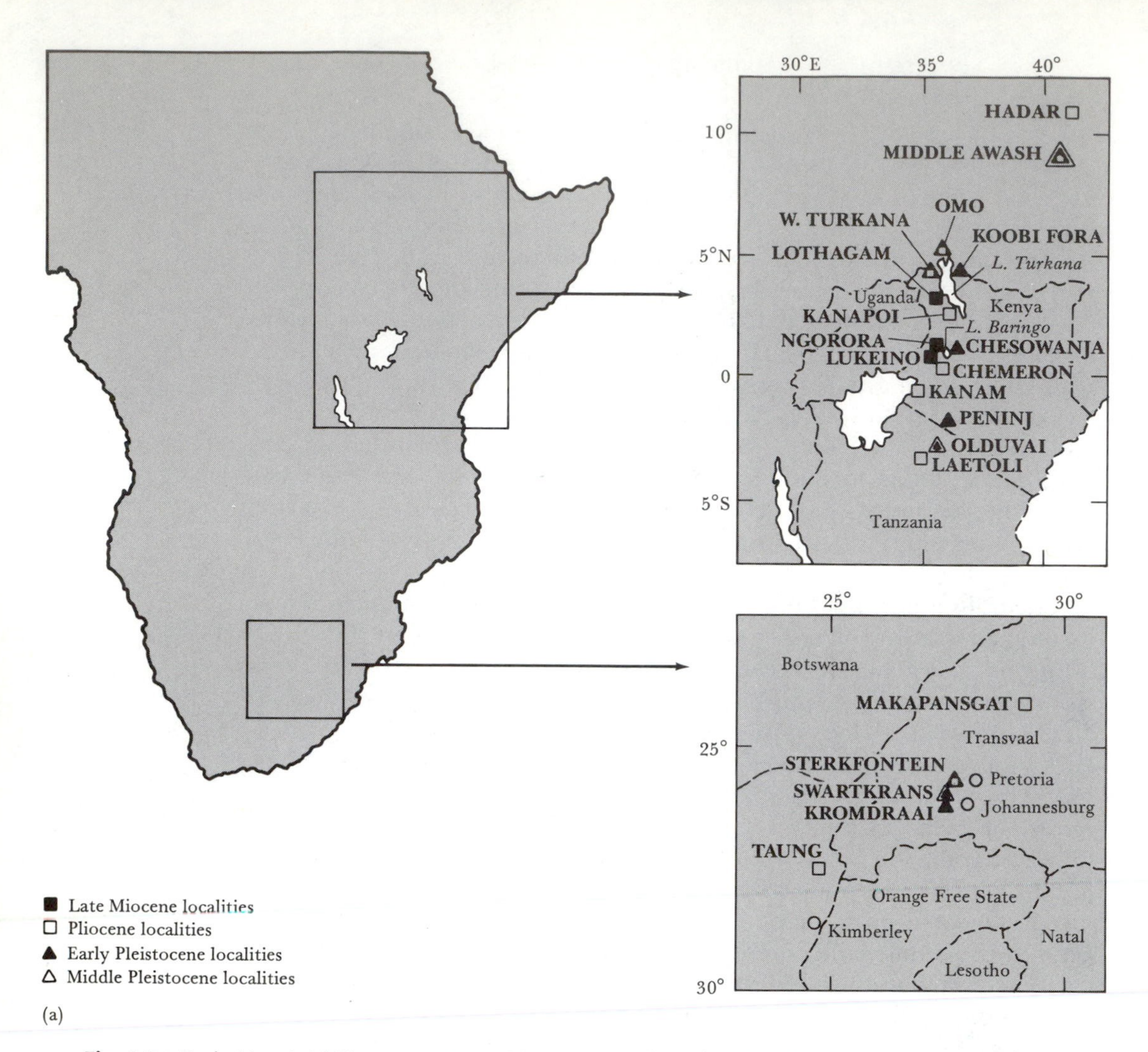

Fig. 6.2 Early Hominid Sites (a) Map of late Miocene to early Pleistocene hominid-bearing localities; (b) chronology of early hominids from major sites. (a, Adapted from Howell, 1978; b, adapted from White et al., 1981, with some age ranges for *H. habilis* and *A. boisei* added and altered to match more recent data)

1. The cavern initially forms by solution of the dolomite in the groundwater, its contours being determined by planes of weakness in the rock.
2. The water level drops in the general area, and the cavern fills with air.
3. Rainwater causes joints or other planes of weakness in the rock to form in the dolomite overlying the cavern.
4. One of the joints breaks through to the surface, providing the first direct link between the cavern and the surface.
5. A talus cone (accumulation of sediment) begins to form beneath the joint, containing the bones of animals living on the surface; if the

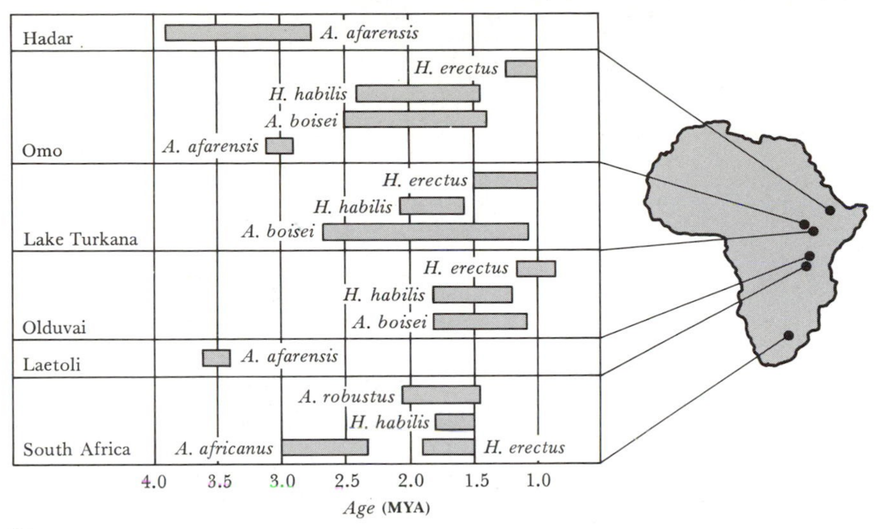

(b)

sediment is calcified by lime-bearing solutions dripping from the roof, the resulting deposit is known as a **cave breccia.**

6. The cavern fills almost completely with breccia.
7. Surface erosion removes much of the roof, exposing the bone-bearing breccia on the surface.

The sites of Sterkfontein, Kromdraai, Swartkrans, and Makapansgat are currently at this latest stage. Unfortunately, the original cave site at Taung was blasted away years ago during lime-quarrying activities.

Following is a brief review of the history and stratigraphy of the South African cave sites. Figure 6.3 shows the main Plio-Pleistocene stratigraphic units yielding early hominid remains in both South and East Africa.

Taung Taung means the place of Tau or Tao, the Lion. Located some 130 km north of Kimberley in the northern Cape Province, it is the most southwesterly site on the African continent yielding fossils of *Australopithecus.*

Sometime in May 1924, a young medical student at the University of the Witwatersrand in Johannesburg, Josephine Salmons, was shown the skull of a fossil baboon by one of her friends that had come from the limeworks at Taung. She in turn brought the fossil to the attention of the new professor of anatomy, Dr. Raymond Dart. Australian by birth, Dart had just assumed the Chair of Anatomy at the new medical school in Johannesburg the year before (after spending a year in the Anatomy Department at Washington University in St. Louis as the first Rockefeller Fellow). The skull piqued his interest because a site yielding fossil baboons might also contain fossil hominids. Dart discussed the fossil find

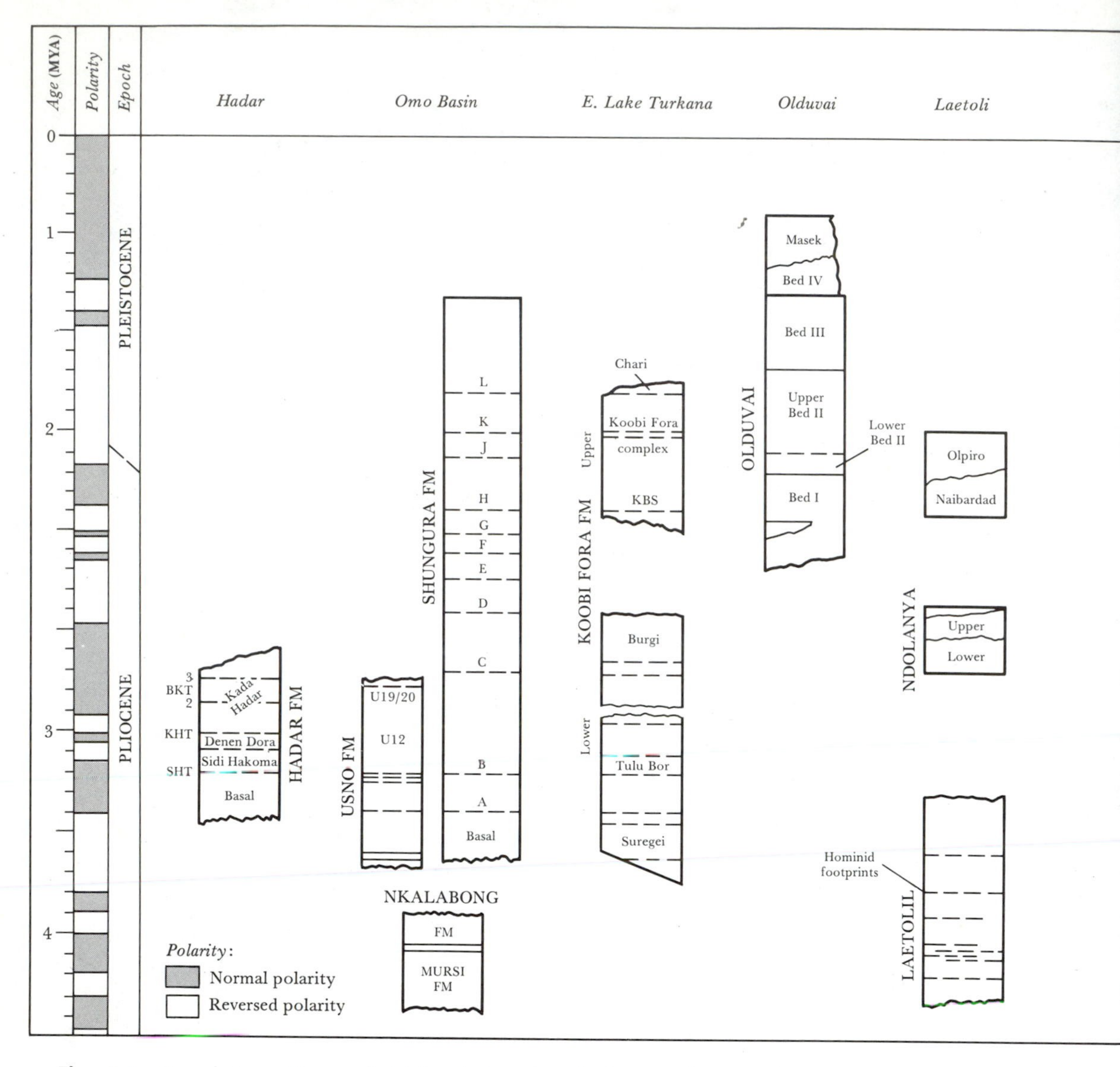

Fig. 6.3 Correlation Chart of Main Early Hominid Sites in East and South Africa Formations (FM) are written vertically, and names within columns identify geological beds and members. The paleomagnetic column is given on the outside left and right margins. The stratigraphy at each site is aligned horizontally with this column so the ages of various beds, members, and formations relative to each other can be seen, with youngest layers on top and oldest layers at the bottom. For example, the youngest hominid-bearing layer at Makapansgat (Member 5) (about 2 to 2.3 MYA) is older than Member H in the Shungura Formation at Omo Basin (about 1.9 to 1.7 MYA). (Adapted from Cooke, 1983)

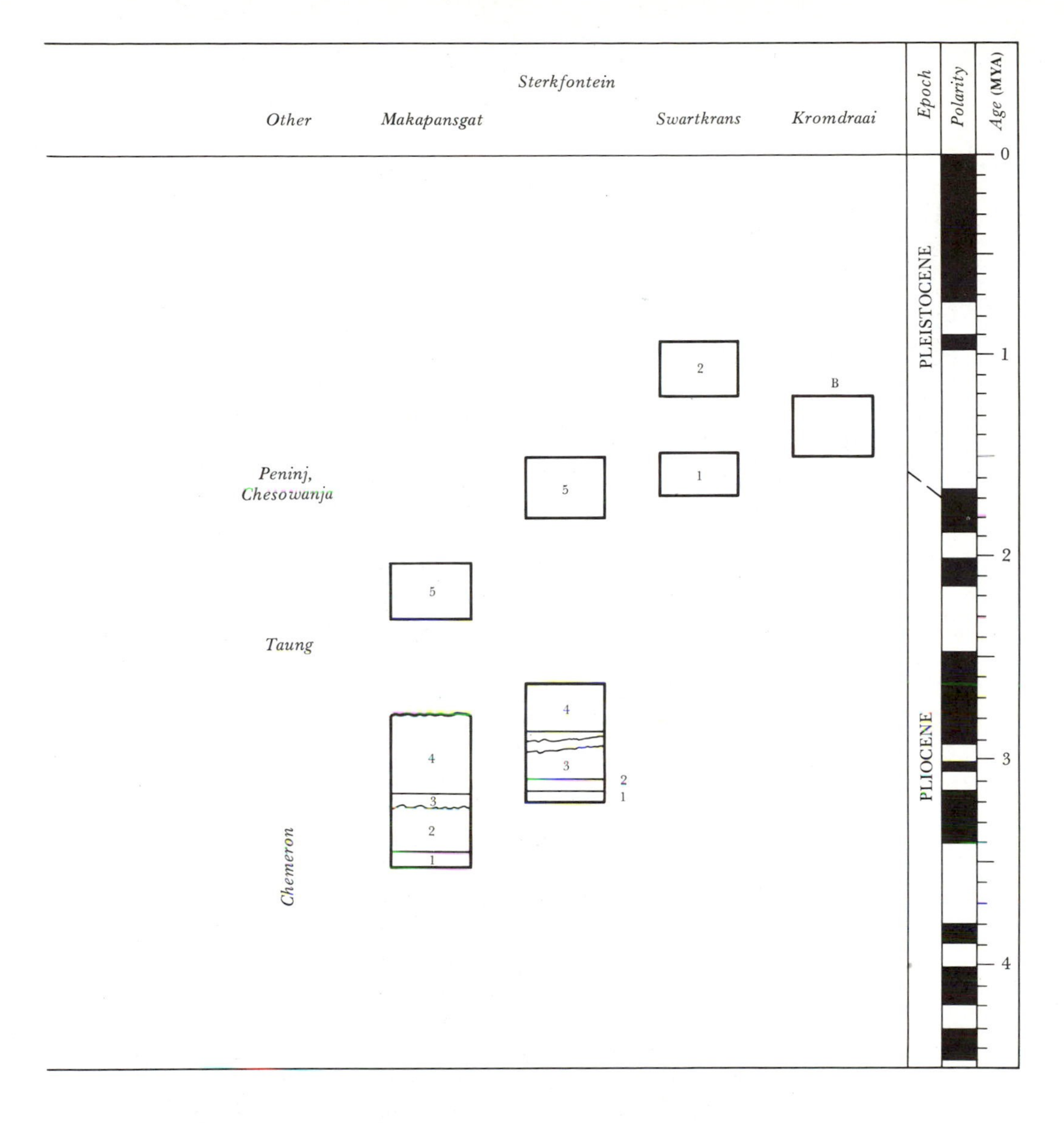

with R. B. Young, a professor of geology who was about to visit the vicinity of Taung to do economic geological research.

In November of that year, one of the limeworkers blasted out a part of another fossilized skull that was lying around the office of the limeworks manager when Young arrived. Knowing Dart's interest in fossil primates, Young brought the fossil back to Johannesburg and handed it over to Dart on November 28, 1924 (Tobias, 1984*b*).

Over the next 40 days Dart extracted the fossil from its rock matrix, analyzed it, and then sent his preliminary conclusions to the British scientific journal *Nature* for publication. The specimen consisted of most of the face, lower jaw, and half of a brain endocast of a young child he named *Australopithecus africanus* ("southern ape of Africa"). Appearing on February 7, 1925, Dart's article triggered an intellectual revolution about human origins that continues unabated to this day (Dart, 1925). Indeed, the popular science magazine *Science 84* included the Taung skull among the "20 discoveries that changed our lives in the 20th century" (Tobias, 1984*a*).

One might have thought that the humanlike qualities of the Taung skull delineated by Dart in 1925 would have been accepted with open arms by a worldwide anthropological community obsessed with finding "missing links" in the evolutionary lineage between apes and humans. However, the reaction to the Taung discovery ranged from total neglect (Hooton, 1931; Leakey, 1935) to extreme criticism (Keith, 1931). Clearly, Dart's little skull did not fulfill the preconceived notions of human evolution prevalent at the time, namely, that early hominids must have had large brains (at least 750 cc according to Keith) and that early man must have evolved in Asia, not Africa. Perhaps Europeans could not accept the view that Africa was the "cradle of mankind," as Dart had referred to it when he exhibited the skull at the 1925 British Empire Exhibition at Wembley. Since the skull was obviously that of an immature creature, critics were quick to point out that if and when adult specimens were found they would certainly look more like apes than hominids, since immature living apes look more "human" than adult apes do. Dart's Taung child remained in taxonomic limbo until adult australopithecines began showing up in other South African sites in the mid-1930s.

Since the exact site of discovery of the only *Australopithecus* specimen from Taung was destroyed by limestone quarrying in the 1920s, precise stratigraphic information for use in dating the fossil remains problematic. Geological considerations and preliminary radiometric dates so far suggest an age of about 1 million years for the hominid specimen (Partridge, 1986), but this is clearly at odds with the faunal dating, particularly of the cercopithecids, which suggests an age of 2.0 to 2.5 MYA (Delson, 1984, 1988). The earlier age has the advantage of being more consistent with the age of other *A. africanus* specimens from Sterkfontein (Member 4) and Makapansgat (Members 3 and 4), which range between 3 and 2.4 MYA.

An analysis of the Taung sediments suggests that they formed under semiarid climatic conditions similar to those of today. The cycle of cave filling that entrapped the hominid seems to have marked the onset of more humid conditions.

Sterkfontein In a 1935 guidebook to places of interest around Johannesburg, the owner of the Sterkfontein caves wrote, "Come to Sterkfontein and find the missing link." On Monday, August 17, 1936, the South African paleontologist and physician Robert Broom did just that.

Earlier in that year two of Dart's former students, G. Schepers and W. Le Riche, alerted Broom to the fact that fossil baboons were known from the caves. Broom, who had joined the staff of the Transvaal Museum in Pretoria only two years before, was understandably excited by these discoveries and paid his first visit to Sterkfontein on August 9, 1936. Eight days later he was rewarded with the discovery of the fossil he at first named *Plesianthropus transvaalensis* (Broom, 1938). It turned out to be the first adult specimen of *Australopithecus africanus* known to science. The find consisted of the anterior two-thirds of a brain cast, the skull base, parts of the parietal bones, and portions of the frontal bone.

Following this initial find, Broom made many discoveries at the site until the outbreak of World War II interrupted his fieldwork. After the war, Broom, assisted by John Robinson, also from the Transvaal Museum at Pretoria, found many other australopithecine remains, including important postcranial remains (Broom and Robinson, 1947). Work continues at the site today under the supervision of Phillip Tobias and Alun Hughes of the University of the Witwatersrand.

Timothy Partridge (1978) has divided the Sterkfontein deposits into Members 1 to 6 (from oldest to youngest) (Fig. 6.4); hominid remains have been found in Members 4 and 5. At least 50 *A. africanus* individuals

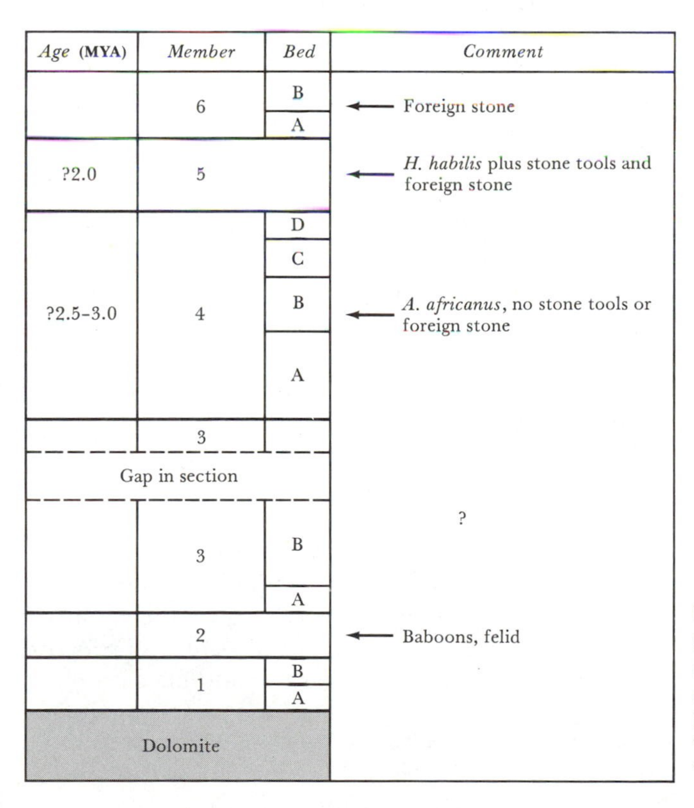

Age (MYA)
Member
Bed
Comment
6
B
A
Foreign stone
?2.0
5
H. habilis plus stone tools and foreign stone
D
C
B
A
?2.5–3.0
4
A. africanus, no stone tools or foreign stone
3
Gap in section
?
3
B
A
2
Baboons, felid
1
B
A
Dolomite

Fig. 6.4 Stratigraphic Sequence of the Sterkfontein Cave Site Note positions of *A. africanus* and *H. habilis* discoveries. Total depth of column is about 27.5 m. (Adapted from Tobias, 1980, after data provided by T. C. Partridge)

are represented at the site, all from Member 4. Included are skulls of varying degrees of completeness, maxillae, mandibles, isolated teeth, and various postcranial bones. Clarke (1988) has recently suggested the presence of two hominid species in Member 4, one more like *A. africanus* and the other more like *A. robustus*. No stone tools have been recovered from Member 4.

Forty years to the day after Broom's first visit to Sterkfontein, A. Hughes recovered the first unequivocal *Homo habilis* specimen from that site (Hughes and Tobias, 1977). (The type specimen, however, was found in the mid-1960s by Jonathan Leakey at Olduvai Gorge; the Latin name *habilis* presumes that it was handy or skillful with tools.) Member 5 contains hominid remains attributed to *H. habilis* along with numerous stone artifacts.

The paleomagnetic stratigraphy of this site is still poorly resolved. Elizabeth Vrba (1982) of the Transvaal Museum in Pretoria, South Africa (and now at Yale University), places the Member 4 fauna between 2.8 and 2.3 MYA and Member 5 between 1.8 and 1.5 MYA. Sedimentological evidence suggests small climatic variations corresponding to Members 2 and 3. Aridification with more seasonal distribution of rainfall is indicated as one goes higher in the section into Members 4 and 5. The onset of this aridification occurs at about 2.5 MYA at both sites. Analysis of the fossil antelope remains suggests a major transition to more open grassland between Members 4 and 5 (Vrba, 1980*b*, 1985*a*).

Vrba (1975) has suggested that the low proportion of juvenile animal bones in Member 5 indicates that some meat was scavenged rather than actively hunted by hominids. The presence of several hundred stone tools combined with indisputable cut marks on the shaft of a small bovid humerus implies that many of the bones in Member 5 represent hominid food remains. At least two bone tools were probably used as instruments for digging up edible bulbs. A number of bovid bones also showed clear evidence of carnivore chewing (Brain, 1981*b*).

Kromdraai In 1938 in somewhat unusual circumstances, the first robust australopithecine was discovered at Kromdraai, situated only 1.5 km east of the Sterkfontein site. The limeworks manager at Sterkfontein, a Mr. Barlow, sold Robert Broom a well-preserved palate with a molar tooth of a fossil hominid but refused to tell him where it had come from. Broom eventually extracted from Barlow the information that the specimen had been found nearby by a schoolboy, Gert Terblanche. Broom took the young boy back to the original finding place, where he soon discovered most of the palate, much of the left side of the face, almost the whole of the left zygomatic arch, the left side of the base of the skull, a portion of the parietal bone, and the greater part of the right mandible with most of the teeth of a hominid. Broom made this skull the type specimen of a new genus and species, *Paranthropus robustus*. Subsequent excavation in

1941 produced a juvenile mandible. After this discovery, further work at Kromdraai was discontinued until the mid-1950s when Brain and later Vrba continued the excavations.

A minimum of six robust australopithecine individuals are now known from Member 3 of the Kromdraai B East (KBE) Formation, also referred to as the Kromdraai Australopithecine Site (Fig. 6.5). The individuals are represented by the partial skull and dentition of the type specimen, by isolated teeth, the proximal end of an ulna, the distal end of a humerus, a second metacarpal, talus, a proximal hand phalanx, and an ilium. The fauna suggests an age of just over 2 MYA to about 1 MYA. This age is consistent with the reversed polarity of this stratigraphic column.

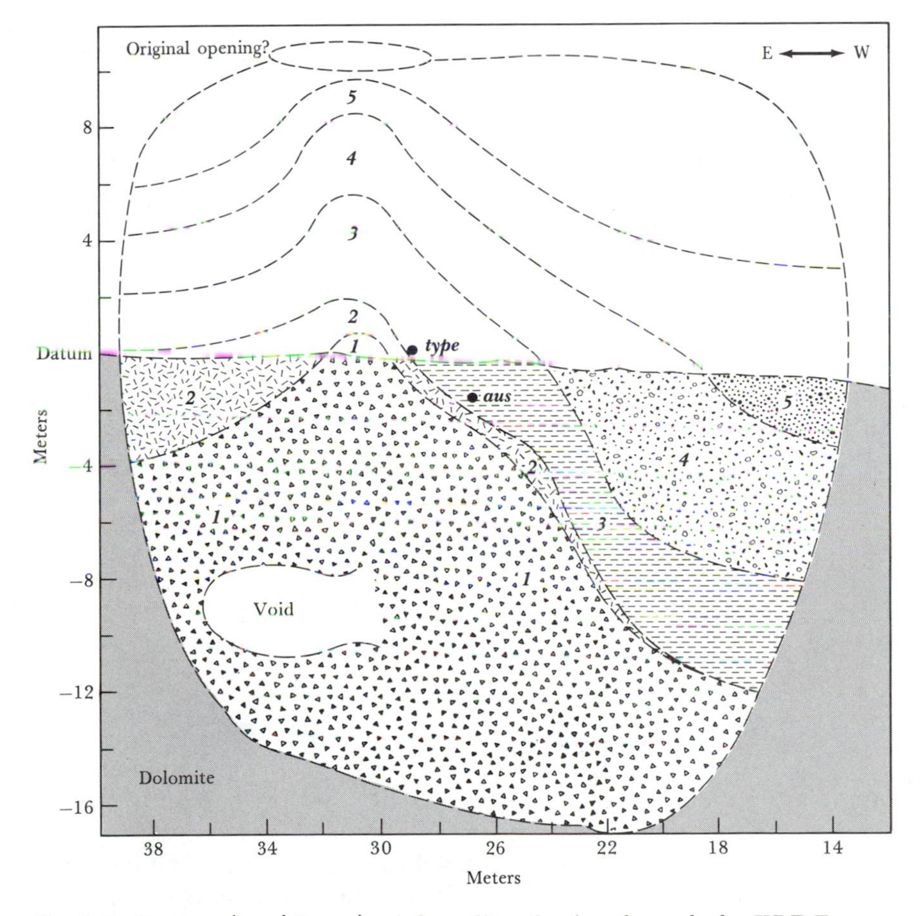

Fig. 6.5 Stratigraphy of Kromdraai Cave Site Section through the KBE Formation (Kromdraai Australopithecine Site). Scale at left gives meters above and below present land surface (datum). Dotted lines above datum show hypothesized original cave contours before erosion. Successive breccia members are labeled 1 to 5. Abbreviations: *type*, possible source of type specimen of *A.* (= *Paranthropus*) *robustus; aus*, mandible of juvenile *A. robustus*. (Adapted from Vrba, 1985*b*)

Only one unquestionable stone tool flake is known from the site, making it unlikely that early hominid scavenging or butchery was responsible for the animal bone accumulations at this site (Brain, 1981*b*).

Sedimentological analyses from Kromdraai indicate a more humid environment than that which prevailed during deposition of Member 4 at both Makapansgat and Sterkfontein. A wooded local environment is evident from the fauna of KBE Member 3, whereas the pollen recovered from the boundary of KBE Members 2 and 3 suggests that a more open savanna characterized the area at that time (Vrba, 1981).

Makapansgat Located in the northern Transvaal, the Makapansgat hominid site is a cavern of almost pure limestone that was mined for a decade starting in 1925. During these mining operations a local teacher of mathematics, W. I. Eitzman, drew Raymond Dart's attention to the abundance of fossil bones preserved in the cave breccia being blasted out by limeworkers. Research and teaching pressures prevented Dart from investigating the site thoroughly until 1945. Many of the vertebrate fossils at the site contained free carbon, which led him to conclude that the bones had been intentionally burned by the original inhabitants of the cave. In September 1947 one of Dart's researchers, J. W. Kitching, discovered the back portion of an australopithecine skull on one of the limeworkers' dumps. Reasoning that this species might have been responsible for some of the burned bones in the deposit, Dart named it *Australopithecus prometheus* (Dart, 1948*a*). By the mid-1960s most workers decided that the majority of fossils previously described as *Plesianthropus* (named for Broom's adult specimen from Sterkfontein) and *A. prometheus* could be included in the single taxon, *A. africanus* (Tobias, 1967).

Other hominid discoveries were made at the site in 1948, including an adolescent mandible, an infant's right parietal bone, several craniofacial fragments and isolated teeth, and two fragments of an adolescent pelvis (Dart, 1948*b*). The morphology of the pelvic remains proved conclusively that *A. africanus* was a biped. Several stone tools have been recovered from the site (Brain et al., 1955). Work at the site continues today under the auspices of the University of the Witwatersrand.

Dart noticed that many of the vertebrate bones seemed to be artificially fractured and that some animal parts were more common than others in the deposit. This suggested to him that the hominids were in some way responsible for the bone accumulation. Dart (1957) proposed the idea that many of the bones and jaws at Makapansgat had been utilized as tools by the early hominids at the cave: dentitions as saws and scrapers, long bones as clubs, and so on. He named this the Osteodontokeratic Culture. However, more recent **taphonomic** studies (concerning the processes by which animal bones become fossilized) have cast doubt on this interpretation, suggesting instead that many of these bone accumulations were the product of carnivore scavengers such as hyenas (Brain, 1981*b*).

A paleomagnetic record for this site extends from the base of Member 1 to the lower levels of Member 4 (Fig. 6.6a), but the calibration of the paleomagnetic record at Makapansgat has been open to several different interpretations. The standard record of polarity reversals during the Plio-Pleistocene is divided into several **polarity epochs** (or **chrons**):

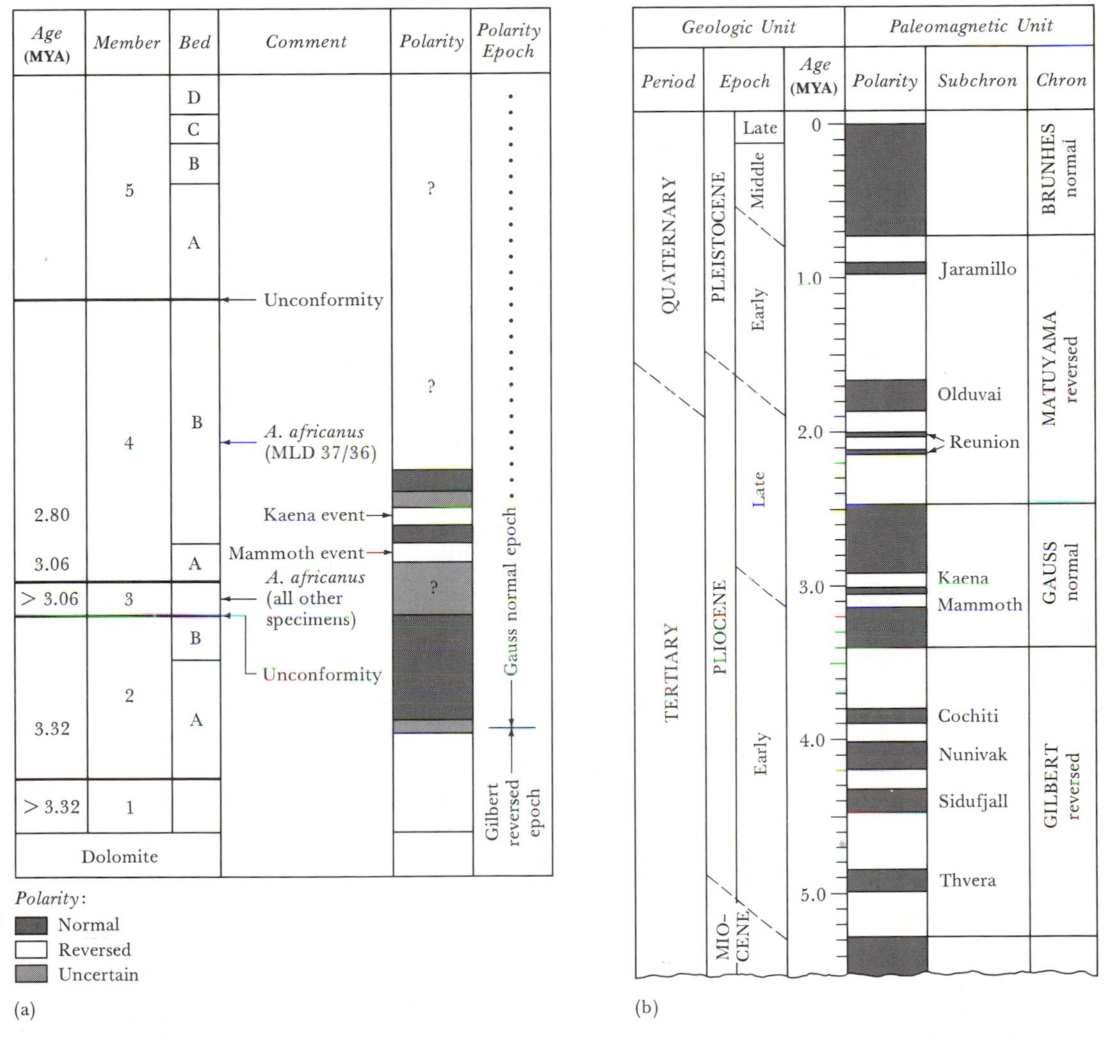

Fig. 6.6 Dating the Makapansgat Cave Site (a) Stratigraphic and paleomagnetic sequence of the Makapansgat Formation showing the location of the major australopithecine discoveries from that site. The partial paleomagnetic column (shown on the right-hand side) can be compared with (b) the standard paleomagnetic column. Subchrons (polarity events) represent short-term magnetic reversals within chrons (long-term polarity epochs). Chrons and subchrons in the Plio-Pleistocene are assigned names, whereas before the Pliocene they are simply numbered. (a, Adapted from Tobias, 1980; b, adapted from Cooke, 1986).

the so-called Gilbert reversed epoch in the early Pliocene; the Gauss normal epoch in the late Pliocene; and the Matuyama reversed epoch, which begins in the late Pliocene, about 2.48 MYA, and covers most of the Pleistocene to about 0.75 MYA. Polarity epochs are in turn divided into **polarity events** (or **subchrons**), which represent short-term magnetic reversals (Fig. 6.6b). Taken together, chrons and subchrons allow one to correlate strata at different sites by age. At Makapansgat the fossil hominids, all assigned to *Australopithecus africanus,* are known only from Members 3 and 4. The australopithecine-bearing strata has been variously dated at about 3 MYA (Delson, 1984; Partridge, 1986) and 2.4 MYA (White et al., 1981). The combination of fossil Bovidae (antelope, buck, buffalo, etc.), suids, and cercopithecoids found in Member 3 is most similar to the faunal assemblages of East African deposits dated at 3 to 2.6 MYA (Vrba, 1985). Such comparisons assume, of course, that East and South Africa had similar enough ecologies to support the same fauna and that similar faunas are truly contemporaneous.

Taphonomic and sedimentological studies suggest that the area had a wetter environment than occurs there today, with a fairly extensive bush cover. Evidence of more episodic rainfall patterns begins to appear after the middle of Member 4. The presence of a higher proportion of grazing mammals signals the onset of more open, drier conditions about this time (Partridge, 1986).

Swartkrans Toward the end of 1948 Robert Broom decided to investigate Swartkrans, a new cave site less than a mile from Sterkfontein. The excavations had an auspicious beginning: the very first dynamite blast yielded an australopithecine tooth. By the end of the first week the mandible of what appeared to be a new species of robust australopithecine, *Paranthropus crassidens,* was discovered. Soon afterward, in April 1949, a very different hominid was found, first named *Telanthropus capensis* but later reclassified as *Homo erectus* (Broom and Robinson, 1950; Clarke et al., 1970). This was the first evidence for the coexistence of *Australopithecus* and *Homo* in the fossil record of Africa (Olson, 1978; White, 1988).

Robert Broom died in Pretoria in April 1951, but the work at Swartkrans was continued by his colleague John Robinson. Since the 1960s the site has been investigated by C. K. Brain and associates from the Transvaal Museum. A large number of robust australopithecines and a lesser number of *Homo* specimens have been recovered.

There are several major rock units defined at this cave site, designated Members 1 to 5 (Fig. 6.7). Member 1, an older orange unit, is preserved as both a basal deposit and as a hanging remnant isolated by erosion; Member 2 is a younger brown unit. Both members contain stone tools and remains of *Australopithecus robustus* and *Homo erectus.* Similar stone tools and *A. robustus* remains are also known from Member 3. Evidence for the earliest use of fire also comes from this member. Middle

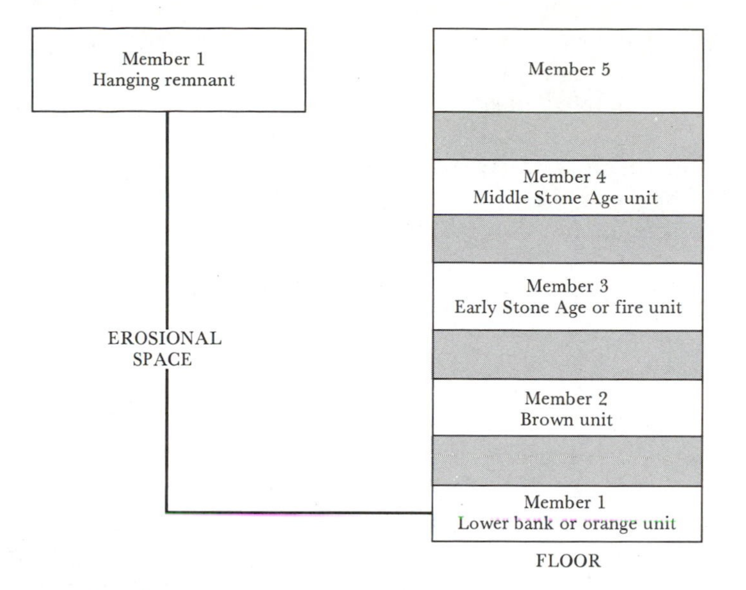

Fig. 6.7 Stratigraphic Sequence of the Swartkrans Cave Site Hominid remains have been found in Members 1 to 3: *A. robustus* and *H. erectus*, both from Members 1 and 2, but only *A. robustus* from Member 3. Stone tools have been found throughout the sequence, and the earliest evidence of fire comes from Member 3. The oldest deposit, Member 1, is divided into two discrete masses: the hanging remnant clinging to the cave's north wall, and the lower bank resting on the cave floor, both thought to date from 1.8 to 1.5 MYA based on faunal considerations. Members 1 to 3 probably fall within a time range of 1.8 to 1.0 MYA, and Members 4 and 5 are probably less than 150,000 years old. Shaded areas indicate erosion. (Adapted from Brain, 1988)

Stone Age artifacts have been recovered from Member 4, although no hominid fossils are known from this level (Partridge, 1986; Brain, 1988). Carnivore tooth marks are fairly common on the mammal bones at the site, suggesting that carnivores were the main agents of bone collection at Swartkrans.

Paleomagnetic results have not been useful at this site so far. However, faunal analysis, again based mainly on the bovids and cercopithecids, suggests that the lower unit ranges in age between 1.8 and 1.5 MYA. Since the Plio-Pleistocene boundary is about 1.8 MYA, the Swartkrans hominids are technically Pleistocene in age. **Thermoluminescence dating** of quartz sands (a technique for determining the timespan between the present and the last heating of the quartz crystal) gives ages of about 1.2 MYA for the brown breccia and 1.65 MYA for the orange breccia (Vogel, 1985).

Analysis of the bovids in the orange unit suggests an open grassland environment compatible with relatively dry conditions, similar to those postulated for Member 5 of Sterkfontein.

Summary of the South African Sites Throughout most of the time during which *A. africanus* was present, greater humidity and thicker bush cover were present than occur in the region today. A major ecological change apparently occurred about 2.5 MYA that led to increasing dryness and the spread of more open grassland conditions. Sedimentological and faunal evidence suggests that a more cyclic (and probably seasonal) rainfall pattern emerged after this time and persisted throughout the Pleistocene. However, none of these subsequent fluctuations between wet and dry climates was as dramatic as the 2.5-MYA event that initiated them. For the most part *A. africanus* was replaced in southern Africa by populations of *A. robustus,* which coexisted at first with *H. habilis* and later with *H. erectus* (Partridge, 1986).

Thus, the three sites near Johannesburg (Sterkfontein, Swartkrans, and Kromdraai) suggest changing patterns of vegetation cover and cave filling over the approximately 2 million years of australopithecine evolution in South Africa. Among fossil bovids, the Alcelaphini (hartebeests) and Antilopini (antelopes) are useful faunal markers for inferring past vegetation cover, since both are indicative of open plains and grassland environments (Vrba, 1974, 1975). These bovids comprise anywhere from about 25 to 90% of the bovid remains from these sites. Although the percentage of these bovids generally increased with time, it appears an abrupt increase occurred somewhere between about 2.5 and 2 MYA. This shift seems to fit well with the idea of a general faunal change in South Africa in the late Pliocene, including the shift from gracile to robust australopithecines about this time in the Sterkfontein valley (Vrba, 1985*a*) (Fig. 6.8).

Thus the five australopithecine-bearing sites in South Africa can be lumped together in the following way: Makapansgat and Sterkfontein as older than 2.5 million years and Swartkrans, Kromdraai, and Taung as younger than 2.5 million years. With the exception of Taung, *A africanus* seems restricted to the older sites and *A. robustus* to the younger ones. As we shall see shortly, this has been used as partial support for the view that *A. africanus* and *A. robustus* were both part of a single evolving robust australopithecine lineage.

East African Sites

As we saw in Chapter 5, the East African Rift System is part of a single tectonic system extending from the Zambesi River in southern Africa to the Dead Sea in the Middle East. It began to develop in the early Miocene when large volcanic domes arose as a result of broad crustal flexing and local faulting in East Africa. Then in the late Miocene floodlike eruptions occurred in the area. But the East African Rift System as we know it today underwent its greatest transformation during the Plio-Pleistocene. The first faulting to affect the entire length of the rift took

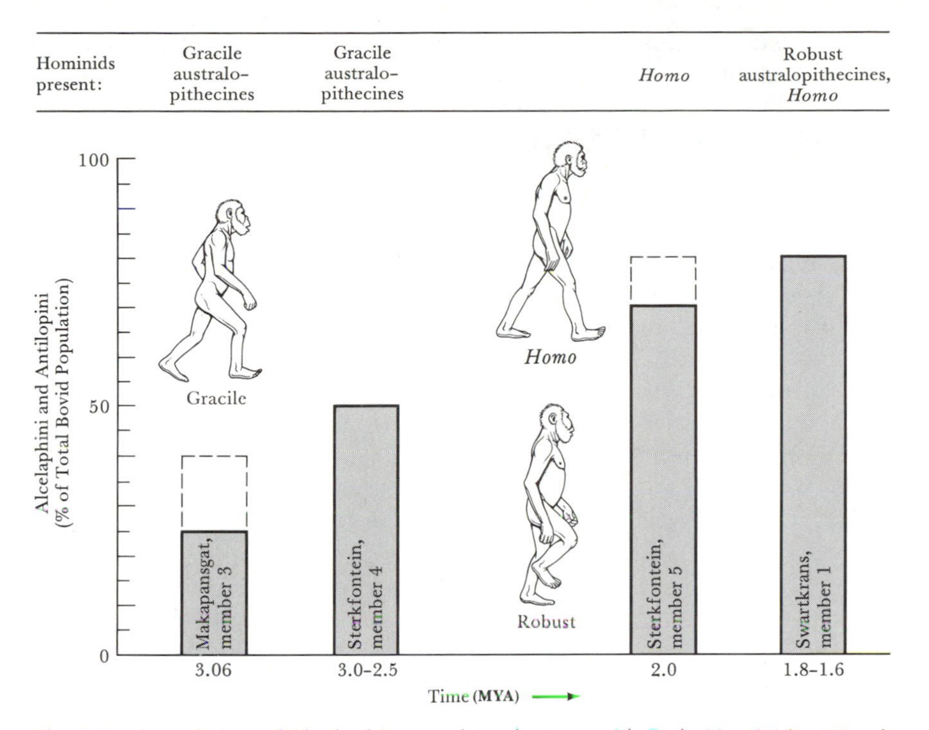

Fig. 6.8 Association of Alcelaphines and Antilopines with Early Hominids at South African Cave Sites Notice that the percentage of the marker species in relation to total bovid population increased with time, suggesting a general ecological shift toward open grassland and savanna. The most abrupt change occurred about 2.5 to 2 MYA. Dotted lines indicate uncertainty due to factors such as mixture from other members. Dates for members come from Figs. 6.4, 6.6, and 6.7. (Graph adapted from Vrba, 1985*a*)

place in the early Pliocene, with the last and major uplift of the area commencing by the end of the Pliocene (Baker and Wohlenberg, 1971).

Australopithecine-bearing deposits in East Africa are concentrated in Ethiopia (Hadar and Omo), Kenya (Lake Turkana and Lake Baringo), and Tanzania (Olduvai and Laetoli). Two species are generally recognized: *A. afarensis* and *A. boisei*. *Homo* fossils (both *H. habilis* and *H. erectus*) have been found at Omo, Lake Turkana, and Olduvai. Following is a brief look at the East African hominid fossil record; the sites are reviewed primarily in chronological order, beginning with the oldest possible evidence for the presence of hominids.

Sites Older than 4 Million Years Although dubious, some evidence exists that hominids may have been present in East Africa earlier than 5.5 MYA, at Lukeino and Ngorora, two sites in the Lake Baringo area of Kenya. An isolated tooth was found at each site. However, these claims are unreliable since it is sometimes very difficult to assign single teeth at the species or even genus level (Hill and Ward, 1988).

The oldest *undisputed* hominid specimen from Africa comes from a locality known as Lothagam Hill in Kenya, which lies some 200 km north of Lake Baringo in the drainage basin of the Kerio River adjacent to Lake Turkana. In 1967 an expedition under the direction of Bryan Patterson from the Museum of Comparative Zoology, Harvard University, investigated fossiliferous deposits at the Lothagam site. This work resulted in the discovery of a right mandibular fragment with the crown of a first lower molar and the broken roots of the second and third lower molars. The specimen comes from the lower layer designated Lothagam-1, which is bracketed by radiometric dates of 8.3 to 3.7 MYA. The associated fauna suggests a date of 5.5 to 5 MYA for the hominid (Patterson et al., 1970; Behrensmeyer, 1976; Brown et al., 1985).

The specimen was originally assigned to *A. africanus* by Patterson et al. (1970), but others have even considered it a pongid (Eckhardt, 1977). Using multivariate analysis, Corruccini and McHenry (1980) have suggested that the Lothagam fossil appears to be closer to the lineage of a hypothetical ancestral hominid. The most recent assessment of the fossil demonstrates its similarity to *A. afarensis,* particularly since it shows signs of adaptations to the increased masticatory power and megadontia characteristic of later australopithecines (Kramer, 1986; Hill and Ward, 1988).

A second nearby site worked by Patterson's group is Kanapoi. A hominid distal left humeral fragment was discovered there that is probably between 4.5 and 4.0 million years old (Patterson and Howells, 1967). It is similar to the distal humerus of both *Australopithecus afarensis* and species of *Homo.* Given its age, the specimen probably belongs to the former taxon.

The next oldest Mio-Pliocene australopithecine specimen from East Africa, a mandibular fragment from Tabarin in the Chemeron Formation near Lake Baringo, is dated to about 4.15 MYA. The specimen is a right mandibular fragment bearing the first and second molars. It too has been classified as *A. afarensis* (Ward and Hill, 1987; Hill and Ward, 1988).

Several other australopithecine specimens from the middle Awash valley south of Hadar in Ethiopia also date to approximately 4 MYA. One of the specimens, a proximal femur, is similar to other specimens known from Hadar. Its anatomy suggests an adaptation to habitual bipedal locomotion and thus provides the earliest fossil evidence documenting the evolution of upright walking. The other fragments from the middle Awash valley come from the frontal bone of an early hominid. The skull fragments are incomplete, but they indicate a skull resembling the chimpanzee in size but not in shape since the bone is slightly thicker and lacks a supratoral sulcus (White, 1984).

Laetoli (Northern Tanzania) The oldest well-documented sample of australopithecines comes from the area known as Laetoli (the correct anglicization of the Masai word for this area) in northern Tanzania. The

Pliocene deposits of this area are known as the Laetolil Beds. All the hominids from this site are classified as *A. afarensis* (Johanson and White, 1979).

Laetoli was the source of the first australopithecine fossils to be discovered in East Africa, an isolated lower canine found in 1935 and a maxillary fragment and isolated third molar found in 1939 (White, 1981). These discoveries were made by a German expedition under the direction of L. Kohl-Larsen in what was formerly German East Africa or Tanganyika (now Tanzania). Laetoli is most renowned, however, for studies conducted there by Mary Leakey and associates from the National Museums of Kenya from 1974 to 1979. The type specimen of the taxon *Australopithecus afarensis* is from this site.

Perhaps the most famous and unusual fossil from the site is a trail of hominid footprints discovered in 1976. These footprints are preserved in fossil form as the result of a unique combination of climatic, volcanic, and mineralogical conditions. Approximately 3.6 MYA a series of light ash falls from a nearby volcano coincided with a series of rain showers. The carbonate in the ash then cemented the ash layer as it dried in the sun. Evidently several individuals crossed the ash layer while it was still wet, leaving their tracks to be preserved (Leakey and Hay, 1979). This is the most incontrovertible evidence that by 3.6 MYA early hominids were bipedal since the footprints are entirely humanlike.

The Laetolil Beds are subdivided into two units, the lower one approximately 70 m thick and the upper one 45 to 60 m thick. Almost all the vertebrate fossils, including the hominids, are from the upper unit, which consists principally of aeolian (wind-blown) and air-fall tuffs. The hominid-bearing strata in the upper unit are well dated at 3.76 to 3.49 MYA (Leakey and Harris, 1987).

Paleoenvironmental reconstructions have been attempted for the site by examining the characteristics of the sediments. The aeolian tuffs are indicative of an environment in which vegetation was at least seasonally insufficient to prevent extensive wind-blown transportation of sand-sized ash particles. In addition, the nature of the weathering of the Laetoli tuffs suggests the presence of a saline, alkaline soil. A present-day analogy to the Pliocene Laetoli is the eastern semiarid–arid part of the modern Serengeti Plain. This area today is a grassland savanna with scattered bush and acacia trees having a seasonal rainfall averaging about 50 cm per year (May, 1981). The ancient Laetoli climate probably had well-defined dry and rainy seasons, as the area still does today.

The vertebrate and invertebrate fauna collected at Laetoli supports the mineralogical evidence for a semiarid climate: the absence of hippopotamus, crocodile, and other water-dwelling remains are characteristic of an upland savanna not necessarily near a permanent water source. The fossil fauna is, in fact, quite similar to the fauna living in the area today. The fossil rodent fauna is a particularly useful paleoenvironmental indicator: rodents do not migrate through the year as larger mammals do, and most species live within quite restricted ecological niches. The fossil rodents suggest an environment of open grassland with occasional acacia trees. The presence of the naked mole rat implies that the climate in the Laetoli area was even warmer then than today (Hay, 1981).

Hadar (Ethiopia) The Afar depression of Ethiopia, in which Hadar lies, is a hot, desolate area of badlands situated at the confluence of three rift systems: the East African, the Red Sea, and the Gulf of Aden rifts. The paleontological potential of the area was first noted by the French geologist Maurice Taieb, who was conducting fieldwork there for his doctoral dissertation on the geological evolution of the Awash River. In 1972 he invited Donald Johanson now of the Institute of Human Origins in Berkeley, Calif., Yves Coppens of the Musée de l'Homme in Paris, and John Kalb of the Rift Valley Research Mission in Ethiopia to participate in a short geological and paleontological survey of the area. The following year Johanson discovered the first hominid remains from Hadar, an associated knee joint (distal femur and proximal tibia) and fragmentary right and left proximal femora.

Numerous australopithecine remains were recovered at Hadar in the period from 1973 to 1977. Over 240 hominid specimens representing a minimum of 35 individuals are known. The most spectacular finds were those of the skeleton nicknamed "Lucy" in 1974 and a collection of at least 13 individuals at site A.L. 333 in 1975 to 1977 (Johanson et al., 1982). To explain why so many individuals were present at the latter locality, Johanson and White (1979) have suggested that it represents a hominid social group that was overcome, buried, and preserved by a single flood event. All of the specimens (including Lucy) are generally identified as *A. afarensis,* although the possibility exists that more than one taxon is being sampled (Falk and Conroy, 1983). While a number of stone tools have been found in the Hadar region, none is definitely associated with *A. afarensis.*

The Hadar Formation contains four parts; from oldest to youngest they are the Basal, Sidi Hakoma, Denen Dora, and Kada Hadar members. All the hominid fossils are from the Sidi Hakoma, Denen Dora, and lowermost Kada Hadar members (Aronson and Taieb, 1981) (Fig. 6.9).

Until early 1982 it was thought that the base of the sequence was probably close to 4 million years old. This conclusion was based on the belief that the basalt near the lower third of the stratigraphic column (the Kadada Moumou Basalt) was securely dated to about 3.65 MYA. Recently, however, it has been determined that this date is probably erroneous, since the Sida Hakoma Tuff near the bottom of the stratigraphic column correlates with other tuffs in East Africa dated to around 3.3 MYA. Thus the Hadar hominids, rather than overlapping the Laetoli hominids in time as recently thought (Johanson et al., 1982; Walter and Aronson, 1982) may actually be between 500,000 and 1 million years later (Boaz et al., 1982; Brown, 1982).

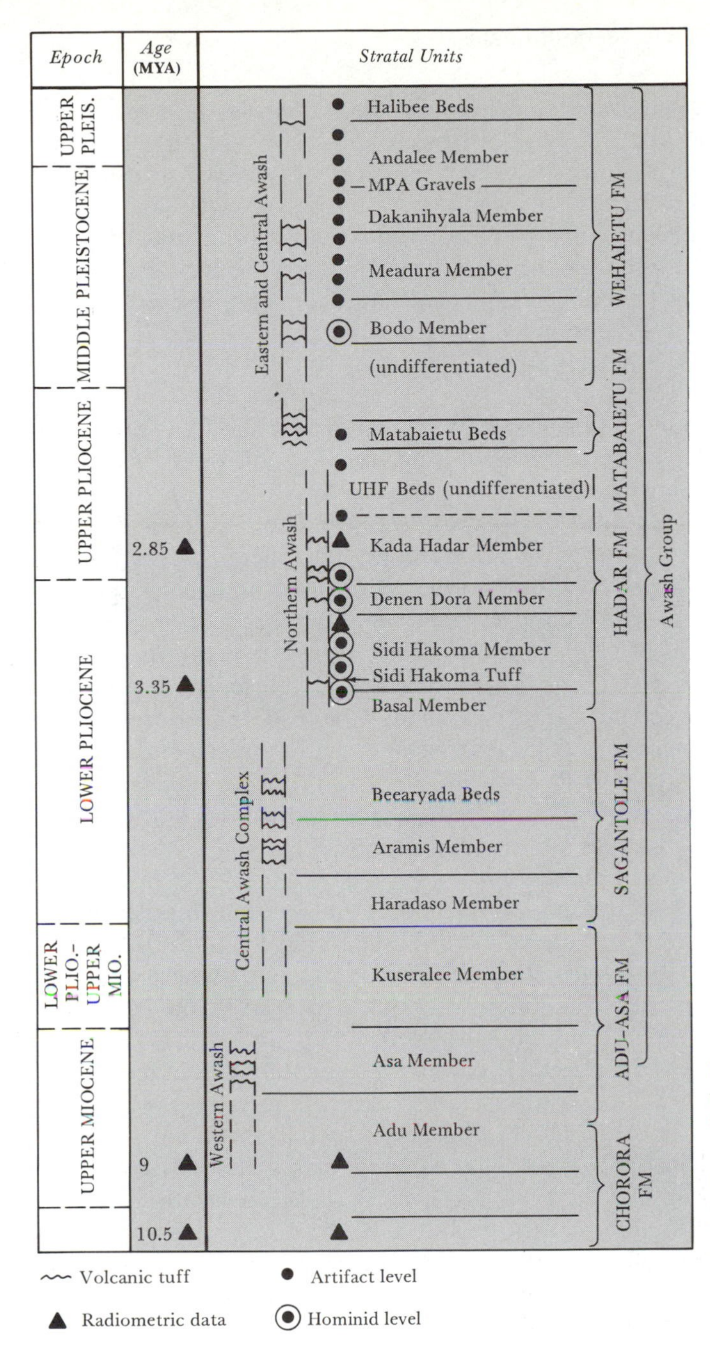

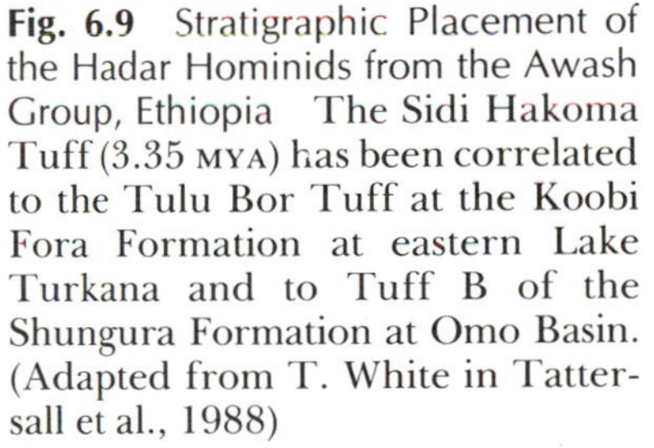

Fig. 6.9 Stratigraphic Placement of the Hadar Hominids from the Awash Group, Ethiopia The Sidi Hakoma Tuff (3.35 MYA) has been correlated to the Tulu Bor Tuff at the Koobi Fora Formation at eastern Lake Turkana and to Tuff B of the Shungura Formation at Omo Basin. (Adapted from T. White in Tattersall et al., 1988)

Examination of these sediments shows that a lake surrounded by marshy environments once existed there and was fed by rivers flowing off the Ethiopian escarpment. In general, the sediments represent marshy, lake-margin, and associated fluvial deposits related to the extensive lake that periodically filled the entire sedimentary basin.

The mammalian fauna of the upper Basal Member and lower Sidi Hakoma Member are indicative of primarily marshy lake-edge environments. A little higher up in the Sidi Hakoma Member a transition to more open habitats occurred in apparent association with lake regression. Clearly, a mosaic of habitats existed through time that included closed and open woodland–bushland and grassland. It appears that the relative proportions of these habitat types varied through time in response to the changes in local climatic conditions and/or lake-basin regression (Johanson et al., 1982).

Omo Basin (Ethiopia) The Omo Basin was first visited by European explorers during Count Teleki's hunting trip through northern Kenya in 1888 (see Chapter 5).

Fossils were first found in the Omo Basin of southwestern Ethiopia in 1902, but it was not until 1933 that the French paleontologist Camille Arambourg established a rudimentary geological sequence for the area and published a paleontological survey of the region.

Detailed studies at Omo began in 1966 with the encouragement of Emperor Haile Selassie. The International Omo Research Expedition was formed one year later and was headed by F. Clark Howell of the University of Chicago (now at U.C., Berkeley), Yves Coppens of the Musée de l'Homme in Paris, and Richard Leakey from the National Museums of Kenya. This was probably the first paleontological expedition to be planned as a multidisciplinary exercise involving geologists, paleontologists, anatomists, and archeologists.

The Kenyan team first explored some of the younger deposits in the area and discovered two hominid skulls of anatomically modern humans at an age of about 100,000 years ago. The French and American teams explored the older deposits dated at about 3 to 1 MYA and recovered several hundred early hominid specimens from nearly 100 separate localities (Feibel et al., 1989). Preliminary assessments suggested the presence of both gracile and robust australopithecines and *Homo* species within the sequence (Howell and Coppens, 1976). The Omo Beds are the longest and best-dated fossil-bearing sequence in East Africa.

The Omo Basin, like the Lake Turkana Basin to the south, is part of the depression of the Eastern Rift valley. The Plio-Pleistocene sediments exposed in the basin have a thickness of nearly 1,100. The sediments outcrop discontinuously over an area of some 200 km in four sectors of the basin. Six formations have been recognized: Mursi, Nkalabong, Loruth Kaado, Usno, Shungura, and Kibish (Butzer, 1976; de Heinzelin et al., 1976), but only two, the Usno and Shungura formations, have yielded hominid fossils.

Most of the hominids come from the river and lake beds of the Shungura Formation. The sequence of deposits has been subdivided into a number of members demarcated by a series of widespread volcanic ash layers (Fig. 6.10).

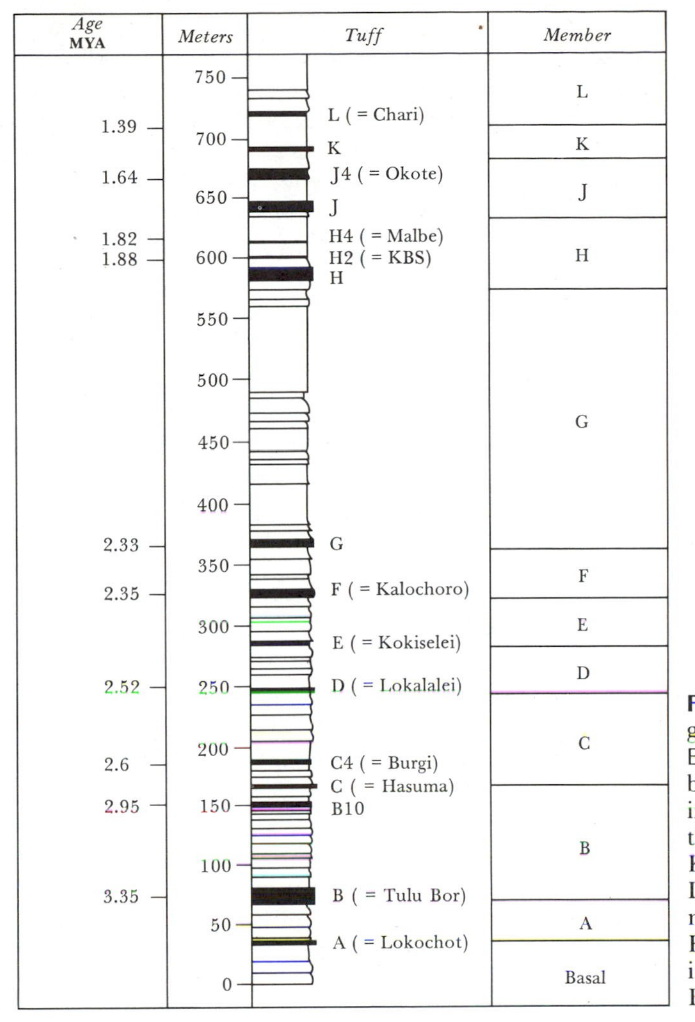

Fig. 6.10 Stratigraphy of the Shungura Formation at Omo Basin, Ethiopia Height of strata above the bottom of the basal member is given in meters; names of tuffs in parentheses refer to field names in the Koobi Fora Formation at eastern Lake Turkana. Hominids have been recovered from Members B through H, K, and L. Stone tools first appear in Members F and G. (From F. Brown in Tattersall et al., 1988)

Fossil hominids have been recovered from two of the several fossiliferous localities within the Usno Formation and from 92 localities in nine members of the Shungura Formation. No hominids are known from the Mursi Formation or from the lowest member of the Shungura Formation. The hominid remains range in age from about 3 to 1 MYA (Feibel et al., 1989).

The two hominid localities of the Usno Formation (called the Brown and White Sands localities) correlate with the lower portion of Member B of the Shungura Formation and are about 3.3 million years old. These sites contain the oldest hominids from Omo. Only teeth are known from these sites: they were provisionally assigned to *A. africanus* by Howell and Coppens (1976).

The earliest hominid remains from the Shungura Formation are from the upper units of Member B. They correspond in age to the middle of the Gauss normal epoch of the paleomagnetic time scale (approximately 2.95 MYA). As with the Usno localities, only dental remains are known and, again, the teeth are most similar to known gracile australopithecine teeth. Howell and Coppens (1976) originally assigned them to *A. africanus*.

Eighteen localities in Member C have yielded hominid remains. Most of these are teeth, but there is a parietal fragment and a phalanx of the hand as well. Member C is capped by Tuff D at about 2.52 MYA. Howell and Coppens again assigned most of the specimens to *A. africanus,* although the parietal bone was said to show some robust australopithecine features such as the beginnings of sagittal cresting.

Member D has 10 hominid localities. All specimens are teeth with the exception of a proximal humerus. Their affinities again seemed to be with *A. africanus*.

Member E has 12 hominid localities yielding teeth, mandibular fragments, a partial juvenile cranium, and a complete ulna. Some of the specimens indicate the presence of robust australopithecines, perhaps *A. boisei*. The maximum length of the ulna suggests an elongated forearm for this particular fossil hominid. The ulna has an accentuated dorsoventral curvature of the shaft, which falls outside the human range and within that of knuckle-walking apes. Cross-sectional profiles of the specimen seem to indicate different patterns of stress from those seen in modern human ulnae. The specimen dates from approximately 2.4 MYA and is classified as *A. boisei* (Howell and Wood, 1974).

Members F and G also have hominids suggestive of robust and gracile australopithecines at an age of slightly less than 2 million years.

Member K has yielded a single tooth and a thickened cranial fragment, indicating the presence of *Homo erectus* around 1.5 MYA.

Howell and Coppens (1976) summarized the Omo hominids as follows: specimens from the Usno Formation, and from the Shungura Formation Members B through G, are similar to specimens of *A. africanus* found in South African sites; they also noted that some of the oldest specimens (notably those from the Usno Formation) may ultimately prove, with additional material, to represent a distinct but related species, possibly *A. afarensis*. Some specimens from Shungura Formation Members E, F, and G were attributed to robust australopithecines and designated *A. boisei;* a very few localities in Shungura Formation Members G and H (and perhaps L) have yielded teeth and cranial parts that are remarkably similar to those attributed to *H. habilis:* some cranial fragments from (the uppermost) Member K are suggestive of *H. erectus* (Hunt and Vitzthum, 1986).

A more recent assessment of the hominids from Omo suggests that the more gracile specimens from the Usno and Shungura formations previously assigned to *A. africanus* may actually belong to *A. afarensis* and/or *H. habilis*. The fragmentary nature of the material makes it difficult to assign many of these specimens with any great degree of confidence. Some of the more robust specimens previously assigned to *A. boisei* are now considered by some to be members of a different species, *A. aethiopicus* (Howell et al., 1987).

There is little consensus on the taxonomic placement of all the Omo hominids, so a summary statement is difficult. The following general points can be made. Australopithecines (possibly *A. afarensis*) first appear at Omo in the Usno Formation between about 3.5 and 3.0 MYA. Gracile specimens (of either *A. africanus* or *H. habilis*) and robust specimens (of either *A. boisei* or *A. aethiopicus*) are present in Members C through G and L of the Shungura Formation, thus spanning a time range of about 2.5 to 1.5 MYA. One specimen referred to *H. erectus* is known from Member K, approximately 1.5 MYA (Hunt and Vitzthum, 1986; Howell et al., 1987; White, 1988).

Stone implements taken from the Shungura Formation at Omo (Members E and F) date from 2.3 to 2.4 MYA and are among the oldest yet known in Africa. Unfortunately, none is in direct association with the hominids, so one cannot say for certain to which taxon or taxa the toolmakers belong (Howell et al., 1987).

What was the paleoenvironment in which these hominids lived? Lacustrine (lake- and swamplike) conditions are apparent in the Basal Member and in some of the higher sections of the Shungura Formation. With these exceptions, most of the Plio-Pleistocene sedimentation and related fossils reflect alluvial landscapes.

Palynological (pollen) data confirm the mosaic character of the vegetation in the lower Omo valley between about 2.5 and 2 MYA. The environment of the hominid sites was more or less wooded savanna with some riverine woodland and forested highlands as well. Strong evidence for a change in vegetation and drier conditions between about 2.5 and 2 MYA is provided by the reduction of arboreal species and the considerable extension of grasslands. This parallels conclusions reached for the South African sites (Bonefille, 1976). The change in the bovid and suid fauna to more grazing and open-country forms also confirms the development of grasslands at about this time (Gentry, 1976).

Lake Turkana (Kenya) During the first field season (1967) of the International Omo Expedition, Richard Leakey chartered a helicopter for a private exploration along the northeastern shores of Lake Turkana (known then as Lake Rudolph). Almost immediately he found some stone tools similar to those from Bed I at Olduvai Gorge (to be described). Buoyed by the prospects of new discoveries, he pulled out of the Omo expedition and organized his own expeditions to Lake Turkana beginning in 1968. These continue to this day on both the east and west sides of the lake and have been enormously successful.

R. Leakey's first hominid discovery at Lake Turkana was the complete skull of a robust australopithecine. Many other specimens were found over the next two decades including representatives of both *H. habilis* and *H. erectus.*

In addition to the great number of australopithecines found at eastern Lake Turkana, a single skull (specimen WT 17000) has been found on the western side of the lake: a hyperrobust australopithecine some workers have called *A. aethiopicus* but which most workers attribute to *A. boisei.* It has been dated to about 2.5 MYA (Walker et al., 1986).

Lake Turkana, in northern Kenya, is one of many African lakes that developed in depressions formed by crustal downwarping and faulting during the Cenozoic. The lake level fluctuated during the Plio-Pleistocene in response to these tectonic movements. This history has resulted in more than 300 m of lake, transitional, and fluvial sediments being deposited that record the gradual shift of the lake's eastern shoreline to the west.

The Plio-Pleistocene strata of the East Turkana Basin consist of a 325-m river-lake complex that rests unconformably on Miocene and Pliocene volcanic strata. (An **unconformity** is a break in the stratigraphic record signifying that deposition was interrupted for a certain length of time.) This complex was subdivided into three formations: the Pliocene Kubi Algi, the Plio-Pleistocene Koobi Fora, and the Pleistocene Guomde formations (Bowen and Vondra, 1973*b*). The fossil hominids are all from the Koobi Fora Formation.

A formal stratigraphic nomenclature for the deposits was proposed by Bruce Bowen and Carl Vondra (1973*b*), both of Iowa State University; an absolute dating system based on radioisotopic analyses of selected tuffs was developed by Frank Fitch of the University of London (Fitch et al., 1974). The radiometric dating has been a matter of some dispute. It seems that most of the hominids are bracketed in age between about 3 and 1 million years.

Several volcanic tuffs suitable for radiometric dating are present in the Kubi Algi Formation. These, in conjunction with faunal analyses, give an age range of 4.9 to 3.8 MYA for this lowermost formation (Maglio, 1972; Fitch and Miller, 1976; White et al., 1981). Prominent tuff marker layers also occur throughout the Koobi Fora Formation (Fig. 6.11). The key stratigraphic marker bed is the KBS Tuff, named for geologist A. Kay Behrensmeyer of the Smithsonian Institution. Dating this tuff is thus critical for understanding temporal relationships among the fossil hominids at eastern Lake Turkana. Average radiometric dates on the various tuffs in the Lower and Upper members of the Koobi Fora Formation were initially determined to be as follows (in MYA): Tulu Bor, 3.18; KBS, 2.42; Okote, 1.56; Koobi Fora, 1.57; Karari, 1.32; and Chari, 1.28 (Fitch et al., 1974; Fitch and Miller, 1976). All the fossil hominids occur between the Tulu Bor and Chari tuffs.

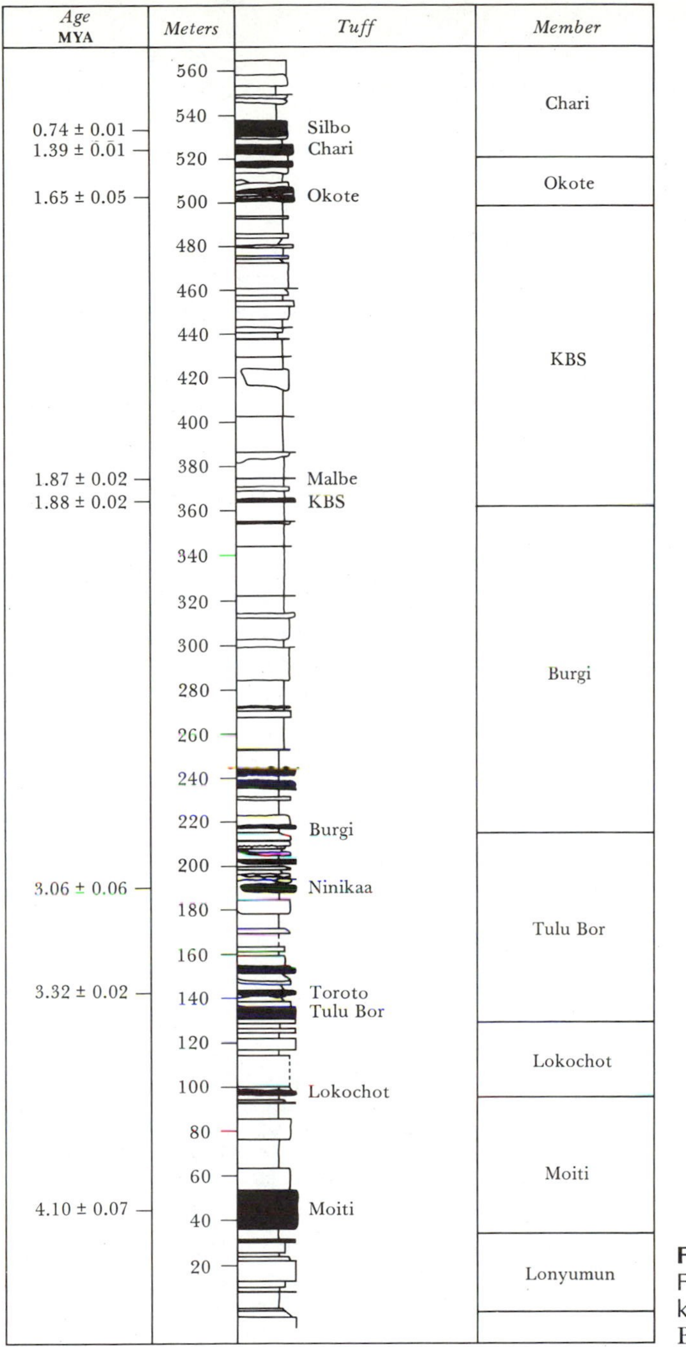

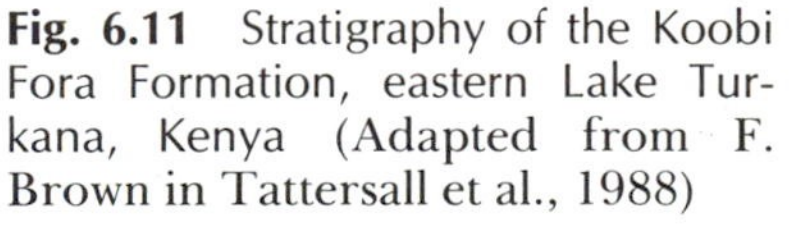
Fig. 6.11 Stratigraphy of the Koobi Fora Formation, eastern Lake Turkana, Kenya (Adapted from F. Brown in Tattersall et al., 1988)

When detailed faunal comparisons were completed between the fossil fauna at Lake Turkana and the well-dated fauna at Omo Basin, it became apparent that some of the Koobi Fora dates were not accurate (Cooke, 1976; White and Harris, 1977; Behrensmeyer, 1978; Brown et al., 1978). In particular, the date of 2.6 to 2.4 MYA for the KBS Tuff seemed too old. A further series of radiometric analyses of this tuff confirmed this suspicion. A more reasonable date for the KBS Tuff is now considered to be around 1.8 MYA (Curtis et al., 1975; Gleadow, 1980; McDougall et al., 1980; McDougall, 1981). This revised date fits both the faunal data and the paleomagnetic stratigraphy (Hillhouse et al., 1977). All that can be said about the hominids taken from strata beneath the KBS Tuff is that their age lies somewhere between about 1.8 and 3.2 MYA.

Over the past few years most of the major tuffs have been radiometrically "fine-tuned"; Fig. 6.11 shows the ages now generally accepted for these tuffs.

Stone tools have also been found in the sediments. The oldest ones, probably around 2.0 MYA, come from within and below the KBS Tuff (Isaac et al., 1976).

Some of the most complete skeletal remains of *Homo erectus* are from sites on both the eastern and western sides of Lake Turkana. The skeleton from the eastern side of the lake is from the Upper Member of the Koobi Fora Formation, just beneath the Koobi Fora Tuff. It is thus around 1.5 to 1.6 millions years old. Interestingly, the partial skeleton shows pathological changes suggesting the individual suffered from hypervitaminosis A, probably caused by a high dietary intake of animal liver (Walker et al., 1982). The specimen from the western side of the lake is the most complete early hominid skeleton ever found and is also dated to around 1.6 MYA. Even though the individual was only about 12 years old, it already stood about 1.68 m tall. Its skeletal morphology is humanlike in many respects, but there are certain anatomical differences between this *H. erectus* and modern humans that we will discuss later in the chapter (Brown et al., 1985).

Lake levels fluctuated significantly during Plio-Pleistocene times, and Lake Turkana was probably a closed, alkaline lake for at least part of this period (Abell, 1982). Thus the setting for the hominids during the period was one of occasional spatial shifts in ecological zones. It seems likely that the river systems, if not ephemeral, suffered greater seasonal fluctuation than rivers with larger drainage basins such as the Omo River. The vertebrate fauna is more comparable to that of the Serengeti Plain and Nairobi Park than to East Turkana today, indicating generally wetter conditions during the Plio-Pleistocene.

The Plio-Pleistocene environment at East Turkana included abundant vegetation, a large and diverse fauna of herbivores and carnivores, and seasonal climatic fluctuations with attendant faunal migrations (Behrensmeyer, 1975*a*, 1975*b*). The terrain was generally flat, sloping upward toward the east, probably with some higher volcanic areas toward the east and northeast. The lake margins were generally swampy, with extensive areas of mudflats, which were seasonally covered with grass.

The fluvial channels were probably bordered by gallery forest, which gave way laterally to grass-covered floodplains.

Of the 84 hominids that Behrensmeyer (1975*a*) assigned to a major depositional environment, 39 were found in fluvial deposits and 45 in lake-margin deposits. *Australopithecus boisei* was more abundant in fluvial environments, whereas *Homo habilis* was rare in such environments. Both are represented in comparable numbers in lake-margin environments, however, *Australopithecus* fossils are more common than *Homo* in both channel and floodplain deposits. The gracile hominids (*H. habilis*) seem to be more restricted ecologically to the lake margin than are the robust forms (*A. boisei*). The stratigraphic and environmental data provide ample evidence that the two hominid taxa were contemporaneous. Both are found from about 10 m below the KBS Tuff to the top of the Kobi Fora Formation. There must have been subtle ecological differences between them since they coexisted for over 1 million years.

Studies of the Lake Turkana region have indicated that Plio-Pleistocene environments were generally comparable to the modern environment of the area even though temperature and rainfall fluctuations have certainly occurred over the past 4 million years. However, the recent recovery of two plant taxa currently characteristic of the central African rain forest from a restricted chronostratigraphic layer dating at 3.3 to 3.4 MYA suggests that the general pattern of environmental stability was punctuated at times by a brief but significant rain-forest extension in East Africa. This brief humid interval apparently coincided with certain major episodes of climatic deterioration in the Northern Hemisphere (Williamson, 1985).

Olduvai Gorge (Tanzania) Olduvai Gorge is a large gash in the Serengeti Plain of northern Tanzania. Louis Leakey of the National Museums of Kenya (and father of Richard Leakey) enjoyed telling the story of its "discovery" in 1911 by one Professor Kattwinkel, an absent-minded German butterfly collector, who nearly plunged to his death down the gorge in pursuit of some elusive specimen. Recovering from his fall, Kattwinkel descended into the gorge and discovered fossil bones along its slopes.

Louis Leakey first visited the site that was to make him world famous in 1931. Within hours of setting up his camp he discovered stone tools in the gorge. During his explorations at Olduvai in the 1930s, Leakey discovered a number of promising sites for the recovery of stone tools. In 1935 a young research assistant, Mary Nicol, discovered fragments of a hominid skull among the remains of antelopes and pigs and some stone tools. The public knew her better as Mary Leakey after she married Louis in 1936. Together they have proved to be one of the most productive husband-and-wife teams in the annals of science.

Constraints of time, money, and other research projects kept the Leakeys away from Olduvai until the early 1950s. During that decade they found and described a number of tool sites and uncovered a few hominid teeth, but nothing very unusual. This all changed on July 17, 1959. That morning Louis remained in Camp recovering from a bout of influenza while Mary revisited the site where the first stone tools had been found in 1931. When she arrived at the site she noticed a skull just eroding through the surface of the slope. After brushing away the dirt to reveal some of the teeth, she knew she had uncovered the most complete hominid fossil ever found at Olduvai. It was a beautifully preserved skull of a robust australopithecine the Leakeys named *Zinjanthropus boisei: Zinj* for the ancient name of East Africa, *anthropus* meaning "man," and *boisei* to honor Charles Boise, one of the Leakeys' financial benefactors. (The public was quick to nickname this hominid "Nutcracker Man" because of its enormous teeth and jaws.) Later it would be attributed to the genus *Australopithecus* along with *A. robustus* and *A. africanus* (Tobias, 1967).

A few hundred meters north of the Zinj site the Leakeys' son Jonathan found another interesting fossil site. Excavations yielded fragments of a rather thin hominid skull, several foot bones, hand bones, and a mandible of a hominid very different from *Zinjanthropus*. In fact, what Jonathan had discovered was the type specimen for *Homo habilis* (Leakey et al., 1964).

Work has continued at Olduvai to the present day. There are four beds (labeled I to IV) plus additional Pleistocene beds above Bed IV (Fig. 6.12). In addition to the numerous *A. boisei* and *H. habilis* fossils taken from Beds I and II, Olduvai has also yielded specimens of *H. erectus* from upper Bed II and lower Bed IV, thus marking the appearance of this species at about 1.2 MYA at this site. It was Louis Leakey who found the first of these *H. erectus* specimens. Considering how strongly his name is associated with early hominid discoveries in the mind of the public, it is ironic that this specimen is the only hominid fossil he personally ever found at Olduvai.

In addition to yielding many australopithecine and *Homo* fossils, Olduvai has also produced an extensive record of Early Stone Age artifacts, some of which we will briefly look at later in the chapter when discussing cultural evidence in early hominids.

Because australopithecines are restricted to Beds I and II (where most *Homo* specimens are found as well), the following comments on the paleoenvironments of the gorge will concentrate on the timeframe corresponding to these two beds. The gorge is between 45 and 90 m deep where the lowest unit of the Olduvai Beds (Bed I) is exposed (Hay, 1973). Varying between 30 and 54 m in thickness, Bed I is probably the most firmly dated of any hominid lower Pleistocene site: the fossils are on the order of 1.7 to 1.8 million years old. Richard Hay (1973) of the University of California at Berkeley, has estimated that the fossiliferous part of Bed I may represent a time span of only 50,000 to 100,000 years.

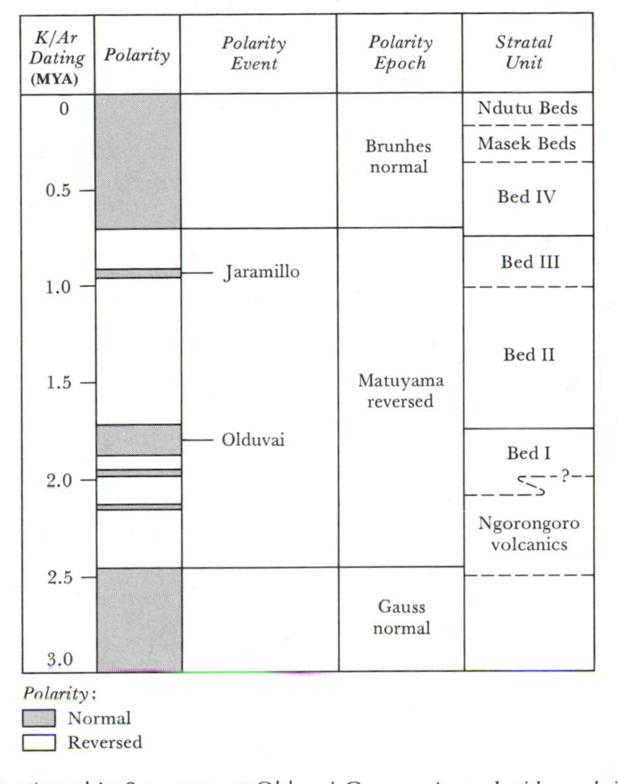

Fig. 6.12 Stratigraphic Sequence at Olduvai Gorge *Australopithecus boisei* is known from Beds I and II, *H. habilis* from Beds I and lower Bed II, and *H. erectus* from upper Bed II and lower Bed IV. Dating of the beds has been inferred from paleomagnetic and potassium–argon studies. The top of the Matuyama reversed epoch may extend into the base of Bed IV. All the *H. erectus* material comes from the Matuyama reversed epoch and is thus older than 700,000 years. (From Hay, 1973)

Bed I can be divided into five lithologically different types: lake deposits, lake-margin deposits, alluvial-fan deposits, alluvial-plain deposits, and lava flows. (An **alluvial fan** is a mass of sediment deposited at a point along a river where there is a decrease in gradient. The mass of the fan is thickest at its point of origin, thinning rapidly in a downstream direction. In time, adjacent fans may coalesce to form **alluvial plains.**)

The lake deposits accumulated in a shallow, perennial lake in the lowest part of the basin at the western foot of the volcanic highlands. The lake either did not have an outlet or overflowed infrequently, resulting in fluctuating lake levels. The perennial part of the lake was saline and alkaline. The lake-margin sediments were laid down on a broad expanse of low-lying, relatively flat terrain periodically flooded by the lake. Hominid fossils in Bed I are concentrated in these lake-margin deposits at the eastern end of the lake (Hay, 1973).

Fossilized leaves and pollen are rare in the sediments of Beds I and II, but swamp vegetation is indicated by abundant vertical root channels and casts possibly made by some kind of reed. Fossil rhizomes of papyrus also suggest the presence of marshland and/or shallow water. Other structures, such as diatoms and traces of algae, point to fluctuating lake levels and saline conditions. The climate was probably semiarid, although wetter than that of Olduvai today, as indicated by the presence of ostracods, freshwater snails, fish, and aquatic birds. Climatic fluctuations may have been seasonal or may have lasted a few tens of years.

The paleogeography of the region represented by Bed II was similar to that of Bed I, at least until faulting began, after which time the perennial lake in the basin was reduced to perhaps a third of its former size. This paleogeographical event accounts for the change in mammalian faunas that occurred about this time whereby swamp-dwelling animals decreased in number and plains-dwelling animals such as horses increased in abundance (Hay, 1971).

Summary of the East African Sites As we have seen, the earliest undisputed hominid appears in the fossil record of East Africa as a single mandibular fragment from the site of Lothagam some 5.5 MYA. Slightly younger than the Lothagam mandible are several tantalizing hominid fragments from Kanapoi in Kenya and the middle Awash valley of Ethiopia that are approximately 4 million years old. The femoral fragment is particularly important in showing that bipedal adaptations of the hindlimb had commenced very early on in hominid evolution. This view was dramatically reinforced by the discovery of fossil hominid footprints at Laetoli dated to about 3.75 MYA. All of these specimens are attributed to *A. afarensis*.

Beginning with the site at Laetoli at about 3.75 MYA and continuing through the sites of Hadar, Omo, Lake Turkana, and Olduvai, the fossil record of early East African hominids is impressive to about 1 MYA. The most common hominid in the earliest part of this fossil record, from about 5.5 to about 2.5 MYA, is *A. afarensis*. By about 2.5 MYA both *H. habilis* and *A. boisei* (or *A. aethiopicus* as it is called by some researchers) were present. In fact, *H. habilis* and *A. boisei* were contemporaneous from about 2.4 to 1.4 MYA at Omo, from about 2.0 to 1.5 MYA at Lake Turkana, and from about 1.8 to 1.2 MYA at Olduvai (White, 1988). *Homo erectus* first appears about 1.5 MYA in East Africa and is known from Lake Turkana, Omo, and Olduvai.

East African early hominid sites are associated with watercourses of one sort or another, from the ancient lake margins at Olduvai to the dense riverine forests along the Omo Valley. Proximity of hominids to water would be expected, especially if the sites represent campsites or home bases where families or groups gathered repeatedly. The cave sites in South Africa, in contrast, have yielded bones that may have been accumulated by hominids and/or carnivores, and thus conclusions regarding proximity to water cannot be drawn.

Having reviewed the paleoclimatological, biogeographical, and geological settings of australopithecines and archaic humans in East and South Africa, we turn now to an examination of the fossils themselves. Two of the most important trends in Plio-Pleistocene primate evolution are the increase in cranial growth and the development of bipedalism. The following two sections focus on distinctions in craniodental morphology in *Australopithecus* and early *Homo;* in the section on Locomotion we will look in detail at the postcranial evidence for the onset of bipedalism.

AUSTRALOPITHECUS

As we have said, most paleoprimatologists consider *A. africanus* to be gracile and *A. robustus* and *A. boisei* robust; historically the term "gracile australopithecine" has not often been applied to *A. afarensis,* but we are considering it gracile in this text. Specimens of *A. afarensis* and *A. boisei* are restricted to East African sites, whereas *A. africanus* and *A. robustus* are found in South Africa. Some gracile hominids in East Africa have also been referred to *A. africanus,* but the presence of that species in East Africa is debatable. Most workers either consider them to be conspecific with *A. afarensis* (Tobias, 1980) or assign most if not all of the East African specimens of *A. africanus* to the taxon *H. habilis.* For the sake of this discussion, I shall adopt the latter view, which is reflected in Fig. 6.2b. Therefore each of the two geographical areas in Africa from which early hominids have so far been found contain one gracile and one robust australopithecine species.

The four species can be ordered by approximate chronological age as *A. afarensis, A. africanus, A. robustus,* and *A. boisei.* In general, the gracile species precede the robust species in the fossil record, with some chronological overlap between early forms of *A. boisei* and later forms of *A. africanus* (see Fig. 6.2b). Not only do the two robust species overlap in time, but their accepted chronological relationship may eventually be reversed based on the recent discovery of the *A. boisei* specimen WT 17000, which is much older than any specimens of *A. robustus.* This point is still open to debate.

We turn now to an examination of the fossils themselves. Most of the discussion in this section will focus on distinctions in craniodental morphology; later, in the section on Locomotion, we will look in detail at the postcranial evidence for the onset of bipedalism.

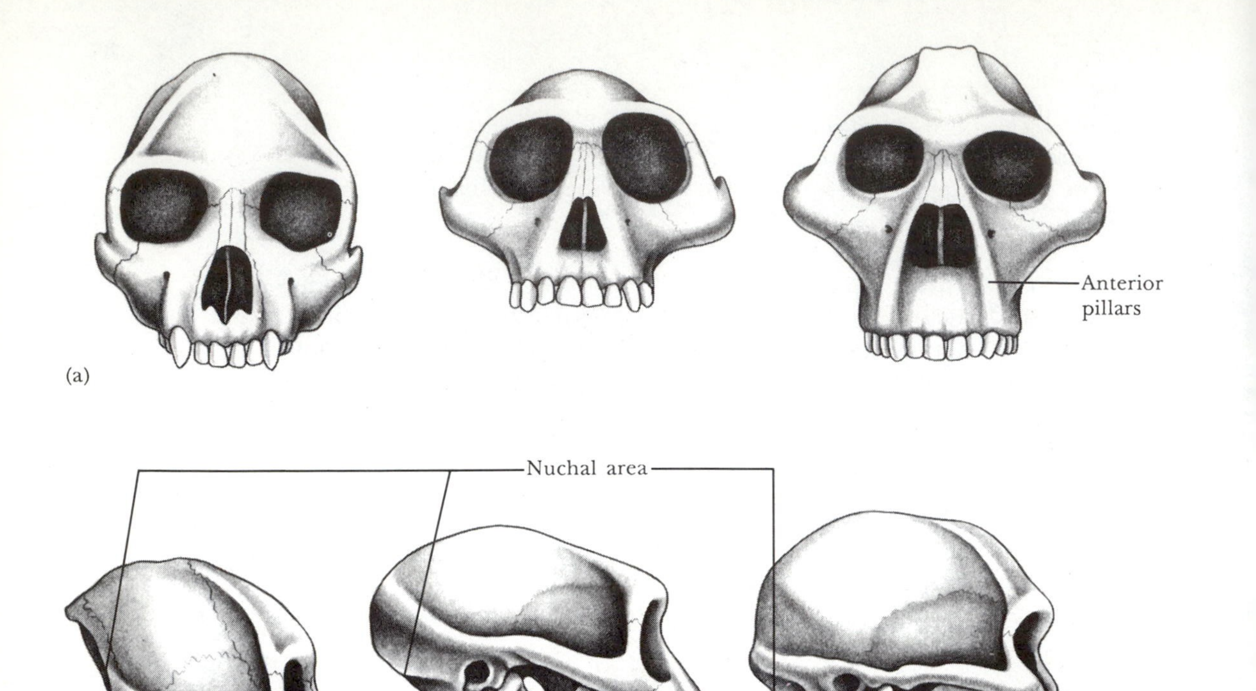

General Craniodental Trends in Australopithecines

Compared with modern humans, australopithecines are characterized by relatively small brains (with average capacities ranging from 413 to 530 cc) housed in skull bones that are quite thin. In spite of this small actual brain size, australopithecine encephalization quotients (EQs) are relatively high (around 3.5 in later species). In addition, trends toward some neural reorganization, like expansion of the temporal lobes of the brain, are evident through time.

As australopithecine skulls become generally more robust with later species, other cranial trends become evident. The robust skull carries forward the adaptive trends already seen, albeit in less-developed form, in the gracile australopithecines, namely, that faces are shorter, deeper, and more massive and that all the structures associated with powerful chewing are accentuated (Fig. 6.13).

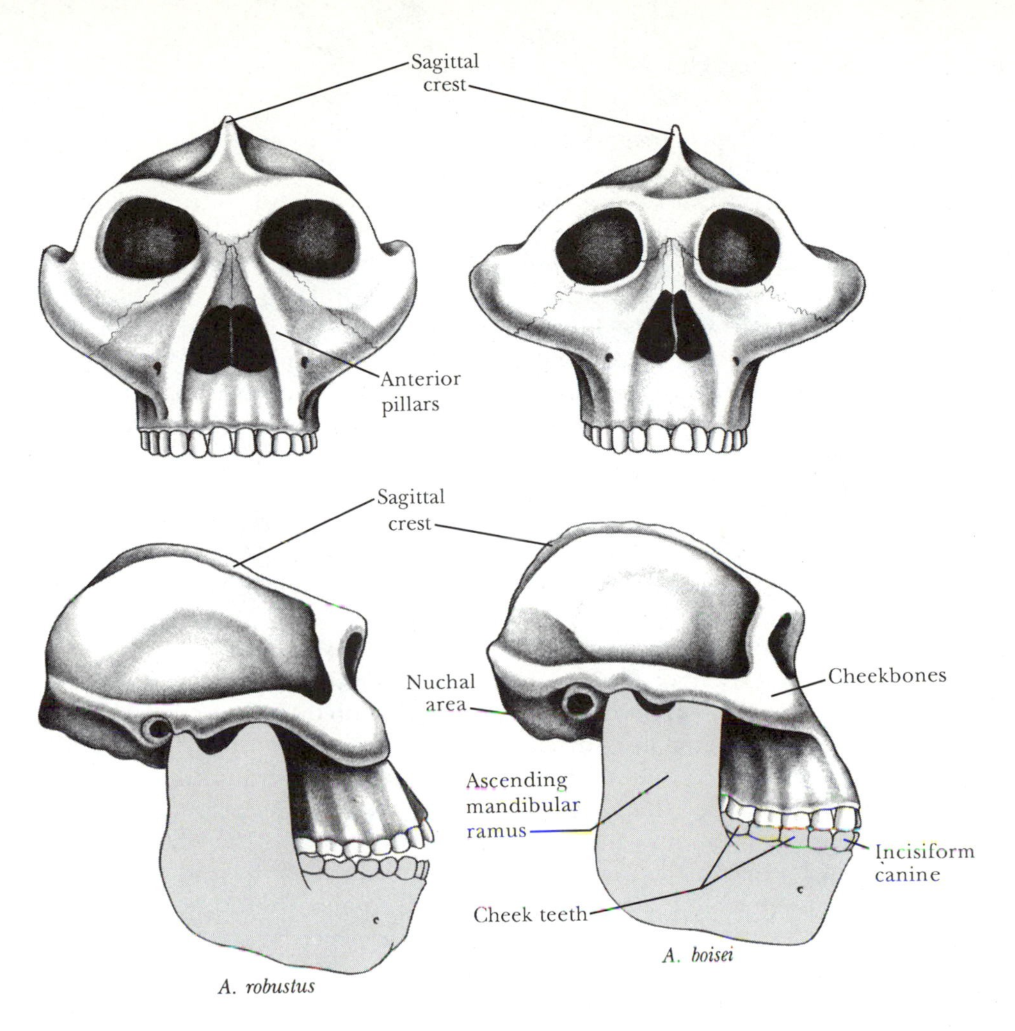

Fig. 6.13 General Craniodental Trends in Australopithecines (a) Frontal and (b) lateral views of a Miocene ape and four australopithecines. Lower jaws are shown in side views only; shaded portions indicate reconstruction. Note that many morphological features associated with powerful chewing and grinding are more accentuated in robust than in gracile australopithecines: faces become shorter, deeper, and more massive, often with strong anterior pillars; skulls may show pronounced sagittal crests; lower jaws become thickened with high ascending mandibular rami; incisors become smaller relative to molars; canines (shaded) become more incisiform; and molars become massive. Figures not drawn to scale. (Redrawn from Walker et al., 1983; Kimbel and White, 1988; Howell, 1978; Rak, 1983; Jolly and Plog, *Anthropology and Archaeology*, McGraw-Hill, 1986; reproduced with permission)

These trends in cranial morphology can be related directly to trends in postcanine tooth size. Studies by Henry McHenry (1984, 1985) show that in *A. afarensis* the cheek teeth are about 1.7 times larger than expected from estimated body weight, in *A. africanus* and *A. robustus* about twice as large, and in *A. boisei* about 2.3 times as large. These studies reveal two important things: (1) all species of *Australopithecus* are characterized by postcanine megadontia; and (2) relative cheek-tooth size tends to increase through time from *A. afarensis* to *A. boisei*.

Larger teeth require larger jaws, which in turn require larger muscles to move them. The larger muscles then need larger bony attachments to the skull and jaw. These include (1) prominent sagittal and nuchal crests for the attachment of powerful chewing and neck muscles respectively, and (2) bony struts in the face to withstand powerful chewing stresses set up through the massive jaws and teeth. One set of struts is called the **anterior pillars** (two massive columns supporting the anterior portion of the palate on both sides of the nasal aperture) (Fig. 6.13c,d). Skull bones also became more **pneumatized** (filled with air spaces) in order to reduce the weight of the enlarged skull. We shall examine these cranial trends in greater detail later as we compare the morphology of the gracile and robust forms.

One general feature of australopithecine cranial morphology deserves some mention here. The upper respiratory system of modern humans is unique among mammals. Although humans share homologous upper respiratory tract features with other mammals, the positional relationship of structures such as the larynx or pharynx has become markedly altered in humans during the course of evolution. These anatomical changes have determined our breathing and swallowing patterns and have provided the physical basis necessary for the production of the full variety of human speech sounds. An analysis of the australopithecine **basicranium** (the base of the skull) indicates that the unique morphology of modern humans had not yet evolved and also that the basicranium, and by extension the australopithecine upper respiratory anatomy, was probably more similar in overall structure and function to that of living apes (Laitman and Heimbuch, 1982). This is evidence that the australopithecines probably were not capable of speech as we know it but were limited to the range of vocalizing seen in modern apes.

Following is an examination of the gracile and robust forms of *Australopithecus*. Throughout we will assume that *A. africanus* is a gracile hominid, although some workers consider it an early robust form, as we shall see in the last section.

Gracile Australopithecines

A. afarensis The oldest australopithecine species is *Australopithecus afarensis*. As discussed earlier, this species is best known from Laetoli and Hadar; fossils that may be of this species have also been found at Omo, Lothagam, Kanapoi, Tabarin, and the middle Awash valley. If the Lothagam mandible is correctly identified, the species would date back to more than 5 MYA. At present, however, the Laetoli specimens are the oldest well-documented sample of the species; they have been dated at about 3.6 million years old.

The morphology of *A. afarensis* is generally considered to be more primitive than that of other hominid species. The following synopsis of morphological features derives from the work of Johanson and White (1979) and from White et al. (1981).

The face of *A. afarensis* is known from several adult and juvenile specimens (Fig. 6.14). The nasoalveolar clivus is convex in order to house the curved incisor roots. The large canine roots form well-developed **canine juga** (vertical ridges of bone that cover the roots) on the lateral walls of the maxilla just above the canine crowns. The anterior margins of the zygomatic processes are situated above the space between the fourth premolar and first molar and are oriented nearly perpendicularly to the tooth rows. The inferior margins of the zygomatic arches flare anteriorly and laterally.

Portions of the adult crania show several primitive features:

1. Strong muscle markings, in particular a compound temporonuchal crest
2. A long, steep nuchal plane
3. Temporal lines that approach the midline and in some cases form a sagittal crest
4. Lateral portions of the cranial base that are heavily pneumatized
5. An external auditory tube (the outer ear canal) that strongly resembles the pongid condition
6. **Mandibular fossae** (depressions in the skull base for articulation with the mandibular condyles) that are broad, have little relief, and are located only partially beneath the braincase
7. A small brain

The mandibular ascending rami are large but not high, and they slope somewhat posteriorly. The symphyseal section usually shows a moderately developed superior transverse torus, whereas the inferior transverse torus is low and round rather than shelflike. There is great disparity among specimens in the size of the mandibles, which suggests a high level of sexual dimorphism within this lineage.

Figure 6.15 shows salient features of the dentition. The upper central incisors are much broader than the lateral incisors. The upper and lower canines both have long, massive roots with crowns projecting beyond the level of adjacent teeth. The upper canine hones against the lower third premolar, producing an exposed strip of dentine along the occlusal edge of the canine after extended wear (Fig. 6.15a) The large, mesiodistally elongated buccal cusp of the lower third premolar often shows vertical wear striations produced by the honing action of the upper canine. In occlusal view the long axis of the third premolar sets at an angle of some 45° to 60 ° to the long axis of the tooth row (Fig. 6.15c).

The first and second lower molars are squarish in outline with cusps arranged in a simple **Y5 pattern** (Fig. 6.16b). The molars decrease in size from front to back. The dental arcades are long, narrow, and straight-

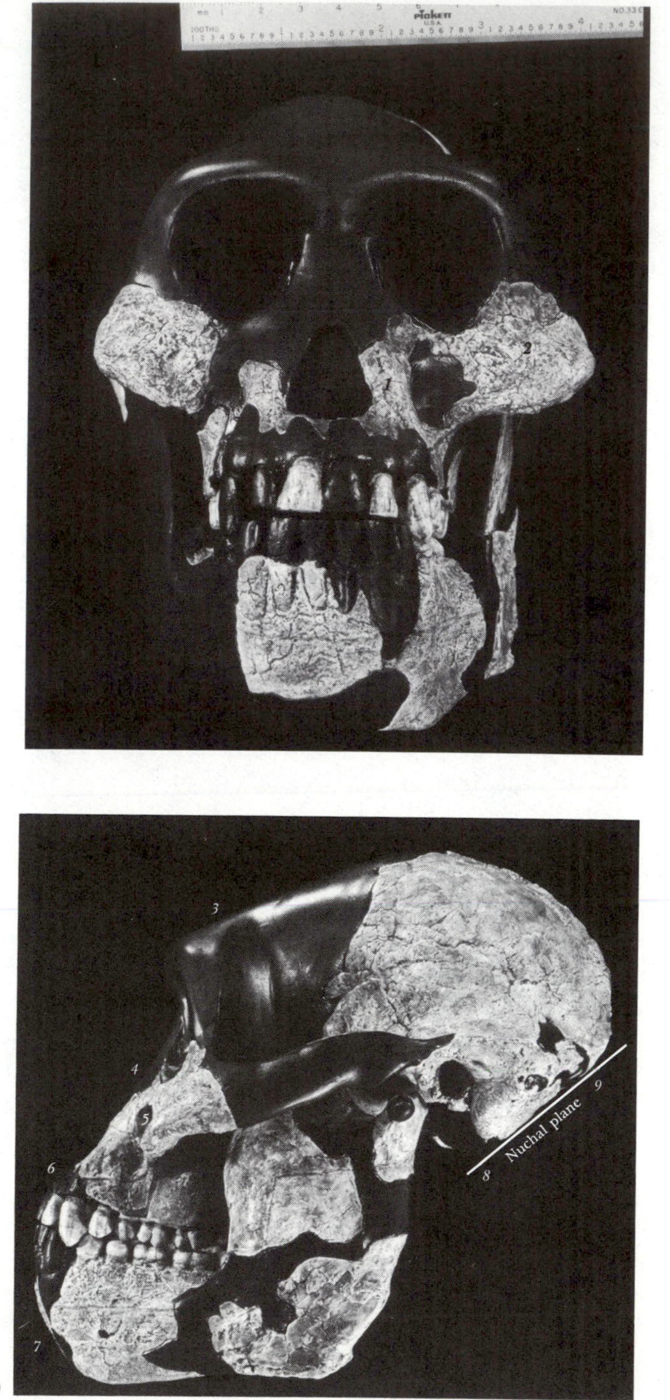

(a)

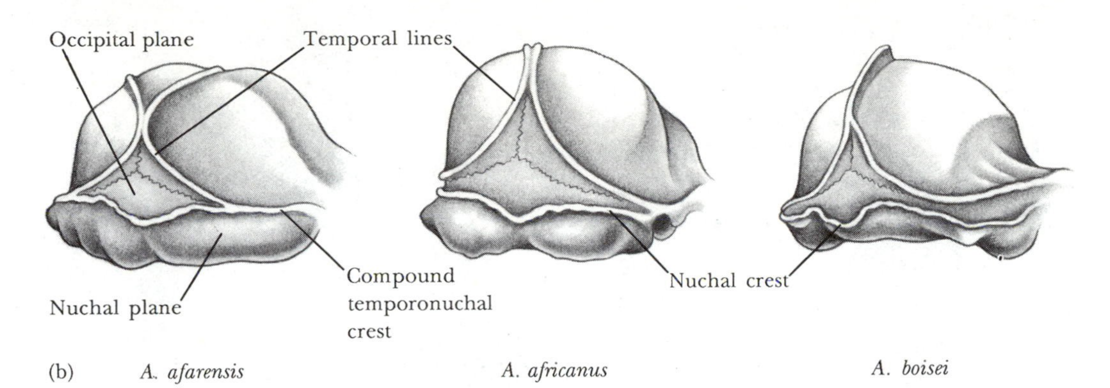

Fig. 6.14 Cranial Features of *A. afarensis* (a) Front and side views of an adult specimen reconstructed by W. Kimbel and T. White (reconstructed portions in black). Note the following features: *1*, the prominent canine jugum; *2*, the widely flaring zygomatic arches; *3*, the relatively flat frontal bone; *4*, the slight concavity of the nasal region; *5*, the root of the zygomatic process between P4 and M1; *6*, the convex nasoalveolar clivus housing the incisor roots; *7*, the sloping contour of the anterior mandibular corpus; *8*, the long, steep slope of the nuchal plane; and *9*, the compound temporonuchal crest. (b) Comparison of posterior view with two other australopithecines. Note the compound temporonuchal crest, absent in all but a few specimens of later australopithecines. (a, Courtesy of the Institute of Human Origins; b, redrawn from Rak, 1983)

sided as in pongids, instead of parabolic as in modern humans. The palates are shallow. A small diastema is sometimes present between the lateral incisors and canines (Fig. 6.15a).

The dentition as a whole seems intermediate between pongid and later hominid dentitions. Neither metric nor morphological data suggest to Johanson and White that more than one hominid lineage is represented by the dental sample from Hadar and Laetoli.

Much of the postcranial skeleton is known and suggests at least some bipedal adaptations. The most complete adult skeleton is "Lucy." This small female was probably 1.1 to 1.2 m in height. All of the postcranial elements indicate high levels of skeletal robustness. The **humerofemoral index** (the ratio of the length of the humerus divided by the length of the femur × 100) is approximately 85 (Jungers, 1982). This value is higher than in modern humans (about 75) and suggests unique body proportions in *A. afarensis* (Jungers and Stern, 1983). The finger and toe bones also differ somewhat from those of modern humans, particularly in their longitudinal curvature. The significance of these features will be discussed later in the section "Morphology of Hominid Locomotion."

A. africanus The next oldest australopithecine species, and the first ever discovered, is *Australopithicus africanus* (Dart, 1925). The type specimen

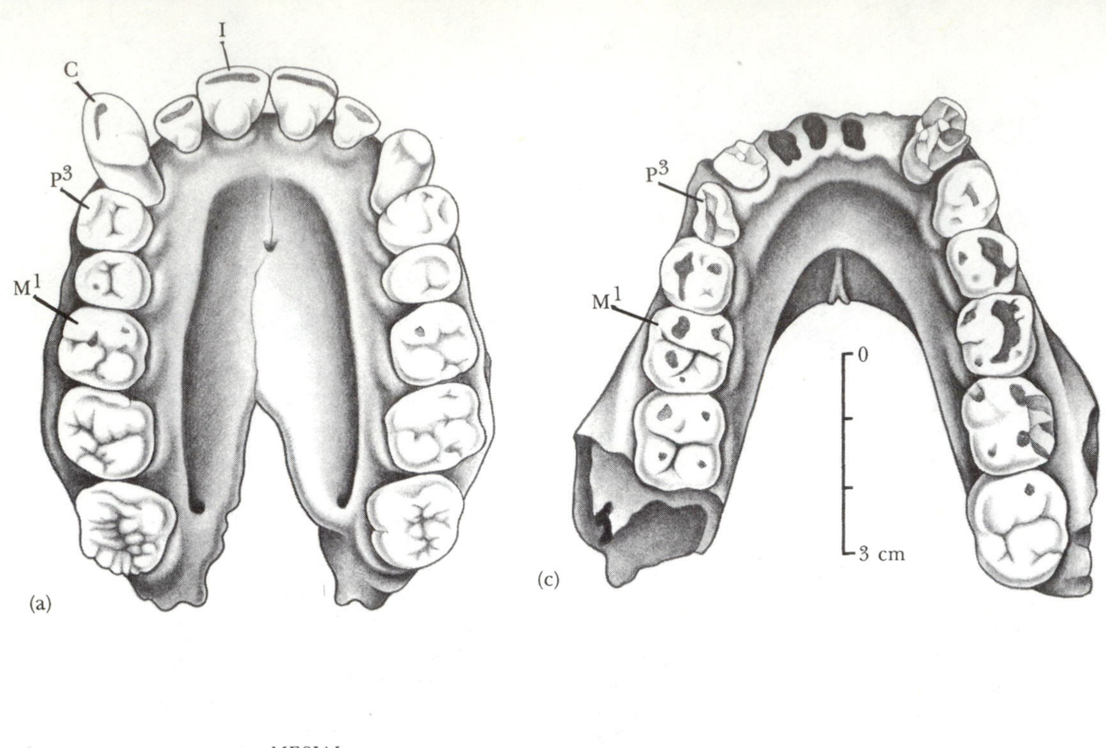

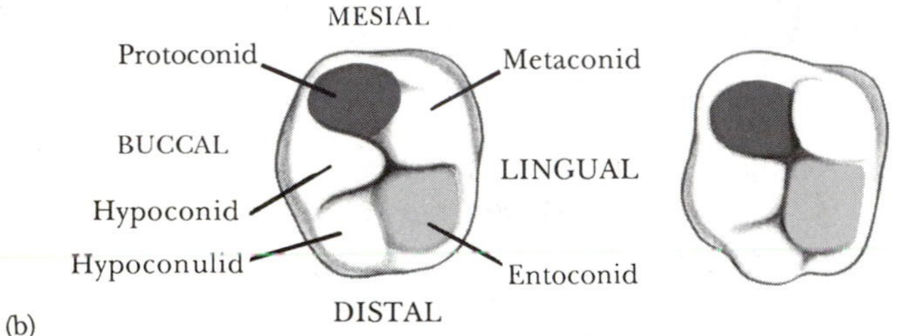

Fig. 6.15 Dentition of *A. afarensis* (a) Upper jaw (specimen A.L. 200-1a from Hadar). Note the broad upper central incisors with worn dentine (black); the slightly projecting canine with a worn area for honing the lower third premolar; the shallow palate with long, narrow, straight-sided dental arcades; and the small diastema between lateral incisors and canines. Vertical line is a crack in the palate. (b) Cusp patterns in hominid lower molars: in the classic *Dryopithecus* or Y5 pattern *(left)* the protoconid does not contact the entoconid, whereas in the +5 pattern *(right)* these two cusps are in contact. (c) Lower jaw (L.H. 4 from Laetolil.) Note the slightly oblique orientation of the lower third premolar, its large buccal cusp, and the squarish first and second lower molars arranged in a simple Y5 pattern. (Redrawn from: a, White et al., 1979; b, Hiiemae, 1981; c, White, 1977*b*)

from Taung consists of the right side of a natural brain endocast, the face, and most of the mandible of an immature "manlike ape." All the deciduous teeth are in place, and the first permanent molars had recently erupted (Fig. 6.16). Based on an analogy with modern children, Dart (1925) considered the fossil child to be about six years old when it died. As we shall see later, this view has been recently challenged (Conroy and Vannier, 1987).

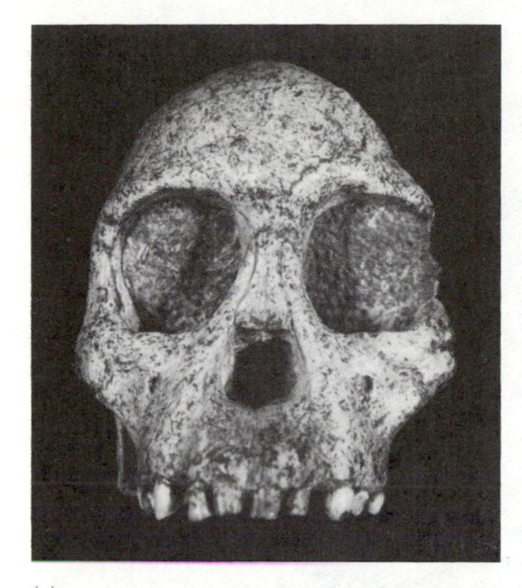

(a)

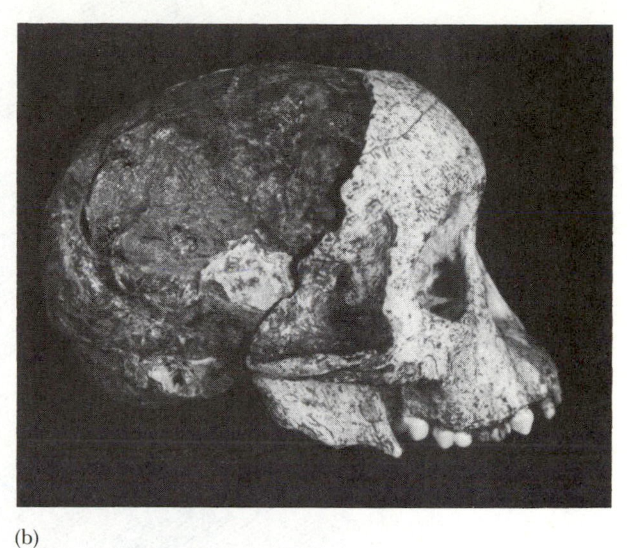

(b)

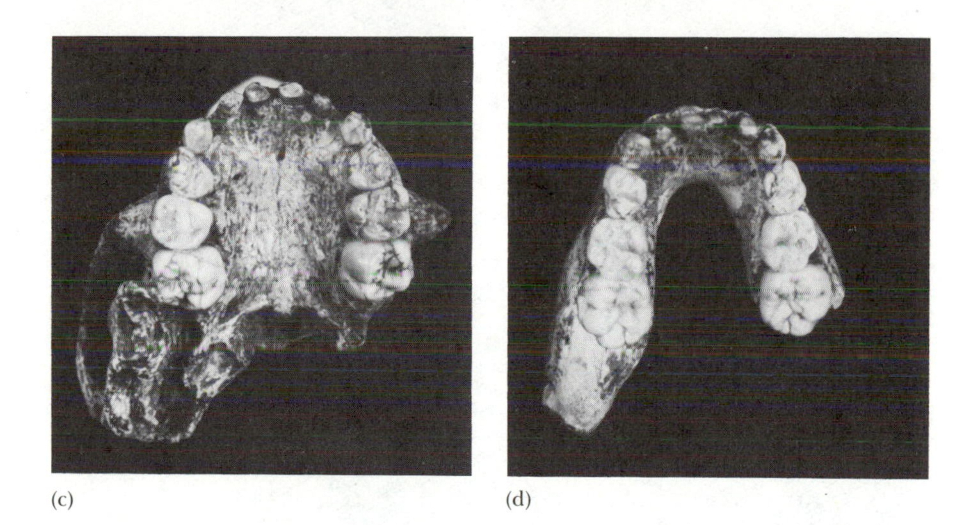

(c) (d)

Fig. 6.16 The Taung Child *(A. africanus)* (a) Frontal and (b) lateral views of the skull and brain endocast; (c) upper and (d) lower jaws. Note the long, narrow cranium, the lack of supraorbital tori, the narrow interorbital distance, the small canines, the parabolic dental arch, and the lack of a simian shelf. (Photos by G. Conroy)

Let's consider what made Dart proclaim he had found "an extinct race of apes intermediate between living anthropoids and man." First, there is the **dolichocephalic** (long and narrow) cranium, which is more humanlike than apelike. Second, there are no supraorbital tori as there are even in young apes. Third, the orbital region is also humanlike: the orbits are circular in outline, not squarish as in apes; the interorbital

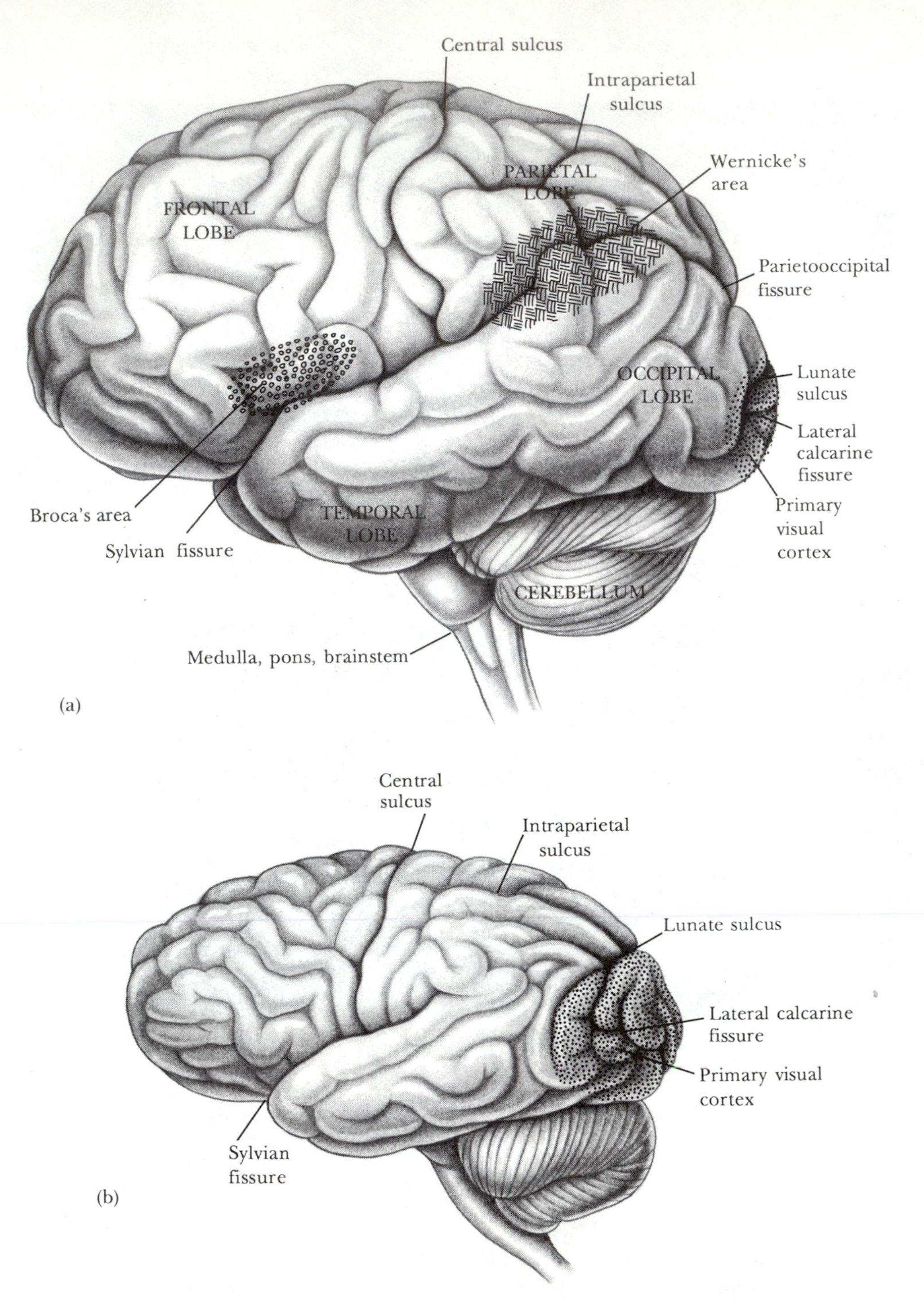

Fig. 6.17 Major Lobes and Sulci in the Hominoid brain Left lateral view of (a) human and (b) chimpanzee brains. The lunate sulcus divides the primary visual cortex posteriorly from the association areas of the parietal and temporal lobes. Note that expansion of the association areas has resulted in a more posterior displacement of the lunate sulcus in the human brain. (b, Redrawn from R. Holloway in Tattersall et al., 1988)

distance is very small; and the ethmoids are not inflated laterally as in modern African apes. Fourth, all the other facial bones are delicate structures, and overall facial prognathism is relatively slight.

Even though the specimen was a juvenile, Dart was able to see that the dentition and mandible are distinctly humanlike. He pointed to (1) the small canines and the absence of gaps between the lower canines and premolars; (2) the parabolic shape of the dental arch; and (3) the lack of a simian shelf.

Dart also remarked on the relatively forward position of the foramen magnum, which to him indicated the "poise of the skull upon the vertebral column" and pointed to "the assumption by this fossil group of an attitude appreciably more erect than that of modern anthropoids." This more vertical posture implied that the hands were freed of their primitive function as accessory organs of locomotion and were becoming better adapted as organs of manipulation and tool using.

Finally, Dart pointed to some unusual features of the endocranial cast. The brain size of this diminutive creature equaled or exceeded that of gorillas many times its overall body size. Other features of the endocast suggested neuronal reorganization as well. For example, the ratio of cerebral to cerebellar matter is greater than that in gorillas. In addition, the association areas of the parietal and temporal lobes (a region of cortex involved with complex functions of comprehension, communication, and consciousness) seem to be relatively enlarged (Figs. 6.16, 6.17). The significance of the enlargement of these so-called association areas in the Taung specimen has become the subject of a heated and mostly inconclusive debate in the recent literature (Holloway, 1984; Falk, 1985).

The discovery at Sterkfontein and Makapansgat of adult australopithecine cranial, dental, and postcranial remains completely vindicated Dart's original claims for *Australopithecus* (Fig. 6.18). It became evident that these early hominids had reduced canines, walked bipedally, made stone tools, and had a large brain relative to body size.

Robust Australopithecines

Robust australopithecines are known from a number of sites in East Africa (Omo, Koobi Fora, Lake Turkana, Olduvai, Peninj, and Chesowanja) and South Africa (Swartkrans and Kromdraai). Fossils attributed to this group range from about 2.5 to 1 million years of age (Howell and Coppens, 1976; Boaz and Howell, 1977; Isaac, 1978; Rak and Howell, 1978; Boaz, 1979; Gowlett et al., 1981; Walker et al., 1986).

A. robustus Only 13 years after Raymond Dart's description of the first gracile australopithecine in 1925, Robert Broom announced the discovery of a much more robust australopithecine from Kromdraai, South Africa. This discovery became the type specimen of a new genus and species of australopithecine that Broom named *Paranthropus robustus*. The

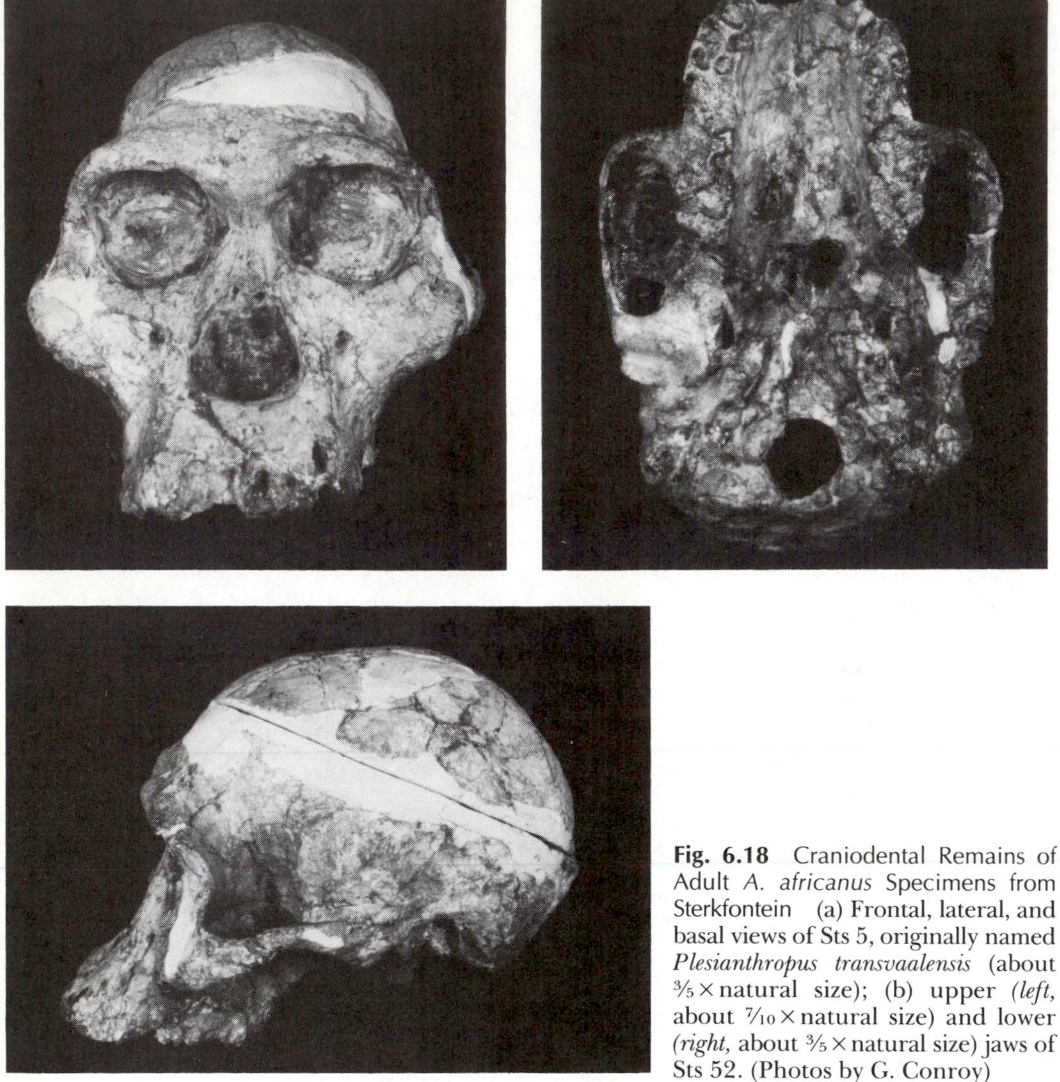

(a)

Fig. 6.18 Craniodental Remains of Adult *A. africanus* Specimens from Sterkfontein (a) Frontal, lateral, and basal views of Sts 5, originally named *Plesianthropus transvaalensis* (about 3/5 × natural size); (b) upper *(left,* about 7/10 × natural size) and lower *(right,* about 3/5 × natural size) jaws of Sts 52. (Photos by G. Conroy)

specimen consisted of most of the palate, much of the left side of the face, almost the whole of the left zygomatic arch, the left side of the base of the skull, portions of the parietal bone, and the greater part of the right mandible with most of the teeth. From the same block of sediment were recovered a number of postcranial bones also belonging to this taxon. This locality was the first association of australopithecine cranial and postcranial remains ever found. A number of australopithecine specimens are now known from Kromdraai, including the remains of at

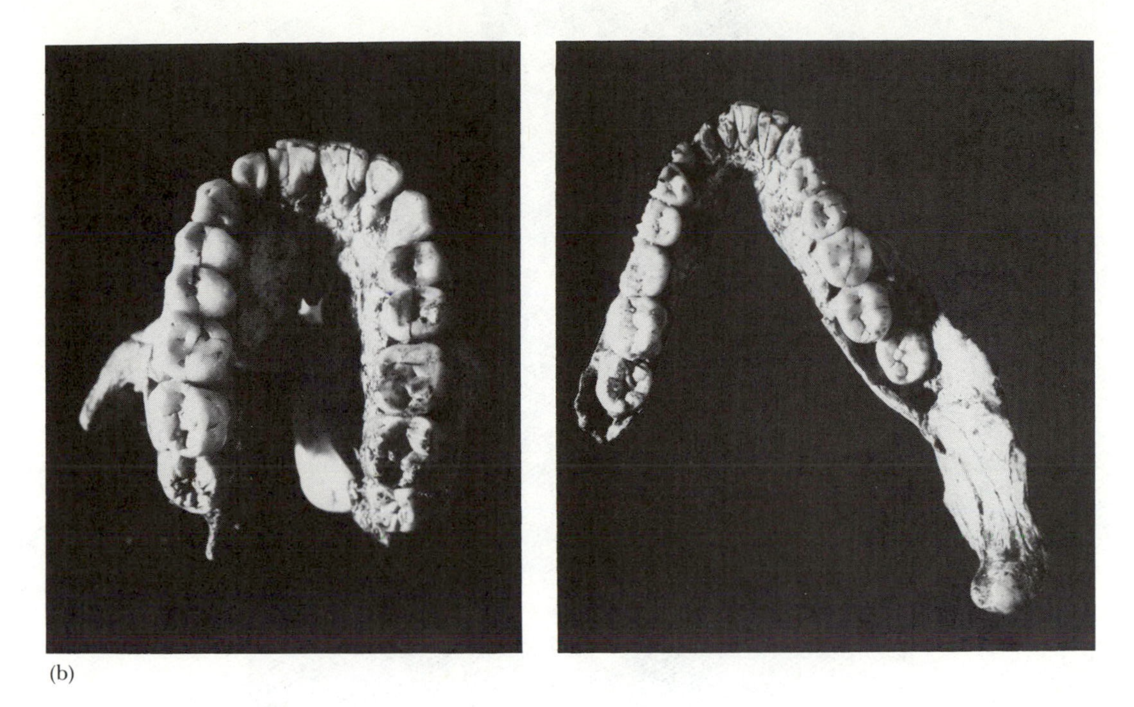

(b)

least two infants, two older children, and two adolescents or young adults (Mann, 1975; Vrba, 1981).

It is clear this species represents a taxon distinct from *A. africanus*. The question of whether or not it represents a distinct genus, however, is still open to debate. Many researchers feel this robust australopithecine is merely a distinct species within the genus *Australopithecus*, appropriately named *Australopithecus robustus*. This is how we shall refer to it here. A current vogue among some (mostly English-trained) anthropologists is to keep the generic distinction and thus retain the name *Paranthropus robustus* for this group of fossils.

Another site in South Africa, Swartkrans, has proved to be an exceptionally rich source of *A. robustus* specimens (Fig. 6.19). Hominids were first discovered there in 1948. Thanks mainly to the work of C. K. Brain, the australopithecine sample from the site now includes over 200 body parts including cranial remains, mandibles, numerous isolated teeth, and limb bones. These represent about 75 individuals, over 40% of which are immature (Mann, 1975).

As the name implies, the major distinction between *A. africanus* and *A. robustus* is one of greater size and overall robustness in the craniofacial and dental features of the latter. While the brain capacity of *A. robustus* (about 530 cc) is slightly larger than that of *A. africanus*, it is still relatively small compared with later hominids of the genus *Homo*. The brain is housed in a skull that is more massive than that of *A. africanus* but less so than that of *A. boisei* (see below). The cranial vault is thin-walled and

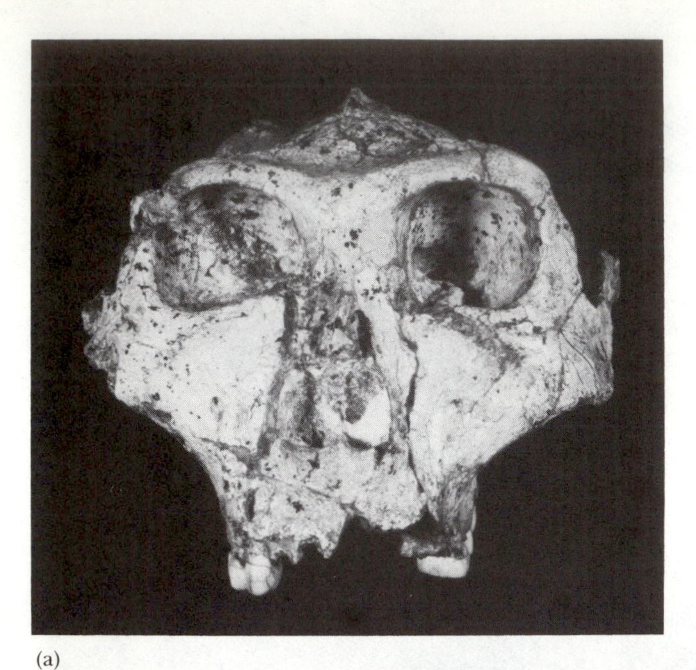

(a)

(b)

Fig. 6.19 Crania of *A. robustus* Specimens from Swartkrans (a) Frontal view (SK 48); (b) lateral view (SK 46). Scale in both photographs about 3/5 × natural size. (Photos by G. Conroy)

normally has well-developed sagittal and nuchal crests. The facial skeleton is hafted onto the **calvaria** (roof of the skull) at a relatively low level; that is, the forehead rises only slightly above the upper margins of the orbit. The brow ridges are well developed, and the face is broad with prominent anterior pillars to withstand large chewing stresses passing through the face (Fig. 6.13).

The dentition is very characteristic of robust australopithecines: the incisors are relatively small, while the premolars and molars are large

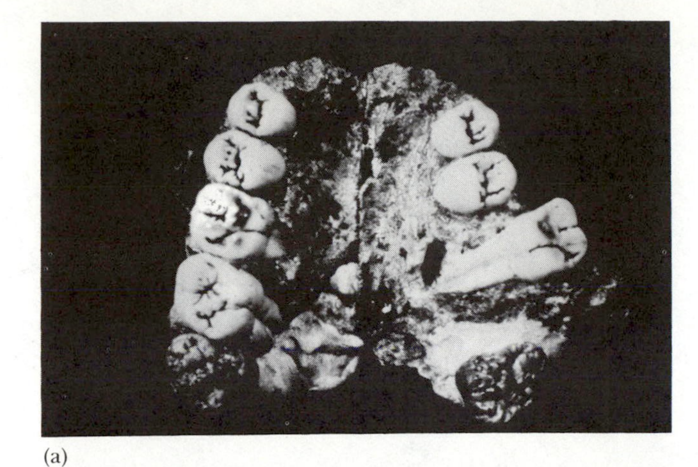

(a)

Fig. 6.20 Dentition of *A. robustus* Specimens from Swartkrans (a) Upper jaw (SK 13); scale about 0.7 × natural size. (b) Lower jaw (SK 12). (Photos by G. Conroy)

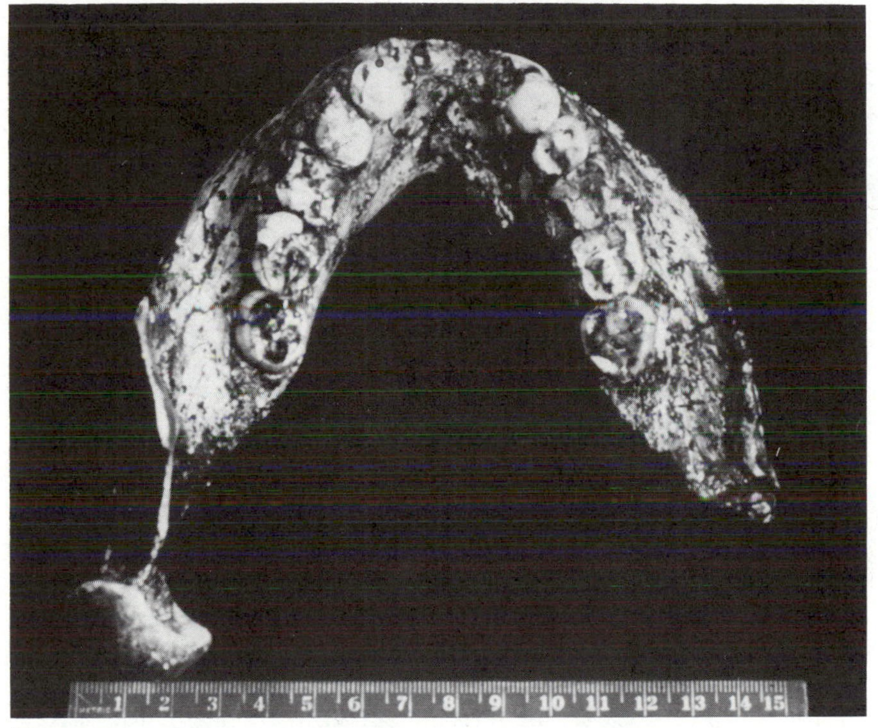

(b)

crushing and grinding teeth with thick enamel and flat surfaces. The canines tend to be functionally incorporated into the incisor row, and thus are not the kind of projecting, pointed stabbing teeth found in so many other primates (Fig. 6.20). As one might expect, the mandibles are also large and robust due to the fact that they (1) house these large teeth; (2) serve as sites of attachment for the powerful chewing muscles; and (3) must withstand the large chewing stresses generated during mastication of tough, fibrous vegetation.

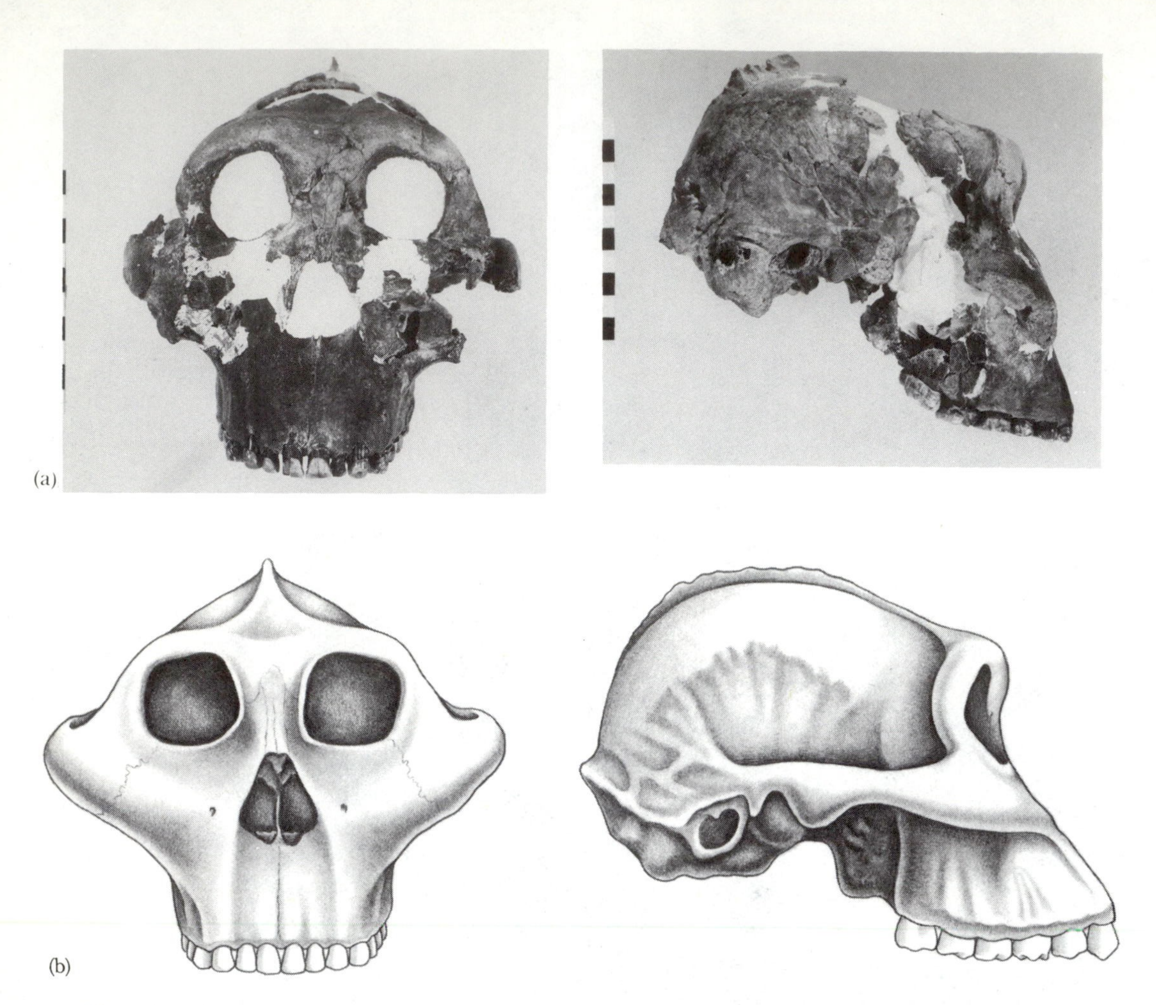

Fig. 6.21 Crania of *A. boisei* (a) *Zinjanthropus* from Bed I of Olduvai Gorge; scale bar in centimeters. (b) Reconstruction based on specimens from Olduvai Gorge and Koobi Fora (Lake Turkana); scale about ⅖ × natural size. (a, Photo by R. G. Klomfass, by permission of P. V. Tobias; b, redrawn from Howell, 1978).

Remains of the pelvis and talus (ankle bone) are more humanlike than apelike, indicating that robust australopithecines were bipedal, although not identical to modern humans in their bipedalism. Recent studies on some hand bones from Swartkrans suggest that *A. robustus* was capable of using and making stone tools (Susman, 1988).

Features such as stature, body weight, and body proportions are very poorly known. Reasonable estimates of stature and body weight are around 155 cm and 50 kg respectively. Roughly speaking, *A. robustus* was probably the size of a modern chimpanzee.

A. boisei The hyperrobust hominid, *Australopithecus boisei,* is well-known from several East African Plio-Pleistocene sites, including Olduvai Gorge (from which the type specimen was taken), eastern Lake Turkana (Koobi

Fora Formation), Omo Basin, and west of Lake Turkana (Fig. 6.21). These specimens are dated at about 2.5 to 1 MYA.

Australopithecus boisei is essentially a more robust version of *A. robustus*. The skull is the most robust of any australopithecine and is characterized by accentuated sagittal and nuchal crests. Pneumatization of the skull is more extensive than in *A. africanus* or *A. robustus*, brow ridges are well developed, the forehead is low or absent, and the face is very long and flat. The jaws and teeth are extremely large, as are the parts of the skull where chewing muscles attach (zygomatic arches and lateral pterygoid plates).

Clearly, selection pressures were favoring increased tooth chewing areas since molars and premolars are enormous compared with incisors and canines (Fig. 6.22). Variations in skull and dental size in specimens attributed to *A. boisei* suggest that the species was markedly sexually dimorphic in body size, with males nearly twice the size of females.

Probably the most spectacular *A. boisei* fossil ever found in terms of overall robustness is the recently discovered specimen (WT 17000) from west of Lake Turkana (Walker et al., 1986) (Fig. 6.23). At 2.5 million years old, it is the oldest representative of the species yet known.

This new specimen reveals a tremendous amount of information about the primitive nature of the skull of this species. The palate is very flat and shallow, and the premaxilla shows pronounced subnasal prognathism. The small size of the brain case, combined with the development of enormous chewing muscles (particularly the posterior fibers of the temporalis muscle), has produced a compound temporonuchal crest and

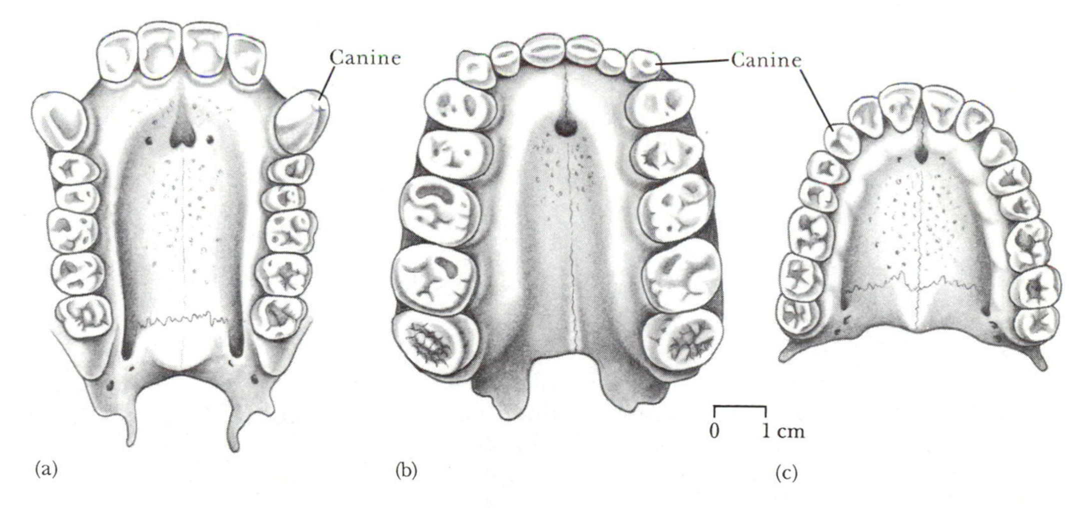

Fig. 6.22 Upper Dentition of (a) Modern Chimpanzee, (b) *A. boisei*, and (c) Modern Human Note the relative size differences between anterior and posterior teeth in *A. boisei* compared with those of the other genera. (Redrawn from Jolly and Plog, *Anthropology and Archaeology*, McGraw-Hill, 1986; reproduced with permission)

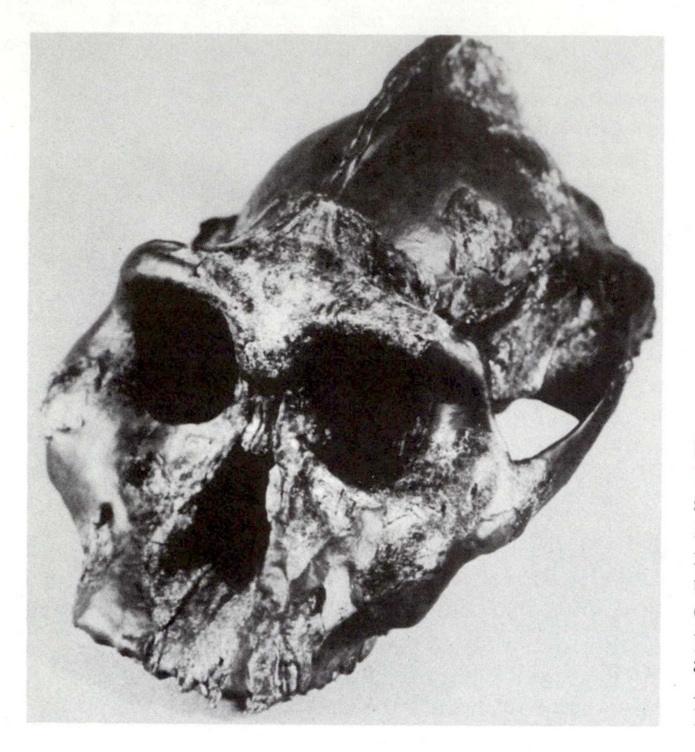

Fig. 6.23 Hyperrobust *A. boisei* (WT 17000) This recently discovered skull came from deposits west of Lake Turkana, dated at about 2.5 MYA. Some researchers consider it to be a distinct species, *A. aethiopicus* (Kimbel et al., 1988). Note the prognathic face, large sagittal crest and low, flat forehead. (Courtesy of A. Walker, © National Museums of Kenya)

a large sagittal crest. Like all robust skulls, the forehead is rather low and flat, and pneumatization of the cranial bones is extensive, particularly in the lateral portions of the cranial base. Later in this chapter we will examine the phylogenetic implications of this new specimen.

Australopithecus boisei is currently represented primarily by craniodental specimens. A few isolated limb bones (partial humerus, femur, tibia) have been attributed to *A. boisei;* however, it is difficult to assign postcranial material to this species with any certainty since skeletal parts have not been recovered in association with any teeth or crania attributed to *A. boisei.*

Comparison of Gracile and Robust Australopithecines

What if anything distinguishes gracile from robust australopithecines? Do they represent two distinct lineages, one ancestral to *Homo (A. africanus)* and the other doomed to extinction *(A. robustus)*? Are they part of one evolving robust australopithecine lineage *(A. africanus–A. robustus–A. boisei),* or are they not separate species but size variants of the same animal? To put these phylogenetic issues into perspective for their examination later in the chapter, we need a clear understanding of the morphological differences between the two groups. The person most responsible for emphasizing the distinctions between them, and for relating these distinctions to inferred behavior and ecology, is John T. Robinson (1956, 1963).

Robinson pointed out that in robust australopithecines the relatively small anterior teeth (incisors and canines) are set in a relatively orthognathous face, whereas the massive postcanine teeth are set in dense bone of the upper and lower jaws. It is obvious that this animal generated large chewing stresses through the skull via these enormous teeth. This is confirmed by the thickened bone in several areas where these forces would have been dissipated throughout the skull: the robust zygomatic process, the anteriorly thickened palate, and (in *A. robustus* at least) the robust anterior pillars (Rak, 1983) (Fig. 6.13).

The large masticatory muscles also left their mark on skull architecture. The presence of a sagittal crest on most robust skills indicates the large size of the temporalis muscle in relation to the overall size of the brain case (on average about 530 cc). The strong development of the lateral **pterygoid plates** (for attachment of the lateral and medial pterygoid muscles) and the zygomatic arch (for attachment of the masseter muscle) indicates the hypertrophy of these chewing muscles as well. The combination of large, strongly flaring zygomatic arches with relatively small anterior dentitions gives most of these robust forms a flattish, even dished, facial appearance. The absence of a true forehead, the great postorbital constriction (again to accommodate a large temporalis muscle), the sagittal crest, and the large brow ridges combine to give this form a skull shape unique among hominids. Muscle markings indicate that the neck muscles were also powerful. Their size undoubtedly contributes to the fact that in posterior view the broadest part of the skull is across the mastoid region.

There are other unique features of the robust australopithecine skull that are related to powerful masticatory stresses. In both juvenile and adult *A. boisei* specimens there is an extraordinarily large overlap of the temporal bone on the parietal. This overlap is a bony adaptation needed to offset the forces produced by the unique combination of certain components of the masticatory system: the massiveness and strength of the temporalis muscle, its relatively anterior location, and the lateral position of the masseter muscle due to the flaring of the zygomatic arches. The effect of the temporalis muscle in chewing is to create excessive pressure on the portion of the **squamosal suture** lying along the parietal bone; the lateral placement of the masseter muscle and the resulting increase of pressure on the temporal bone via the zygomatic arch tends to loosen the contact between the temporal and parietal bones (Rak, 1978).

According to Robinson (1963), the total morphological pattern in gracile australopithecines is very different from the robust australopithecine pattern. In these animals the anterior teeth are relatively larger, the postcanine teeth relatively smaller, and the face more prognathous than in robust forms. Chewing stresses were apparently less, since the great ruggedness of the bones, including sagittal cresting, is for the most part absent. The cranial vault is also higher than that seen in robust forms (Figs. 6.13b). It is interesting to note that even though the overall appearance of the skull in both forms is of a rather rugged creature, the actual skull bones themselves are quite thin.

To Robinson these craniodental differences reflected a considerable difference in diet and behavior between the two species. The larger anterior teeth in the gracile forms were seen as an adaptation to more omnivorous diets that included meat eating (and consequently hunting), whereas the emphasis on the postcanine dentition in the robust forms was seen as a dental adaptation to tough vegetarian diets utilizing roots, tubers, bulbs, and so forth.[1] Robinson considered *A. africanus* so human-like in both morphology and (implied) hunting behaviors that he eventually reclassified it as *Homo africanus.*

Body Size and Encephalization Quotients

Throughout this book, the biological importance of overall body size has been emphasized, not only for implications about diet and habitat, but also for derivation of other measures such as encephalization quotients.

How can we attack the problem of determining australopithecine body size? Returning to the principles of allometry discussed in Chapter 1, we could perform a regression analysis of some osteological variable against body weight in modern animals to see how good a body-weight predictor that variable might be when applied to the fossil record. Since in bipedal animals the body weight must be transmitted through the vertebral column, a good test might be to measure the cross-sectional area of several vertebrae as a predictor of body weight. Using a formula relating vertebral cross-sectional area to body weight in the modern human, the body weight of *A. africanus* was calculated as ranging from about 28 to 43 kg and that of *A. robustus* from about 36 to 53 kg (McHenry, 1975*a*, 1976). Different techniques give varying results: estimates of body weight have ranged between 18 and 43 kg for *A. africanus* and 36 to 80 kg for *A. robustus.* An acceptable average is probably 35 kg for the gracile australopithecines and 44 kg for the robust ones (McHenry, 1982).[2]

If these body-weight estimates are reasonable, then we could calculate encephalization quotients for early hominids in exactly the same way one does for nonhuman primates. Using an average endocranial volume of 441 cc for *A. africanus* and 530 cc for *A. robustus,* the ence-

[1] These distinctions were highlighted in Robert Ardrey's *African Genesis* (Bell, New York, 1961), in which *A. africanus* was immortalized as the "killer ape," while *A. robustus* was pictured as a plodding, vegetarian hominid that eventually became extinct about 1 MYA.

[2] More recent estimates of body-weight ranges predicted from morphological features of the hindlimb are as follows: about 30 to 80 kg for *A. afarensis* (average = 50 kg); 33 to 67 kg for *A. africanus* (average = 45 kg); 37 to 88 kg for *A. robustus* (average = 48 kg); and 33 to 88 kg for *A. boisei* (average = 48kg) (Jungers, 1988; McHenry, 1988).

phalization quotient (EQ) becomes 3.4 for the former and 3.5 for the latter (compared with 7.6 for modern humans and 2.6 for chimpanzees) (McHenry, 1975*a*, 1976). Table 6.2 provides comparative information on mean endocranial capacity, estimated body mass, and EQs for a number of early hominids and compares them with the same data for the chimpanzee.

Estimates of height can be carried out using the same principles. Stature estimates range from about 129 to 160 cm for gracile australopithecines and about 147 to 157 cm for robust australopithecines (McHenry, 1974).

What can we conclude from the trend in EQs? One obvious difference between modern humans and apes is the relative size of the brain. *Homo sapiens* has a cranial capacity about three times that of the greater apes. It is quite clear from the fossil record, however, that expansion in the size of the brain was not one of the earliest hominid evolutionary trends, since the brain size of australopithecines was quite similar to that seen in modern apes. Marked expansion of the hominid cerebral cortex took place only during the last 1 to 2 million years of human evolution. Brain size in the oldest australopithecine, *A. afarensis,* was well within the range for extant apes. As we shall see shortly, the pelvis of the *A. afarensis* specimen nicknamed Lucy has a birth canal whose shape and dimensions show little or no effect of selection for passage of an enlarged fetal cranium, adaptations that so clearly dominate the form of the modern human pelvis (Lovejoy, 1981, 1988).

Having considered size and craniodental morphology in australopithecines, we now look briefly at the same in *Homo* before moving on to examine the postcranial evidence in both groups.

Table 6.2 Mean Endocranial Capacity, Estimated Body Mass, and Encephalization Quotients for a Series of Hominoids

	Mean Endocranial Capacity (cc)	*Estimated Body Mass (kg)*	*EQ (Actual Value)*	*EQ (as % of H. sapiens Value)*
P. troglodytes	395.0	45.0	2.6	34
A. afarensis	413.5	37.1	3.1	41
A. africanus	441.2	35.3	3.4	45
A. robustus	530.0	44.4	3.5	46
H. habilis	640.2	48.0	4.0	53
H. erectus (Asia and Africa)	937.2	53.0	5.5	72
H. e. pekinensis	1,043.0	53.0	6.1	80
H. sapiens	1,350.0	57.0	7.6	100

Source: Tobias (1987).

HOMO

The first appearance of our own genus, *Homo,* occurred about 2.5 to 2.0 MYA in East Africa. Three species are generally recognized: *H. habilis, H. erectus,* and *H. sapiens.* It is often difficult to distinguish early specimens of *Homo* from some of the more gracile australopithecines, but some general characteristics of the genus as a whole include the following.

The trend of generally increasing stature seen in the australopithecines continues in *Homo.* Some *H. habilis* specimens indicate a stature similar to that of australopithecines, but by 1.6 MYA some populations of *H. erectus* were 170 to 180 cm (nearly 6 feet) tall (Brown et al., 1986). Similarly, the trend of increasing average body weight continues throughout the *Homo* line: 48 kg for *H. habilis,* 53 kg. for *H. erectus,* and 57 kg for *H. sapiens* (Table 6.2).

Brain size in *Homo* is greater than in the australopithecines, being at least 600 cc in *H. habilis* and two to three times that size in *H. sapiens.* This compares with average cranial capacities of 413 cc for A. *afarensis,* 441 cc in *A. africanus,* 530 cc in *A. robustus,* and 513 cc in *A. boisei* (Tobias, 1987).

The whole facial skeleton and masticatory apparatus of *Homo* is reduced in size compared with *Australopithecus,* although brow ridges become prominent in later species. Muscular markings on the skull, including crest development, are reduced relative to those in *Australopithecus.*

Early species of *Homo* are the first hominids to show anatomical evidence of the capacity for speech and language, and dependence on stone tools and culturally patterned behavior is characteristic of the genus.

H. habilis Specimens identified as *Homo habilis* are known from a number of sites in East Africa, including eastern Lake Turkana (Fig. 6.24), Omo Basin, and Olduvai Gorge, and from Sterkfontein in South Africa. As mentioned earlier, the East African specimens were contemporaries of australopithecines from about 2.4 to 1.2 MYA. At Sterkfontein the *H. habilis* specimens are from Member 5, thus postdating the *A. africanus* specimens from Member 4 (Hughes and Tobias, 1977).

The average cranial capacity of *H. habilis* is about 650 cc. The first indication of possible speech capacity in hominids occurs in this species with the endocranial evidence for **Broca's area** (the speech area in the left cerebral cortex) (see Fig. 6.17).

In general, the skull in *H. habilis* is more gracile than in *Australopithecus,* particularly in the less well developed muscular crests. Facial heights and breadths are reduced relative to those in *Australopithecus,* and prominent anterior pillars are absent. In size (particularly in width) the pre-

(a)

Fig. 6.24 Crania of *H. habilis* from Koobi Fora Formation (Eastern Lake Turkana) (a) Specimen ER-1470; (b,c) reconstructions based on (b) the same skull, presumably a male, and on (c) specimen ER-1813, presumably a female. Scale in (a) about ½ and in (b,c) about ⅓ × natural size. Note the reduced nuchal and sagittal crests and lack of anterior pillars. (a, Courtesy of Alan Walker, © National Museums of Kenya; b,c redrawn from Howell, 1978)

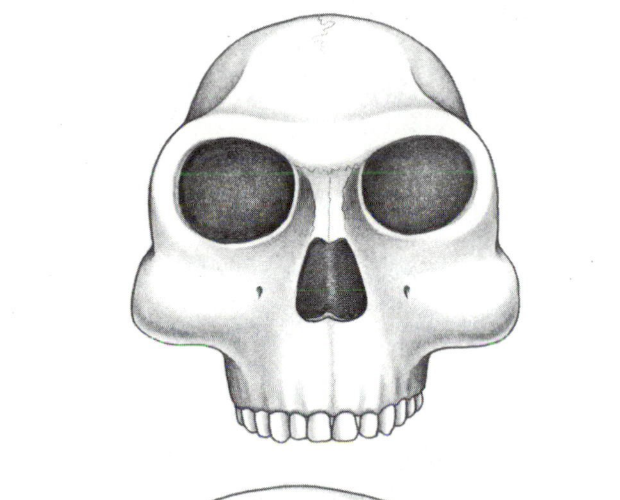

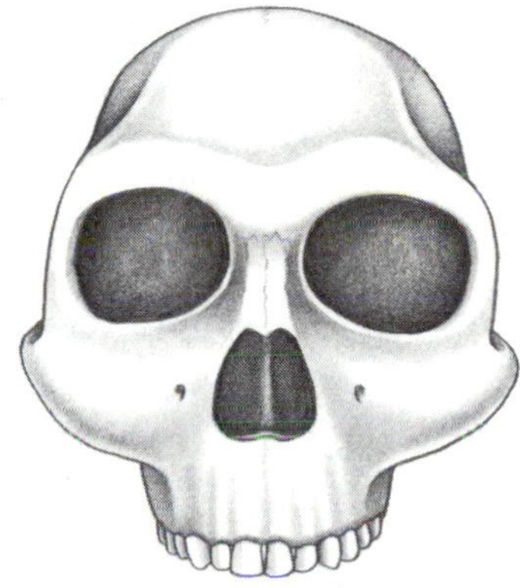

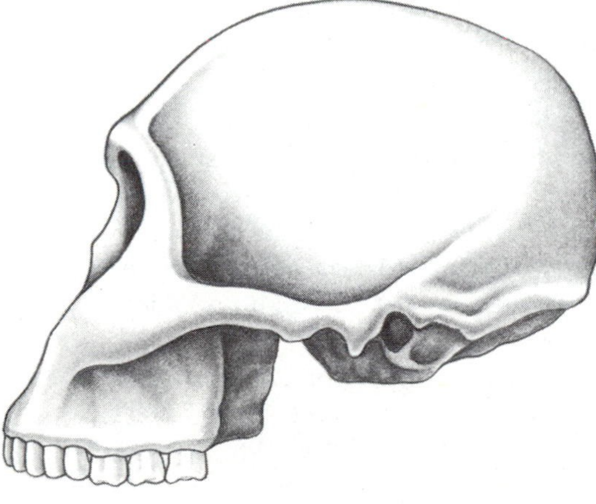

(b)

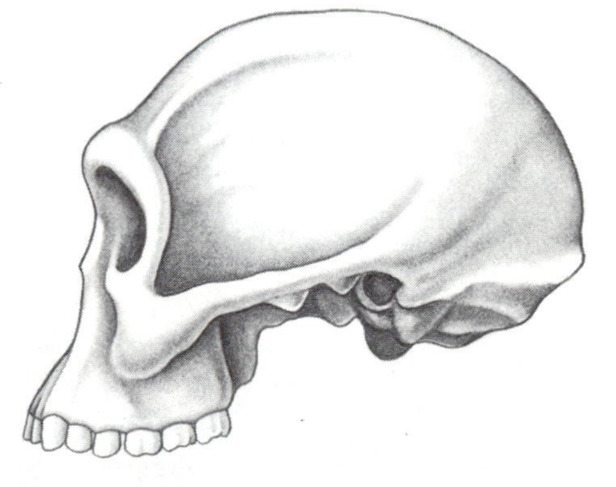

(c)

molars and molars of *H. habilis* are within or below the range found in *A. africanus;* the incisors are relatively larger.

Parts of an associated skeleton recently discovered at Olduvai shed some light on the postcranial anatomy of this species. Of particular interest is the similarity in estimated stature and overall morphology between *H. habilis* and *Australopithecus.* It now appears that the evolutionary differentiation of *Australopithecus* and early *Homo* species is based mostly on cranial, mandibular, and dental characters rather than on postcranial ones (Johanson et al., 1987). We shall return to this issue in a subsequent section.

H. erectus The next species of *Homo, H. erectus,* first appears about 1.5 MYA in several East African sites including Lake Turkana (Fig. 6.25), Omo, and Olduvai. It also appears at about the same time in South Africa at Swartkrans. This is the first hominid species to spread throughout Africa and eventually into the rest of the Old World: remains attributable to this species have been found in Europe, Indonesia, and China.

The cranial capacity of *H. erectus* averages 937 cc for both Asian and African specimens, with a range of about 900 to 1,000 cc; this is above the range for *H. habilis* but falls within the low end of the range of modern humans (Tobias, 1987). The distinctive skull of *H. erectus* shows a low braincase and very thick bones of the cranial vault. At the front of the skull the supraorbital tori are often massive ledges, while at the back of the skull there is a thickened shelf of bone called the **occipital torus** (Fig. 6.25b). The forehead is receding, and the front part of the cranium behind the brow ridges is constricted from side to side. The nose is wide, and the jaws and palate are broad. The teeth are larger than in modern humans but much smaller than in australopithecines.

Evidence of the developing capacity for human speech and language continues with this species: in addition to Broca's area in the cortex, *H. erectus* apparently had a larynx positioned as it is in modern humans, that is, shifted downward relative to the base of the skull so that there was increased air space above the vocal cords. (Laitman, 1985).

H. sapiens Not surprisingly, *Homo sapiens* was the first and for a long time the only hominid in primate classifications; our species name, which in Latin means "wise man," comes from Linnaeus (1758). Fossils of early *H. sapiens* have been recovered from sites throughout Europe, Africa, and Asia (Fig. 6.26), but it is still unclear where or when the first representatives of our species developed, a point we shall consider in more detail in the final section of the chapter.

Because of the tremendous diversity of modern human groups, only a very general definition of *H. sapiens* can be given. Brain size averages about 1,300 cc but varies from about 1,000 to 2,000. The brain and vocal tract are obviously fully adapted for speech. This includes cerebral

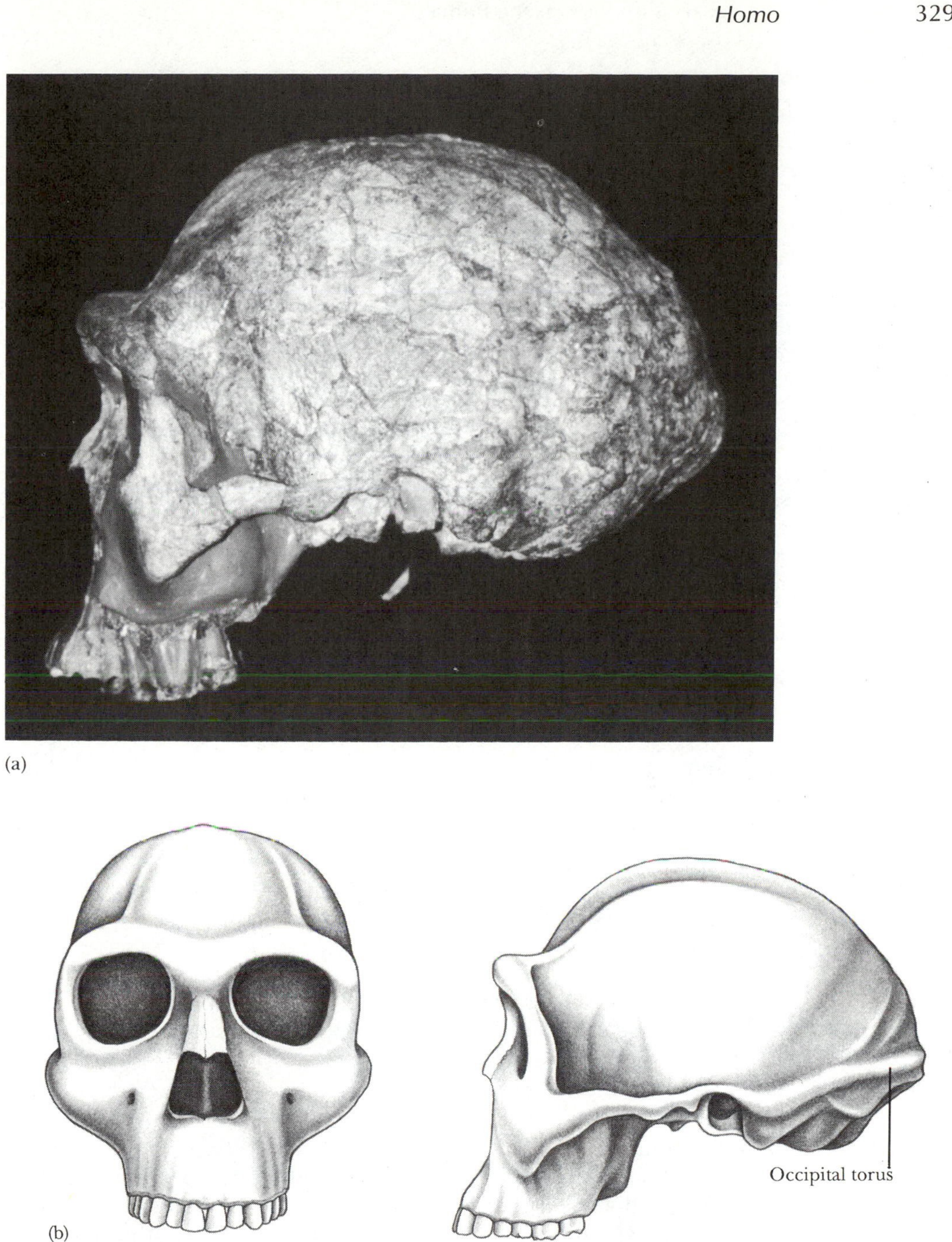

Fig. 6.25 Cranium of *H. erectus* from Koobi Fora (eastern Lake Turkana) (a) Specimen ER-3733, probably a male; (b) reconstruction based on the same specimen. Note the prominent brow ridges and the occipital torus. Scale in (a) about ½ and in (b) about ⅓ × natural size. (a, Courtesy of A. Walker, © National Museums of Kenya; b, redrawn from Howell, 1978)

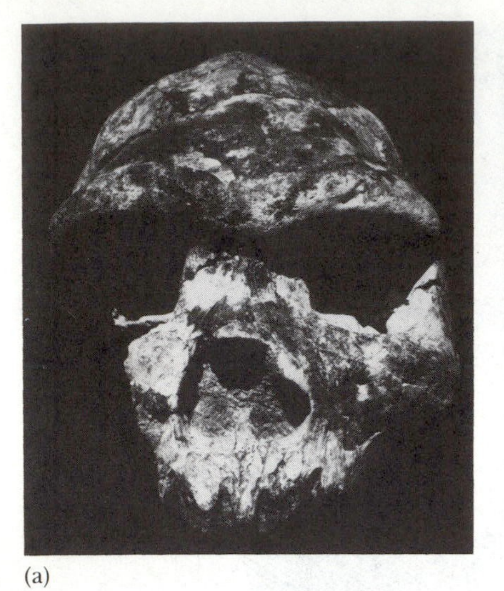
(a)

(b)

Fig. 6.26 Cranium of Early *H. sapiens* from East Africa (a) Frontal and (b) lateral views. This skull, from the Bodo Member of the Wehaietu Formation in the middle Awash valley of Ethiopia (Fig. 6.9), is one of the oldest specimens of "archaic" *H. sapiens* found in Africa. It dates from sediments thought to be 500,000 to 200,000 years old. (Photo by and courtesy of Tim White)

asymmetry, with language centers predominantly in the left cerebral hemisphere. The cranial bones are much thinner than the corresponding bones in *H. erectus;* the whole structure of the face and dentition is more gracile than in *H. erectus* as well.

Some of the synapomorphic features distinguishing *H. sapiens* from other species of *Homo* include the following:

1. A more gracile skeleton
2. A more voluminous cranium that is typically short, high, and domed
3. Reduced or absent brow ridges and cranial buttressing
4. Reduced size of the teeth and their supporting structure
5. A more orthognathous face
6. A **mental eminence** (chin) that develops at a young age. (Stringer and Andrews, 1988)

Modern humans have an unusually prolonged lifespan for mammals of the same size, a protracted period of childhood dependency, and a pronounced adolescent growth spurt. The skeleton is fully adapted to a striding bipedal gait, and there is very fine motor control over movements of the hand and thumb.

Perhaps the most distinguishing feature of the species is a complete reliance on material culture for survival.

MORPHOLOGY OF HOMINID LOCOMOTION

It is obvious that at some early stage in hominid evolution our ancestors evolved from quadrupedal to bipedal movement. The fossil evidence shows significant adaptations to bipedalism very early in hominid evolution, by at least 3 to 4 MYA. This conclusion receives unequivocal support from the presence of hominid footprints discovered at Laetoli (White and Suwa, 1987) (Fig. 6.27). In this section we will discuss the biomechanics of bipedal posture and examine the morphological evidence for bipedalism in australopithecines and archaic humans; in the following section we will consider behavioral pressures that may have contributed to the development of this mode of locomotion.

The exact nature of the hominid quadrupedal ancestor is still open to debate. Was the immediate quadrupedal ancestor a knuckle walker like the African apes? Was it more a fist walker like the orangutan? Or was it something quite different from either (Tuttle, 1974)? No matter what type of quadrupedalism was practiced by these "prehominids," it is important to understand what morphological changes were necessary to evolve into a bipedal creature before we review the actual fossil evidence.

(a)

(b)

Fig. 6.27 Laetoli Hominid Footprints These footprints, dated to approximately 3.6 MYA, are entirely humanlike in form. (a) They show a well-developed medial arch and an adducted hallux (big toe). (b) Foot size and stride length suggest that the larger individual *(right)* was about 140 cm tall and the smaller one *(center)* about 120 cm tall. (Photo by Tim White)

Biomechanical Principles of Bipedalism

There are at least two basic requirements for bipedal posture to be effective and efficient: (1) the body must be able to balance itself effectively not only while standing but also while striding when only one limb is supporting the trunk; and (2) the legs must be able to move quickly through a relatively wide arc, which requires that the limb can be extended past the vertical to provide propulsion to the stride. We will shortly see whether or not such adaptations were alreadly underway in the Pliocene.

The ability of humans to maintain trunk balance is due in large part to the large muscles of the vertebral column (**erector spinae** group), the abdomen **(external** and **internal obliques),** and the gluteal region **(gluteus medius** and **minimus).** It is important to note that all these muscles attach to the pelvis, specifically to the ilium. If we were to take an imaginary cross section through the pelvis, the resulting picture would look something like a wheel-and-axle, the axle being the vertebral column at the center and the steering wheel being the iliac blades on the periphery (Fig. 6.28).

A wheel-and-axle is an example of a simple machine (in a class with levers, pulleys, inclined planes, and the like). The gluteal and abdominal muscles act on the wheel (the ilium), and the erector spinae group acts directly on the axle (the vertebral column). From simple physics it is obvious that the leverage of the forces acting on the wheel (gluteal and abdominal muscles) will be greater the further their line of force is from the center of rotation (the axle, or vertebral column). Thus a clear adaptation to improved stability in both lateral and front-to-back directions is to "wrap" the ilium around the vertebral column at an ever increasing distance.

If we compare the pelves of modern hominids and pongids, this is exactly what we see. The pongid ilium is rather long and thin, and the iliac blades flare out to the side so that their flat inner surface almost faces forward. Hominids, on the other hand, have short, broad ilia that curve around the vertebral column, so that the flat inner surface faces more medially.

The most critical muscles for lateral stability in bipedal walking are the gluteus medius and gluteus minimus, which run from the lateral surface of the ilium and attach to the greater trochanter of the femur. Their attachments to the femur allow them to function as abductors of the lower limb. Everytime a bipedal hominid takes a step, these muscles hold the hip steady. If these muscles were not in a position to do this, walking would require lurching over sideways to maintain the center of gravity over the leg that is on the ground; otherwise the hip would collapse to the opposite side of each stride. This is exactly the type of gait seen in patients with paralysis of these muscles or with dislocated hips

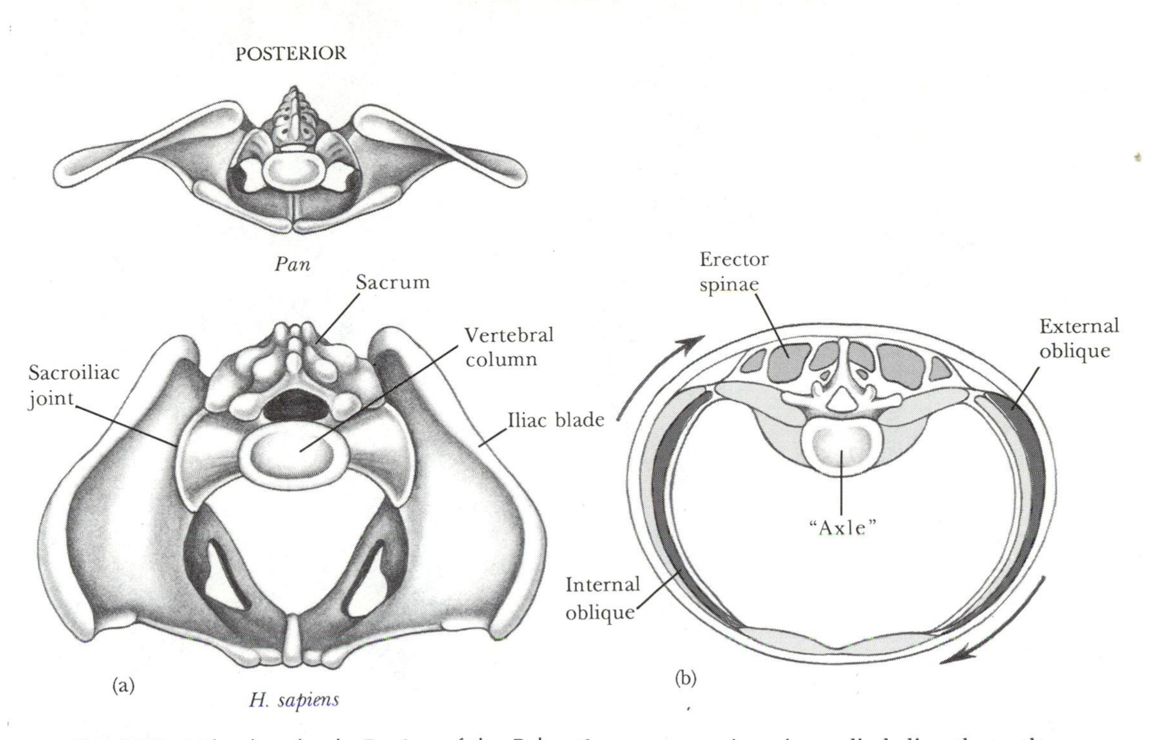

Fig. 6.28 Wheel-and-axle Design of the Pelvic Region in Bipeds (a) Overhead view of chimpanzee and human pelves. In chimpanzees the external surfaces of the ilia face dorsally (backwards), whereas in modern humans they face laterally and "wrap around" the vertebral column, resulting in a more wheel-like morphology for muscle attachment. (b) Cross section of the human trunk shown at about the first lumbar vertebra. The force of the erector spinae is applied directly to the vertebral column (axle), and the force of the oblique abdominal muscles is applied to the ilia (wheel); arrows indicate the direction of force of each oblique muscle. (a, Redrawn from "The Evolution of Human Walking," by C. Owen Lovejoy. Copyright © 1988 by Scientific American, Inc. All rights reserved. b, Adapted from Wells, 1971)

and to a lesser extent it is also evident in apes walking bipedally because in the latter gluteal muscles act not as abductors but as extensors of the hip (Fig. 6.29).

During the lifetime of an organism its bones will respond to stresses passing through them by remodeling themselves. The ilium is no exception. If the large stresses generated by body weight are counterbalanced by the gluteal muscles (as we saw earlier), the structure of the ilium will come to reflect it. Indeed, this strain results in the formation of the **iliac pillar** (a supporting bony buttress in the form of a thickening of the outer table of bone along the ilium between the iliac crest and the acetabulum). The presence of this pillar is important evidence of the efficient lateral balance control required by bipedalism. In addition, the ilium becomes broader from front to back to increase the leverage of hip flex-

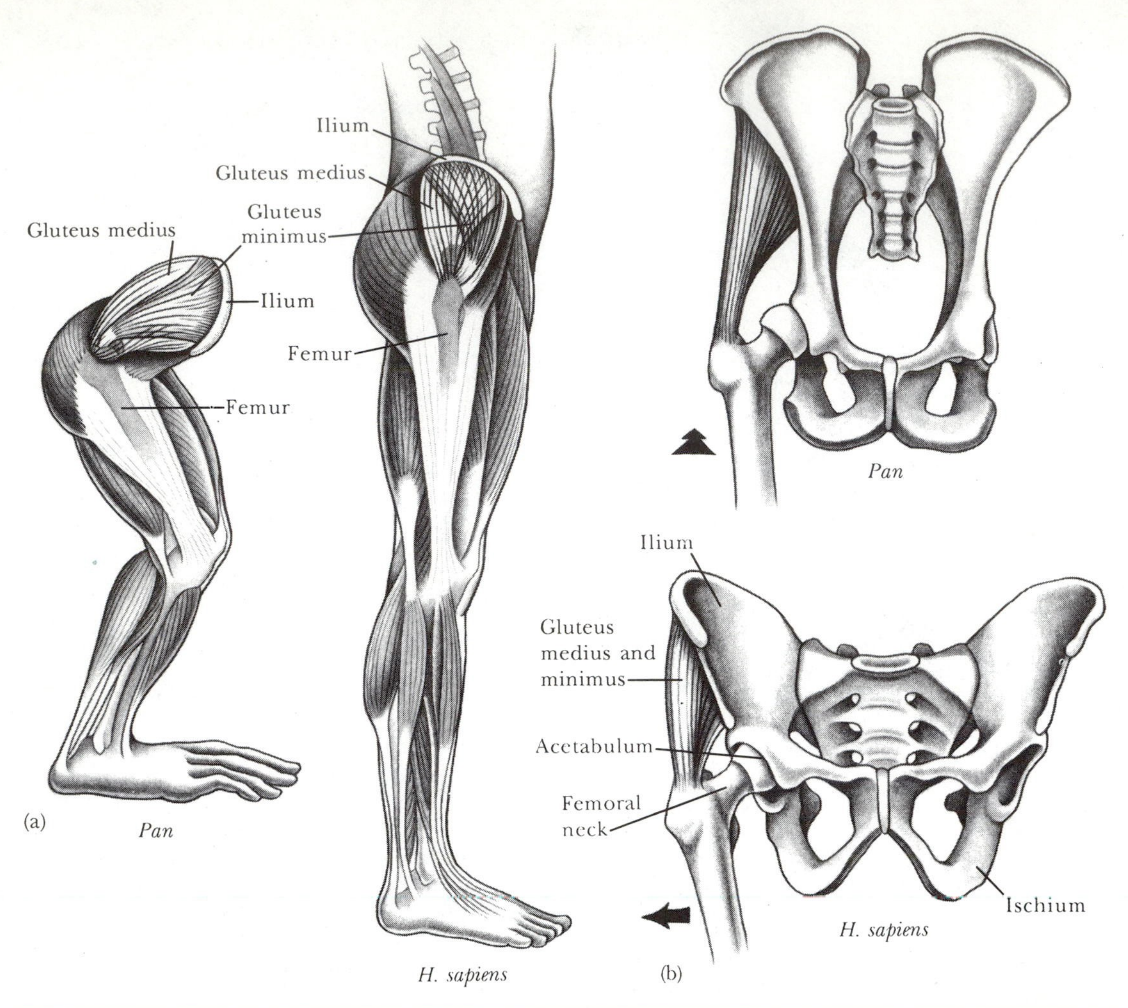

Fig. 6.29 Muscle Configuration in the Pelvic Region of Apes and Humans (a) Lateral and (b) frontal views of the gluteal muscles in *H. sapiens* and *Pan*. In chimpanzees the gluteus medius and minimus are extensors of the hip (pulling the thigh backward), whereas in humans they are abductors (pulling the thigh out to the side of the trunk). In humans these muscles are also critical for maintaining lateral stability during walking: the abductors on the side of the body supporting the body during a stride contract to prevent the hip joint from collapsing to the opposite (unsupported) side. (All views except frontal view of *Pan* redrawn from "The Evolution of Human Walking," by C. Owen Lovejoy.)

ors and extensors. These morphological changes result in the development of a **sciatic notch** and prominent **anterior and posterior iliac spines** (Fig. 6.30).

The **sacroiliac joint** also changes in orientation in the bipedal hominid. The surface for articulating with the sacrum becomes more posterior relative to the **acetabulum** (hip socket for the femoral head). This shifts the line of gravity of the trunk from being far in front of the

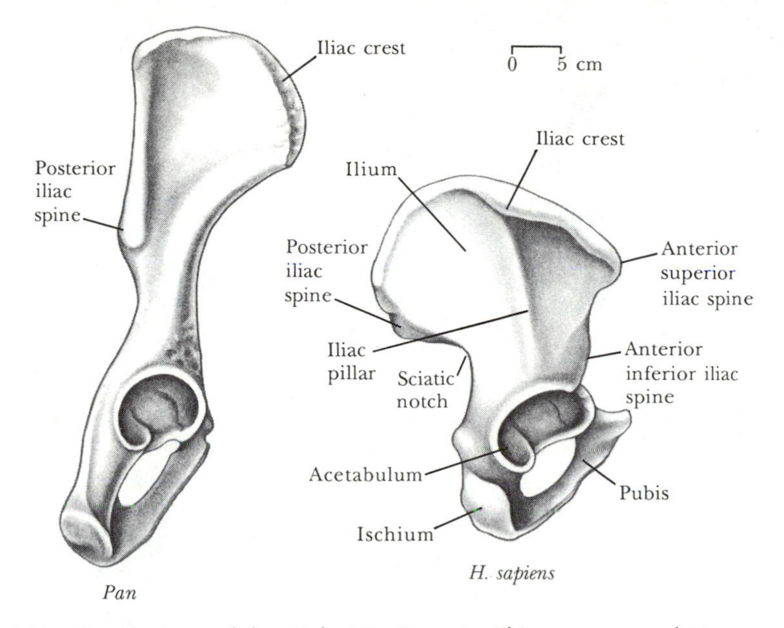

Fig. 6.30 Comparison of the Right Hip Bone in Chimpanzees and Humans In response to increased stresses passing through the bipedal hip joint, the human pelvis has developed a prominant iliac pillar and anterior and posterior spines not evident in the chimpanzee. In addition, the human ilium is shorter (to help lower the center of gravity) and wider (to enlarge the attachment area of flexor and extensor muscles of the hip). This alteration in shape results in a distinctive sciatic notch. (Redrawn from Jolly and Plog, 1976)

femoral heads (as in the chimpanzee) to being closer to them. The result of this shift is to reduce the muscular effort needed to control balance of the vertical trunk on the femoral heads during bipedal locomotion (Stern and Susman, 1983).

Another important aspect of balance control in hominids concerns lowering the center of gravity of the body. In apes the center of gravity is well forward of the hip joint. Thus the movement of a pongid walking bipedally is very inefficient because the natural tendency for the body to fall forward over the hip joints has to be resisted by powerful muscular activity. In modern hominids the center of gravity has shifted downward and backward relative to overall height, so that it is slightly above and behind the hip joint. In fact, there is even a slight tendency for the human trunk to fall backward over the hip joint in erect posture; this tendency is easily resisted by the **iliofemoral ligament** (which runs from the anterior inferior spine of the ilium to the proximal femur). Thus hominids expend little muscular energy in maintaining an erect posture (Fig. 6.31).

There are several important factors in hominid evolution that have led to lowering the center of gravity. One has been to increase the size

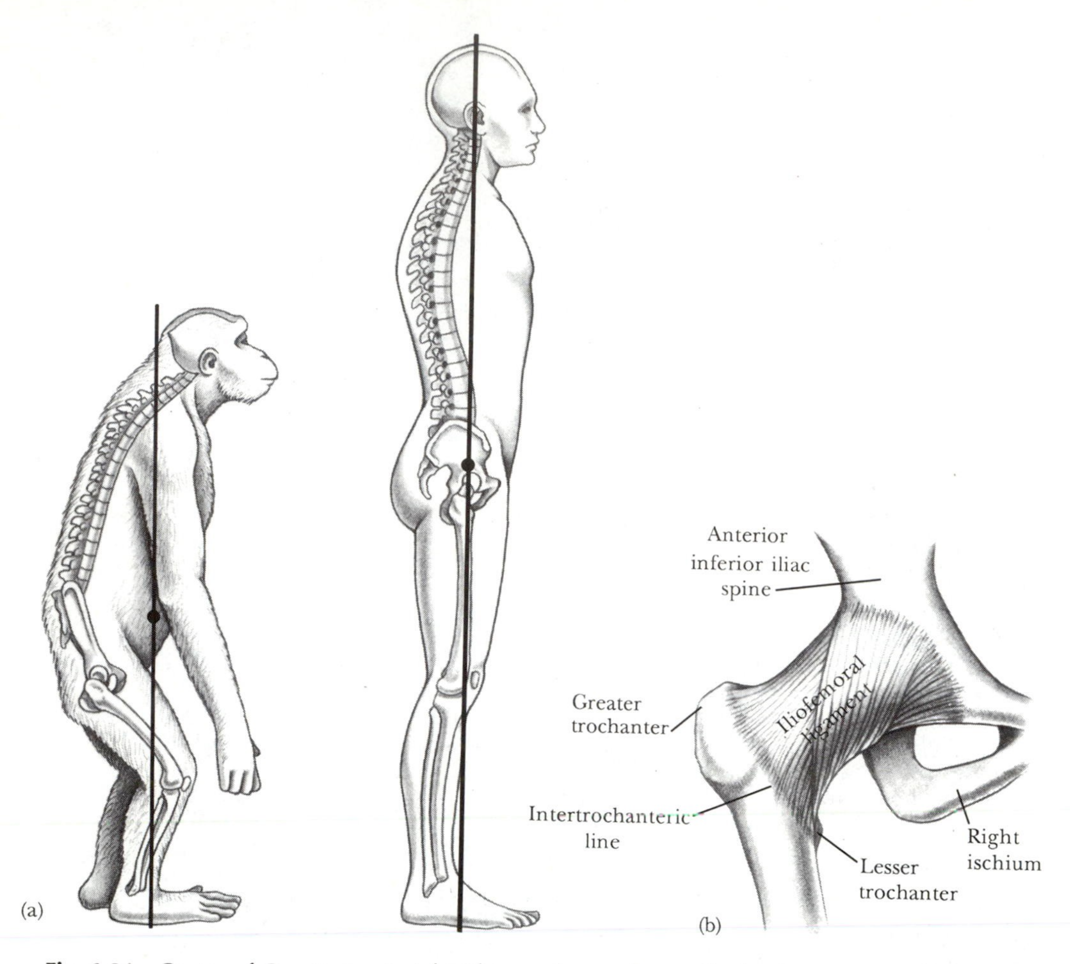

Fig. 6.31 Center of Gravity in Upright Chimpanzees and Modern Humans (a) The human's center of gravity lies near the hip joint, and the line of gravity falls slightly anterior to the ankle joint. In contrast, the chimpanzee's center of gravity lies above and well forward of the hip joint. In the human, note the lumbar curve in the spine and the larger lower limb—factors contributing to lowering the center of gravity. (b) Right human hip joint seen from the front; tension in the iliofemoral ligament helps keep the trunk from falling backward when standing upright. (a, Redrawn from Zihlman and Brunker, 1979; b, redrawn from Warwick and Williams, 1973)

of the lower limbs in order to increase the mass of the lower portion of the body. With the assumption of erect posture, more body weight must be supported and propelled by the hind limbs, increasing the stresses on the pelvic and limb joints. The lower limbs, which account for less than one-third of the total body weight, must alone carry the remaining two-thirds. As a percentage of total body weight, the upper limbs of humans are only 60% as heavy as those of chimpanzees, whereas the lower limbs

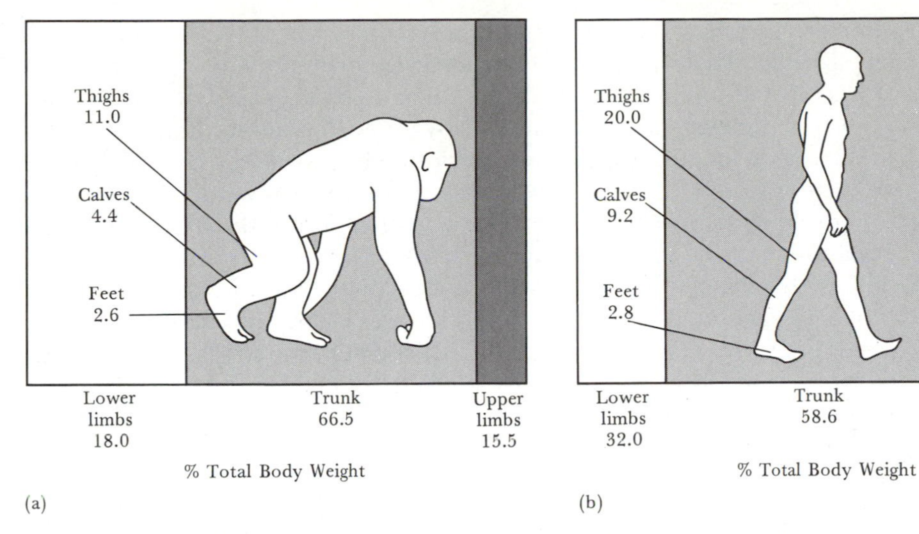

Fig. 6.32 Body Weight Distribution in (a) Chimpanzees and (b) Modern Humans (Adapted from Zihlman and Brunker, 1979)

are nearly twice as heavy (Zihlman and Brunker, 1979). This arrangement of body weight contributes to a more stable posture, since the center of gravity is low in the pelvis. (Fig. 6.32).

Two other factors in lowering the center of gravity are reducing the height of the ilium (see Fig. 6.30) and developing a lumbar curve to bring the upper portion of the trunk back over the pelvis without blocking the birth canal.

Evidence of Bipedalism in Australopithecines and Archaic Humans

Precise information about the body proportions of early hominids is crucial for accurate functional and phylogenetic interpretations of early human evolution. We shall now examine in detail the features of *Australopithecus* and *Homo* that have a bearing on their modes of locomotion.

A. afarensis The partial skeleton of *A. afarensis* recovered from Hadar ("Lucy") permits the first direct assessment of body size, limb proportions, and skeletal allometry in ancestral hominids that predate 3 MYA (Fig. 6.33). Using allometric relationships between limb lengths and body weights in nonhuman catarrhine primates and African apes as empirical baselines for comparison, William Jungers (1982) of SUNY at Stony Brook showed that the limb proportions of *A. afarensis* are unique among hominoids. *A. afarensis* had already attained forelimb proportions similar to those of modern humans but had hindlimbs that were relatively much shorter. Lucy had very short hindlimbs proportioned most like those of small-bodied African apes and outside the known human range. By contrast, relative humerus length in relation to body size is similar to that seen in a skeletal sample of human pygmies. A similar combination of relative limb lengths appears to exist in the larger individuals of *A. afarensis* and probably other gracile australopithecines as well.

It follows that relative and absolute elongation of the hindlimbs represents one of the major evolutionary changes in later human evolution. The body proportions of Lucy are not incompatible with some form of bipedal locomotion, but that locomotion was clearly not identical to the bipedal gait of modern humans. Reduced relative stride length in *A. afarensis* probably implies both greater relative energy cost and relatively lower peak velocities of bipedal locomotion in *A. afarensis* than in later hominids (Jungers and Stern, 1983)

Jack Stern and Randall Susman (1983), also of SUNY at Stony Brook, have argued that *A. afarensis* had many primitive (pongidlike) postcranial features indicating a significant adaptation to or retention of movement in the trees. Several of the wrist bones are markedly chimpanzeelike, and the fingers are slender and curved as much as in apes and far beyond the degree seen in the human hand. In addition, the overall morphology of the metacarpals is similar to that of chimpanzees and might be interpreted as evidence of well-developed grasping capabilities associated with suspensory behavior. The fossil hand bones lack features associated with knuckle walking, however.

The glenoid cavity of the scapula is directed in a more cranial orientation than is typical of modern humans, and Stern and Susman suggest that this trait also implies use of the upper limb in elevated positions, as is common during climbing behavior.

The pelvis of *A. afarensis* shows some distinctly humanlike traits such as (1) a low, broad ilium; (2) a deep sciatic notch, (3) a prominent anterior inferior iliac spine, and (4) an ischial surface for the origin of the hamstring muscles (biceps femoris, semimembranosus, and semitendinosis) that is divided by a vertical ridge into lateral and medial portions. Stern and Susman suggest that *A. afarensis* had a relatively long hamstring moment arm but one that did not fall outside the range of human variation. The pelvis also shows the typical australopithecine trait of a laterally flaring anterior superior iliac spine that is far removed from the anterior inferior iliac spine.

The articular surface of the acetabulum in *A. afarensis* lacks the large contribution from the pubic bone that characterizes modern humans. This suggests the absence of full, humanlike extension at the hip joint during bipedal locomotion. Other structural features of the femur also point to a mode of terrestrial bipedalism that involved less extension at the hip and knee than occurs in modern humans. However, the pelvis shows some major changes toward the human condition regarding location of the sacroiliac joint.

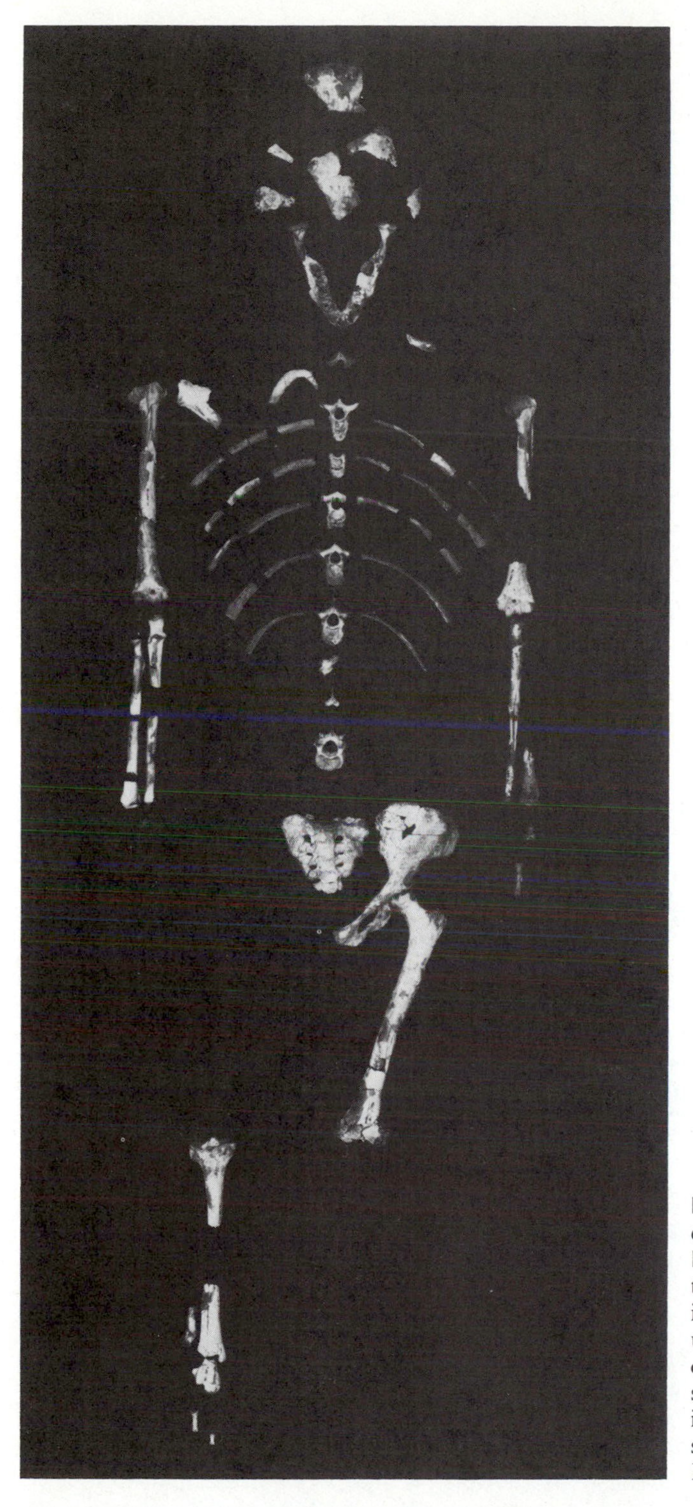

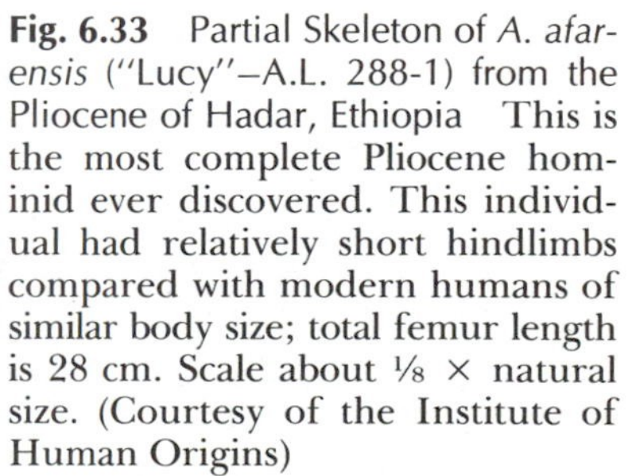

Fig. 6.33 Partial Skeleton of *A. afarensis* ("Lucy"–A.L. 288-1) from the Pliocene of Hadar, Ethiopia This is the most complete Pliocene hominid ever discovered. This individual had relatively short hindlimbs compared with modern humans of similar body size; total femur length is 28 cm. Scale about ⅛ × natural size. (Courtesy of the Institute of Human Origins)

Striking features of the *A. afarensis* sacrum are that the ventral concavity is only slightly developed and that the first segment of the sacrum lacks well-developed transverse processes. These features suggest poorly developed **sacrotuberous ligaments** (which bind together the sacrum and the ischium), which in turn implies a less well developed mechanism for sacroiliac stabilization when the trunk is in an erect position. The orientation of the iliac blade is very similar to that seen in chimpanzees, that is, its external surface faces more posterolaterally. The pelvis of *A. afarensis* reveals that the process of wrapping the iliac blades around the vertebral column had already begun but had not reached the degree of curvature seen in modern hominids (Lovejoy, 1975; Stern and Susman, 1983) (Fig. 6.34). This feature suggests that in *A. afarensis* the mechanism of lateral pelvic balance during bipedalism was more similar to apes than to humans. The anterior placement of the iliac pillar suggests a bent-kneed gait, since electromyography (recording of electrical activity in active muscles) has shown that during such a gait the medial rotators of the hip are important for balance, and they arise from the more anterior aspect of the ilium (Stern and Susman, 1983).

Evidence suggests that the toes from *A. afarensis* at Hadar are both longer and much more curved than those in modern humans. Stern and Susman (1983) interpret these features as signs of adaptation to or retention of arboreal activity (since it is easier for feet to grip tree branches with long, curved digits). This observation has led to some recent debate over *A. afarensis*. Some workers have suggested that the feet typified by the remains at Hadar could not have belonged to the same species that made the Laetoli footprints (Tuttle, 1985).

A comparison of the smaller (presumably female) specimens to the larger (presumably male) specimens suggests sexual differences in locomotor behavior linked to marked size dimorphism. The males were probably less arboreal and engaged more frequently in terrestrial bipedalism (Stern and Susman, 1983).

A very different viewpoint is expressed by Lovejoy (1975, 1988), who considers *A. afarensis* to be fully adapted to bipedalism. He has identified several adaptations that characterize human bipedalism:

1. A knee positioned close to the midline to minimize side-to-side shifting of the center of gravity during locomotion
2. Raised lateral lip of the patellar groove to prevent lateral displacement of the patella (kneecap) during extension of the leg
3. Increased cartilage contact in the knee joint
4. A well-developed iliofemoral ligament, which as we have seen helps keep the erect trunk from falling backwards
5. A long abductor-muscle moment arm for the gluteus medius and minimus for lateral stability in the hip joint during walking
6. A posterior position of the gluteus maximus, which acts as an extensor of the lower limb

He argues that these adaptations, which were present in later australopithecine species, were already evident in *A. afarensis*.

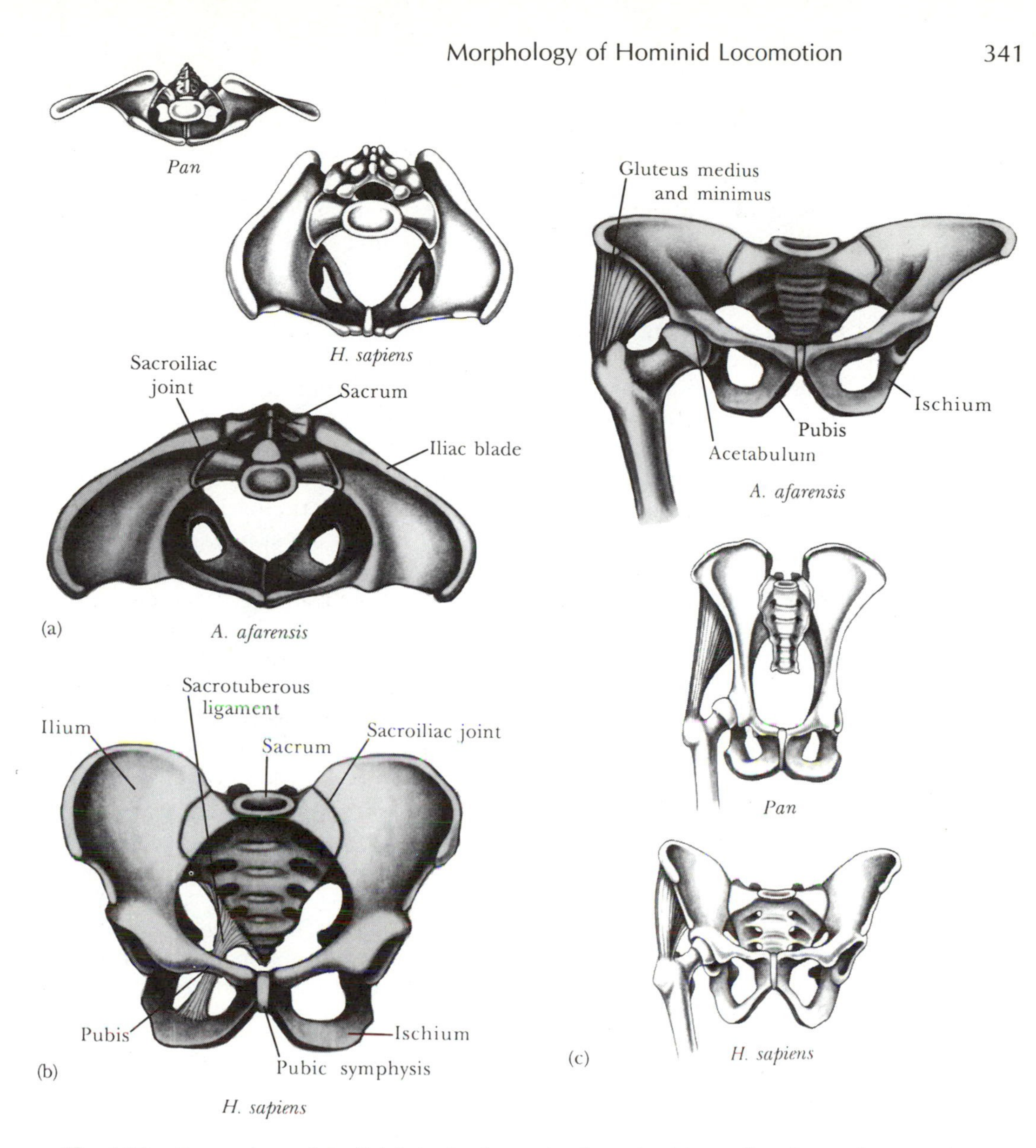

Fig. 6.34 Comparison of the Pelvis in *A. afarensis, Pan,* and *H. sapiens* (a) Overhead view. Note that the iliac blades in *A. afarensis* are beginning to curve around the vertebral column, as in the modern human, but the blade angle is still closer to that seen in the chimpanzee. (b) Anterior view of the human pelvis. Strong sacrotuberous ligaments in the human (shown on one side only) help stabilize the sacroiliac joint when the body is erect; features of the sacrum in *A. afarensis* suggest it lacked well-developed sacrotuberous ligaments. (c) Anterior view of the full pelvis in *A. afarensis* with smaller views of *Pan* and *H. sapiens* for comparison. The gluteal muscles are acting as partial abductors in *A. afarensis,* providing some lateral stability; note, however, that the ball-and-socket joint in the australopithecine is shallower than in the human, due to less contribution to the acetabulum from the pubis. (a,c, Redrawn from "The Evolution of Human Walking," by C. Owen Lovejoy. Copyright © 1988 by Scientific American, Inc. All rights reserved. b, Redrawn from Wells, 1971)

Other Australopithecines A good deal of postcranial material of *A. africanus* is available from which to extrapolate locomotor behavior in that species (Lovejoy et al., 1973; Lovejoy, 1975). The most complete specimen (Sts 14) is from a mature adult female from Sterkfontein; it consists of much of the vertebral column (including the lumbar vertebrae and sacrum), a few ribs, almost the complete pelvis, and much of the femur. The ribs, vertebral bodies, and the femoral head, neck, and shaft are all slender, indicating a small, gracile animal. The stature of this individual is estimated at 122 to 137 cm (Robinson, 1972). The presence of six lumbar vertebrae is interesting in that modern humans normally have five and African apes no more than three or four (Schultz, 1968). Importantly, the specimen had a well-developed lumbar curve.

All postcranial evidence from Sterkfontein and Makapansgat suggests that *A. africanus* was a gracile animal. On the basis of an estimate of femur length from one individual (Sts 14), it appears the femur is about the same length relative to overall body length as the femur in modern humans. From this information Robinson (1972) has concluded that in *A. africanus* the lower limbs were about the same length relative to upper limbs as in modern humans. If true, this would be in clear contrast to the proportions calculated for *A. afarensis*, with its relatively reduced lower limb length (Jungers, 1982). However, if the Sts 14 individual had a humerus as long as that in another *A. africanus* specimen (Sts 7), it would indicate that the upper limb–lower limb ratio was greater than in modern humans but not as great as in pongids.

The *A. africanus* pelvis, similar in many respects to that described for *A. afarensis*, is clearly approaching the modern human pelvis in morphology (Fig. 6.35a). Moreover, *A. africanus* has been described as showing a well-developed, humanlike lumbar curve, which suggests that the posture was more like that found in modern humans than in pongids (Robinson, 1972). Another bipedal feature is the angle of the femoral shaft, which demonstrates that the knees, and therefore the feet, were close together during standing and walking.

The postcranial evidence is not as satisfactory for *A. robustus*. Nothing definite is known about lumbar curvature or body proportions for this australopithecine. The ilium, however, does show some important adaptations to bipedalism, such as the reduction of iliac height and the development of an iliac pillar. It is certainly clear that *A. robustus* was a more robust and heavily built creature than *A. africanus* (Fig 6.35b). For example, the pelvis of *A. robustus* from Swartkrans is about 22% larger than the specimen of *A. africanus* from Sterkfontein. A robust specimen from Kromdraai seems just as large as the one from Swartkrans. The width of the acetabulum is about 11% larger in the robust pelvis. The situation is the same for the few robust femoral remains. For example, one robust femur has a shaft thickness greater than that of the Sterkfontein specimen by 50% of the thickness of the latter, and a second specimen is thicker by 75% (Robinson, 1972). As in all australopithecines, the femo-

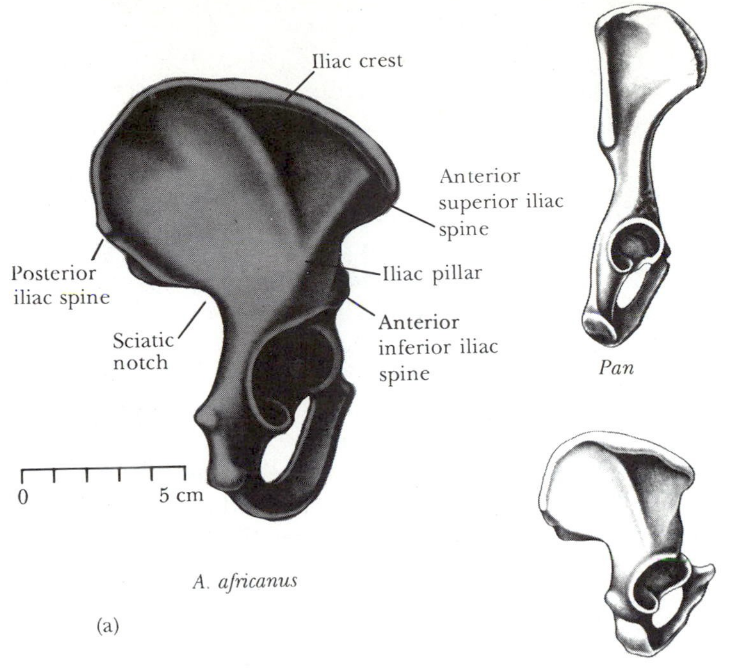

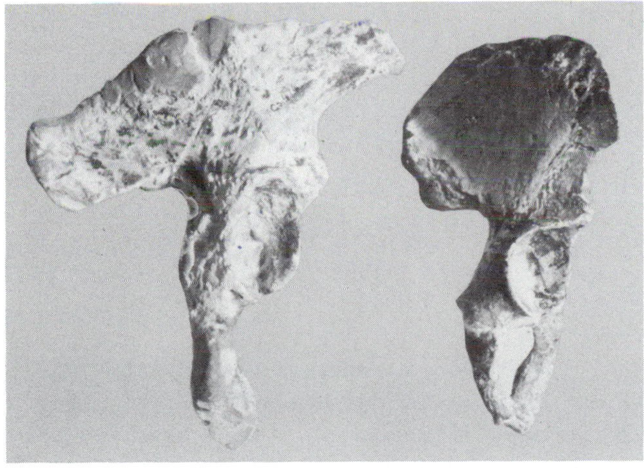

Fig. 6.35 Pelvis of South African Australopithecines (a) Hip bone in *A. africanus;* smaller views of *Pan* and *H. sapiens* are provided for comparison. (b) Lateral view of the right pelvis of *A. robustus* from Swartkrans *(left)* and *A. africanus* from Sterkfontein *(right)*. Note the larger overall size of the *A. robustus* specimen. Both pelves have broad, short ilia with well-developed sciatic notches. Scale about ⅓ × natural size. (a, Redrawn from Jolly and Plog, 1976; b, from Robinson, 1972)

ral neck is relatively long and the femoral head is small. This suggests greater moment arms for the hips abductors (gluteus medius and minimus) and relatively less weight being borne directly on the femoral heads.

Even less can be said about the postcranial evidence of bipedalism in *A. boisei* than in *A. robustus;* as mentioned earlier, few skeletal parts have been found, and none in association with teeth or crania, making attribution tentative at best. However, because the two species are so closely related, most workers assume bipedalism for *A. boisei,* as well as for *A. robustus.*

Homo With the emergence and evolution of *Homo,* the postcranial skeleton began to develop features leading to the morphology typical of modern humans.

The earliest species, *H. habilis,* shows an interesting mosaic of australopithecine and *Homo*-like features. For example, overall body size and limb proportions are strikingly similar between *H. habilis* (specimen OH 62 from Olduvai) and *A. afarensis* (specimen A.L. 288–1 from Hadar). The high humerofemoral index of around 95 in OH 62 clearly indicates that the arms were long and powerful in this early *Homo* species (Johanson et al., 1987).

Other morphological features of *H. habilis* are more modern. For example, the foot and lower leg are humanlike in terms of metatarsal robustness and in having a nonopposable big toe. The middle phalanges of the hand are somewhat apelike in being robust and curved with well-marked insertions for powerful muscles of finger flexion (for grasping behavior); however, the overall length and morphology of the distal phalanges are similar to those of modern humans. In addition, the **carpometacarpal joint** (thumb joint) is distinctly humanlike. It would thus seem that *H. habilis* was certainly bipedal but retained a hand capable of powerful grasping. To Susman and Stern (1982) this combination of traits reflects some residual arboreal activity in this species.

While skeletons of *H. erectus* are clearly more humanlike than those of *Australopithecus* or *H. habilis,* they still differ in some respects from those of modern humans. For example, the spinous processes on the vertebrae are relatively longer and less inclined, resulting in less overlapping of spines of adjacent vertebrae than is typical of *H. sapiens.* In addition, the vertebral canal (for passage of the spinal cord) is smaller relative to vertebral body size than in modern humans.

In the *H. erectus* pelvis a strong iliac pillar has not yet developed, and there was a marked degree of iliac flare. This latter feature is concordant with the femoral necks, which are as long relative to femoral length as those of robust australopithecines. It also appears that birth canal diameters in early specimens of *H. erectus* would have been significantly smaller than in *H. sapiens* and that passage of a modern human-sized, full-term fetus would not have been possible (Brown et al., 1985). This conclusion is consistent with the evidence that *H. erectus* had an average brain size still only two-thirds that of *H. sapiens.*

Summary of Trends in Hominid Pelvic Morphology Since the anatomy of the pelvis is so critical for interpreting locomotor behavior in fossil hominids, let us summarize the salient points. Distinctive features of the modern human pelvis include the following: (1) a low, broad ilium; (2) a short distance from the center of the acetabulum to the articular surface of the sacrum; (3) a deep sciatic notch; (4) a more lateral orientation of the iliac blades, and (5) the presence of an iliac pillar. Fossil evidence from both robust and gracile australopithecines clearly indicates that pelvic reorientation from a typically apelike configuration to a more humanlike one had already commenced several million years ago (Lovejoy et al., 1973; McHenry and Corruccini, 1975*b;* McHenry, 1975b, 1975c; Lovejoy, 1988).

That this pelvic reorganization was still incomplete in australopithecines is evidenced by the fact that they still had (1) a relatively small area on the sacrum for articulation with the ilium; (2) relatively small femoral heads; and (3) incomplete rotation of the iliac blades so that they still faced in a more dorsolateral position than is characteristic of modern

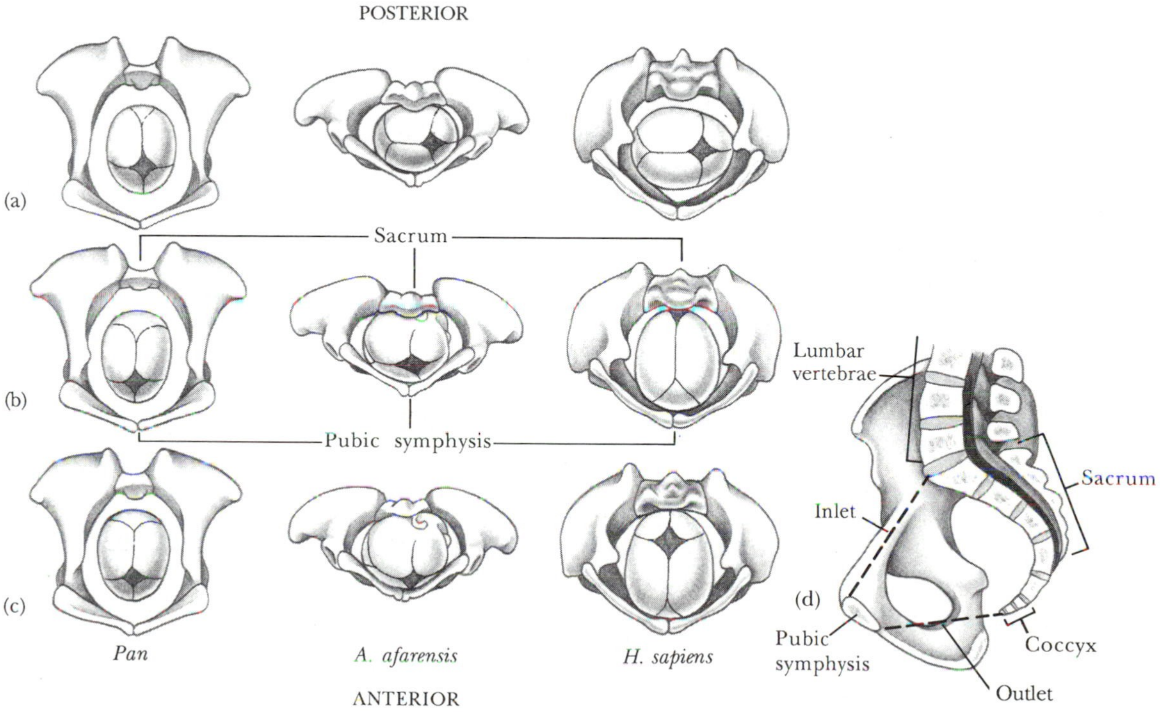

Fig. 6.36 Pelvic Dimensions, Encephalization, and Parturition in Pongids and Hominids Fetal head is shown passing through (a) inlet, (b) midplane, and (c) outlet of the birth canal seen from below; the anterior fontanelle (membrane-covered soft spot in the infant's incompletely ossified skull) is marked in black to indicate fetal orientation. (d) Side view of the human pelvic area identifying inlet and outlet planes; the midplane is approximately halfway in between. In *A. afarensis* the birth process was probably more difficult than in the chimpanzee because, although the birth canal was broad, it was constricted from front to back; the infant's cranium could pass through only if it first turned sideways and then tilted. During the human birth process the canal is even more constricted, and a second rotation of the fetal cranium within the birth canal is required. (a–c, Redrawn from "The Evolution of Human Walking," by C. Owen Lovejoy. Copyright © 1988 by Scientific American, Inc. All rights reserved. d, Adapted from Warwick and Williams, 1973)

humans. This last feature seems to indicate that the hip abductor mechanism discussed earlier did not yet function as it does in modern humans, although this inference is still a point of contention (Stern and Susman, 1983; Lovejoy, 1988).

What explains some of the morphological rearrangements of the australopithecine pelvis, like the reduction of lateral flare of the ilium, and the commensurate changes in such features as the anterior position of the iliac pillar, longer biomechanical femoral neck length, and anterior iliac spine position? Owen Lovejoy of Kent State University suggests that the rapid encephalization that took place during the Pleistocene might be the cause (Lovejoy et al., 1973; Lovejoy, 1988). As he sees it, there would have been significant selective pressure on the female to maintain adequate pelvic breadth to allow parturition in the face of the constantly increasing size of the fetal cranium (Fig. 6.36). There is certainly a maximum total pelvic breadth that can be efficiently maintained in a biped. The pelvic breadth of *Australopithecus* was already great because of the laterally flared ilium and a broad interacetabular distance relative to that in quadrupeds. Because of increasing fetal cranial dimensions, adjustments that enlarged the pelvic opening but did not increase total pelvic breadth would have been favored. If stature and total pelvic breadth were to remain unchanged relative to one another and the dimensions of the parturition canal were to increase, the only avenue of change would have been to increase the interacetabular width and reduce the length of the moment arm of the hip abductors. Such changes would have led to increased pressures on the femoral heads, and one response would have been enlargement of the femoral heads. In addition, the resulting increase in the muscle force of the abductor mechanism (because of its reduced moment arm) would have required a commensurate increase in the iliac pillar and general robustness of the ilium. It is precisely these changes, in addition to those discussed earlier, that can be seen between the hip complex of *Australopithecus* and *H. sapiens.*

BEHAVIORAL AND CULTURAL TRENDS IN AUSTRALOPITHECINES AND ARCHAIC HUMANS

Thus far we have concentrated on the importance of fossils for unraveling human origins, but equally important is the recognition of a close genetic relationship between humans and the other extant hominoids (especially *Pan* and *Gorilla*) (Sarich, 1971; Kohne et al., 1972; King and Wilson, 1975; Goodman, 1976). Experiments involving DNA hybridization indicate at least 98% identity in nonrepeated DNA between human and chimpanzee. These molecular studies serve to reinforce more traditional comparative anatomical studies that have emphasized the similarities between humans and apes for over a century (Huxley, 1863).

It is also important to recognize that models of human origins must directly address both the morphological and behavioral differences separating humans and apes (Lovejoy, 1981). In this section we look at how characteristics of diet and/or reproductive behavior may have served as the impetus for and acceleration of hominid evolution. We shall also examine briefly the emergence of culture: tool use, tool making, and the use of fire.

Behavioral Theories for the Evolution of Bipedalism

As we have seen, australopithecines were showing unmistakable adaptations to bipedalism nearly 4 MYA. Why did bipedalism develop at all? Are two legs better than four? Bipedalism is an unusual mode of locomotion in mammals because it has the disadvantages of reducing speed, agility, and energy efficiency.

Many theories have been advanced to explain the evolution of bipedalism: it frees the hands for tool use; it allows for new feeding adaptations; it allows for other behaviors, such as carrying food and infants, or using the hands for display; or some combination of these. However, there is no doubt from the fossil record that bipedalism predated the use of stone tools, the modern hominid dental complex, and even increased encephalization above the average levels for other hominoids.

Could bipedalism be more energy efficient than quadrupedalism? Empirical studies have shown that at maximum speed human bipedalism costs twice as much energy per kilogram per kilometer as is predicted for a true mammalian quadruped of the same size (Taylor et al., 1970). It has also been shown that for chimpanzees and capuchin monkeys the energy costs of traveling quadrupedally and bipedally are about the same. Does this mean that energy efficiency is not a good argument to explain the evolution of human bipedalism (Taylor and Rowntree, 1973)? Peter Rodman and Henry McHenry of the University of California, Davis, have made the important point that bipedalism in humans today is much more efficient than quadrupedalism is in extant apes. They hypothesize a scenario in which energy efficiency in terms of food-gathering strategies would actually have increased through bipedalism. The climatic fluctuations in the Miocene resulted in changing distributions of forests and open country. In places where the forests were receding, food sources would no longer have been as concentrated; ancestral australopithecine populations in these marginal areas would thus have had to travel farther to forage for food. One might argue that morphological changes to improve quadrupedalism might have evolved instead, in order to increase efficiency at moving between food sites; however, this would have lessened the ease of gathering food at the food sites themselves. By modifying only the hind limbs, the evolution of bipedalism provided for the possibility of improved efficiency of travel, while still allowing for arboreal feeding (Rodman and McHenry, 1980).

In the last two decades two of the most interesting and influential models for the evolution of bipedalism have been Clifford Jolly's (1970) seed-eating hypothesis and Owen Lovejoy's (1981) sexual-and-reproductive-strategy model. Both have had their critics, yet both stand as landmark attempts to synthesize anatomical and behavioral observations on living primates into plausible scenarios of human origins and adaptations.

Jolly's (1970) seed-eating hypothesis makes an analogy to anatomical and behavioral characters shared by living gelada baboons *(Theropithecus gelada)* and some early hominids. He suggests that early hominid populations relied on small-object feeding, that this dietary specialization resulted in a suite of adaptations to the grassland savanna, and that bipedalism developed in response to feeding posture. This model fits quite well some of the dental specializations seen in the robust australopithecines, but the more generalized dentition of *A. afarensis* is more difficult to fit, unless of course *A. afarensis* represents the beginning of this new adaptation to savanna living and to small-object feeding, a scenario that is quite possible.

Lovejoy (1981) presents a somewhat different view of human origins based on an obvious trend he has observed toward prolonged lifespan in the primate evolutionary record. An increased primate lifespan has implications both for primate physiology and population dynamics, which bear, as we shall see, on the issue of hominid origins. The trends in primate physiology we have studied correlate well with a longer period of infant dependency, prolonged gestation, single rather than multiple births, and successively greater periods between pregnancies. This progressive slowing of life phases can be accounted for by an evolutionary strategy in which populations devote a greater proportion of their reproductive energy to the care of young and greater investment in the survival of fewer young. The increase in primate lifespan is accompanied by both a proportionate delay in the onset of reproductive readiness and greater spacing between births. This requires a female to survive to an older age in order to maintain the same reproductive potential. Increased longevity, according to Lovejoy, can occur only if features such as strong social bonds, high levels of intelligence, intense parenting, and long periods of learning are in evidence to reduce environmentally induced mortality. The emergence of successful hominid clades in the Pliocene strongly suggests that a major change in reproductive strategy occurred as these hominids came to occupy new environments. Yet neither brain expansion nor significant material culture appear at this time and were presumably not responsible for the shift (unless the material culture was not preserved).

Lovejoy (1981) suggests a novel behavioral pattern that could have evolved from typical primate survival strategies, in which males and females had nonoverlapping feeding areas, with the females (and infants) having a much-reduced day range to minimize injury to the infants. Such a division of feeding areas, however, would not genetically favor males unless it specifically reduced competition with their own biological offspring and did not reduce their opportunities for having consort relationships. Polygynous mating (several males with one female) would not be favored by this adaptive strategy, but monogamous pair bonding would be. Sexual division of labor would develop, with the male provisioning the female and the male's biological offspring at some type of **home base** (an area to which hominids return to meet other group members, share food, and make tools). In this model such male provisioning is also seen as an impetus to the evolution of bipedalism.

A critical component of Lovejoy's model concerns the notion of extended periods of learning. In modern humans the course of development from conception to maturity requires nearly twice the time as in modern apes. This extended period of maturation in humans is usually regarded as a major evolutionary advance since it enhances the importance of learning in reducing environmentally induced mortality. This view has best been expressed by Theodosius Dobzhansky (1962), who stated that,

> although a prolonged period of juvenile helplessness and dependency would, by itself, be disadvantageous to a species because it endangers the young and handicaps their parents, it is a help to man because the slow development provides time for learning and training, which are far more extensive and important in man than in any other animal.

Lovejoy's model assumes that long periods of learning already existed in the early lifetimes of Pliocene hominids. This assumption relies on the seminal work by Alan Mann (1975) of the University of Pennsylvania, who concluded that "a long childhood dependency period . . . which the evidence . . . seems to indicate for *Australopithecus,* would provide the time necessary for the skills associated with tool-making to be developed in the young."

How can we determine whether long learning periods really occurred in australopithecines? As Mann (1975) pointed out, it would be crucial to know how old individual australopithecines were when they died. He estimated their age by examining their teeth. Nearly all mammals have one set of deciduous teeth that are first formed and then replaced by permanent teeth in a predictable sequence. A jaw with some deciduous teeth in it reveals fairly precisely at which point along the way to adulthood the individual died. The key is translating the relative ages of the fossil into actual time spans in years and months. The crucial question is whether australopithecines developed on a humanlike or a pongidlike timetable. The difference is significant. Based on an analysis of the timing of tooth eruptions in mainly *A. robustus* teeth, Mann concluded that australopithecines showed a humanlike pattern of dental development, and so he based his age estimates on the human timetable. Thus it was thought that alteration in the timing of growth and development occurred very early in hominid evolution, indicating that a long maturational period had already evolved in Pliocene hominids older than *A. robustus*.

Because of the widespread influence of Mann's (1975) work, it has been anthropological dogma for the past decade or so that humanlike traits of sociality, enhanced intelligence, elaborate communication, and so on, arose very early in the history of the hominid lineage. Recent evidence, however, suggests that the situation may be more complicated than previously thought (Bromage and Dean, 1985; Smith, 1986; Conroy, 1988; Conroy and Vannier, 1987).

For example, B. Holly Smith (1986) of the University of Michigan, using many more specimens than were available to Mann, dated all the teeth separately for each fossil jaw she studied. She then statistically compared the dental age ranges of the jaw as determined by the individual teeth to chimpanzee and human calibration standards to see which group the fossil most closely resembled. Theoretically, all the teeth in the jaw should follow a standard pattern and rate of development for that species. For example, if the incisors are in a state of maturation that is typical of a six-year-old human, the molars should be as well. Surprisingly, her results showed that only the robust australopithecines followed a humanlike pattern, while *A. afarensis* and *A. africanus* (and *H. habilis*) showed an apelike pattern.

Actually, a more humanlike pattern of dental eruption in robust australopithecines was first suggested by Broom and Robinson (1951) and has been the subject of periodic debate ever since (Wallace, 1972; Dean, 1985; Grine, 1987). This alleged humanlike pattern concerns the relative rate of eruption of the first permanent central incisors and molars. In *H. sapiens* permanent central incisors and permanent first molars erupt with little time in between. In pongids, on the other hand, eruption of the first permanent incisors is often delayed more than 2.5 years after permanent first-molar eruption. M. C. Dean (1985) has suggested that robust australopithecines show the *H. sapiens* pattern, and gracile australopithecines the pongidlike pattern. Part of the difficulty in resolving this debate is that the developing teeth (particularly the incisors) in some of the specimens most crucial to solving this debate have never been clearly visualized by conventional radiographic techniques because of heavy mineralization in the fossils. However, recent studies by Michael Vannier of Washington University and myself using high-resolution CT scans suggest that both gracile and robust australopithecines had a more apelike dental maturation pattern (Conroy, 1988, Conroy and Vannier, 1987, in press *a,* in press *b*).

Timothy Bromage and M. C. Dean (1985) of the University of London have used a totally independent method of analysis to age the australopithecines at death. Their study employed a technique of absolute age determination that relies on counting the periodic increments of enamel laid down as teeth are formed. Their results essentially confirm Smith's conclusions. Both studies conclude that many of the ages at death determined by Mann were too old by a factor of about 2; in other words, australopithecines generally followed a pongidlike maturational process. Both studies suggest that many of the immature australopithecines apparently died between the ages of 2.5 and 3.5 years old. An interesting avenue for future research would be to see whether these new estimates for ages at death might be related to weaning stresses in australopithecine populations (Shipman, 1987).

It is difficult to be certain just when the modern human maturational pattern finally emerged. Very preliminary studies suggest that even some specimens of *H. habilis* and *H. erectus* show a more pongidlike pattern of dental development, but few individuals have been examined. A fully modern pattern has been discerned in at least one Neanderthal *(H. sapiens)* child, a specimen called Gibraltar 2 (Smith, 1986).

We now turn our attention away from bones and teeth to the perhaps more sublime realm of human culture.

Beginnings of Culture

What is culture, and when did it begin to play a role in hominid evolution? Definitions of culture have always been elusive and changeable as ethologists and anthropologists learn more about present-day and fossil hominids. Culture, in a broad sense, can be defined as a system of shared meanings, symbols, customs, and beliefs that are learned in a society and used to cope with the environment, to communicate with others, and to transmit information through the generations. Culture also defines the rules that guide the production of human artifacts (Jolly and Plog, 1986).

Most people accept tool use, tool making, and use of fire as three of the many kinds of cultural behavior. Certainly they are among the earliest evidence for culture among fossil hominids. It is important to recognize, however, that tool use and even rudimentary tool making have been documented among other living primates besides humans. For example, chimpanzees are known to adapt thin branches for extracting termites from the ground or to utilize leaf "sponges" for sopping up water in the hollow of trees (Goodall, 1976). It is not that *H. sapiens* is unique in showing cultural behavior; but we *are* unique in the complexity and extent of the cultural behavior on which we depend for our survival.

When did hominids begin using tools, and what materials were these tools made of? Figure 6.37 provides a general overview of stone and other tool technologies associated with African fossil sites. Use of simple **opportunistic tools** (unaltered objects found nearby) presumably predated tool making (physically altering an object for one or more specific uses). It is difficult to make inferences about opportunistic tool use from artifacts found at fossil sites because many perishable objects such as

Age and Main Division	*Stone Technology*		*Some Other Tools*[a]	*Behavioral Traits*
	General	*Specific*		
Present Late Stone Age 0.035 MYA	Specialized composite tools Abstractions Cave art	Many regional traditions Many functionally specific tool sets		Population expansion Small bands Intensive seasonal resource use in smaller areas Cohesion for ceremonies and exchange
0.04 MYA Middle Stone Age 0.15 MYA	Simple compound tools Hafting begins	Several regional traditions and chronological sets		Seasonal herding of livestock Resource and task-specific camps Regular reoccupation by large bands in large territories
0.2 MYA Early Stone Age II Acheulian and Developed Oldowan 1.8 MYA	First standard tool forms	Large cutting bifacial tools Chopper/light-duty complex		Organized hunting of large animals Scavenging and planned collecting Widely adapted sites reoccupied
2.0 MYA Early Stone Age I Oldowan 2.5 MYA	Selected opportunistic tools	Choppers Spheroids Light duty Heavy duty Small cutting tools	? – Digging sticks ? – Carrying trays, etc. Spears Throwing sticks Bone tools: opportunistic — Shaped Fire ?	Scavenging and hunting of small animals Collecting insects, eggs, and plant foods from brief stopovers
No known flaked stone tools 5.0 MYA	Simple opportunistic tools (sticks, etc.)?	Manuports?		Collecting plant foods, insects, and small mammals

[a] - - - Inferred ——— Confirmed

Fig. 6.37 Summary of Stone Age Tool Use in Africa Period covers the early Pliocene to the present. (Adapted from Clark, 1985)

wood may have been used long before permanent materials were. Even preserved objects such as stones and fossilized animal bone found in association with fossil hominids are sometimes difficult to assess as opportunistic tools; for example, bones may have been refuse from eating, and stones foreign to the site may have been deposited there through geological activity rather than human intervention.

There is no evidence of tool use or tool making associated with any fossil hominoids in the Miocene, nor is it definite which if any australopithecine species made the stone tools found in the numerous Plio-Pleistocene tool sites in East and South Africa. This uncertainty is due to the fact that early species of *Homo* are also found in the same areas. However, it is probably safe to assume that in the Pliocene, from about 5.0 to 2.5 MYA, early australopithecines were using simple opportunistic tools (such as sticks, bones, or stones) and **manuports** (unaltered objects carried some distance before use) for collecting plant foods, insects, scavenging, and killing small animals.

The period of 2.5 to 2.0 MYA (Early Stone Age I) provides the first incontrovertible evidence for the use of stone tools in Africa. Several sites from the Shungura Formation at Omo have yielded stone artifacts. Since both early *Homo* and *A. boisei* are represented in these sediments, it is not certain if one or both groups made the tools. Several sites slightly younger than 2.0 MYA at eastern Lake Turkana have also yielded stone tools. Again, there is no certainty as to the identity of the toolmakers. Some of these sites may have been butchery sites; all of them are distinguished by a low density of artifacts, a high frequency of flakes, and a rarity of larger shaped core tools (Howell, 1978). (Cores are simply lumps of stone from which flakes have been removed; sometimes a core is simply the byproduct of tool making, but it may also be shaped and modified to serve as a tool in its own right.) Most of the early stone tools were made from quartz, quartzite, lava, and chert. The stone-tool industry from this time period is referred to as the **Oldowan** tradition, named after Olduvai Gorge, where it was first identified (Fig. 6.38).

Undoubtedly the finest archeological record for early hominids comes from Olduvai Gorge (Leakey, 1981). Dozens of occupational sites are found in Beds I and II, which yield artifacts not only of the Oldowan tradition but also of the **Developed Oldowan** industry. The latter succeeds the former in lower Bed II and occurs thereafter in Beds II and IV. Oldowan and Developed Oldowan tools have been found at several sites in North Africa (coastal Morocco and Tunisia), East Africa (Olduvai, Koobi Fora, Omo, Hadar), and South Africa (Swartkrans and Sterkfontein). After 2.0 MYA these two traditions often overlap with each other and with later Stone Age industries.

The basic forms of stone tools that predominate from 2.0 to 1.5 MYA in the Oldowan tradition include an assortment of choppers (usually made from cobblestones of quartz or quartzite), hammerstones made from pebbles or cobblestones, simple cores, choppers, polyhedrals (angular tools having three or more intersecting working edges), and flake scrapers (Toth, 1987). The Developed Oldowan tradition continues the flake-and-chopper industry seen in the Oldowan tradition. In addition it includes **proto-handaxes** (simple pointed choppers) and crude bifacial (two-sided) forms (Fig. 6.39). Rarely are more sophisticated bifacial forms, such as handaxes, present. (A **handaxe** is a bifacial core with one end pointed for cutting and the other end rounded for hand-holding comfort). These artifacts provide clear evidence of a rudimentary knowledge of working stone for the production of flakes and chipping edges. Some of these archeological occurrences suggest that the species making the tools may have established home bases at some of these sites.

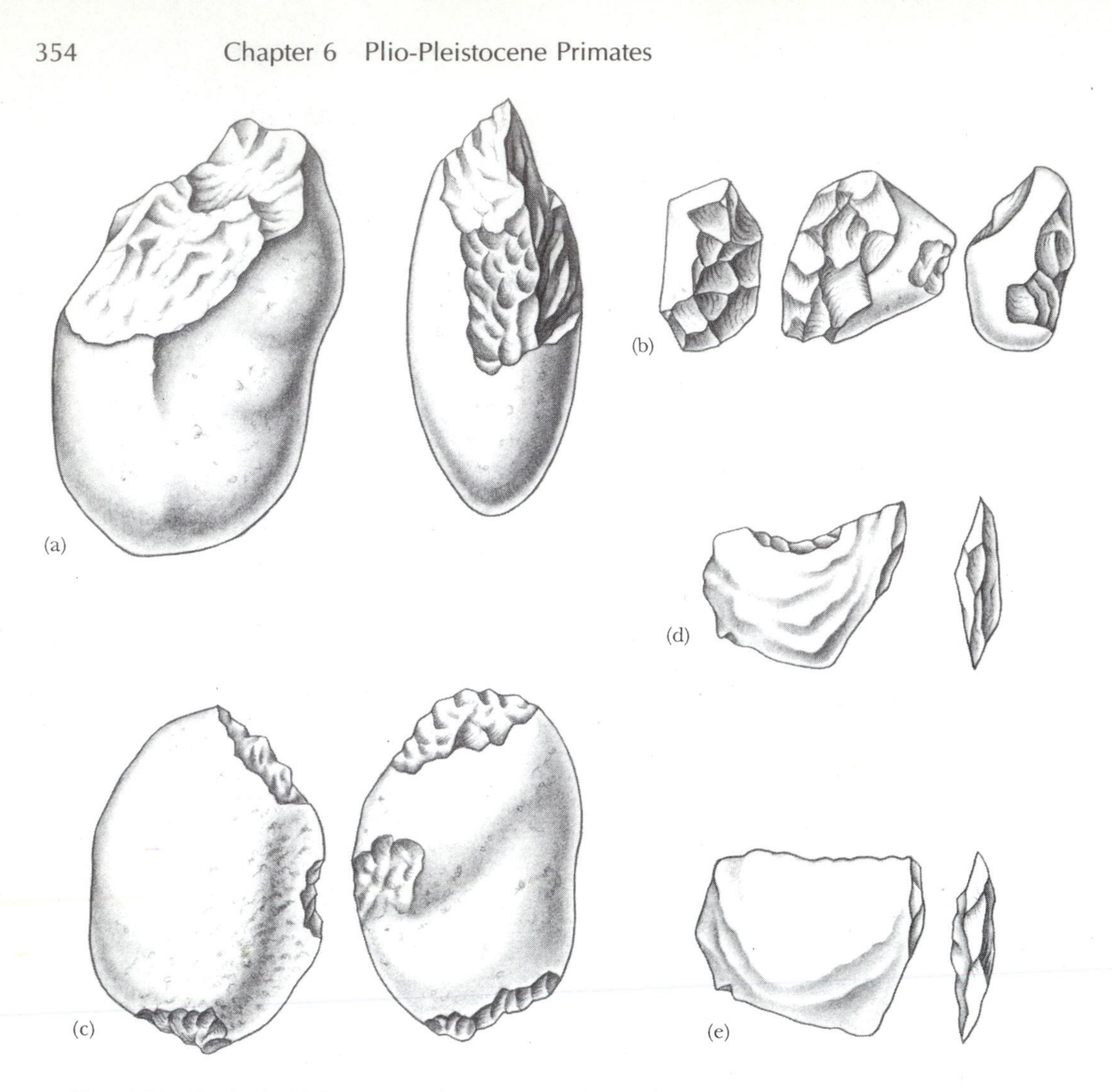

Fig. 6.38 Typical Oldowan Artifacts from Olduvai Gorge (a) Chopper; (b) polyhedron; (c) hammerstone; (d) utilized flake (stone fragment removed from a larger core by percussion or pressure); (e) debitage (unmodified or waste) flake. Multiple views of each tool are shown. (Redrawn from R. Potts in Tattersall et al., 1988)

All the sites in East and South Africa yielding Oldowan and Developed Oldowan stone tools also contain remains of *H. habilis* and/or *H. erectus*. However, one cannot definitely conclude that *Homo* was the sole maker of these tools, since australopithecines have also been found in association with these artifacts: *A. boisei* at Lake Turkana (Koobi Fora), Omo, and Olduvai (Beds I and II) and *A. robustus* at Swartkrans (Members 1 to 3) in South Africa.

The most widespread and longest-lived cultural tradition on record is the **Acheulian** industrial complex, a stone-tool tradition predominat-

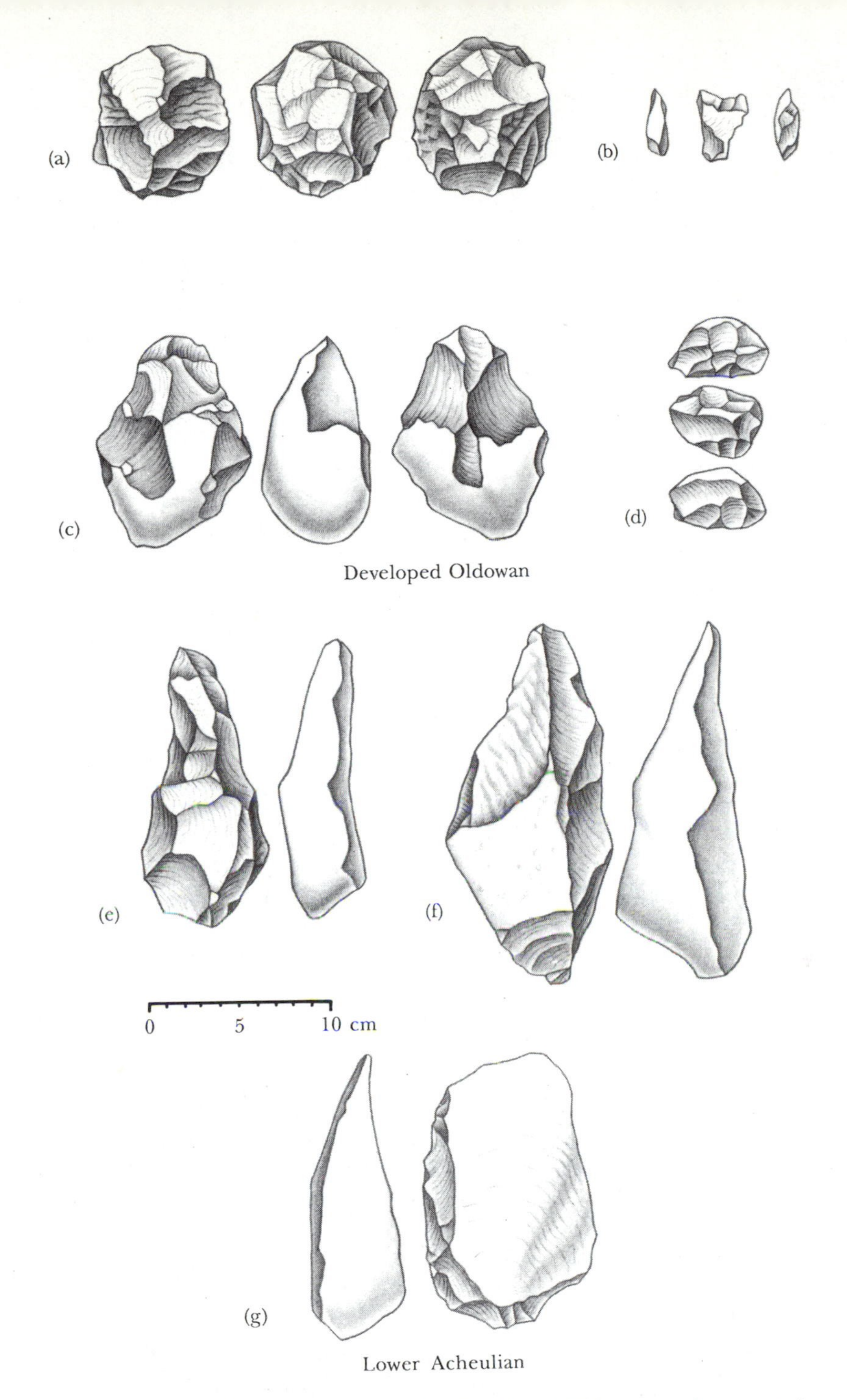

Fig. 6.39 Stone Tools Characteristic of the Developed Oldowan and Lower Acheulian Traditions of Africa *Developed Oldowan:* (a) Multifaceted polyhedral from North Africa; (b) small flake scraper from Olduvai; (c) core scraper from Olduvai; (d) proto-handaxe from Sterkfontein. *Lower Acheulian:* (e) handaxe from North Africa; (f) trihedral pick from North Africa; (g) cleaver from North Africa. Multiple views of each tool are shown. (Redrawn from Clark, 1970)

ing in late Early Stone Age. This tradition was named for St. Acheul, a site near Amiens, France, where artifacts of this style were first identified in the 19th century. Its earliest occurrence goes back at least 1.5 MYA (at Olduvai Gorge), and it continues in some areas (for example at Kalambo Falls in northern Zambia) to about 60,000 years ago (Clark, 1970). More sophisticated than the Oldowan and Developed Oldowan traditions, it is sometimes found concurrently with the latter (for example, from the middle of Bed II up through Beds III and IV at Olduvai). The Acheulian tradition is characterized by large cutting tools like true handaxes, **cleavers** (axelike stone implements with a sharp, somewhat straight cutting edge on one end), and **trihedrals** (three-sided picklike handaxes) (Fig. 6.39 e–g).

The Acheulian industry is often associated with *Homo erectus*. This association is sometimes difficult to confirm, however. For example, *H. erectus* is present at Lake Turkana from at least 1.6 to 1.2 MYA, but there is no trace of an Acheulian industry there. In South Africa some early Acheulian artifacts have been found at Sterkfontein (Member 5), which has yielded *H. habilis*.

Evidence from China indicates that *H. erectus* clearly had controlled mastery over fire by about 500,000 years ago. It has recently been reported, however, that fire was in use at least 1 MYA at Swartkrans. The remains consist of altered bones that laboratory analysis indicates were burned at temperatures consistent with campfires rather than natural conflagrations. These bones are found in Member 3 dated at about 1.5 to 1.0 MYA (Brain and Sillen, 1988). Although remains of *A. robustus* and *H. erectus* are found in Members 1 and 2, only *A. robustus* is so far known from Member 3; however, *Homo* is presumed to have been present as well. Tools of the Developed Oldowan tradition, usually associated with *Homo*, are found throughout the sequence. It is not clear, therefore, which hominid was using fire at Swartkrans.

The Middle Stone Age in Africa, generally considered to have lasted from about 100,000 to 20,000 years ago, is characterized by the use of carefully prepared stone cores for the production of flake tools. It is also the time in Africa when regional variation in stone-tool technologies became most evident. The Late Stone Age in Africa, which generally postdates 20,000 years ago, is characterized by a predominance of blade tools and microliths (tiny, geometrically shaped blades often set in handles made of bone or wood). The Middle and Late Stone ages are both associated exclusively with *H. sapiens*.

In summary, we can say that deliberate stone tool making had evolved by about 2.5 MYA but that the identity of the toolmakers is inconclusive. Stone tools have not been found in direct association with either *A. afarensis* at Laetoli or Hadar or with *A. africanus* at Sterkfontein or Taung. Although these two species were presumably using some perishable tools, we have no evidence that they were actually making stone tools. The earliest stone tools of the Oldowan and Developed Oldowan tradition have been found at a number of sites yielding *A. robustus* and/or *A. boisei* and *H. habilis,* including Koobi Fora, Omo, and Olduvai. *Homo habilis* is also associated with an early Acheulian industry in both East Africa (e.g., Olduvai) and South Africa (Sterkfontien, Member 5). The more developed Acheulian is generally associated with *H. erectus* and the Middle and Late Stone Age industries with *H. sapiens.*

PHYLOGENY AND CLASSIFICATION OF EARLY HOMINIDS

As we saw in Chapter 5, attempts to identify direct lineages between Miocene hominoids and present-day apes and hominids have not been successful. The reason for this is quite straightforward: there is simply no decent fossil evidence for hominoid evolution in Africa during the critical time period of about 12 to 5 MYA. Presumably this is when various African Miocene hominoids were differentiating into the African great apes and early hominids.

Although no likely Miocene candidates for direct ancestry of the first hominids can be advanced, a number of plausible scenarios for relationships among the various hominid groups have been proposed, and these we turn to next.

Phylogenetic Relationships within *Australopithecus*

Throughout the chapter we have recognized four australopithecine species, grouping them into gracile *(A. afarensis* and *A. africanus)* and robust *(A. robustus* and *A. boisei)* forms. There has been much debate over how many separate species these fossils truly represent and over what their phylogenetic and ecological relationships are to one another and to *Homo* (Fig. 6.40). The significance of the robust–gracile distinction and its role in hominid evolution has also been controversial. All of these issues will be examined in this section.

The relationship between the two oldest australopithecines—*A. afarensis* and *A. africanus*—is anything but clear. Are they geographic variants of the same species (Tobias, 1980), or is *A. afarensis* the more primitive of the two (White et al., 1981)? Owing to their antiquity, *A. afarensis* and *A. africanus* not surprisingly, share several primitive craniofacial characteristics. Some have argued that *A. africanus* displays a derived morphological composite superimposed on this primitive craniofacial plan, making it an excellent structural and phylogenetic intermediate between *A. afarensis, A. robustus,* and *A. boisei* (Johanson and White, 1979; White et al., 1981; Rak, 1985). These authors would exclude *A. africanus* from direct ancestry of *Homo habilis* by the apparent fact that it is specialized in the

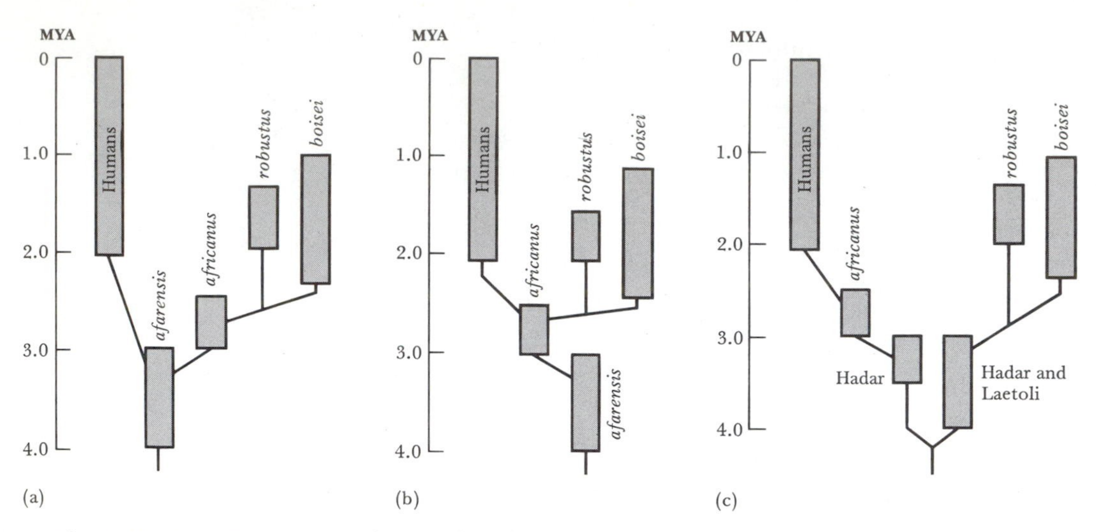

Fig. 6.40 Five Recent Hypotheses of Evolutionary Relationships among the Australopithecines (a) *A. afarensis* represents the stem hominid from which both the *Homo* and robust australopithecine lineages diverged; here *A. africanus* is part of the robust lineage. (b) *A. afarensis* represents the stem hominid from which *A. africanus* evolved; here *A. africanus* represents the last common ancestor of the *Homo* and robust lineages. (c) The robust lineage diverged very early from the *Homo* line, and the ancestors of each lineage can be found in the Hadar and Laetoli specimens. (d) *A. afarensis* represents the common ancestor from which the East African robust lineage stemmed, together with another lineage that led ultimately to the South African robust form and to *Homo* through a common ancestor *(A. africanus)*. (e) As in hypothesis c, the robust taxa from South and East Africa represent a single evolutionary branch; however, only one species *(A. afarensis)* is recognized in the earlier fossils from Hadar and Laetoli, and *A. afarensis* is the last common ancestor of both the robust australopithecine line and the lineage leading to humans. Unlike hypothesis d, *A. africanus* here represents a direct forbear of *Homo*. (Adapted from F. Grine in Tattersall et al., 1988)

direction of robust australopithecines (Fig. 6.40a). Such specializations revolve around the stronger molarization of its premolars, increased relative size of its postcanine dentition, and increased buttressing and robustness of its mandibular body. Other features are all closely related to the increased size of the posterior dentition. Larger postcanine teeth are functionally associated with greater mandibular buttressing, more strongly developed muscle attachment areas on the cranium, a shift forward of the center of action of the temporalis muscle, and midfacial buttressing (McHenry, 1985).

McHenry (1985) has also pointed out that although *A. afarensis* is remarkably primitive in many features, *A. africanus, A. robustus,* and *A. boisei* more closely resemble *H. habilis* in many of the same traits. In fact, *H. habilis* and the later species of *Australopithecus* share an extensive suite of derived traits not present in *A. afarensis*. This implies that the imme-

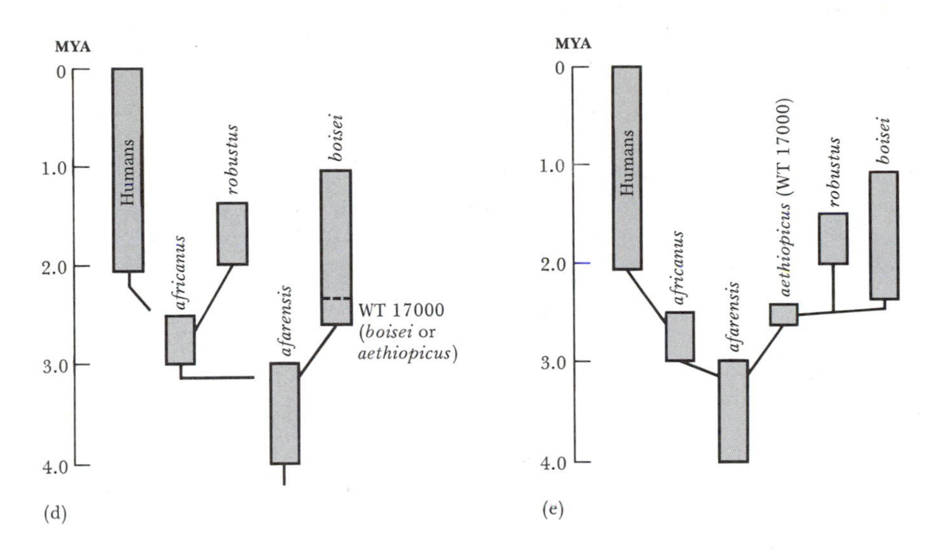

diate ancestor of *H. habilis* also probably shared these traits. If *A. afarensis,* rather than *A. africanus,* is the immediate ancestor of *H. habilis,* then a great deal of parallel evolution would have to be postulated. From this point of view it is clearly more parsimonious to assume that the hominid species immediately predating the appearance of *Homo, A. africanus,* is also the immediate common ancestor not only of *H. habilis* but also of *A. robustus* and *A. boisei* (Fig. 6.40b).

What is the relationship between *A. africanus* and *A. robustus?* Both are found in South African sites. Robinson (1963), as we have seen, viewed these two australopithecines as occupying separate ecological niches based on dietary preference, the former being omnivorous and the latter vegetarian. Presumably he considered the dating of the fossils to be indeterminate enough that the two species could have overlapped in time as well as geographic area; however, most proposals for dating the South African sediments place the earliest *A. robustus* specimen at least a half million years after the last *A. africanus* specimen. In any case, Robinson considered these two groups distinct enough to assign them separate genera, renaming the gracile hominid *H. africanus* because of its similarity to other *Homo* species.

Some have advocated an ancestor–descendant relationship between *A. africanus* and *A. robustus,* with the latter then giving rise to *A. boisei.* This view is based in large part on an analysis by Yoel Rak (1985) of Tel Aviv University of the structure and function of the australopithecine face. He has followed the evolutionary scheme mentioned shown in Fig. 6.40a in which the origins of the robust clade are in *A. africanus*, which is thereby removed from consideration as a human ancestor. To those who see *A. africanus–A. robustus–A. boisei* as a single evolving lineage, the beginnings of specialization characterizing the robust australopithecines are already manifest in almost every aspect of the masticatory system of *A. africanus*. Evidence for this viewpoint includes the presence of anterior pillars in both forms (Fig. 6.41). This buttressing is viewed as a structural response to the greater occlusal load arising from the beginning stages of molarization of the premolars. In *A. africanus* this molarization process was just beginning, but the still considerable protrusion of the palate relative to the more peripheral facial frame increased the need for such incipient pillars. The anterior pillars and the advancement of the inferior part of the infraorbital region (where the masseter muscle originated) played a major role in molding the facial topography of *A. africanus*. The presence of such pillars in *A. africanus* is considered evidence linking it to the robust australopithecine clade. The absence of the pillars in *A. afarensis* has led Rak (1985) to conclude that its face was the most primitive among the australopithecines.

This scenario was gaining widespread acceptance before the recent discovery of the 2.5 million-year-old. *A. boisei* skull (specimen WT 17000) from west of Lake Turkana (Walker et al., 1986). This specimen shows that the *A. boisei* lineage was established prior to the well-dated *A. robustus* specimens from East and South Africa and that, in robustness and tooth size, at least some members of the early *A. boisei* population were as large or larger than the later ones. Previously, most workers have suggested that within the robust australopithecine lineage there was a trend toward increasing size and robustness of the skull and jaws. This view is no longer tenable. Two possible phylogenetic schemes involving this new specimen are shown in Fig. 6.40d,e.

Australopithecus robustus shares with younger examples of *A. boisei* several features that are apparently derived from the condition seen in specimen WT 17000:

1. The cresting pattern with emphasis on the anterior and middle parts of the temporalis muscle
2. The orthognathism
3. The deep **temporomandibular joint** (where the lower jaw articulates with the skull) with a strong **articular eminence** (a bony projection anterior to the mandibular fossa on which the mandibular condyle slides as the jaw opens)

Thus this new specimen shows that *A. robustus* is a related, smaller species that was derived from ancestral forms earlier than 2.5 MYA and/or had evolved independently in southern Africa, perhaps from *A. africanus*.

Although the new *A. boisei* specimen shows characteristics of robustness and tooth size usually thought to be typical of later "advanced" robust australopithecines, it also has a number of craniofacial features that are more primitive than those in *A. africanus* and similar to those in *A. afarensis*. Some of the primitive features it shares with *A. afarensis* include the following:

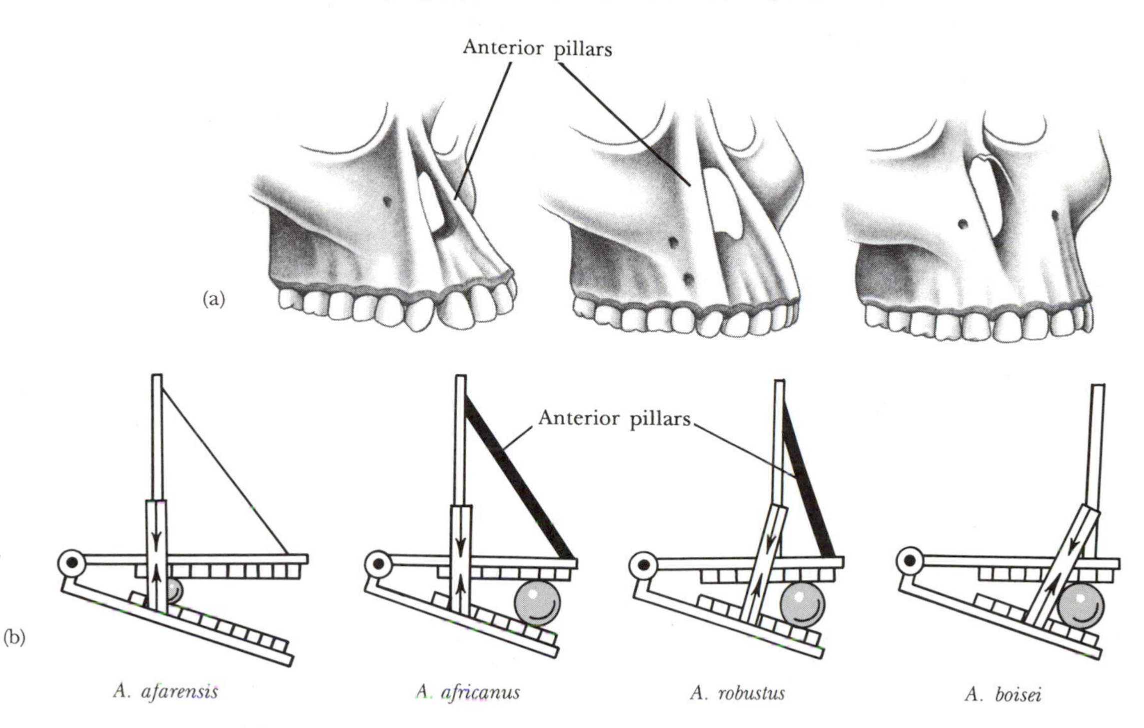

Fig. 6.41 Facial Buttressing in Australopithecines (a) Three-quarter view of the facial region in *A. africanus*, *A. robustus*, and *A. boisei*. Note the increased extent of the bony buttressing along the lateral side of the nasal cavity; in both *A. africanus* and *A. robustus* this buttressing takes the form of distinct anterior pillars. (b) Diagrams of the mechanical forces requiring facial buttressing. The vertical columns represent the infraorbital region, and the opposing arrows the contracting masseter muscle. When the occlusal load (food to be crushed by the molars and represented by the circles) not only increases but is also extended forward from the infraorbital region, additional buttressing in the form of anterior pillars becomes necessary. In *A. boisei* the whole infraorbital region has moved forward sufficiently to replace the anterior pillars functionally. (a, Redrawn from Rak, 1983; b, from Rak, 1985)

1. Strong upper facial prognathism
2. A flat cranial base
3. A flat temporomandibular fossa
4. Extensive pneumatization in the temporal bone
5. A large anterior tooth row
6. A flat, shallow palate
7. A maxillary dental arch that converges posteriorly (Kimbel et al., 1988)

In spite of these primitive features, WT 17000 seems clearly to be a member of the *A. boisei* clade. Might it be possible that the Hadar sample (from which *A. afarensis* comes) actually consists of two species, one of which gave rise to *A. boisei*? Todd Olson (1985*a*, 1985*b*) of the City University of New York has identified features of the nasal region that suggest that at least one of the juvenile crania from Hadar has features associated with robust australopithecines, and Dean Falk, of SUNY at Albany, and I (1983) have shown that the pattern of cranial venous circulation in some Hadar crania leads to the same conclusion. The phylogenetic implications of this conclusion can be seen in Fig. 6.40c. Both of these lines of evidence have been recently challenged, however (Kimbel, 1984; Eckhardt, 1987).

In analyzing the differences between gracile and robust australopithecines, two related questions come to mind that return us to principles introduced in Chapter 1. First, how much intra- and interspecific variation should we expect to find in the two forms? Second, are the differences between gracile and robust forms the products of real functional and biological differences, or are they merely the differences one would predict in two similar animals of different body size? In other words, are robust australopithecines merely "blown-up" versions of gracile australopithecines, or are they morphologically, behaviorally, and/or ecologically different, as Robinson suggested? These are critical questions too often ignored in taxonomic arguments about early hominids.

As mentioned earlier, Robinson believes the two forms to be distinct at the generic level and has developed a dietary hypothesis to explain these differences. But how distinct in quantitative terms are they? How does their morphological variability stack up against known variability in living hominoids? In postcanine dental areas the gracile species are about 88% the size of the robust species. Is this a great difference? In terms of both absolute and percentage differences, gracile and robust australopithecines are more similar in this dimension than many modern human populations are. According to Robinson's dietary hypothesis, gracile australopithecines are supposed to have had reduced posterior dentitions, because of their presumably more omnivorous diet. However, gracile australopithecine cheek teeth are not really as small as Robinson implies. After all, their size lies completely within the range of variation in modern gorillas, in spite of the fact that gracile australopithecine body size was only a fraction of that of gorillas (Wolpoff, 1971).

As noted earlier, another way to look at cheek-tooth size is to ask the question, What is the relative size of the cheek teeth in *Australopithecus* in comparison to body size? It turns out that all australopithecine species are megadont: as mentioned earlier, *A. afarensis* has a cheek-tooth size about 1.7 times larger than expected for modern hominoids of equivalent body size, and *A. africanus* and *A. robustus* both have more than twice the expected cheek-tooth size. The sequence *A. afarensis–A. africanus–A. robustus–A. boisei* shows strong positive allometry indicating increasing megadontia through time (McHenry, 1984, 1985).

The question of whether gracile species and robust species were merely allometric variants of one another was directly addressed by Pilbeam and Gould (1974). They argued that the three species, *A. africanus, A. robus-*

tus, and *A. boisei,* simply represented the "same" animal expressed over a wide range of body size. "In other words, size increase may be the only independent adaptation of these animals, changes in shape simply preserving the function of the smaller prototype at larger sizes." They came to this conclusion after examining the relationship between brain size and body size and between tooth size and body size in the three species. Intraspecific plots of brain size versus body size have been carried out for numerous birds and mammals. (Here "intraspecific" includes not only individual adults or races within a species but also very closely related species displaying the "same" body plan over a wide size range.) The slopes of these plots range from 0.2 to 0.4. The slope for the three australopithecine species calculated by Gould and Pilbeam is about 0.33, well within this intraspecific range.

When tooth areas are plotted against body weights, a slope of ⅔ is the expected value of geometric scaling for constant proportions throughout the size range. Any value greater than ⅔ indicates that large animals have relatively larger postcanine teeth than smaller members of the series. The value calculated by Pilbeam and Gould (1974) for *A. africanus, A. robustus,* and *A. boisei* is 0.7 and for the sequence pygmy chimpanzee–chimpanzee–gorilla, 0.85. This would confirm that in both these groups the larger species has relatively larger cheek teeth. Based on similar data from other mammals, including rodents, pigs, deer, and monkeys, however, they concluded that such positive allometry would be expected in sequences of related mammals that vary in size but not in basic design. Perhaps cheek-tooth size scales more to metabolic rate than to geometric similarity.

Phylogenetic Relationships within *Homo*

With the last of the australopithecines, we are on the brink of humanity. As we have seen, it is unclear which if any of the currently recognized species of *Australopithecus* gave rise to the earliest known species of *Homo (H. habilis),* although it was most probably a gracile form (see Fig. 6.40). Within the genus, however, the three species of *Homo* appear to bear an approximate ancestor–descendant relationship to one another, with one species grading morphologically and chronologically into the next: *H. habilis–H. erectus–H. sapiens.*

It is difficult to decide at what point we became truly human. Many of the biological and environmental forces that helped shape the genetic makeup of modern humans were operating by at least the Pliocene. Whatever the actual selection pressures were for human traits—bipedal locomotion, reduced canine teeth, enlarged brains, parental care of the young, home bases, food sharing, verbal communication skills, savanna-adapted diets, cooperative hunting and scavenging, tool use, opposable thumbs, and division of labor, among others—they were certainly in evidence by the mid-Pleistocene in *H. erectus*. By that time *H. erectus* had emerged from the continent of Africa to populate much of the Old World.

It is easy to assume that the more recent the time period under examination, the more confident paleontologists necessarily can be about the lines of descent of particular taxa. It may therefore surprise the reader to learn that the source, timing, and area of differentiation—in short, the origin—of our own species, *H. sapiens,* is still largely unknown. However, we can probably safely say that *H. erectus* is the direct ancestor of *H. sapiens*. Most of the differences between the two species are confined to the skull and teeth; below the neck the two species are morphologically very similar.

There is no unanimity of opinion on the timing of the *H. erectus–H. sapiens* transition, but it is probably fair to say that it took place sometime in the middle Pleistocene, most likely less than 500,000 years ago. Because there is morphological continuity between the two species, the dividing line is somewhat arbitrary.

Presently there are two competing models to explain the evolution of modern *H. sapiens:* the multiregional model and the single-origin model. In the **multiregional model,** (also called the regional continuity model) recent human variation is seen as the product of the early-to-middle Pleistocene radiation of *H. erectus* from Africa. These regional populations gradually evolved into the modern populations now seen in the different parts of the world. Obviously, gene flow between these regional populations must have been such as to maintain overall grade similarities (that is, prevent further speciation) yet at the same time allow regional morphological characteristics to persist. According to this theory *H. sapiens* appeared as the result of continual long-term evolutionary trends within the genus *Homo*.

The **single-origin model** (also called the "Noah's Ark" or "Garden of Eden" hypothesis) is diametrically opposed to the multiregional model. It assumes that there was a relatively recent common ancestral population of *H. sapiens* in Africa that had already evolved most of the anatomical characters typical of modern humans. This population regionally differentiated in Africa and then spread out of Africa to replace all other populations of *Homo* in the Old World. Based on data derived from mitochondrial DNA studies of modern human populations, some geneticists have concluded that the mitochondrial DNA of all modern humans could be derived from a single African ancestor living approximately 140,000 to 290,000 years ago (Cann et al., 1987).

Each of the two models leads to different predictions concerning the evolution of *Homo sapiens* (Table 6.3). At present, each model has its defenders (Stringer and Andrews, 1988; Wolpoff, in press). One thing seems clear, however. Anatomically modern *H. sapiens* must have appeared long before 100,000 years ago, since their remains are known from deposits of approximately that age from such geographically diverse sites as the southern tip of Africa and the Near East (Stringer et al., 1989).

Table 6.3 Theoretical Predictions from Models of *Homo sapiens* Evolution

Aspect	*Multiregional Model*	*Single-Origin Model*
1. Geographic patterning of human evolution	Continuity of pattern from middle Pleistocene to present	Continuity of pattern only from late Pleistocene appearance of *H. sapiens* to present
	Interpopulation differences high; greatest between each peripheral area	Interpopulation differences relatively low; greatest between African and non-African populations
	Intrapopulation variation greatest at center of human range	Intrapopulation variation greatest in African populations
2. Regional continuity and the establishment of *H. sapiens*	Transitional fossils widespread	Transitional fossils restricted to Africa, population replacement elsewhere
	Modern regional characters of high antiquity at peripheries	Modern regional characters of low antiquity at peripheries (except Africa)
	No consistent temporal pattern of appearance of *H. sapiens* characters between areas	Phased establishment of *H. sapiens* suite of characters: (i) Africa, (ii) southwestern Asia, (iii) other areas
3. Selective and behavioral factors involved in the origin of *H. sapiens*	Factors varied and widespread, perhaps related to technology; local behavioral continuity expected	Factors special and localized in Africa; behavioral discontinuities expected outside Africa

Source: Stringer and Andrews (1988).

Aside from the difficulty of determining when we became anatomically human, there is the issue of when we became human in the larger sense of our culture and behavior, a question we leave to other texts to explore in detail. On a lighter note, one wag has observed that the current ubiquitous bipedal primate we call *Homo sapiens* is simply the missing link between apes and humanity! Perhaps the first truly human act, at least in a symbolic sense, was when some bereaved Neanderthal sprinkled flowers over a grave at Shanidar, Iraq, some 50,000 years ago.

EPILOG

In the preceding six chapters we have reviewed nearly 70 million years of primate history focusing on the major paleontological evidence and the inferences drawn from it. It seems fitting now to look back over this span of time, especially the latest human era, to ask a few more probing philosophical questions.

Although 70 million years might seem like an immense time span, in terms of geological time it is just a brief moment. This humbling fact calls to mind Thoreau's observation that "time is but a stream I go fishin' in." In effect, the primate fossil record consists of a few fish out of the millions that have swum by undetected over the millennia. We know so much, yet there is still so much to know.

Our own genus, *Homo,* has existed for only about 2 million years, yet it is often said that we are the most successful primate. On what is this anthropomorphic judgment based? There is no question that *H. sapiens* is dominant for the present; the sudden extinction of all other primates and most other organisms is well within our collective grasp to accomplish or prevent. *Homo sapiens* has shaped its environment more than any other primate, but only time will tell if our influence constitutes a positive environmental impact.

What other grounds do we have at this time for saying that our species is an evolutionary success? As we have seen, many obscure plesiadapiforms, adapids, and omomyids survived far longer. Even the small-brained dinosaur, often touted as the epitome of lumbering stupidity, survived tens of millions of years longer than our own genus has. We would be hard pressed to calculate the odds that a Human Age will continue another three generations, much less another 100 million years to rival the Age of Dinosaurs.

Those who see *Homo sapiens* rightfully occupying the pinnacle of primate evolution point to our relatively large brain as the *sine qua non* of human success. Whether the consequences of bearing and employing this hypertrophied mass of tissue ultimately redeems that view or proves us as inviable a species as the dodo is yet another fascinating but unresolvable question. With our perspective of 70 million years of primate evolution behind us, we are prepared to ask it with special understanding.

APPENDIX A

Geological Time Scale

The time scales used in paleoprimatology all have their origin, directly or indirectly, with sedimentary geology. The basic starting point for understanding the stratigraphy of any region in the world is the identification of its **rock units,** such as formations, groups, and members. **(Groups** are subdivided into **formations,** which are subdivided into **members,** and these are further subdivided into **beds.)**

Rock units carry with them no connotation of time and exist only as three-dimensional bodies of rock. As attempts are made to correlate particular rock units with those of other geographical areas, two problems emerge. First, it may be impossible to physically correlate the two rock units because of the outcrop pattern of the rock in the intervening areas. For example, the formation in question may have eroded away in this area or become submerged beneath the ground surface. Second, rock units are often **diachronous** (formed at different times) over large geographical areas. For example, if a sandstone deposit originated as a beach that developed in a westward direction across a shallow marine environment, the eastern portion of the sandstone unit would be younger than the western portion (Matthews, 1974). Thus in order to study large areas of earth's history we must closely consider the dimension of time.

By the early 19th century William Smith in England had discovered the Law of Faunal Succession from which he showed that succeeding rock units could be identified by their distinctive fossils. These major sedimentary units, each representing an important segment of geological time, became known as the **geological systems.** Each system, at least in its **type section** (where it was first named and described), was deposited during a discrete portion of geological time and is represented by a considerable thickness of rock. Any rock anywhere in the world formed during the same time range belongs to the same geological system. The three systems covered by this text are the Paleogene, the Neogene, and the Quaternary. Rocks formed during a specified interval of time are

Paleoenvironmental Changes

Neogene

Age	Interval	Paleoenvironmental Changes
Recent–1.8 MYA	Quaternary	Further increase in upwelling and biogenic productivity at Antarctic Convergence; increase in ice rafting
2.5–3 MYA	Late Pliocene	Continued global cooling; development of Northern Hemisphere ice caps
3.5 MYA	Middle Pliocene	First major glaciation in South America
4–6 MYA	Late Miocene–early Pliocene	Expansion of Antarctic ice cap; global cooling; northward expansion of Antarctic surface water; major increase in upwelling and biogenic productivity; increase in ice rafting
12–14 MYA	Middle Miocene	Development of major ice cap on east Antarctica; increased ice rafting; drop in surface-water temperatures
22 MYA	Early Miocene	Initial development of Antarctic Convergence—but low degree of upwelling and biogenic productivity; clear differences between Antarctic and sub-Antarctic microfossil assemblages; increased sub-Antarctic surface-water temperatures to 10°C

Paleogene

Age	Interval	Paleoenvironmental Changes
22–25 MYA	Late Oligocene	Development of unrestricted Circum-Antarctic Current
22–38 MYA	Oligocene	Prolonged Antarctic glaciation—no ice cap; sub-Antarctic surface-water temperatures of 7°C; cool temperate vegetation disappearing; increase in ice rafting

Time (MYA): 5, 10, 15, 20, 25, 30, 35

Paleomagnetic Polarity: *

Polarity Event (Subchron): Jaramillo, Olduvai, Kaena, Cochiti, C_1, C_2, Nunivak, Mammoth

Polarity Epoch (Chron): Brunhes, Matuyama, Gauss, Gilbert

Series / Epoch		*System / Period*		*Era*
PLEISTOCENE	U, L; L, E	QUATERNARY		CENOZOIC
PLIOCENE	U, L; L, E	NEOGENE	TERTIARY	
MIOCENE	Upper, Late; Middle; Lower, Early			
OLIGOCENE	Upper, Late; Lower, Early	PALEOGENE		

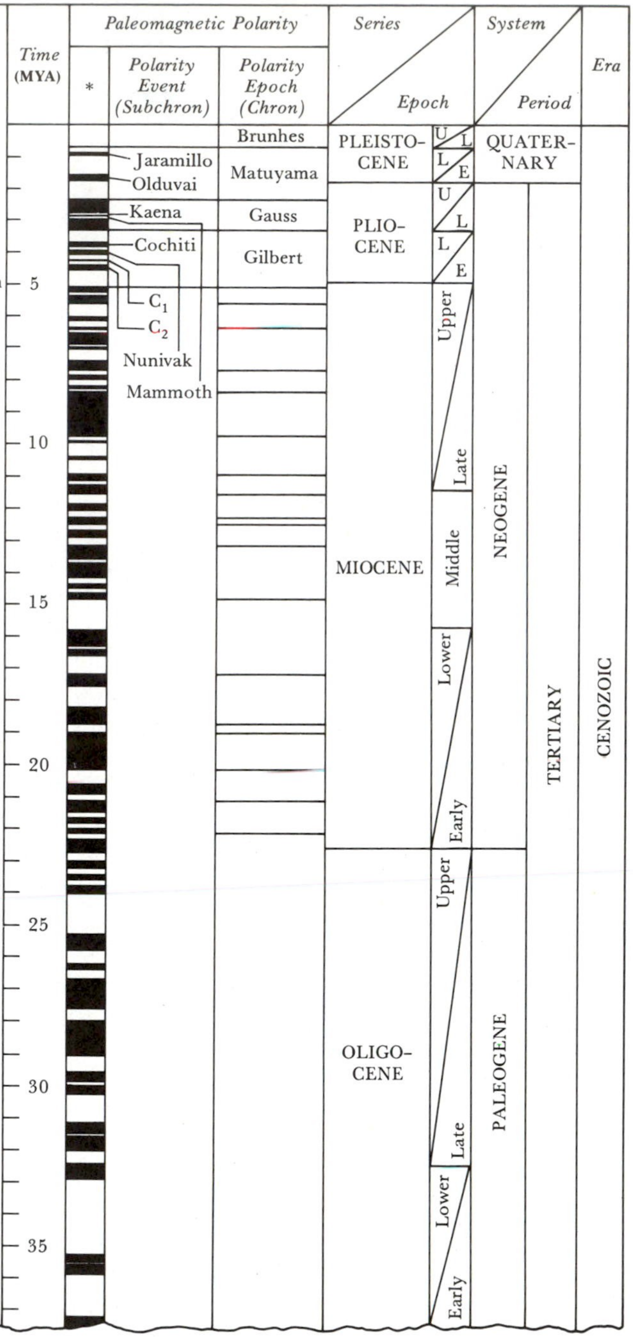

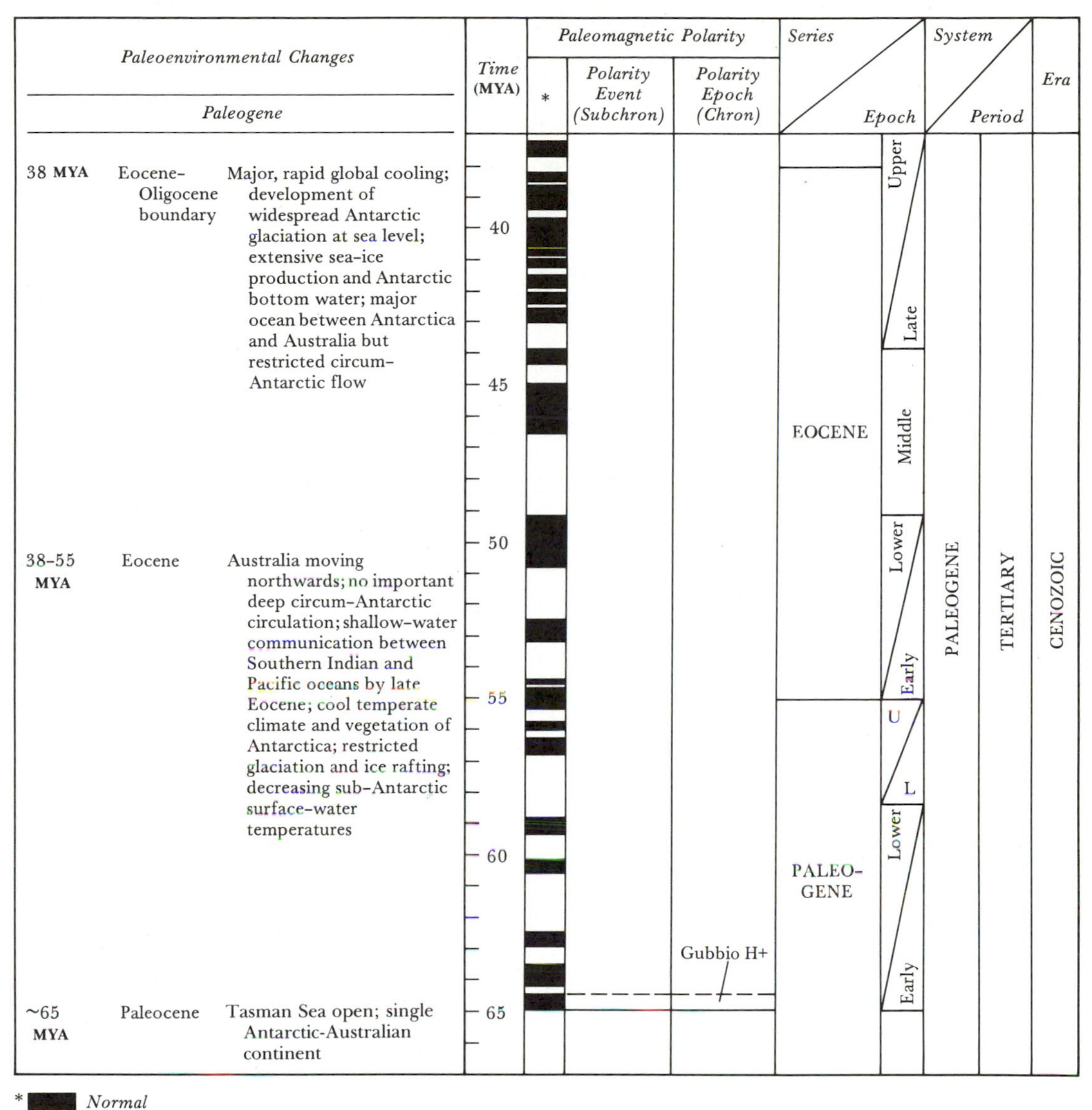

Paleoenvironmental Changes
Paleogene
38 MYA
Eocene-Oligocene boundary
Major, rapid global cooling; development of widespread Antarctic glaciation at sea level; extensive sea-ice production and Antarctic bottom water; major ocean between Antarctica and Australia but restricted circum-Antarctic flow
38–55 MYA
Eocene
Australia moving northwards; no important deep circum-Antarctic circulation; shallow-water communication between Southern Indian and Pacific oceans by late Eocene; cool temperate climate and vegetation of Antarctica; restricted glaciation and ice rafting; decreasing sub-Antarctic surface-water temperatures
~65 MYA
Paleocene
Tasman Sea open; single Antarctic-Australian continent
Time (MYA)
40
45
50
55
60
65
Paleomagnetic Polarity
*
Polarity Event (Subchron)
Polarity Epoch (Chron)
Gubbio H+
Series
Epoch
EOCENE
PALEO-GENE
Upper
Late
Middle
Lower
Early
U
L
Lower
Early
System
Period
PALEOGENE
TERTIARY
Era
CENOZOIC
* Normal
Reversed

Fig. A.1 Correlation of Time-Stratigraphic Units, Geological Time Units, Magnetic Reversals, and Environmental Changes in the Paleogene and Neogene (Adapted from LaBrecque et al., 1977, and Kennett, 1978)

MYA	*Epoch*	*Subepoch*	*Standard Stage*	*European Land-Mammal Age*	*N. American Land-Mammal Age*
0	PLEISTOCENE	Late	Flandrian		Rancholabrean
			Tyrrhenian		
0.25		Middle	Sicilian	Oldenburgian	
0.5				Biharian (Cromerian)	
0.75					
1.0		Early	Calabrian	(Villanyian) Villafranchian (Csarnotian)	Irvingtonian
1.5	PLIOCENE	Late	Piacenzian		
2					
4		Early	Zanclean		Blancan
				Ruscinian	
6	MIOCENE	Late	Messinian	Turolian (Pikermian)	Hemphillian
8			Tortonian		
10				Vallesian	Clarendonian
12		Middle	Serravallian	Astaracian (Vindobonian)	Barstovian
14					
16			Langhian	Orleanian (Burdigalian)	
18		Early	Burdigalian		Hemingfordian
20					
22			Aquitanian	Agenian (Aquitanian)	Arikareean
24					

called **time-stratigraphic (chronostratigraphic) units,** and the geological systems are the basic time-stratigraphic units of historical geology. Systems are subdivided into smaller time-stratigraphic units called **series** and **standard stages.**

A time-stratigraphic unit is independent of rock type or thickness. It cannot be recognized by rock type except in one place, the **type locality,** where it exists by definition. Everywhere else it must be recognized by time correlation. One difficulty in time correlation occurs when rocks

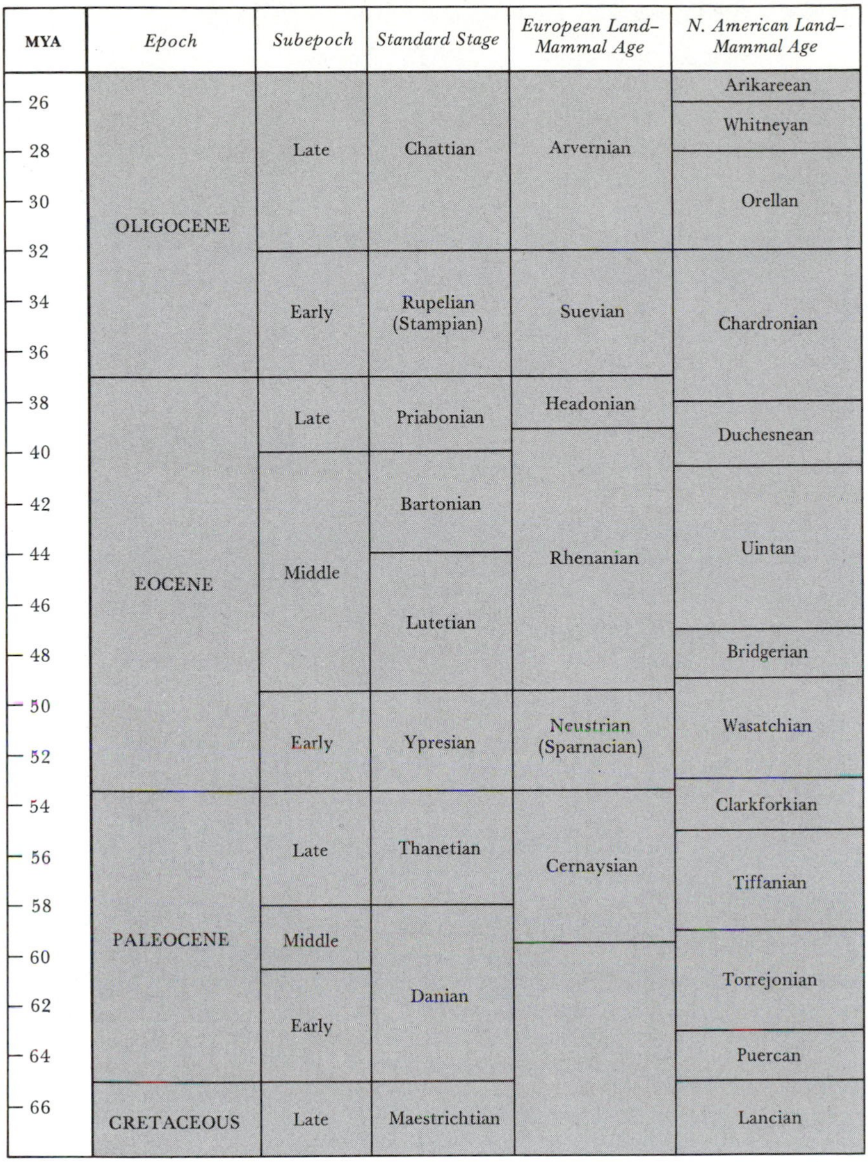

MYA	*Epoch*	*Subepoch*	*Standard Stage*	*European Land-Mammal Age*	*N. American Land-Mammal Age*
	OLIGOCENE	Late	Chattian	Arvernian	Arikareean
26					Whitneyan
28, 30					Orellan
32, 34, 36		Early	Rupelian (Stampian)	Suevian	Chardronian
	EOCENE	Late	Priabonian	Headonian	
38					Duchesnean
				Rhenanian	
40		Middle	Bartonian		
42					Uintan
44, 46			Lutetian		
48					Bridgerian
					Wasatchian
50, 52		Early	Ypresian	Neustrian (Sparnacian)	
					Clarkforkian
54	PALEOCENE	Late	Thanetian	Cernaysian	
56					Tiffanian
58		Middle	Danian		
					Torrejonian
60					
62		Early			
64					Puercan
66	CRETACEOUS	Late	Maestrichtian		Lancian

Fig. A.2 Correlation of Tertiary Epochs, Standard Stages, and Land-mammal Ages (Adapted from Szalay and Delson, 1979)

representative of the system are missing as a result of nondeposition or erosion. In order to have a continuous timeline that could be used consistently everywhere in the world, abstract geological time units were devised. Corresponding to the time-stratigraphic units of systems, series, and standard stages are, respectively, the geological time units of periods, epochs, and land-mammal ages (Fig. A.1).

The **epochs** of the **Tertiary period** were originally defined in the 1800s by Charles Lyell (Eocene, Miocene, and Pliocene), August von

Beyrich (Oligocene), and Wilhelm Schimper (Paleocene). They were generally defined according to the percentage of modern-day taxa (mostly invertebrates) they contained. For example, the Eocene was defined as containing 3% extant species, the Miocene 17%, and the Pliocene 50% to 67%.

A longstanding goal of paleontologists has been to describe criteria by which time intervals of the Cenozoic era could be recognized on the basis of fossil mammals. These **land-mammal ages** are intended to represent divisions of the Cenozioc era based on characteristic groups of fossil mammals whose temporal relationships and overall stage of evolution are indicative of a particular interval of geological time. Land-mammal ages are often defined on the basis of first and last occurrences of particular fossils or by characteristic faunal assemblages (Fig. A.2).

A major advance in geology has come from magnetostratigraphy. The earth's magnetic field has reversed its polarity numerous times over the course of its history, leaving evidence of polar direction in rocks at the time of their formation. Because reversals of the earth's magnetic field are geologically instantaneous, boundaries between reversals are considered to be **isochronous** (occurring at the same time). This results in a consistent pattern that can be used to calibrate the dating of rocks throughout the world. The more recent part of the Cenozoic magnetic time scale (back to about 5 MYA) is generally well calibrated by directly associated potassium–argon dates. However, the dating of much of the earlier time scale is based on assumptions about the rates of seafloor spreading since these determine the magnetic patterns of this time scale (Woodburne, 1987).

The record of reversals is divided into **polarity epochs** (or **chrons**), which correspond to specific time spans during which the earth's magnetic field had a characteristic normal or reversed polarity. The polarity epochs (chrons) of the last 5 million years have been given names (for example, the Gauss normal polarity epoch or the Matuyama reversed polarity epoch), while the earlier polarity epochs (chrons) have been assigned numbers (for example, magnetic polarity chron 5N, or 5 normal). Polarity epochs are further divided into **polarity events** (or **subchrons**), which correspond to short-term polarity reversals (Fig. A.1).

APPENDIX B

Basic Primate Skeletal Morphology

Basic osteological landmarks of the primate skeleton are illustrated in Fig. B.1 using the modern prosimians (sifaka), a modern anthropoid ape (chimpanzee), and human as examples.

CRANIAL MORPHOLOGY

The primate skull serves several important functions: it houses and protects the brain and special sense organs; it forms the anchoring structure for the dentition; and it provides the bony surface area for attachment of the muscles of mastication. To a great extent the overall shape of the skull in various primates reflects the relative degree of enlargement or specialization of these various functions. For example, primates that rely more heavily on olfaction (the sense of smell) tend to have longer snouts than those that rely mainly on vision. Likewise, those primates with well-developed muscles of mastication (chewing) often have accentuated bony crests on the skull that serve to increase the surface area for the muscle attachments and so withstand the increased stresses of these chewing muscles.

The eyes of all modern primates are protected by bone in two different ways. In the lower primates (prosimians) a **postorbital bar** (ring of bone) formed by portions of the **frontal** and **zygomatic** bones surrounds the orbit. In higher primates (anthropoids) the orbital contents are protected within a bony cup formed from a number of bones including the frontal, **maxilla,** zygomatic, **sphenoid, lacrimal, ethmoid,** and **palatine** bones. A skull with this morphology is said to show complete **postorbital closure.**

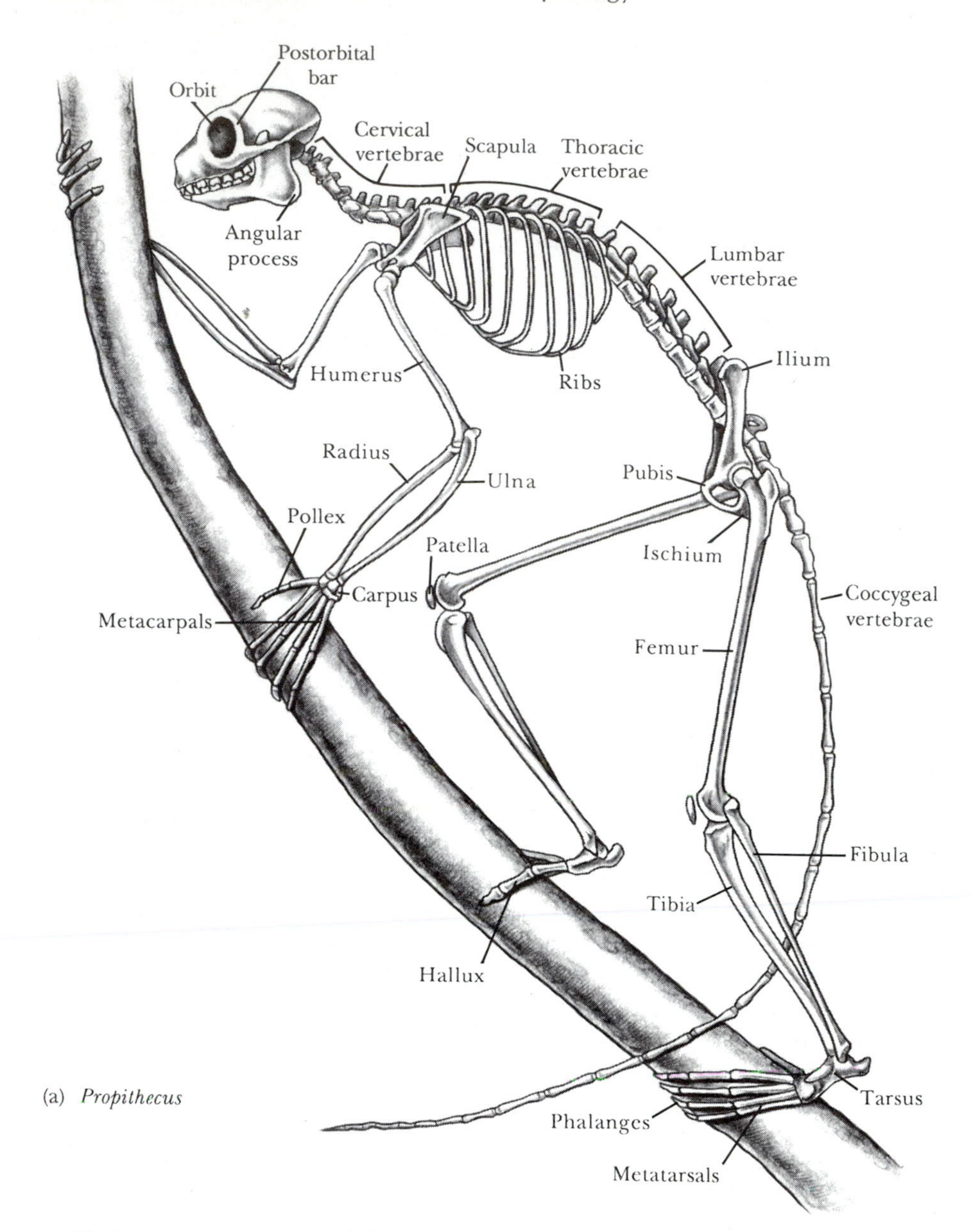

Fig. B.1 Major Features of the Modern Primate Skeleton Shown are a prosimian (a, the sifaka *Propithecus*) and two anthropoids (b, *Pan,* and c, *Homo*). Note in the human that the right hand is pronated (palm facing backward), whereas the left hand is supininated (palm facing forward). Note also that the three skeletons are not to scale. (Redrawn from: a, Gregory, 1920; b, Warwick and Williams, 1973; c, Le Gros Clark, 1959)

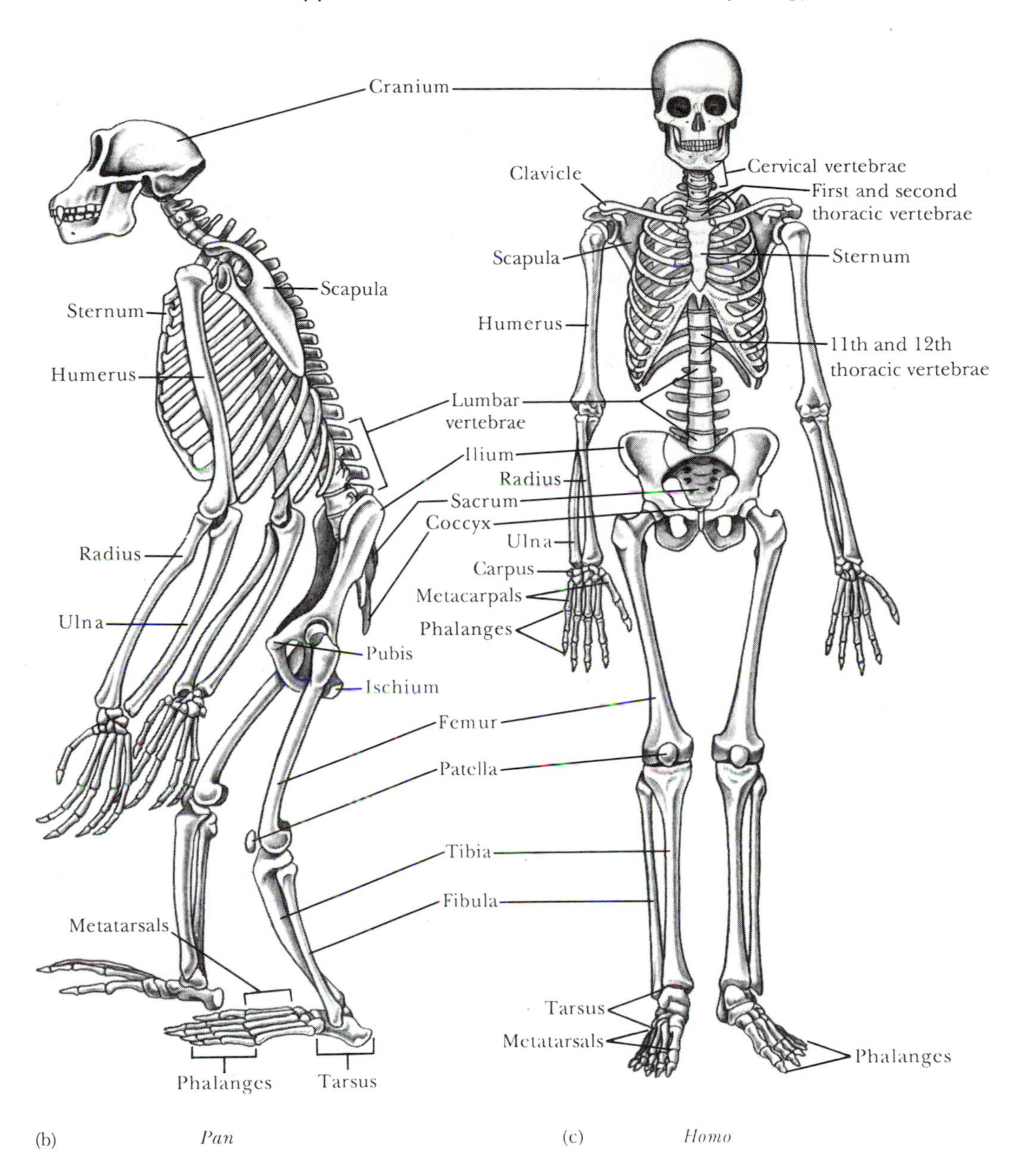
Cranium
Scapula
Sternum
Humerus
Radius
Ulna
Lumbar vertebrae
Ilium
Sacrum
Coccyx
Pubis
Ischium
Femur
Patella
Tibia
Fibula
Metatarsals
Phalanges
Tarsus
Clavicle
Cervical vertebrae
First and second thoracic vertebrae
Scapula
Sternum
Humerus
11th and 12th thoracic vertebrae
Radius
Ulna
Carpus
Metacarpals
Phalanges
Tarsus
Metatarsals
Phalanges

(b) *Pan* (c) *Homo*

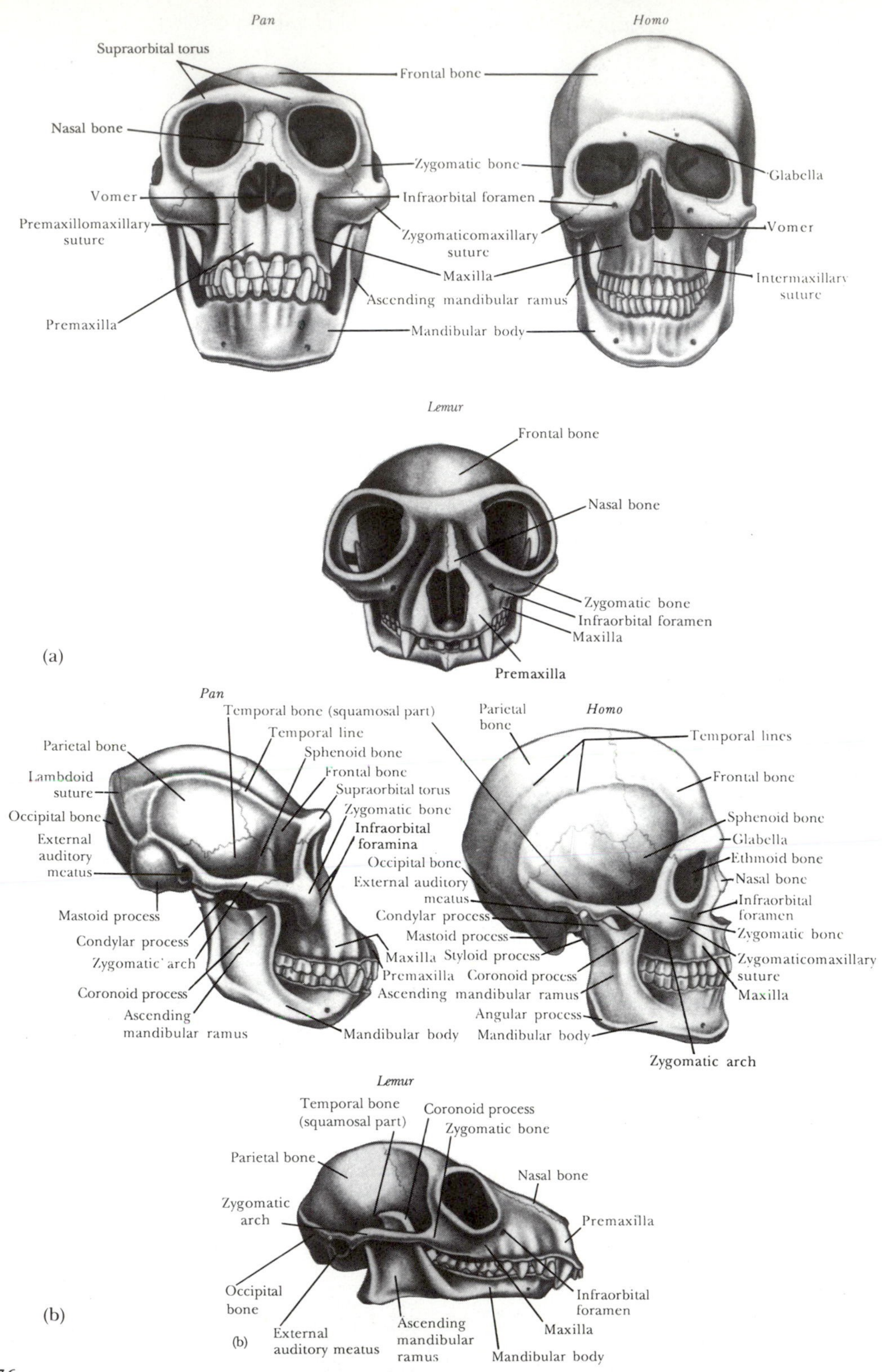
Pan
Homo
Supraorbital torus
Frontal bone
Nasal bone
Zygomatic bone
Glabella
Vomer
Infraorbital foramen
Premaxillomaxillary suture
Zygomaticomaxillary suture
Maxilla
Intermaxillary suture
Ascending mandibular ramus
Premaxilla
Mandibular body
Lemur
Frontal bone
Nasal bone
Zygomatic bone
Infraorbital foramen
Maxilla
Premaxilla
(a)
Pan
Temporal bone (squamosal part)
Temporal line
Parietal bone
Sphenoid bone
Lambdoid suture
Frontal bone
Supraorbital torus
Occipital bone
Zygomatic bone
External auditory meatus
Infraorbital foramina
Mastoid process
Condylar process
Zygomatic arch
Maxilla
Premaxilla
Coronoid process
Ascending mandibular ramus
Mandibular body
Homo
Parietal bone
Temporal lines
Frontal bone
Sphenoid bone
Glabella
Occipital bone
Ethmoid bone
External auditory meatus
Nasal bone
Condylar process
Infraorbital foramen
Mastoid process
Zygomatic bone
Styloid process
Zygomaticomaxillary suture
Coronoid process
Maxilla
Ascending mandibular ramus
Angular process
Mandibular body
Zygomatic arch
Lemur
Temporal bone (squamosal part)
Coronoid process
Zygomatic bone
Parietal bone
Nasal bone
Zygomatic arch
Premaxilla
Occipital bone
Infraorbital foramen
External auditory meatus
Ascending mandibular ramus
Maxilla
Mandibular body
(b)

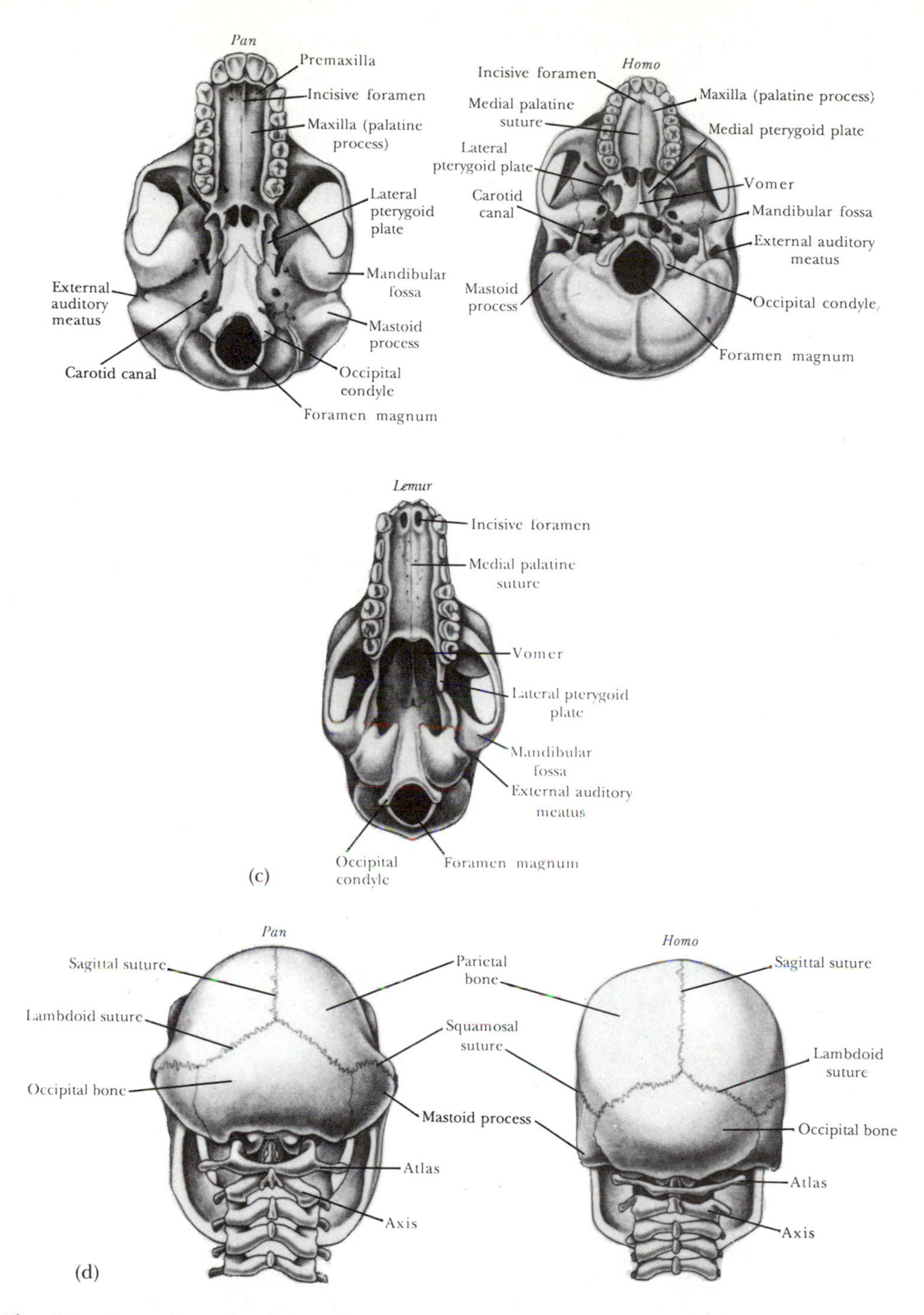

Fig. B.2 Cranial Landmarks in the Modern Primate Here prosimians are represented by *Lemur* and anthropoids by *Pan* and *Homo*. (a) Frontal, (b) lateral, (c) basal, and (d) posterior views. Note that the basal view excludes the lower jaw. Skulls are not to the same scale. (*Lemur* redrawn from Elliot, 1913; *Pan* and *Homo* redrawn from Swindler and Wood, 1973)

Posture sometimes affects important cranial landmarks. For example, there is a general tendency for the **foramen magnum** (the hole at the bottom of the skull for passage of the spinal cord) to face more directly downward, rather than backward, in primates that routinely hold their trunk more erect in both resting and locomotor postures. However, other factors may also affect the relative position of the foramen magnum: (1) the flexion of the cranial base so that the face comes to lie below rather than in front of the brain case; and (2) the relative enlargement of the occipital lobes (those at the back of the brain), which results from the expansion of the visual area of the cerebral cortex (Le Gros Clark, 1959).

The thin bones of the **calvaria** (the roof of the skull which protects the brain) are the frontal bone anteriorly, the occipital bone posteriorly, and the parietal, temporal, and sphenoid bones laterally. These bones are joined to one another by fibrous joints called **sutures.** All of the upper teeth are housed in the **maxilla** except for the incisors, which are housed in the **premaxilla.** All the lower teeth are contained in the **mandible** (lower jaw). In many lower primates the two halves of the mandible are joined at a midline suture called the **mandibular symphysis.**

There is a great deal of variation in the skull of different primates, particularly when we include fossil primates as well. The main osteological features of the primate skull are identified in Fig. B.2.

PRIMATE DENTITION

Anatomical orientation when referring to teeth is indicated by particular terminology. Figure B.3a is a lateral view of the jaw and teeth, while Figs. B.3b,c show occlusal views. In a crown or **occlusal view,** one is looking at the occlusal (chewing) surfaces of the teeth of the lower jaw from above or the upper-jaw teeth from underneath. The four directions in an occlusal view are **buccal** (facing laterally, or toward the cheek), **lingual** (facing medially, or toward the tongue), **mesial** (facing anteriorly, or toward the front of the mouth), and **distal** (facing posteriorly, or toward the back of the jaw).

A major characteristic of the primate dentition, as in all mammals, is its **heterodonty** (teeth that are regionally differentiated in form so as to serve special functions) (Table B.1). Thus on each side of the upper and lower jaws at the the mesial end are cutting and cropping teeth **(incisors),** then a pointed and projecting stabbing tooth **(canine),** and distal to this the **postcanine** dentition consisting of more complicated chewing teeth, called **premolars** and **molars.** It is generally accepted that primates evolved from primitive mammals having a **dental formula** of 3.1.4.3/

3.1.4.3: three incisors, one canine, four premolars, and three molars on each side of the upper and lower jaws (Fig. B.3a). This dental formula has been modified in the evolutionary history of many primate lineages by a reduction in the number of teeth. For example, no primate, living or fossil, has ever had more than two incisors on each side of the upper and lower jaws.

In most primates the incisors are relatively small, simple teeth that are somewhat spatulate in form. However, in some primate lineages they become greatly enlarged and/or specialized, as in the rodentlike incisors of plesiadapiforms and the tooth combs of lemurs. The upper incisors are implanted in the **premaxilla** bone.

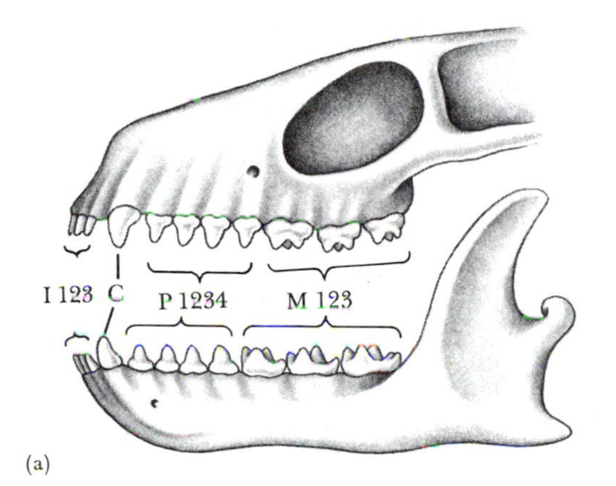

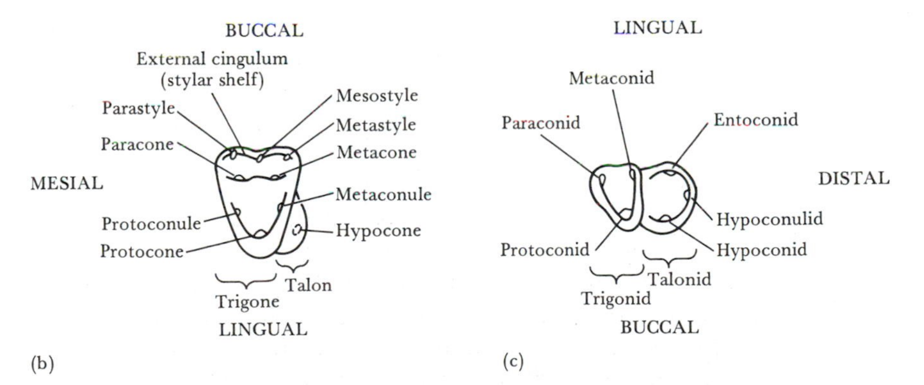

Fig. B.3 Primitive Eutherian Dentition (a) Jaws of a hypothetical primitive eutherian with a dental formula of 3.1.4.3/3.1.4.3; upper teeth are usually identified with superscripts for example, (M^1 = upper first molar) and lower teeth with subscripts (P_3 = lower third premolar). (b) Basic cusp pattern of the upper left molar. (c) Basic cusp pattern of the lower left molar. (a, Redrawn from Le Gros Clark, 1959; b,c, adapted from Simpson, 1937)

Table B.1 Characteristics of Primate Dentitions

	Lemurs and Lorises	*Tarsier*	*New World Monkeys*
Dental formula	2.1.3.3/2.1.3.3 except: (a) Indriidae 2.1.2.3/1.1.2.3 (b) *Daubentonia* 1.0.1.3/1.0.0.3	2.1.3.3/1.1.3.3	Cebidae 2.1.3.3/2.1.3.3 Callithrichidae 2.1.3.2/2.1.3.2
Shape of arch	V-shaped	V-shaped	Variable: V-shaped, parabolic, rectangular
	Median diastema between upper central incisors		Diastema between I^2
Incisors	Uppers: small subconical Lowers: comb (with canines)	Subconical	Spatulate ———
Canines	Uppers: large, projecting Lowers: incisiform (comb)	Pointed, caniniform	Large, caniniform Usually projecting. Some sexual dimorphism
Premolars	Lower first caniniform Conical with lingual cingulum	Last molariform	Bicuspid; lower first often sectorial ———
Molars	Uppers: tribosphenic[a] incipient hypocone or four-cusped Lowers: tribosphenic[a] with reduction or loss of paraconid	Tribosphenic[a]	Four-cusped Uppers: oblique ridge Lowers: paraconid lost

[a]Tribosphenic = triangular-shaped *Source:* Hiiemae (1981).

The canines are usually large, pointed teeth, although they have become reduced in some early primates and are incorporated into the dental comb in some lemurs. By definition, the upper canine is the first tooth immediately behind the suture between the premaxilla and the maxilla, and the lower canine is the tooth immediately in front of the upper canine when the upper and lower jaws are **occluded** (brought together).

The premolars are often of simple construction with either one or two main cusps. Those with two are called **bicuspid.** There is often a thickened ring of enamel around the base of the crown called the **cingulum.** In some primates, particularly Old World monkeys, the anterior premolar acts as a **honing** (sharpening) stone to sharpen the posterior edge of the upper canine every time the two teeth come into contact. Such as adaptation is referred to as a **sectorial** premolar.

The upper molars of primates all derive from a **tritubercular** (triangular-cusp) pattern. The crown of each tooth bears three main cusps: protocone, metacone, and paracone. The **protocone** is the main cusp on the lingual side, and the **paracone** and **metacone** are on the buccal side. Enamel crests often connect these cusps, and **cuspules** (subsidiary cusps) may develop on these crests. Two of the more prominent cuspules are

Table B.1 (continued)

Old World Monkeys	*Apes*	*Humans*
2.1.2.3/2.1.2.3 ——→		
Variable, generally rectangular	Rectangular	Parabolic, continuous
and upper canine for occlusion of lower canine		
——→		
Large, caniniform Sexual dimorphism especially baboons	Stout, large, projecting Sexual dimorphism	Incisiform
——→	Bicuspid ——→	
Upper first: canine fossa on mesial surface ——→ Lower first: sectorial		
Bilophodont	*Dryopithecus* (Y5) pattern ——→ Uppers: four cusps, oblique ridge ——→ Lowers: four or five cusps. ——→	
$M1 < M2 \leq M_3^{(3)}$	$M1 < M2 > M3$	$M1 > M2 > M3$

the **protoconule** and the **metaconule.** Cingula may develop around the inner side of the base of the crown and/or on the buccal side, where it may form a flattened shelf called the **external cingulum** or **stylar shelf.** Little pointed cuspules may also develop from the stylar shelf along the side of the tooth: from front to back these are the **parastyle, mesostyle,** and **metastyle.** The three main cusps of the upper molars form the points of a triangle; this formation is called the **trigone.** In many primates a fourth cusp is added at the posterolingual border to change the tritubercular tooth into a more **quadritubercular** (four-sided) one. This additional cusp is the **hypocone** (Fig. B.3b). In some primates transverse ridges **(lophs)** develop between cusps; for example, Old World monkeys are said to be **bilophodont** because their molars each have two lophs.

The crowns of lower molars consist of two parts: the **trigonid** (an anterior portion bearing three main cusps) and the **talonid** (a posterior, heellike projection). In early primates the trigonid is elevated well above the level of the talonid, but this height differential decreases throughout primate evolution. The three main cusps of the trigonid are the mirror images of the cusps of the upper-molar trigone: the **protoconid** on the buccal side, and the **paraconid** and **metaconid** on the lingual side. Note that the lower-molar cusp and cuspule names end in **–id,** while those for

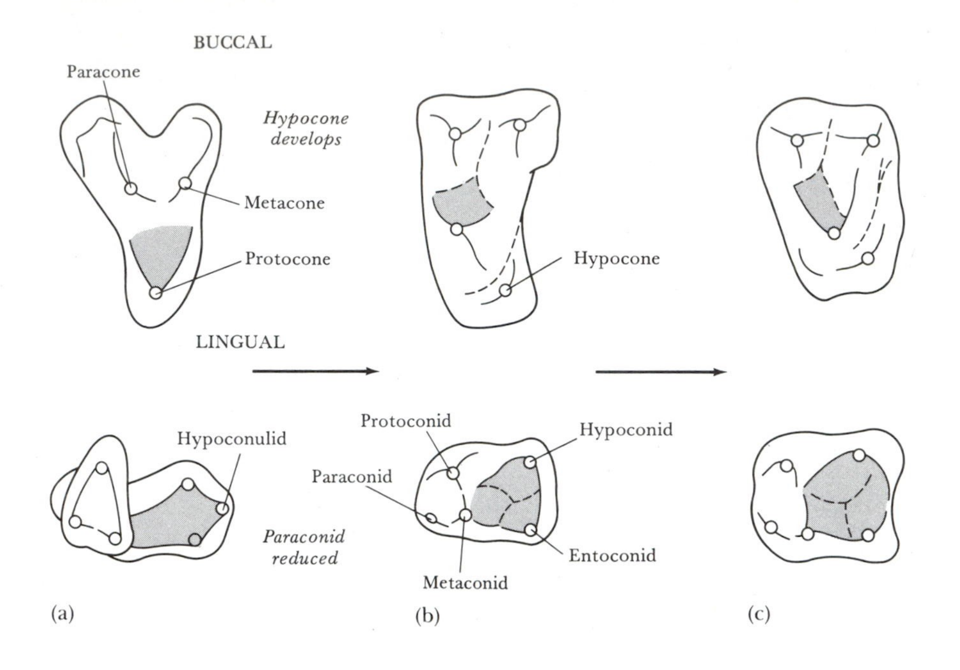

the upper molars end in **–e.** The talonid is usually a basinlike structure surrounded by a raised enamel rim with two main cusps: the **hypoconid** laterally and the **entoconid** medially. Often there is an additional cusp, the **hypoconulid,** toward the middle of the posterior margin of the rim; it is often well developed on lower third molars. Otherwise, the terminology is very similar for upper and lower molars (Fig. B.3c).

During chewing, the protocones of the upper molars fit into the talonid basins of the corresponding lower molars, like the action of a pestle in a mortar.

Figure B.4 shows some of the main changes in the evolution of primate molar teeth from primitive Cretaceous mammals through modern Old World monkeys.

POSTCRANIAL ADAPTATIONS

The general structure of the primate postcranial skeleton is basically rather primitive. In very few primate groups do we see the number of functional specializations that typify other mammalian groups. This is particularly true of adaptations to terrestrial locomotion. Among the terrestrial adaptations seen in some other mammals are (1) reduction in mobility at the shoulder joint, (2) atrophy or disappearance of the **clavicle** (col-

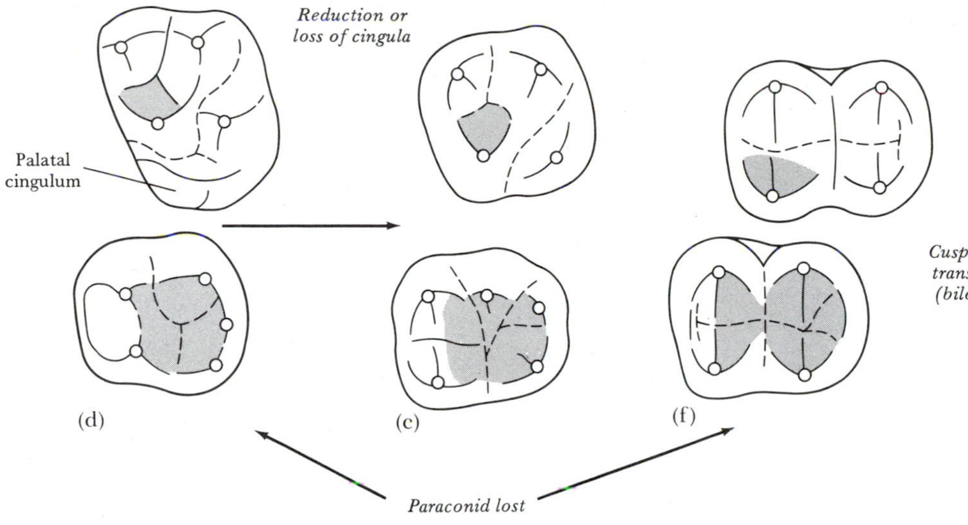

Fig. B.4 Evolution of Primate Molar Teeth Upper left molar *(above)* and lower right *(below)* molar shown for each group, with mesial (anterior) surfaces to the left and buccal (lateral) surfaces to the top. Shading shows where the lingual side of the protocone occludes the talonid basin. (a) A Cretaceous nonprimate mammal (*Pappotherium,* an insectivoran); (b) a Paleocene plesiadapiform *(Palenochtha);* (c) an Eocene adapid *(Cantius);* (d) an Oligocene anthropoid *(Aegyptopithecus);* (e) a Miocene catarrhine *(Dryopithecus);* (f) a modern cercopithecoid (Old World monkey). (Redrawn from Hiiemae, 1981)

larbone), (3) fusion of the **radius** and **ulna** (forearm bones) with loss of the ability to rotate between them, (4) fusion of the **tibia** and **fibula** (lower leg bones) and (5) atrophy or loss of one or more of the digits in the hand or foot. In contrast, postcranial adaptations associated with fully terrestrial locomotor behavior are not present in the primates. For example, primates retain five fingers and toes, a relatively mobile shoulder joint for free movement of the upper limb in all directions, prehensile (grasping) capability in both hands and feet, a well-developed clavicle, and completely separate bones of the forearm (radius and ulna) and leg (tibia and fibula).

Primates have developed highly efficient friction pads on the hands and feet, and in all modern primates the distal phalanges of at least some digits are covered by flattened nails instead of sharp, curved claws. These nails and friction pads provide an efficient mechanism for both grasping arboreal supports and manipulating objects.

The skeleton of all primates consists of the same basic elements, although in shape and size the bones naturally vary among the different genera. Figure B.5 illustrates postcranial skeletal features of the chimpanzee and the human.

Fig. B.5 Primate Postcranial Skeleton Separate parts are shown throughout for chimpanzee *(left)* and human *(right)*. All views except (a) and (g) are from the right side of the body. (a) Bony thorax, (b) scapula, (c) humerus, (d) ulna, (e) radius, (f) hand, (g) articulated pelvis (frontal view), (h) pelvis (lateral and medial views), (i) femur, (j) tibia, (k) fibula, (1) foot (superior view). Note that each digit in the hand and foot is assigned a roman numeral. (Redrawn from Swindler and Wood, 1973)

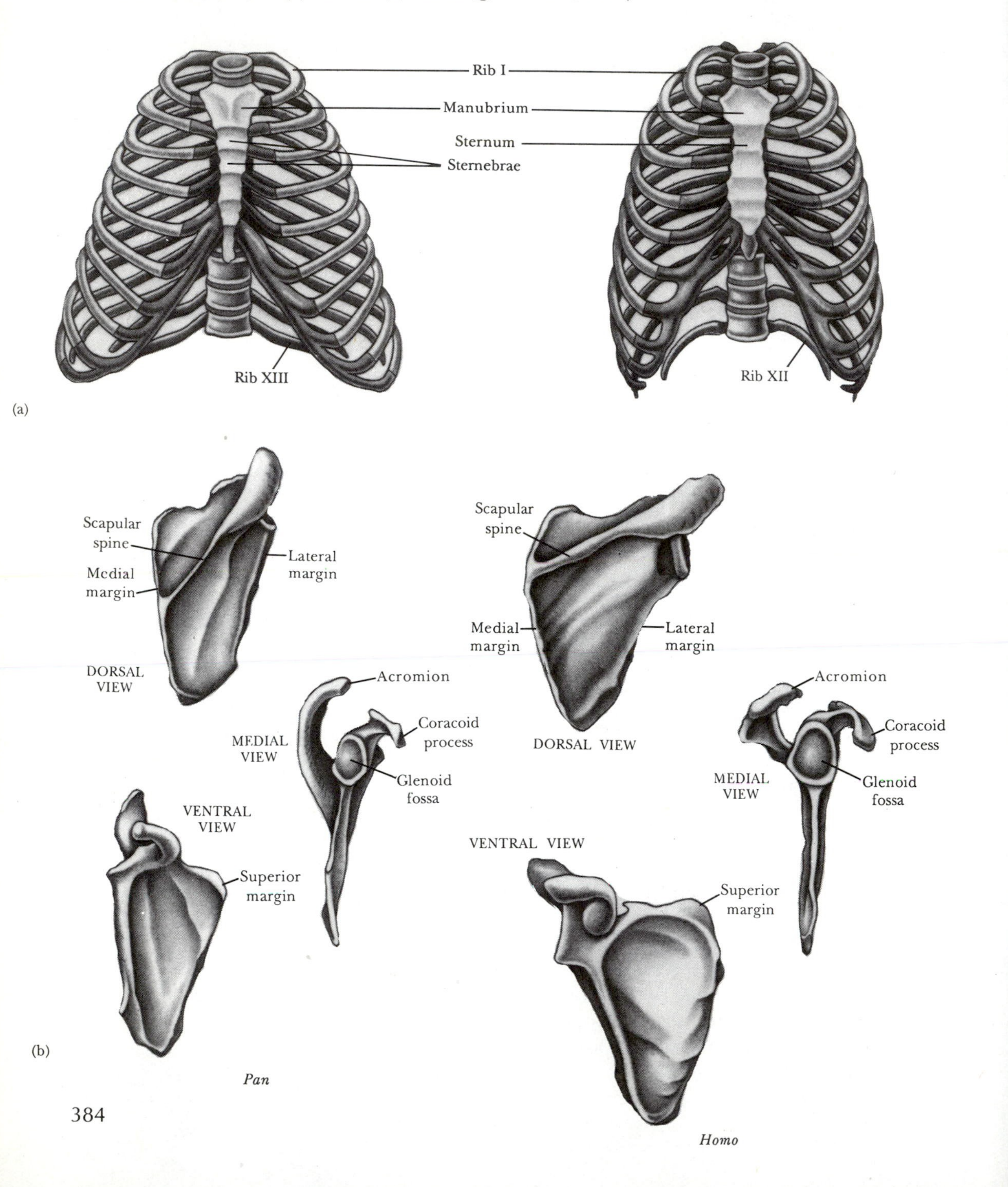

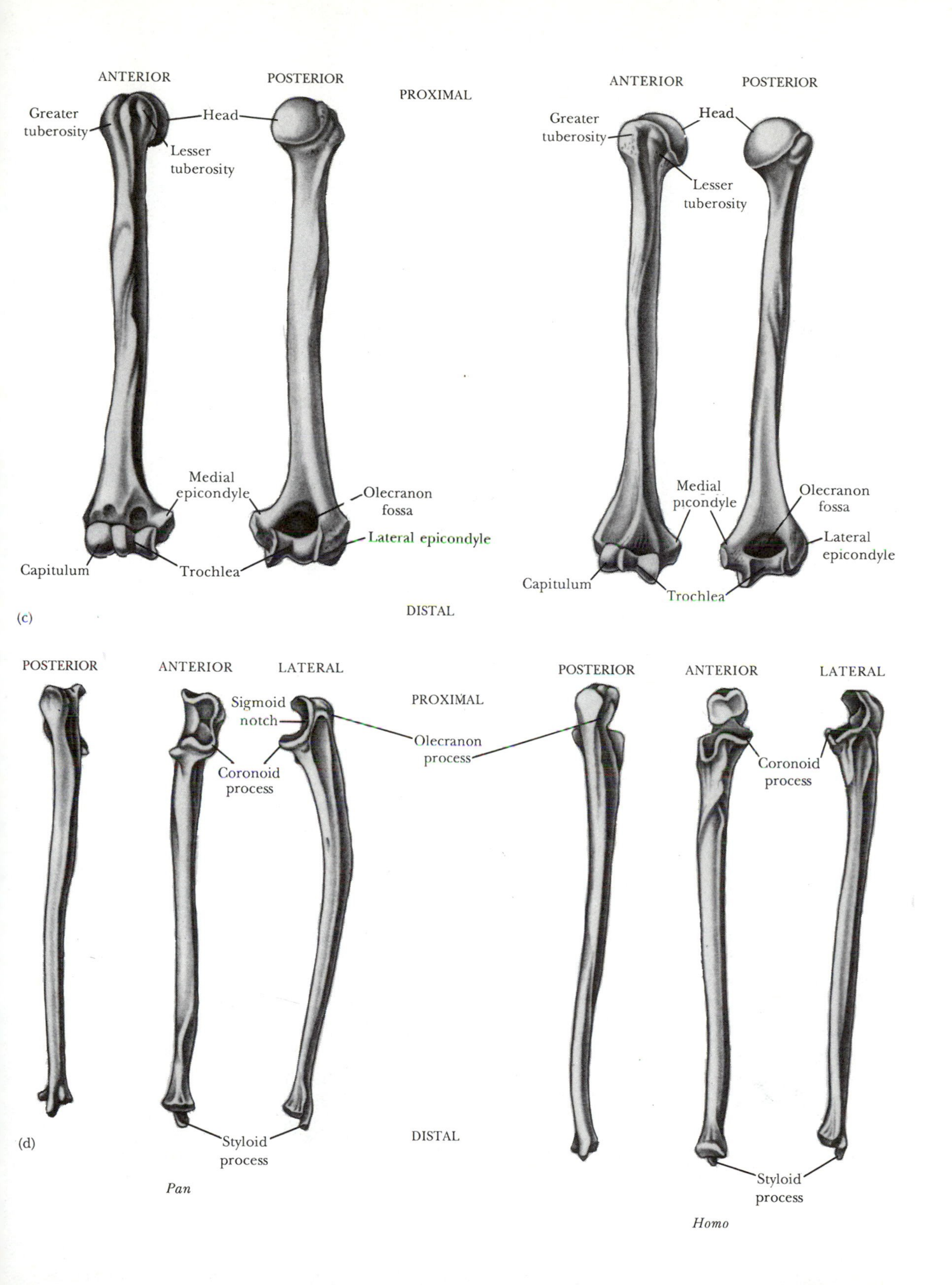

ANTERIOR
POSTERIOR
PROXIMAL
Greater tuberosity
Head
Lesser tuberosity
Medial epicondyle
Olecranon fossa
Lateral epicondyle
Capitulum
Trochlea
ANTERIOR
POSTERIOR
Greater tuberosity
Head
Lesser tuberosity
Medial picondyle
Olecranon fossa
Lateral epicondyle
Capitulum
Trochlea
(c)
DISTAL
POSTERIOR
ANTERIOR
LATERAL
Sigmoid notch
PROXIMAL
Olecranon process
Coronoid process
POSTERIOR
ANTERIOR
LATERAL
Coronoid process
(d)
Styloid process
DISTAL
Styloid process
Pan
Homo

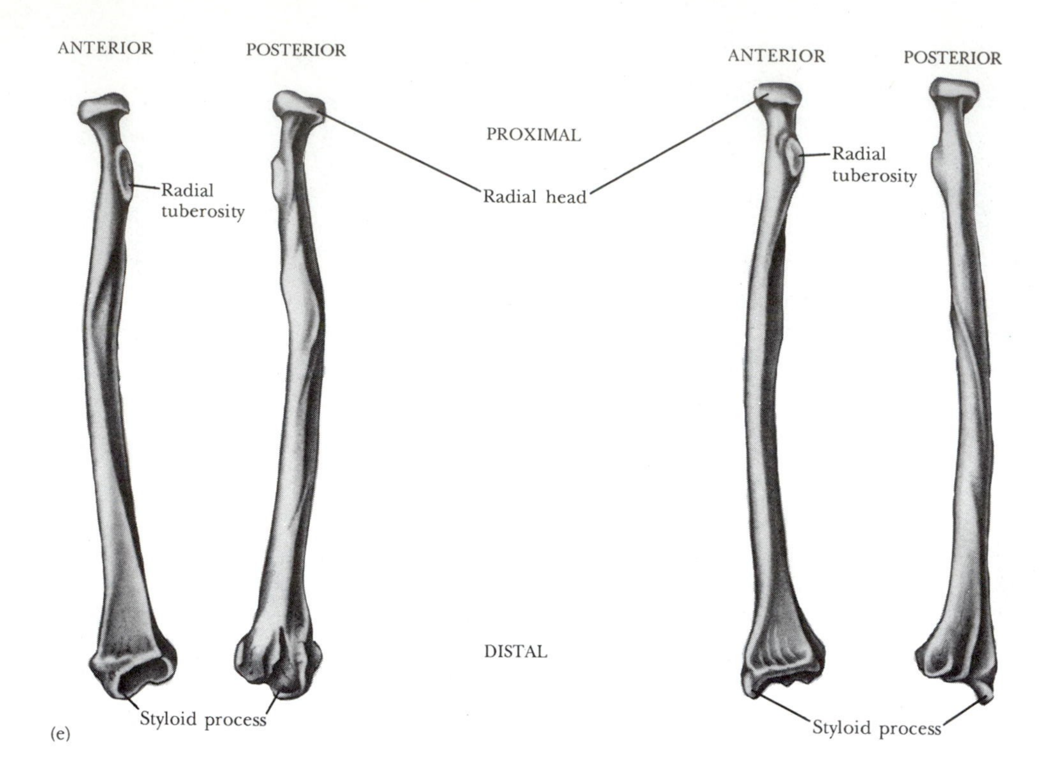
ANTERIOR
POSTERIOR
ANTERIOR
POSTERIOR
PROXIMAL
Radial tuberosity
Radial head
Radial tuberosity
DISTAL
Styloid process
Styloid process
(e)

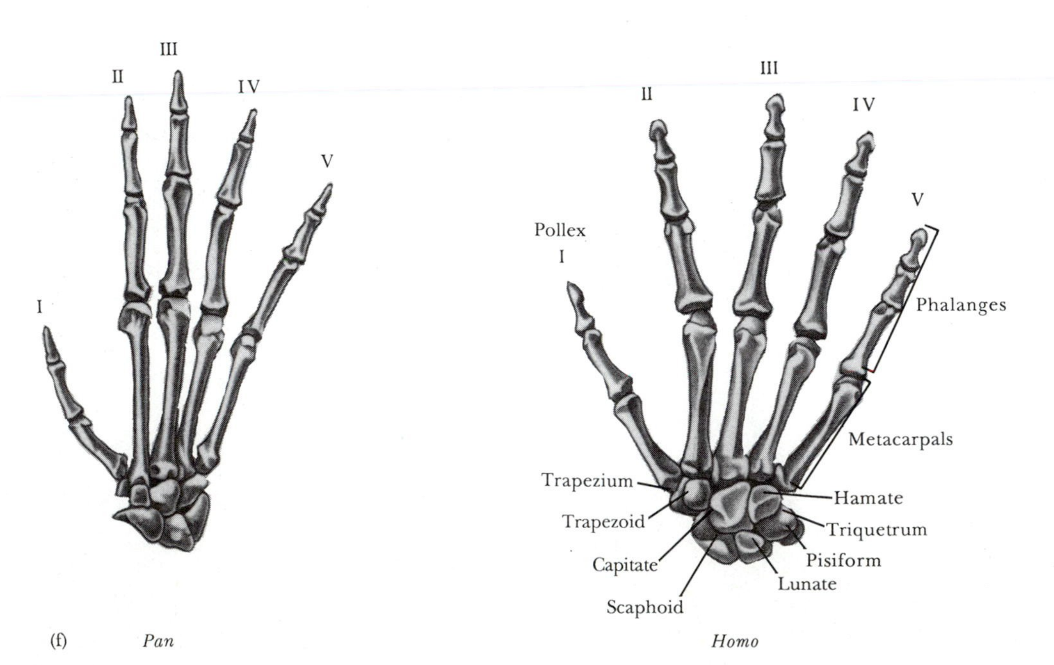
III
II
IV
V
I
III
II
IV
V
Pollex
I
Phalanges
Metacarpals
Trapezium
Hamate
Trapezoid
Triquetrum
Capitate
Pisiform
Lunate
Scaphoid
(f)
Pan
Homo

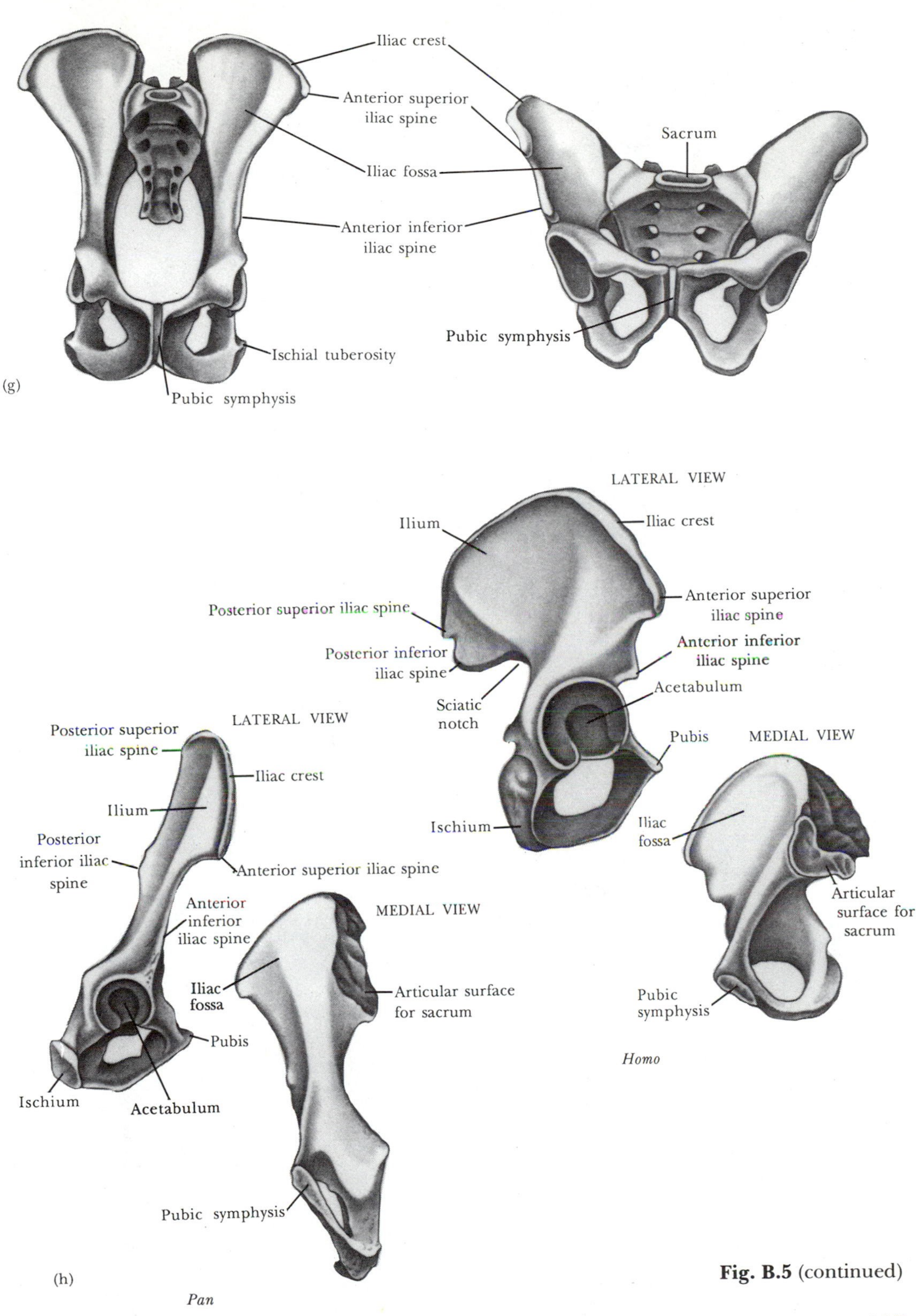

Fig. B.5 (continued)

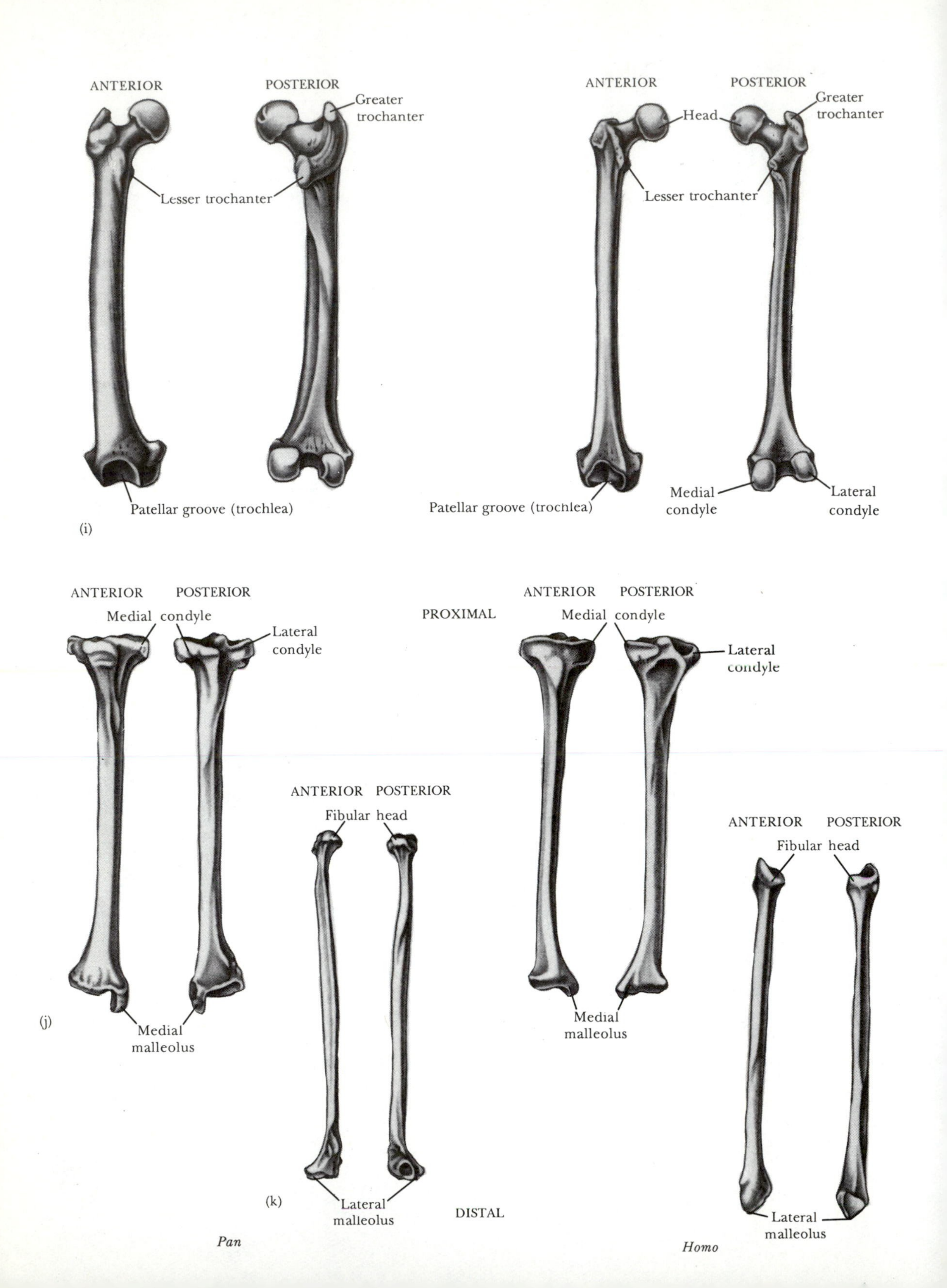
ANTERIOR
POSTERIOR
Greater trochanter
Lesser trochanter
Patellar groove (trochlea)
(i)
ANTERIOR
POSTERIOR
Head
Greater trochanter
Lesser trochanter
Patellar groove (trochlea)
Medial condyle
Lateral condyle
ANTERIOR
POSTERIOR
Medial condyle
Lateral condyle
(j)
Medial malleolus
PROXIMAL
ANTERIOR
POSTERIOR
Medial condyle
Lateral condyle
Medial malleolus
ANTERIOR
POSTERIOR
Fibular head
(k)
Lateral malleolus
DISTAL
ANTERIOR
POSTERIOR
Fibular head
Lateral malleolus
Pan
Homo

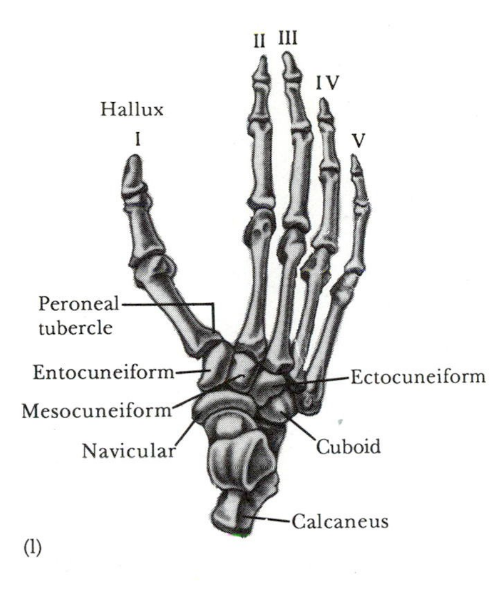
Hallux
I
II
III
IV
V
Peroneal tubercle
Entocuneiform
Mesocuneiform
Navicular
Ectocuneiform
Cuboid
Calcaneus
(l)

Pan

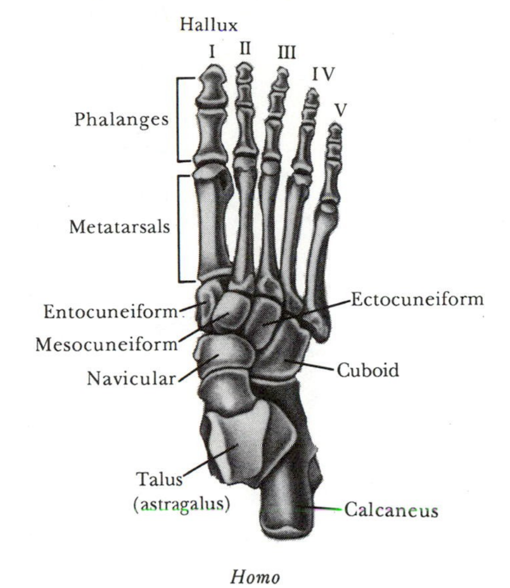
Hallux
I
II
III
IV
V
Phalanges
Metatarsals
Entocuneiform
Mesocuneiform
Navicular
Ectocuneiform
Cuboid
Talus (astragalus)
Calcaneus

Homo

APPENDIX C

Correlates of Feeding Behavior in Modern Primates

In order to integrate some basic features of primate biology with various dietary preferences found in extant primates (and by analogy in fossil primates as well), it is necessary to consider features such a positional behavior, general morphology of jaws and teeth, food types, plant chemistry, and digestive anatomy.

Figure C.1 arranges a number of variables (and their ranges of variation) in relation to diet ranging from 100% **faunivory** (eating vertebrates and invertebrates) through **frugivory** (eating fruit) to 100% **folivory** (eating leaves). This figure provides a good first approximation to the correlations found among the many variables associated with particular dietary preferences in primates (Chivers et al., 1984).

Let us consider faunivorous primates as an example. We would expect most of them to share the following attributes:

1. Have a low biomass density (that is, be rare)
2. Be small in body size.
3. Live in small social groups (or be solitary)
4. Be nocturnal
5. Progress by leaping or quadrupedal walk-run, often clinging when stationary
6. Have light jaws, a large gape, and little lateral jaw movement when chewing
7. Have a small (but often modified) anterior dentition
8. Have molars with thin enamel and postcanine teeth with low shearing blades
9. Have pointed cusps on their molars for puncture-crushing hard and

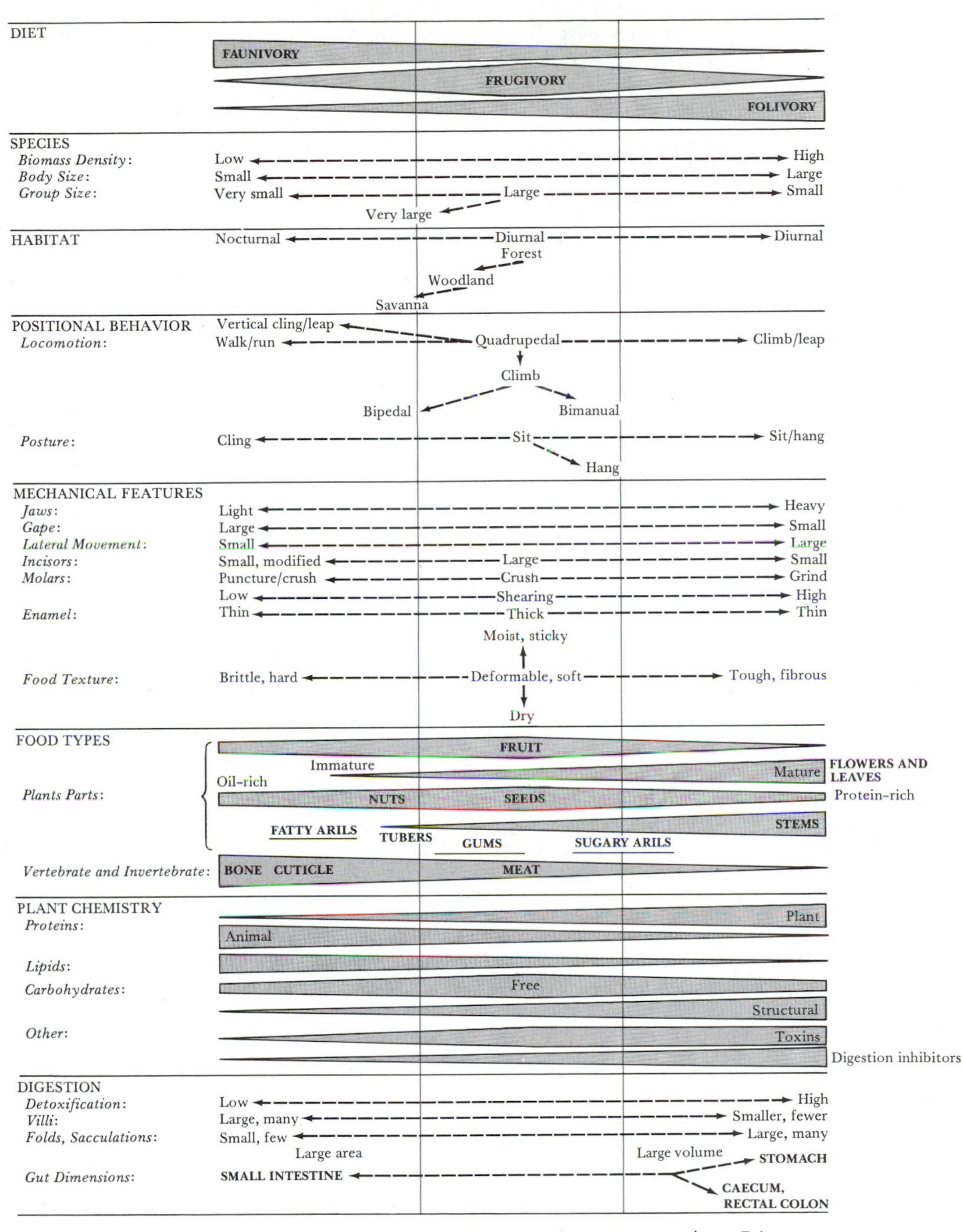

Fig. C.1 Correlations among Feeding, Ecology, and Behavior in Modern Primates. (Adapted from Chivers et al., 1984)

brittle foods (such as the bone and cartilage of vertebrates or the cuticle of invertebrates)

10. Eat foods high in animal protein and lipids but low in structural carbohydrates and plant protein
11. Have gastrointestinal tracts that are small relative to body size but dominated by the small intestine.

By using this chart, we can infer many of the attributes of fossil primates just by knowing about their dentition.

APPENDIX D

Classifications of the Order Primates

One of the difficulties in the study of primate evolution is the lack of consensus concerning the classification of living and extinct members of the order Primates. Numerous taxonomies have been devised since Linnaeus first named the order in 1758, especially in the last two decades as workers have attempted to account for new evidence in paleoanthropology and molecular biology. Included here are four current taxonomies: Simons (1972), Schwartz et al. (1978), Szalay and Delson (1979), and Fleagle (1988). They are presented in chronological order without comment on the efficacy of one over another. Extinct genera are shown in boldface.

Table D.1 Simons (1972) Taxonomy of the Primates

- Suborder PROSIMII
  - Infraorder PLESIADAPIFORMES
    - Superfamily PLESIADAPOIDEA
      - Family PLESIADAPIDAE
        - Subfamily PLESIADAPINAE
          - ***Platychoerops***
          - ***Plesiadapis***
          - ***Chiromyoides***
          - ***Pronothodectes***
        - Subfamily SAXONELLINAE
          - ***Saxonella***
      - Family CARPOLESTIDAE
        - ***Elphidotarsius***
        - ***Carpodaptes***
        - ***Carpolestes***
    - Superfamily PLESIADAPOIDEA, *incertae sedis*
      - Family PAROMOMYIDAE
        - Subfamily PAROMOMYINAE
          - ***Paromomys***
          - ***Palaechthon***
          - ***Plesiolestes***
          - ***Palenochtha***
          - ***Purgatorius***
        - Subfamily PHENACOLEMURINAE
          - ***Phenacolemur***
      - Family PICRODONTIDAE
        - ***Picrodus***
        - ***Zanycteris***
  - Infraorder LEMURIFORMES
    - Superfamily ADAPOIDEA

Simons (1972) Taxonomy (continued)

Suborder PROSIMII (cont.)
Infraorder LEMURIFORMES (cont.)
Superfamily ADAPOIDEA (cont.)
Family ADAPIDAE
Subfamily ADAPINAE
Adapis
Pronycticebus
Protoadapis
Anchomomys
Caenopithecus
Lantianius
Agerinia
Subfamily NOTHARCTINAE
Notharctus
Pelycodus
Smilodectes
Superfamily LEMUROIDEA
Family LEMURIDAE
Subfamily LEMURINAE
Lemur
Hapalemur
Lepilemur
Subfamily CHEIROGALEINAE
Cheirogaleus
Microcebus
Phaner
?*Allocebus*
Family INDRIIDAE
Subfamily INDRIINAE
Palaeopropithecus
Mesopropithecus
Propithecus
Archaeoindris
Indri
Avahi
Subfamily HADROPITHECINAE
Hadropithecus
Subfamily ARCHAEOLEMURINAE
Archaeolemur
Family DAUBENTONIIDAE
Daubentonia
Family MEGALADAPIDAE
Megaladapis
Infraorder LORISIFORMES
Superfamily LORISOIDEA
Family LORISIDAE
Subfamily LORISIDAE
Indraloris
Loris
Nycticebus
Arctocebus
Perodicticus
Subfamily GALAGINAE
Galago
Family LORISIDAE, *incertae sedis*
Progalago
Komba
Infraorder TARSIIFORMES
Superfamily TARSIIOIDEA
Family TARSIIDAE
Subfamily TARSIINAE
Tarsius
Subfamily MICROCHOERINAE
Nannopithex
Necrolemur
Microchoerus
Pseudoloris
Superfamily TARSIOIDEA, *incertae sedis*
Family ANAPTOMORPHIDAE
Subfamily ANAPTOMORPHINAE
Absarokius
Tetonius
Tetonoides
Uintalacus
Anemorhysis
Trogolemur
Anaptomorphus
Uintanius
Subfamily OMOMYINAE
Omomys
Loveina
Hemiacodon
Washakius
Shoshonius
Macrotarsius
Teilhardina
Utahia
Stockia
Ourayia
Rooneyia
Ekgmowechashala
Family ?OMOMYIDAE, *incertae sedis*
Periconodon
Hoanghonius
Lushius
Suborder ANTHROPOIDEA
Infraorder PLATYRRHINI
Superfamily CEBOIDEA
Family CEBIDAE
Subfamily AOTINAE
Homunculus
Aotus
Callicebus
Dolichocebus

Simons (1972) Taxonomy (continued)

Suborder ANTHROPOIDEA (cont.)
Infraorder PLATYRRHINI (cont.)
Superfamily CEBOIDEA (cont.)
Family CEBIDAE (cont.)
Subfamily PITHECINAE
Cacajao
Pithecia
Chiropotes
Subfamily ALOUATTINAE
Alouatta
Subfamily CEBINAE
Cebus
Saimiri
Neosaimiri
Stirtonia
Subfamily CEBUPITHECINAE
Cebupithecia
Subfamily ATELINAE
Ateles
Brachyteles
Lagothrix
Subfamily CALLIMICONINAE
Callimico
Family CALLITHRICIDAE
Callithrix
Saguinus
Family XENOTHRICIDAE
Xenothrix
Superfamily ?CEBOIDEA
Branisella
Infraorder CATARRHINI
Superfamily CERCOPITHECOIDEA
Family CERCOPITHECIDAE
Subfamily CERCOPITHECINAE
Macaca
Libypithecus
Cercocebus
Parapapio
Dinopithecus
Gorgopithecus
Papio
Procynocephalus
Theropithecus
Cercopithecus
Allenopithecus
Erythrocebus
Miopithecus
Subfamily PARAPITHECINAE
Parapithecus
Apidium
Subfamily COLOBINAE
Mesopithecus
Dolichopithecus
Presbytis
Pygathrix
Rhinopithecus
Simias
Nasalis
Colobus
Procolobus
Family CERCOPITHECIDAE, *incertae sedis*
Cercopithecoides
Paracolobus
Prohylobates
Victoriapithecus
Superfamily OREOPITHECOIDEA
Family OREOPITHECIDAE
Oreopithecus
Mabokopithecus
Superfamily HOMINOIDEA
Family HYLOBATIDAE
Subfamily PLIOPITHECINAE
Pliopithecus
Limnopithecus
Aeolopithecus
Subfamily HYLOBATINAE
Hylobates
Symphalangus
Family PONGIDAE
Subfamily DRYOPITHECINAE
Dryopithecus
Aegyptopithecus
Propliopithecus
Subfamily ?DRYOPITHECINAE, *incertae sedis*
Oligopithecus
Subfamily PONGINAE
Pongo
Pan
Gorilla
Subfamily GIGANTOPITHECINAE
Gigantopithecus
Family HOMINIDAE
Ramapithecus
Australopithecus
Homo
Superfamily ?HOMINOIDEA
Amphipithecus
Pondaungia

Table D.2 Schwartz et al. (1978) Taxonomy of the Primates

Semiorder PLESITARSIIFORMES
Suborder MICROSYOPIDA
Cynodontomys
Microsyops
Craseops
Alsaticopithecus
Suborder PLESITARSIIDA
Infraorder TARSIIFORMES
Family UINTASORICIDAE
Tinimomys
Uintasorex
Niptomomys
Family OMOMYIDAE
Subfamily OMOMYINAE
Omomys
Tarsius
"Uintanius ameghini"
Chumashius
Pseudoloris
Subfamily MACROTARSIINAE
Macrotarsius
Ourayia
Mytonius
Infraorder PLESIADAPIFORMES
Superfamily ANAPTOMORPHOIDEA
Plesion ***Teilhardina***
Family MICROCHOERIDAE
Plesion ***Loveina***
Plesion ***Shoshonius***
Tribe WASHAKIINI
Hemiacodon
Dyseolemur
Washakius
Tribe MICROCHOERINI
Nannopithex
Rooneyia
Necrolemur
Microchoerus
Family ANAPTOMORPHIDAE
Anemorhysis
Altanius
Pseudotetonius
Trogolemur
Anaptomorphus
Tetonius
Absarokius
Superfamily PLESIADAPOIDEA
Family PLESIADAPIDAE
Plesion ***Pronothodectes***
Subfamily CARPOLESTINAE
Elphidotarsius
Carpodaptes
Carpolestes
Subfamily PLESIADAPINAE
Nannodectes
Plesiadapis
Chiromyoides
Platychoerops
Family PAROMOMYIDAE
Palenochtha
Palaechthon (= Plesiolestes)
Talpohenach
Torrejonia
Paromomys (= Stockia)
Phenacolemur
Picrodus
Zanycteris
Semiorder SIMIOLEMURIFORMES
Suborder STREPSIRHINI
Infraorder LORISIFORMES
Family LORISIDAE
Loris
Nycticebus
Perodicticus
Arctocebus
? ***Indraloris***
Family GALAGIDAE
Subfamily GALAGINAE
Galago
Euoticus
Subfamily CHEIROGALEINAE
Phaner
Cheirogaleus
Allocebus
Microcebus
Infraorder LEMURIFORMES
Superfamily LEMUROIDEA
Family LEMURIDAE
Lemur
Varecia
Family ADAPIDAE
Subfamily ADAPINAE[a]
Pelycodus
Notharctus
Smilodectes
Adapis
Pronycticebus
Protoadapis
Anchomomys
Agerinia
Caenopithecus
Mahgarita
Cercamonius
Periconodon
Subfamily LEPILEMURINAE
Hapalemur
Prolemur
Megaladapis
Lepilemur
Superfamily INDRIOIDEA

Semiorder SIMIOLEMURIFORMES (cont.)
Suborder STREPSIRHINI (cont.)
Infraorder LEMURIFORMES (cont.)
Superfamily INDRIOIDEA (cont.)
Plesion *Daubentonia*
Family ARCHAEOLEMURIDAE
Archaeolemur
Hadropithecus
Family INDRIIDAE
Subfamily PALAEOPROPITHECINAE
Palaeopropithecus
Archaeoindris
Subfamily INDRIINAE
Indri
Avahi
Mesopropithecus
Propithecus
Suborder ANTHROPOIDEA
Plesion ***Amphipithecus***
Plesion ***Oligopithecus***
Infraorder PLATYRRHINI
Plesion ***Branisella***
Family CEBIDAE[a]
Tremacebus
Stirtonia
Neosaimiri
Saimiri
Aotus
Callicebus
Alouatta
Cebupithecia
Pithecia
Chiropotes
Cacajao
Cebus
Lagothrix
Ateles
Brachyteles
Dolichocebus
Homunculus
Xenothrix
Family CALLITRICHIDAE[a]
Cebuella
Callithrix
Saguinus
Leontopithecus
Callimico
Infraorder CATARRHINI
Plesion PARAPITHECIDAE
Parapithecus
Apidium
Plesion ***Oreopithecus***
Superfamily CERCOPITHECOIDEA
Plesion ***Prohylobates***
Plesion ***Victoriapithecus***
Family CERCOPITHECIDAE[a]
Subfamily CERCOPITHECINAE
Tribe CERCOPITHECINI
Cercopithecus
Erythrocebus
Allenopithecus
Tribe PAPIONINI
Papio
Cercocebus
Dinopithecus
Gorgopithecus
Macaca
Procynocephalus
Paradolichopithecus
Theropithecus
Subfamily COLOBINAE
Colobus
Libypithecus
Cercopithecoides
Paracolobus
Presbytis
Pygathrix
Nasalis
Mesopithecus
Dolichopithecus
Superfamily HOMINOIDEA
Plesion ***Propliopithecus***
Plesion ***Aeolopithecus***
Family HYLOBATIDAE
Hylobates
Symphalangus
Plesion ***Dendropithecus***
Plesion ***Pliopithecus***
Family HOMINIDAE
Subfamily DRYOPITHECINAE
Plesion ***Aegyptopithecus***
Dryopithecus
Limnopithecus
Subfamily HOMININAE
Supertribe PONGINU
Pongo
Supertribe HOMININU
Tribe PANINI
Pan
Gorilla
Tribe HOMININI
Plesion ***Sivapithecus***
Plesion ***Gigantopithecus***
Ramapithecus
Australopithecus
Homo

[a] Indicates that order of taxa within the clade is arbitrary, not phylogenetic.

Table D.3 Szalay and Delson (1979) Taxonomy of the Primates

- Suborder PLESIADAPIFORMES
  - Superfamily PAROMOMYOIDEA
    - Family PAROMOMYIDAE
      - Tribe PURGATORIINI
        - ***Purgatorius***
      - Tribe PAROMOMYINI
        - Subtribe PALAECHTHONINA
          - ***Palaechthon***
          - ***Plesiolestes***
          - ***Palenochtha***
        - Subtribe PAROMOMYINA
          - ***Paromomys***
          - ***Ignacius***
          - ***Phenacolemur***
      - Tribe MICROMOMYINI
        - ***Micromomys***
        - ***Tinimomys***
      - Tribe NAVAJOVIINI
        - ***Navajovius***
        - ***Berruvius***
    - Family PICRODONTIDAE
      - ***Picrodus***
      - ***Zanycteris***
  - Superfamily PLESIADAPOIDEA
    - Family PLESIADAPIDAE
      - ***Pronothodectes***
      - ***Plesiadapis***
      - ***Chiromyoides***
      - ***Platychoerops***
    - Family SAXONELLIDAE
      - ***Saxonella***
    - Family CARPOLESTIDAE
      - ***Elphidotarsius***
      - ***Carpodaptes***
- Suborder STREPSIRHINI
  - Infraorder ADAPIFORMES
    - Family ADAPIDAE
      - Subfamily NOTHARCTINAE
        - ***Pelycodus***
        - ***Notharctus***
        - ***Smilodectes***
        - ***Copelemur***
      - Subfamily ADAPINAE
        - Tribe PROTOADAPINI
          - ***Protoadapis***
          - ***Agerinia***
          - ***Europolemur***
          - ***Mahgarita***
          - ***Pronycticebus***
        - Tribe ANCHOMOMYINI
          - ***Anchomomys***
          - ***Huerzeleris***
          - ***Periconodon***
        - Tribe MICROADAPINI
          - ***Microadapis***
        - Tribe ADAPINI
          - Subtribe ADAPINA
            - ***Leptadapis***
            - ***Adapis***
          - Subtribe CAENOPITHECINA
            - ***Caenopithecus***
        - Tribe INDRALORISINI
          - ***Indraloris***
    - Family ADAPIDAE, *incertae sedis*
      - ***Amphipithecus***
      - ***Lushius***
  - Infraorder LEMURIFORMES
    - Superfamily LEMUROIDEA
      - Family LEMURIDAE
        - *Lemur*
        - *Lepilemur*
        - *Hapalemur*
        - *Varecia*
      - Family MEGALADAPIDAE
        - ***Megaladapis***
    - Superfamily INDRIOIDEA
      - Family INDRIIDAE
        - *Indri*
        - *Propithecus*
        - *Avahi*
        - ***Mesopropithecus***
      - Family DAUBENTONIIDAE
        - *Daubentonia*
      - Family ARCHAEOLEMURIDAE
        - ***Archaeolemur***
        - ***Hadropithecus***
      - Family PALAEOPROPITHECIDAE
        - ***Palaeopropithecus***
        - ***Archaeoindris***
    - Superfamily LORISOIDEA
      - Family CHEIROGALEIDAE
        - *Cheirogaleus*
        - *Phaner*
        - *Allocebus*
        - *Microcebus*
      - Family LORISIDAE
        - Subfamily GALAGINAE
          - *Galago*
          - *Galagoides*
          - *Euoticus*
          - ***Progalago***
          - ***Komba***
        - Subfamily LORISINAE
          - *Loris*
          - *Nycticebus*
          - *Arctocebus*

Szalay and Delson (1979) Taxonomy (continued)

Suborder STREPSIRHINI (cont.)
Infraorder LEMURIFORMES (cont.)
Superfamily LORISOIDEA (cont.)
Family LORISIDAE (cont.)
Subfamily LORISINAE (cont.)
Perodicticus
Mioeuoticus
Suborder HAPLORHINI
Infraorder TARSIIFORMES
Family OMOMYIDAE
Subfamily ANAPTOMORPHINAE
Tribe ANAPTOMORPHINI
Subtribe TEILHARDININA
Teilhardina
Chlororhysis
Subtribe ANAPTOMORPHINA
Anaptomorphus
Subtribe TETONIINA
Tetonius
Absarokius
Anemorhysis
Altanius
Mckennamorphus
Tribe TROGOLEMURINI
Trogolemur
Subfamily OMOMYINAE
Tribe OMOMYINI
Subtribe OMOMYINA
Omomys
Chumashius
Subtribe MYTONIINA
Ourayia
Macrotarsius
Tribe WASHAKIINI
Loveina
Shoshonius
Washakius
Dyseolemur
Hemiacodon
Tribe UINTANIINI
Uintanius
Tribe UTAHIINI
Utahia
Stockia
Tribe ROONEYIINI
Rooneyia
Subfamily EKGMOWECHASHALINAE
Ekgmowechashala
Subfamily MICROCHOERINAE
Nannopithex
Necrolemur
Microchoerus
Pseudoloris
Family OMOMYIDAE, *incertae sedis*
Donrussellia
Hoanghonius
Family TARSIIDAE
Tarsius
Infraorder PLATYRRHINI
Family CEBIDAE
Subfamily CEBINAE
Cebus
Saimiri
Neosaimiri
Dolichocebus
Subfamily BRANISELLINAE
Branisella
Subfamily CALLITRICHINAE
Tribe CALLITRICHINI
Callithrix
Cebuella
Saguinus
Leontopithecus
Tribe CALLIMICONINI
Callimico
Family ATELIDAE
Subfamily ATELINAE
Tribe ATELINI
Ateles
Lagothrix
Brachyteles
Tribe ALOUATTINI
Alouatta
Stirtonia
Subfamily PITHECIINAE
Tribe PITHECIINI
Subtribe CALLICEBINA
Callicebus
Subtribe PITHECIINA
Pithecia
Chiropotes
Cacajao
Cebupithecia
Tribe XENOTRICHINI
Xenothrix
Tribe HOMUNCULINI
Aotus
Homunculus
Tremacebus
Infraorder CATARRHINI
Superfamily PARAPITHECOIDEA
Family PARAPITHECIDAE
Parapithecus
Apidium
Superfamily CERCOPITHECOIDEA
Family CERCOPITHECIDAE

Szalay and Delson (1979) Taxonomy (continued)

Suborder HAPLORHINI (cont.)
Infraorder CATARRHINI (cont.)
Superfamily CERCOPITHECOIDEA (cont.)
Family CERCOPITHECIDAE (cont.)
Subfamily CERCOPITHECINAE
Tribe CERCOPITHECINI
Cercopithecus
Erythrocebus
Allenopithecus
Tribe PAPIONINI
Subtribe PAPIONINA
Papio
Cercocebus
Parapapio
Dinopithecus
Gorgopithecus
Subtribe MACACINA
Macaca
Procynocephalus
Paradolichopithecus
Subtribe THEROPITHECINA
Theropithecus
Subfamily COLOBINAE
Subtribe COLOBINA
Colobus
Libypithecus
Cercopithecoides
Paracolobus
Colobina, new genus
Colobina, gen. indet.
Subtribe SEMNOPITHECINA
Presbytis
Pygathrix
Nasalis
Subfamily COLOBINAE, *incertae sedis*
Mesopithecus
Dolichopithecus
Family CERCOPITHECIDAE, *incertae sedis*
Prohylobates
Victoriapithecus
Family OREOPITHECIDAE
Oreopithecus
Superfamily HOMINOIDEA
Family PLIOPITHECIDAE
Propliopithecus
Pliopithecus
Dendropithecus
Family ?PLIOPITHECIDAE, *incertae sedis*
Family HOMINIDAE
Subfamily HYLOBATINAE
Hylobates
Subfamily PONGINAE
Tribe PONGINI
Pongo
Pan
Tribe DRYOPITHECINI
Dryopithecus
Tribe SUGRIVAPITHECINI
Sivapithecus
Gigantopithecus
Subfamily HOMININAE
Ramapithecus
Australopithecus
Homo
Homo or ***Australopithecus,*** sp(p). indet.
HOMININAE, gen. et sp(p). indet.
Infraorder CATARRHINI, *incertae sedis*
Pondaungia
Oligopithecus

Table D.4 Fleagle (1988) Taxonomy of the Primates

Suborder PLESIADAPIFORMES
Family MICROSYOPIDAE
Palaechthon
Plesiolestes
Talpohenach
Torrejonia
Palenochtha
Berruvius
Navajovius
Micromomys
Tinimomys
Niptomomys
Uintasorex
Microsyops
Arctodontomys
Craseops
Alveojunctus
Family PLESIADAPIDAE
Pronothodectes
Nannodectes

Fleagle (1988) Taxonomy (continued)

Suborder PLESIADAPIFORMES (cont.)
Family PLESIADAPIDAE (cont.)
Plesiadapis
Chiromyoides
Platychoerops
Family CARPOLESTIDAE
Elphidotarsius
Carpodaptes
Carpolestes
Family SAXONELLIDAE
Saxonella
Family PAROMOMYIDAE
Paromomys
Ignacius
Phenacolemur
Elwynella
Arcius
Family PICRODONTIDAE
Picrodus
Zanycteris
Draconodus
Family *incertae sedis*
Purgatorius
Suborder PROSIMII
Infraorder LEMURIFORMES
Superfamily LEMUROIDEA
Family LEMURIDAE
Lemur
Varecia
Pachylemur
Hapalemur
Family LEPILEMURIDAE
Subfamily LEPILEMURINAE
Lepilemur
Subfamily MEGALADAPINAE
Megaladapis
Family INDRIIDAE
Subfamily INDRIINAE
Avahi
Propithecus
Indri
Mesopropithecus
Subfamily ARCHAEOLEMURINAE
Archaeolemur
Hadropithecus
Subfamily PALAEOPROPITHECINAE
Palaeopropithecus
Archaeoindris
Family DAUBENTONIIDAE
Daubentonia
Superfamily LORISOIDEA
Family CHEIROGALEIDAE
Microcebus
Mirza
Cheirogaleus
Phaner
Allocebus
Family GALAGIDAE
Otolemur
Galago
Galagoides
Euoticus
Progalago
Komba
Family LORISIDAE
Perodicticus
Arctocebus
Nycticebus
Loris
Mioeuoticus
Nycticeboides
Infraorder ADAPIFORMES
Family ADAPIDAE
Subfamily NOTHARCTINAE
Cantius
Copelemur
Notharctus
Smilodectes
Pelycodus
Subfamily ADAPINAE
Donrussellia
Protoadapis
Europolemur
Periconodon
Agerinia
Caenopithecus
Pronycticebus
Cercamonius
Cryptadapis
Microadapis
Anchomomys
Adapis
Leptadapis
Mahgarita
Subfamily SIVALADAPINAE
Indraloris
Sivaladapis
Sinoadapis
Subfamily *incertae sedis*
Azibius
Panobius
Hoanghonius
Lushius
Infraorder TARSIIFORMES
Family TARSIIDAE
Tarsius

Fleagle (1988) Taxonomy (continued)

Suborder PROSIMII (cont.)
Infraorder TARSIIFORMES (cont.)
Family TARSIIDAE (cont.)
Afrotarsius
Infraorder OMOMYIFORMES
Family OMOMYIDAE
Subfamily ANAPTOMORPHINAE
Teilhardina
Anemorhysis
Chlororhysis
Pseudotetonius
Absarokius
Anaptomorphus
Tetonius
Trogolemur
Aycrossia
Strigorhysis
Gazinius
Steinius
Loveina
Subfamily OMOMYINAE
Arapahovius
Omomys
Chumashius
Ourayia
Shoshonius
Washakius
Utahia
Hemiacodon
Dyseolemur
Stockia
Macrotarsius
Uintanius
Jemezius
Rooneyia
Ekgmowechashala
Subfamily MICROCHOERINAE
Nannopithex
Pseudoloris
Necrolemur
Microchoerus
Subfamily *incertae sedis*
Altanius
Kohatius
Suborder ANTHROPOIDEA
Infraorder *incertae sedis*
Amphipithecus
Pondaungia
Oligopithecus
Infraorder PARAPITHECOIDEA
Family PARAPITHECIDAE
Qatrania
Apidium
Parapithecus
Infraorder PLATYRRHINI
Superfamily CEBOIDEA
Family CEBIDAE
Subfamily CEBINAE
Cebus
Saimiri
Neosaimiri
Subfamily AOTINAE
Aotus
Tremacebus
Callicebus
Homunculus
Family ATELIDAE
Subfamily PITHECIINAE
Pithecia
Chiropotes
Cacajao
Mohanamico
Cebupithecia
Subfamily ATELINAE
Alouatta
Stirtonia
Lagothrix
Brachyteles
Ateles
Family CALLITRICHIDAE
Subfamily CALLITRICHINAE
Callimico
Saguinus
Leontopithecus
Callithrix
Cebuella
Micodon
Family *incertae sedis*
Subfamily *incertae sedis*
Branisella
Dolichocebus
Soriacebus
Xenothrix
Infraorder CATARRHINI
Superfamily HOMINOIDEA
Family PROPLIOPITHECIDAE
Propliopithecus
Aegyptopithecus
Family PLIOPITHECIDAE
Pliopithecus
Crouzelia
Laccopithecus
Family PROCONSULIDAE
Proconsul

Fleagle (1988) Taxonomy (continued)

Suborder ANTHROPOIDEA (cont.)
- Infraorder CATARRHINI (cont.)
  - Superfamily HOMINOIDEA (cont.)
    - Family PROCONSULIDAE (cont.)
      - ***Limnopithecus***
      - ***Dendropithecus***
      - ***Simiolus***
      - ***Rangwapithecus***
      - ***Micropithecus***
      - ***Dionysopithecus***
      - ***Platydontopithecus***
    - Family *incertae sedis*
      - ***Turkanapithecus***
      - ***Afropithecus***
      - ***Kenyapithecus***
    - Family OREOPITHECIDAE
      - ***Nyanzapithecus***
      - ***Oreopithecus***
    - Family HYLOBATIDAE
      - *Hylobates*
    - Family PONGIDAE
      - Subfamily DRYOPITHECINAE
        - ***Dryopithecus***
        - ***Lufengpithecus***
      - Subfamily PONGINAE
        - *Pongo*
        - ***Sivapithecus***
        - ***Gigantopithecus***
      - Subfamily GORILLINAE
        - ***Graecopithecus***
        - *Gorilla*
        - *Pan*
    - Family HOMINIDAE
      - ***Australopithecus***
      - *Homo*
  - Superfamily CERCOPITHECOIDEA
    - Family VICTORIAPITHECIDAE
      - Subfamily VICTORIAPITHECINAE
        - ***Victoriapithecus***
        - ***Prohylobates***
    - Family CERCOPITHECIDAE
      - Subfamily CERCOPITHECINAE
        - *Macaca*
        - ***Procynocephalus***
        - ***Paradolichopithecus***
        - *Cercocebus*
        - ***Parapapio***
        - *Papio*
        - *Mandrillus*
        - ***Dinopithecus***
        - ***Gorgopithecus***
        - *Theropithecus*
        - *Cercopithecus*
        - *Allenopithecus*
        - *Miopithecus*
        - *Erythrocebus*
      - Subfamily COLOBINAE
        - ***Mesopithecus***
        - ***Dolichopithecus***
        - ***Microcolobus***
        - ***Libypithecus***
        - ***Cercopithecoides***
        - ***Paracolobus***
        - ***Rhinocolobus***
        - *Colobus*
        - *Piliocolobus*
        - *Procolobus*
        - *Presbytis*
        - *Simias*
        - *Nasalis*
        - *Pygathrix*
        - *Rhinopithecus*

References

Abel, O. 1902. Zwei neue Menschenaffen aus den Leithakalkbildungen des Wiener Bekkens. *S. Ber. Akad. Wiss. Wien* **111:**1171–1207.

Abell, P. 1982. Palaeoclimates at Lake Turkana, Kenya, from oxygen isotope ratios of gastropod shells. *Nature* **297:**321–323.

Allman, J. 1982. Reconstructing the evolution of the brain in primates through the use of comparative neurophysiological and neuroanatomical data. In *Primate brain evolution: methods and concepts,* ed. E. Armstrong and D. Falk, pp. 13–28. Plenum, New York.

Alvarez, L.; Alvarez, W.; Asaro, F.; and Michel, H. 1980. Extraterrestrial cause for the Cretaceous-Tertiary extinction. *Science* **208:**1095–1108.

Alvarez, W.; Kauffman, E.; Surlyk, F.; Alvarez, L.; Asaro, F.; and Michel, H. 1984. Impact theory of mass extinctions and the invertebrate fossil record. *Science* **223:**1135–1141.

Anapol, F. 1983. Scapula of *Apidium phiomense:* a small anthropoid from the Oligocene of Egypt. *Folia Primatol.* **40:**11–31.

Andrews, P. 1974. New species of *Dryopithecus* from Kenya. *Nature* **249:**188–190.

Andrews, P. 1978*a*. A revision of the Miocene Hominoidea of East Africa. *Bull. Br. Mus. (Nat. Hist.) Geol.* **30:**85–224.

Andrews, P. 1978*b*. Taxonomy and relationships of fossil apes. In *Recent advances in primatology,* 4 vols., ed. D. Chivers and K. Joysey, vol. 3, pp. 43–56. Academic Press, London.

Andrews, P. 1981. Species diversity and diet in monkeys and apes during the Miocene. In *Aspects of human evolution,* ed. C. Stringer, pp. 25–61. Taylor & Francis, London.

Andrews, P. 1982. Ecological polarity in primate evolution. *Zool. J. Linn. Soc.* **74:**233–244.

Andrews, P. 1983. The natural history of *Sivapithecus*. In *New interpretations of ape and human ancestry,* ed. R. Ciochon and R. Corruccini, pp. 441–464. Plenum, New York.

Andrews, P. 1985. Family group systematics and evolution among catarrhine primates. In *Ancestors: the hard evidence,* ed. E. Delson, pp. 14–22. Alan Liss, New York.

Andrews, P. 1986. Molecular evidence for catarrhine evolution. In *Major topics in primate and human evolution,* ed. B. Wood, L. Martin, and P. Andrews, pp. 107–129. Cambridge Univ. Press.

Andrews, P., and Aiello, L. 1984. An evolutionary model for feeding and positional behavior. In *Food acquisition and processing in primates,* ed. D. Chivers, B. Wood, and A. Bilsborough, pp. 429–466. Plenum, New York.

Andrews, P., and Cronin, J. 1982. The relationships of *Sivapithecus* and *Ramapithecus* and the evolution of the orang-utan. *Nature* **297:**541–546

Andrews, P., and Evans, E. 1979. The environment of *Ramapithecus* in Africa. *Paleobiology* **5:**22–30.

Andrews, P., and Simons, E. 1977. A new African Miocene Gibbon-like genus, *Dendropithecus,* with distinctive postcranial adaptations: its significance to origin of Hylobatidae. *Folia Primatol.* **28:**161–169.

Andrews, P., and Tekkaya, I. 1980. A revision of the Turkish Miocene hominoid *Sivapithecus meteai. Palaeontology* **23:**85–95.

Andrews, P., and Tobien, H. 1977. New Miocene locality in Turkey with evidence on the origin of *Ramapithecus* and *Sivapithecus. Nature* **268:**699–701.

Andrews, P., and Van Couvering, J. 1975. Paleoenvironments in the East African Miocene. In *Approaches to primate paleobiology,* ed. F. Szalay, pp. 62–103, Karger, Basel.

Andrews, P., and Walker, A. 1976. The primate and other fauna from Fort Ternan, Kenya. In *Human origins: Louis Leakey and the East African evidence,* ed. G. Isaac and E. McCown, pp. 279–304. Benjamin-Cummings, Menlo Park, Calif.

Andrews, P.; Lord, J.; and Evans, E. 1979. Patterns of ecological diversity in fossil and modern mammalian faunas. *Biol. J. Linn. Soc.* **11:**177–205.

Andrews, P.; Harrison, T.; Martin, L.; and Pickford, M. 1981. Hominoid primates from a new Miocene locality named Meswa Bridge in Kenya. *J. Hum. Evol.* **10:**123–128.

Ankel, F. 1965. Der Canalis sacralis als indikator für die lange der caudalregion der Primaten. *Folio Primatol.* **3:**263–276.

Ankel-Simons, F. 1983. *A survey of living primates and their anatomy.* Macmillan, New York.

Archer, M.; Flannery, T.; Ritchie, A.; and Molnar, R. 1985. First Mesozoic mammal from Australia, an early Cretaceous monotreme. *Nature* **318:**363–366.

Archibald, J. 1977. Ecotympanic bone and internal carotid circulation of eutherians in reference to anthropoid origins.*J. Hum. Evol.* **6:**609–622.

Archibald, J., and Clemens, W. 1984. Mammal evolution near the Cretaceous-Tertiary boundary. In *Catastrophes and earth history: the new uniformitarianism,* ed. W. Berggren and J. Van Couvering, pp. 339–371. Princeton University Press, Princeton, N.J.

Aronson, J., and Taieb, M. 1981. Geology and paleogeography of the Hadar hominid site, Ethiopia. In *Hominid sites: their geologic settings,* ed. G. Rapp and C. Vondra, pp. 165–196. Westview Press, Boulder, Colo.

Axelrod, D. 1975. Evolution and biogeography of Madrean-Tethyan sclerophyll vegetation. *Ann. Mo. Bot. Gard.* **62:**280–334.

Axelrod, D., and Raven, P. 1978. Late Cretaceous and Tertiary vegetation history of Africa. In *Biogeography and ecology of southern Africa,* ed. M. Werger, pp. 77–130. W. Junk, The Hague.

Badgely, C., and Behrensmeyer, A. K. 1980. Paleoecology of Middle Siwalik sediments and faunas, northern Pakistan. *Palaeogeogr., Palaeoclimatol., Palaeoecol.* **30:**133–155.

Baker, B., and Wohlenberg, J. 1971. Structure and evolution of the Kenya Rift Valley. *Nature* **229:**538–542.

Barry, J.; Behrensmeyer, A.; and Monaghan, M. 1980. A geologic and biostratigraphic framework for Miocene sediments near Khaur Village, northern Pakistan. *Postilla* **183:**1–19.

Barry, J.; Lindsay, E.; and Jacobs, L. 1982. A biostratigraphic zonation of the middle and upper Siwaliks of the Potwar Plateau of northern Pakistan. *Palaeogeogr., Palaeoclimatol., Palaeoecol.* **37:**95–130.

Barry, J.; Johnson, N.; Raza, S.; and Jacobs, L. 1985. Neogene mammalian faunal change in southern Asia: correlations with climatic, tectonic, and eustatic events. *Geology* **13:**637–640.

Beard, K.; Teaford, M.; and Walker, A. 1986. New wrist bones of *Proconsul africanus* and *P. nyanzae* from Rusinga Island, Kenya. *Folia Primatol.* **47:**97–118.

Behrensmeyer, A. 1975*a*. Taphonomy and paleoecology in the hominid fossil record. *Yearb. Phys. Anthropol.* **19:**36–50.

Behrensmeyer, A. 1975*b*. The habitat of Plio-Pleistocene hominids in East Africa: taphonomic and microstratigraphic evidence. In *Early hominids of Africa,* ed. C. Jolly, pp. 165–189. Duckworth, London.

Behrensmeyer, A. 1976. Lothagam Hill, Kanapoi, and Ekora: a general summary of stratigraphy and faunas. In *Earliest man and environments in the Lake Rudolf Basin,* ed. Y. Coppens, F. Howell, G. Isaac, and R. Leakey, pp. 163–170. Univ. Chicago Press, Chicago.

Behrensmeyer, A. 1978. Correlation of Plio-Pleistocene sequences in the northern Lake Turkana Basin: a summary of evidence and issues. In *Geological background to fossil man: recent research in the Gregory Rift Valley, East Africa,* ed. W. Bishop, pp. 421–440. Scottish Academic Press, Edinburgh.

Bell, R. 1971. A grazing ecosystem in the Serengeti. *Sci. Am.* **225:**86–93.

Benefit, B., and Pickford, M. 1986. Miocene fossil cercopithecoids from Kenya. *Am. J. Phys. Anthropol.* **69:**441–464.

Berggren, W., and Van Couvering, J. 1974. The late Neogene. *Palaeogeogr., Palaeoclimatol., Palaeoecol.* **16:**1–216.

Bernor, R. 1983. Geochronology and zoogeographic relationships of Miocene Hominoidea. In *New interpretations of ape and human ancestry,* ed. R. Ciochon and R. Corruccini, pp. 21–66. Plenum, New York.

Bernor, R.; Flynn, L.; Harrison, T.; Hussain, S.; and Kelley, J. 1988. *Dionysopithecus* from southern Pakistan and the biochronology and biogeography of early Eurasian catarrhines. *J. Hum. Evol.* **17:**339–358.

Biknevicius, A. 1986. Dental function and diet in the Carpolestidae. *Am. J. Phys. Anthropol.* **71:**157–172.

Bishop, W. 1964. More fossil primates and other Miocene mammals from northeast Uganda. *Nature* **203:**1327–1331.

Bishop, W. 1967. The later Tertiary in East Africa: volcanics, sediments, and faunal inventory. In *Background to evolution in Africa,* ed. W. Bishop and J. D. Clark, pp. 31–56. Univ. Chicago, Press, Chicago.

Bishop, W., ed. 1978. *Geological background to fossil man: recent research in the Gregory Rift Valley, East Africa.* Scottish Academic Press, Edinburgh.

Bishop, W.; Miller, J.; and Fitch, F. 1969. New potassium–argon age determinations relevant to the Miocene fossil mammal sequence in East Africa. *Am. J. Sci.* **267:**669–699.

Boaz, N. 1979. Hominid evolution in Eastern Africa during the Pliocene and early Pleistocene. *Annu. Rev. Anthropol.* **8:**71–85.

Boaz, N., and Howell, F. 1977. A gracile hominid cranium from upper member G of the Shungura Formation, Ethiopia. *Am. J. Phys. Anthropol.* **46:**93–108.

Boaz, N.; Howell, F.; and McCrossin, M. 1982. Faunal age of the Usno, Shungura B and Hadar formations, Ethiopia. *Nature* **300:**633–635.

Bonnefille, R. 1976. Palynological evidence for an important change in the vegetation of the Omo Basin between 2.5 and 2 million years. In *Earliest man and environments in the Lake Rudolf Basin,* ed. Y. Coppens, F. Howell, G. Isaac, and R. Leakey, pp. 421–431. Univ. Chicago Press, Chicago.

Bosler, W. 1981. Species groupings of early Miocene Dryopithecine teeth from East Africa. *J. Hum. Evol.* **10:**151–158.

Bowen, B. 1971. Paleoenvironmental interpretations of the Oligocene Jebel el Qatrani Formation, Fayum Depression, Egypt. *Abs. Geol. Soc. Am. Bull. 1971,* p. 254.

Bowen, B., and Vondra, C. 1973*a*. Paleoenvironmental interpretation of the Oligocene Jebel el Qatrani Formation, Fayum Depression, Egypt. Unpublished ms.

Bowen, B., and Vondra, C. 1973*b*. Stratigraphical relationships of the Plio-Pleistocene deposits, East Rudolf, Kenya. *Nature* **242:**391–393.

Bown, T. 1982. Ichnofossils and rhizoliths of the fluvial nearshore Jebel Qatrani Formation, Fayum Province, Egypt. *Palaeogeogr. Palaeoclimatol. Palaeoecol.* **40:**255–309.

Bown, T., and Gingerich, P. 1973. The Paleocene primate *Plesiolestes* and the origin of Microsyopidae. *Folia Primatol.* **19:**1–18.

Bown, T., and Rose, K. 1976. New early Tertiary primates and a reappraisal of some Plesiadapiformes. *Folia Primatol.* **26:**109–138.

Bown, T., and Rose, K. 1987. Patterns of dental evolution in the Early Eocene anaptomorphine primates (Omomyidae) from the Bighorn basin, Wyoming. *J. Paleo. Soc. Mem.* **23:**1–162.

Bown, T.; Krause, M.; Wing, S.; Fleagle, J.; Tiffany, B.; Simons, E.; and Vondra, C. 1982. The Fayum primate forest revisited. *J. Hum. Evol.* **11:**603–632.

Brain, C. 1981*a*. Hominid evolution and climatic change. *S. Afr. J. Sci.* **77:**104–105.

Brain, C. 1981*b*. *The hunters or the hunted? An introduction to African cave taphonomy.* Univ. Chicago Press, Chicago.

Brain, C. 1983. The terminal Miocene event: a critical environmental and evolutionary episode. In *SASQUA Intl. Symp.,* ed. J. Vogel, pp. 491–498. A. Balkema, Rotterdam.

Brain, C. 1985. Cultural and taphonomic comparisons of hominids from Swartkrans and Sterkfontein. In *Ancestors: the hard evidence,* ed. E. Delson, pp. 72–75, Alan Liss, New York.

Brain, C. 1988, New information from the Swartkrans cave of relevance to "robust" australopithecines. In *Evolutionary History of the "robust" australopithecines,* ed. F. Grine, pp. 311–316. Aldine de Gruyter, New York.

Brain, C., and Sillen, A. 1988. Evidence from the Swartkrans cave for the earliest use of fire. *Nature* **336:**464–466.

Brain, C.; van Riet Lowe, C.; and Dart, R. 1955. Kafuan stone artefacts in the post australopithecine breccia at Makapansgat. *Nature* **175:**16.

Branco, W. 1898. Die menschenahnlichen Zahne aus dem Bohnerz der schwabischen Alb. *Jh. Ver. vaterl. naturk. Wurtt.* **54:**1–144.

Britten, R. 1986. Rates of DNA sequence evolution differ between taxonomic groups. *Science* **231:**1393–1398.

Bromage, T., and Dean, C. 1985. Re-evaluation of the age at death of immature fossil hominids. *Nature* **317:**525–527.

Broom, R. 1938. More discoveries of *Australopithecus. Nature* **141:**828–829.

Broom, R., and Robinson, J. 1947. Further remains of the Sterkfontein ape-man, *Plesianthropus*. *Nature* **160:**430–431.

Broom, R., and Robinson, J. 1950. Man contemporaneous with the Swartkrans ape-man. *Am. J. Phys. Anthropol.* **8:**151–156.

Broom, R., and Robinson, J. 1951. Eruption of the permanent teeth in South African fossil ape-men. *Nature* **167:**443.

Brown, B.; Gregory, W.; and Hellman, M. 1924. On three incomplete anthropoid jaws from the Siwaliks, India. *Am. Mus. Novit.* **130:**1–9.

Brown, F. 1982. Tulu Bor Tuff at Koobi Fora correlated with the Sidi Hakoma tuff at Hadar. *Nature* **300:**631–633.

Brown, F., and Shuey, R. 1976. Magnetostratigraphy of the Shungura and Usno formations, lower Omo Valley, Ethiopia. In *Earliest man and environments in the Lake Rudolf Basin,* ed. Y. Coppens, F. Howell, G. Isaac, and R. Leakey, pp. 64–78. Univ. Chicago Press, Chicago.

Brown, F.; Howell, F.; and Eck, G. 1978. Observations on problems of correlation of late Cenozoic hominid-bearing formations in the North Lake Turkana Basin. In *Geological background to fossil man: recent advances in the Gregory Rift Valley, East Africa,* ed. W. Bishop, pp. 473–498. Scottish Academic Press, Edinburgh.

Brown, F.; McDougall, I.; Davies, T.; and Maier, R. 1985. An integrated Plio-Pleistocene chronology for the Turkana Basin. In *Ancestors: the hard evidence,* ed. E. Delson, pp. 82–90. Alan Liss, New York.

Brown, F.; Harris, J.; Leakey, R.; and Walker, A. 1985. Early *Homo erectus* skeleton from west Lake Turkana, Kenya. *Nature* **316:**788–792.

Buchardt, B. 1978. Oxygen isotope palaeotemperatures from the Tertiary period in the North Sea area. *Nature* **275:**121–123.

Butzer, K. 1976. The Mursi, Nkalabong, and Kibish formations, lower Omo Basin, Ethiopia, In *Earliest man and environments in the Lake Rudolf Basin,* ed. Y. Coppens, F. Howell, G. Isaac, and R. Leakey, pp. 12–23. Univ. Chicago Press, Chicago.

Butzer, K., and Hansen, C. 1968. *Desert and river in Nubia.* Univ. of Wisconsin Press, Madison.

Cachel, S. 1975. The beginning of primates. In *Primate functional morphology and evolution,* ed. R. Tuttle, pp. 23–36. Mouton, The Hague.

Cachel, S. 1981. Plate tectonics and the problem of anthropoid origins. *Yearb. Phys. Anthropol.* **24:**139–172.

Cann, R.; Brown, W.; and Wilson, A. 1987. Mitochondrial DNA and human evolution. *Nature* **325:**31–36.

Cartmill, M. 1971. Ethmoid component in the orbit of primates. *Nature* **232:**566–567.

Cartmill, M. 1972. Arboreal adaptations and the origin of the order Primates. In *Functional and evolutionary biology of primates,* ed. R. Tuttle, pp. 97–122. Aldine, Chicago.

Cartmill, M. 1974. Rethinking primate origins. *Science* **184:**436–443.

Cartmill, M. 1975. Strepsirhine basicranial structures and the affinities of the Cheirogaleidae. In *Phylogeny of the primates: a multidisciplinary approach,* ed. W. Luckett and F. Szalay, pp. 313–354. Plenum, New York.

Cartmill, M. 1978. The orbital mosaic in prosimians and the use of variable traits in systematics. *Folia Primatol.* **30:**89–114.

Cartmill, M. 1979. The volar skin of primates: its frictional characteristics and their functional significance. *Am. J. Phys. Anthropol.* **50:**497–510.

Cartmill, M. 1981. Morphology, function, and evolution of the anthropoid postorbital septum. In *Evolutionary biology of New World monkeys and continental drift,* ed. R. Ciochon and A. Chiarelli, pp. 243–273. Plenum, New York.

Cartmill, M., and Gingerich, P. 1978. An ethmoid exposure (os planum) in the orbit of *Indri indri. Am. J. Phys. Anthropol.* **48:**535–538.

Cartmill, M., and Kay, R. 1978. Craniodental morphology, tarsier affinities, and primate suborders. In *Recent advances in primatology,* 4 vols., ed. D. Chivers and K. Joysey, **3:**205–214. Academic Press, London.

Cartmill, M.; MacPhee, R.; and Simons, E. 1981. Anatomy of the temporal bone in early anthropoids, with remarks on the problem of anthropoid origins. *Am. J. Phys. Anthropol.* **56:**3–21.

Cartmill, M.; Pilbeam, D.; and Isaac, G. 1986. One hundred years of paleoanthropology. *Am. Sci.* **74:**410–420.

Cave, A. 1961. Frontal sinus of the gorilla. *Proc. Zool. Soc. Lond.* **136:**359–373.

Cave, A., and Haines, R. 1940. The paranasal sinuses of the anthropoid apes. *J. Anat.* **72:**493–523.

Charlesworth, E. 1854. *Report of the British Association for the Advancement of Science for 1854.* p. 80. Brit. Assn. Adv. Sci. Liverpool.

Chesters, K. 1957. The Miocene flora of Rusinga Island, Lake Victoria, Kenya. *Palaeontographica* **101:**30–67.

Chivers, D., and Hladik, C. 1984. Diet and gut morphology in primates. In *Food acquisition and processing in primates,* ed. D. Chivers, B. Wood, and A. Bilsborough, pp. 213–230. Plenum, New York.

Chivers, D.; Andrews, P.; Preuschoft, H.; Bilsborough, A.; Wood, B. 1984. Food acquisition and processing in primates: concluding discussion. In *Food Acquisition and Processing in Primates,* ed. D. Chivers, B. Wood, and A. Bilsborough, pp. 545–556. Plenum, New York.

Churcher, C. 1970. Two new upper Miocene giraffids from Fort Ternan, Kenya, East Africa. In *Fossil Vertebrates of Africa,* ed.. L. Leakey and R. Savage, **2:**1–106. Academic Press, London.

Ciochon, R. 1983. Hominoid cladistics and the ancestry of modern apes and humans. In *New interpretations of ape and human ancestry,* ed. R. Ciochon and R. Corruccini, pp. 781–844. Plenum, New York.

Ciochon, R., and Chiarelli, A. 1980. Paleobiogeographic perspectives on the origin of the Platyrrhini. In *Evolutionary biology of the New World monkeys and continental drift,* ed. R. Ciochon and A. Chiarelli, pp. 459–493. Plenum, New York.

Ciochon, R., and Corruccini, R. 1975. Morphometric analysis of platyrrhine femora with taxonomic implications and notes on two fossil forms. *J. Hum. Evol.* **4:**193–217.

Ciochon, R., and Corruccini, R. 1977. The phenetic position of *Pliopithecus* and its phylogenetic relationship to the Hominoidea. *Syst. Zool.* **26:**290–299.

Ciochon, R.; Savage, D.; Tint, T.; and Maw, B. 1985. Anthropoid origins in Asia? New discovery of *Amphipithecus* from the Eocene of Burma. *Science* **229:**756–759.

Clark, J. 1970. The prehistory of Africa. Praeger, New York.

Clark, J. 1985. Leaving no stone unturned: archaeological advances and behavioral adaptation. In *Hominid Evolution: Past, Present, and Future,* ed. P. Tobias, pp. 65–88. Alan Liss, New York.

Clarke, R. 1988. A new *Australopithecus* cranium from Sterkfontein and its bearing on the ancestry of *Paranthropus*. In *Evolutionary history of the "robust" australopithecines,* ed. F. Grine, pp. 285–292. Aldine de Gruyter, New York.

Clarke, R.; Howell, F.; and Brain, C. 1970. More evidence of an advanced hominid at Swartkrans. *Nature* **225:**1217–1220.

Clemens, W. 1974. *Purgatorius,* an early paromomyid primate. *Science* **184:**903–905.

Coe, M. 1984. Primates: their niche structure and habitats. In *Food acquisition and processing in primates,* ed. D. Chivers, B. Wood, and A. Bilsborough, pp. 1–32. Plenum, New York.

Colbert, E. 1935. Siwalik mammals in the American Museum of Natural History. *Trans. Am. Philos. Soc.* **26:**1–401.

Colbert, E. 1937. A new primate from the upper Eocene Pondaung formations of Burma. *Am. Mus. Novit.* no. 951, pp. 1–18.

Collins, E. T. 1921. Changes in the visual organs associated with the adoption of arboreal life with the assumption of the erect posture. *Trans. Ophthalmol. Soc. U. K.* **41:**10–90.

Conroy, G. 1974. Primate postcranial remains from the Fayum province, Egypt. Ph.D. dissertation, Yale University.

Conroy, G. 1976*a*. Primate postcranial remains from the Oligocene of Egypt. *Contrib. Primatol.* **8:**1–134.

Conroy, G. 1976*b*. Hallucal tarsometatarsal joint in an Oligocene anthropoid, *Aegyptopithecus zeuxis. Nature* **262:**684–686.

Conroy, G. 1978. Candidates for anthropoid ancestry: some morphological and paleozoogeographical considerations. In *Recent advances in primatology,* 4 vols., ed. D. Chivers and K. Joysey, **3:**27–41. Academic Press, London.

Conroy, G. 1980. Ontogeny, auditory structures and primate evolution. *Am. J. Phys. Anthropol.* **52:**443–451.

Conroy, G. 1981. A review of the Torrejonian (middle Paleocene) primates from the San Juan Basin, New Mexico. In *Advances in San Juan Basin paleontology,* ed. S. Lucas, K. Rigby, and B. Kues, pp. 161–176. Univ. New Mexico Press, Albuquerque.

Conroy, G. 1987. Problems in body weight estimation in fossil primates. *Int. J. Primatol.* **8:**115–137.

Conroy, G. 1988. Alleged synapomorphy of the M1/I1 eruption pattern in robust australopithecines and *Homo:* evidence from high-resolution computed tomography. *Am. J. Phys. Anthropol.* **75:**487–492.

Conroy, G., and Fleagle, J. 1972. Locomotor behavior in living and fossil pongids. *Nature* **237:**103–104.

Conroy, G., and Pilbeam, D. 1975. *Ramapithecus,* a review of its hominid status. In *Paleoanthropology: morphology and paleoecology,* ed. R. Tuttle, pp. 59–86. Mouton, The Hague.

Conroy, G., and Rose, M. 1983. Evolution of the primate foot from the earliest primates to the Miocene hominoids. *Foot and Ankle* **3:**342–364.

Conroy, G., and Vannier, M. 1987. The Taung skull revisited: new evidence from high-resolution computed tomography. *Nature* **329:**625–627.

Conroy, G., and Vannier, M. (in press, *a*). Dental development in South African australopithecines. Part I. Problems of pattern and chronology. *Am. J. Phys. Anthropol.*

Conroy, G. and Vannier, M. (in press, *b*). Dental development in South African australopithecines. Part II. Dental stage assessment. *Am. J. Phys. Anthropol.*

Conroy, G., and Wible, J. 1978. Middle ear morphology of *Lemur variegatus:* some implications for primate paleontology. *Folia Primatol.* **29:**81–85.

Conroy, G.; Schwartz, J.; and Simons, E. 1975. Dental eruption patterns in Parapithecidae. *Folia Primatol.* **24:**275–281.

Cooke, H. 1976. Suidae from Pliocene-Pleistocene strata in the Rudolf Basin. In *Earliest man and environments in the Lake Rudolf Basin,* ed. Y. Coppens, F. Howell, G. Isaac, and R. Leakey, pp. 251–263. Univ. Chicago Press, Chicago.

Cooke, H. 1983. Human evolution: the geological framework. *Can. J. Anthropol.* **3:**143–161.

Cooke, H. 1986. *Changing perspectives on the age of man.* Raymond Dart Lecture no. 21. Witwatersrand Univ. Press, Johannesburg.

Cope, E. 1872. On a new vertebrate genus from the northern part of the Tertiary basin of Green River. *Proc. Am. Philos. Soc.* **12:**554.

Corliss, B.; Aubry, M.; Berggren, W.; Fenner, J.; Keigwin, L.; and Keller, G. 1984. The Eocene/Oligocene boundary event in the deep sea. *Science* **226:**806–810.

Corruccini, R., and McHenry, H. 1980. Cladometric analysis of Pliocene hominids. *J. Hum. Evol.* **9:**209–221.

Corruccini, R.; Ciochon, R.; and McHenry, H. 1975. Osteometric shape relationships in the wrist joint of some anthropoids. *Folia Primatol.* **24:**250–274.

Corruccini, R.; Ciochon, R.; and McHenry, H. 1976. The postcranium of Miocene hominoids: were dryopithecines merely "dental apes"? *Primates* **17:**205–223.

Covert, B. 1985. The skeleton of *Smilodectes gracilis. Am. J. Phys. Anthropol.* **66:**159.

Cox, A. 1972. Geomagnetic reversals: their frequency, their origin, and some problems of correlation. In *Calibration of hominoid evolution: recent advances in isotopic and other dating methods applicable to the origin of man,* ed. W. Bishop and J. Miller, pp. 93–106. Scottish Academic Press, Edinburgh.

Cronin, J. 1983. Apes, humans, and molecular clocks. In *New interpretations of ape and human ancestry,* ed. R. Ciochon and R. Corruccini, pp. 115–149. Plenum, New York.

Cronin, J.; Boaz, N.; Stringer, C.; and Rak, Y. 1981. Tempo and mode in human evolution. *Nature* **292:**113–122.

Crook, J., and Gartlan, S. 1966. Evolution of primate societies. *Nature* **210:**1200–1203.

Curtis, G.; Drake, R.; Cerling, T.; and Hampel, J. 1975. Age of KBS tuff in Koobi Fora Formation, East Rudolf, Kenya. *Nature* **258:**395–398.

Cuvier, G. 1821. Discours sur la théorie de la terre, servant d'introduction aux recherchés sur les ossements fossiles. Paris.

Dagosto, M. 1983. Postcranium of *Adapis parisiensis* and *Leptadapis magnus. Folia Primatol.* **41:**49–101.

Dagosto, M. 1985. The distal tibia of Primates with special reference to the Omomyidae. *Int. J. Primatol.* **6:**45–75.

Dart, R. 1925. *Australopithecus africanus*: the man-ape of South Africa. *Nature* **115:**195–199.

Dart, R. 1948*a*. An adolescent promethean australopithecine mandible from Makapansgat. *S. Afr. J. Sci.* **45:**73–75.

Dart, R. 1948*b*. The adolescent mandible of *Australopithecus prometheus. Am. J. Phys. Anthropol.* **6:**391–412.

Dart, R. 1948*c*. The Makapansgat proto-human *Australopithecus prometheus Am. J. Phys. Anthropol.* **6:**259–281.

Dart, R. 1957. The Osteodontokeratic Culture of *Australopithecus prometheus. Transvaal Mus. Mem.* no. 10.

Darwin, C. 1859. *The origin of species by means of natural selection or the preservation of favored races in the struggle for life.* Murray, London.

Darwin, C. 1871. *The descent of man, and selection in relation to sex.* Murray, London.

Dashzeveg, D., and McKenna, M. 1977. Tarsioid primate from the early Tertiary of the Mongolian People's Republic. *Acta Palaeontol. Pol.* **22:**119–137.

Dawson, M.; West, R.; Langston, W.; and Hutchinson, J. 1976. Paleogene terrestrial vertebrates: northernmost occurrence, Ellesmere Island, Canada. *Science* **192:**781–782.

Dean, M. 1985. The eruption pattern of the permanent incisors and first permanent molars in *Australopithecus robustus. Am. J. Phys. Anthropol.* **67:**251–257.

de Bonis, L., and Melentis, J. 1977. Les primates hominoides du Vallesien de Macedoine. Etude de la machoire inferieure. *Geobios (Lyon)* **10:**849–885.

de Bonis, L., and Melentis, J. 1978. Les primates hominoides du Miocène superieur de Macedoine. Etude de la machoire superieure. *Ann. Paleontol.* **64:**185–202.

de Bonis, L., and Melentis, J. 1980. Nouvelles remarques sur l'anatomie d'un primate hominoide du Miocène: *Ouranopithecus macedoniensis. C. R. Acad. Sci. Paris, Ser. D* **290:**755–758.

de Bonis, L.; Bouvrain, G.; Geraads, D.; and Melentis, J. 1974. Première decouverte d'un primate hominoide dans le Miocène superieur de Macedoine. *C. R. Acad. Sci. (Paris)* **278:**3063–3066.

de Groot, J., and Chusid, J. 1988. *Correlative neuroanatomy.* Appleton & Lange, East Norwalk, Conn.

de Heinzelin, J.; Haesaerts, P.; and Howell, F. 1976. Plio-Pleistocene formations of the lower Omo Basin, with particular reference to the Shungura Formation. In *Earliest man and environments in the Lake Rudolf Basin,* ed. Y. Coppens, F. Howell, G. Isaac and R. Leakey, pp. 24–49. Univ. Chicago Press, Chicago.

Delson, E. 1975*a*. Toward the origin of the Old World monkeys. *Colloq. Int. Cent. Natl. Rech. Sci.,* **218:**839–850.

Delson, E. 1975*b*. Evolutionary history of the Cercopithecidae. *Contrib. Primatol.* **5:**167–217.

Delson, E. 1977. Catarrhine phylogeny and classification: principles, methods, and comments. *J. Hum. Evol.* **6:**433–459.

Delson, E. 1979. *Prohylobates* from the early Miocene of Libya: a new species and its implications for cercopithecid origins. *Geobios (Lyon)* **12:**725–733.

Delson, E. 1984. Cercopithecoid biochronology of the African Plio-Pleistocene: correlation among eastern and southern hominid-bearing localities. *Cour. Forschungsinst. Senckenb.* **69:**199–218.

Delson, E. 1986. Human phylogeny revised again. *Nature* **322:**496–497.

Delson, E. 1988. Chronology of South African australopith site units. In *Evolutionary history of the "robust" australopithecines,* ed. F. Grine, pp. 317–324. Aldine de Gruyter, New York.

Delson, E., and Andrews, P. 1975. Evolution and interrelationships of the catarrhine primates. In *Phylogeny of the primates: a multidisciplinary approach,* ed. W. Luckett and F. Szalay, pp. 405–446. Plenum, New York.

Delson, E., and Rosenberger, A. 1980. Phyletic perspectives on platyrrhine origins and anthropoid relationships. In *Evolutionary biology of the New World monkeys and continental drift,* ed. R. Ciochon and A Chiarelli, pp. 445–458. Plenum, New York.

Demment, M. 1983. Feeding ecology and the evolution of body size of baboons. *Afr. J. Ecol.* **21:**219–233.

Dene, H.; Goodman, M.; and Prychodko, W. 1976. Immunodiffusion evidence on the phylogeny of the primates. In *Molecular anthropology: genes and proteins in the evolutionary ascent of the primates,* ed. M. Goodman and R. Tashian, pp. 171–190. Plenum, New York.

Dietz, R., and Holden, J. 1970. The breakup of Pangaea. *Sci. Am.* **223:**30–41.

Dobzhansky, T. 1962. *Mankind evolving.* Yale Univ. Press, New Haven.

Eckhardt, R. 1977. Hominid origins: the Lothagam mandible. *Curr. Anthropol.* **18:**356.
Eckhardt, R. 1987. Hominoid nasal region polymorphism and its phylogenetic significance. *Nature* **328:**333–335.
Eicher, D. 1976. *Geologic time.* Prentice-Hall, Englewood Cliffs, N.J.
Eldredge, N., and Gould, S. 1972. Punctuated equilibria: an alternative to phyletic gradualism. In *Models in paleobiology,* ed. T. Schopf, pp. 82–115. Freeman, Cooper, San Francisco.
Elliot, D. 1913. A review of the Primates. *Am. Mus. Nat. Hist.* Monograph 1, vol. 1.
Etler, D. 1984. The fossil hominoids of Lufeng, Yunnan Province, The Peoples Republic of China: a series of translations. *Yearb. Phys. Anthropol.* **27:**1–56.
Evans, J.; Van Couvering, J.; and Andrews, P. 1981. Paleoecology of Miocene sites in western Kenya. *J. Hum. Evol.* **10:**99–116.

Falconer, H. 1868. *Palaeontological memoirs* **2; 8:**675.
Falk, D. 1983. Cerebral cortices of East African early hominids. *Science* **222:**1072–1074.
Falk, D. 1985. Apples, oranges, and the lunate sulcus. *Am. J. Phys. Anthropol.* **67:**313–315.
Falk, D., and Conroy, G. 1983. The cranial venous sinus system of *Australopithecus afarensis. Nature* **306:**779–781.
Feibel, C.; Brown, F.; and McDougall, I. 1989. Stratigraphic context of fossil hominids from the Omo Group deposits: Northern Turkana Basin, Kenya, and Ethiopia. *Am. J. Phys. Anthropol.* **78:**595–623.
Fitch, R. 1972. Selection of suitable material for dating and the assessment of geological error in potassium-argon age determination. In *Calibration of hominoid evolution: recent advances in isotopic and other dating methods applicable to the origin of man,* ed. W. Bishop and J. Miller, pp. 77–92. Scottish Academic Press, Edinburgh.
Fitch, F., and Miller, J. 1976. Conventional potassium–argon and argon-40/argon-39 dating of volcanic rocks from East Rudolf. In *Earliest man and environments in the Lake Rudolf Basin,* ed. Y. Coppens, F. Howell, G. Isaac, and R. Leakey, pp. 123–147. Univ. Chicago Press, Chicago.
Fitch, F.; Findlater, I.; Watkins, R.; and Miller, J. 1974. Dating of the rock succession containing fossil hominids at East Rudolf, Kenya. *Nature* **251:**213–215.
Fleagle, J. 1974. Dynamics of a brachiating siamang. *Nature* **248:**259–260.
Fleagle, J. 1975. A small gibbon-like hominoid from the Miocene of Uganda. *Folia Primatol.* **24:**1–15.
Fleagle, J. 1980. Locomotor behavior of the earliest anthropoids: a review of the current evidence. *Z. Morphol. Anthropol.* **71:**149–156.
Fleagle, J. 1988. *Primate adaptation and evolution.* Academic Press, New York.
Fleagle, J., and Bown, T. 1983. New primate fossils from Late Oligocene localities of Chubut Province, Argentina. *Folia Primatol.* **41:**240–266.
Fleagle, J., and Kay, R. 1983. New interpretations of the phyletic position of Oligocene hominoids. In *New interpretations of ape and human ancestry,* ed. R. Ciochon and R. Corruccini, pp. 181–210. Plenum, New York.
Fleagle, J., and Kay, R. 1987. The phyletic position of the Parapithecidae. *J. Hum. Evol.* **16:**483–532.
Fleagle, J., and Rosenberger, A. 1983. Cranial morphology of the earliest anthropoids. In *Morphologie, evolutive, morphogenèse du crane et origine de l'homme,* ed. M. Sakka, pp. 141–153, Centre National de la Recherche de Scientifique, Paris.

Fleagle, J., and Simons, E. 1978. *Micropithecus clarki,* a small ape from the Miocene of Uganda. *Am. J. Phys. Anthropol.* **49:**427–440.
Fleagle, J., and Simons, E. 1979. Anatomy of the bony pelvis in parapithecid primates. *Folia Primatol.* **31:**176–186.
Fleagle, J., and Simons, E. 1982*a*. Skeletal remains of *Propliopithecus chirobates* from the Egyptian Oligocene. *Folia Primatol.* **39:**161–177.
Fleagle, J., and Simons, E. 1982*b*. The humerus of *Aegyptopithecus zeuxis,* a primitive anthropoid. *Am. J. Phys. Anthropol.* **59:**175–193.
Fleagle, J., and Simons, E. 1983. The tibio-fibular articulation in *Apidium phiomense,* an Oligocene anthropoid. *Nature* **301:**238–239.
Fleagle, J.; Kay, R.; and Simons, E. 1980. Sexual dimorphism in early anthropoids. *Nature* **287:**328–330.
Fleagle, J.; Simons, E.; and Conroy, G. 1975. Ape limb bone from Oligocene of Egypt. *Science* **189:**135–137.
Fleagle, J.; Bown, T.; Obradovich, J.; and Simons, E. 1986. Age of the earliest African anthropoids. *Science* **234:**1247–1249.
Fleagle, J.; Powers, D.; Conroy, G.; and Watters, J. 1987. New fossil primates from Santa Cruz Province, Argentina. *Folia Primatol.* **48:**65–77.
Flint, R. 1971. *Glacial and quaternary geology.* Wiley, New York.
Ford, S. 1986*a*. Comment on the evolution of claw-like nails in callitrichids. *Am. J. Phys. Anthropol.* **70:**25–26.
Ford, S. 1986*b*. Systematics of the New World monkeys. In *Comparative primate biology,* 4 vols., ed. D. Swindler and J. Erwin, **1:**73–135. Alan Liss, New York.
Fourtau, R. 1918. *Contribution à l'étude des vertèbres miocène de l'Egypt.* Egypt Survey Dept., Cairo.
Frakes, L., and Kemp, E. 1972. Influence of continental positions on early Tertiary climates. *Nature* **240:**97–100.
Frisch, J. 1973. The hylobatid dentition. In *Gibbon and Siamang,* ed. D. Rumbaugh, **2:**56–95. Karger, Basel.

Gantt, D. 1983. The enamel of Neogene hominoids: structural and phyletic implications. In *New interpretations of ape and human ancestry,* ed. R. Ciochon and R. Corruccini, pp. 249–298. Plenum, New York.
Garber, P. 1984. Proposed nutritional importance of plant exudates in the diet of the Panamanian tamarin, *Saquinus oedipus geoffroyi. Int. J. Primatol.* **5:**1–15.
Gartner, S., and McGuirk, J. 1979. Terminal Cretaceous extinction scenario for a catastrophe. *Science* **206:**1272–1276.
Gaudry, A. 1890. Le dryopithèque. *Mem. Soc. Geol. Fr.* **1:**1–11.
Gaulin, S. 1979. A Jarman/Bell model of primate feeding niches. *Hum. Ecol.* **7:**1–20.
Gaulin, S., and Sailer, L. 1984. Sexual dimorphism in weight among the primates: the relative impact of allometry and sexual selection. *Int. J. Primatol.* **5:**515–535.
Gazin, C. 1958. A review of the middle and upper Eocene primates of North America. *Smithson. Misc. Collect.* **136:**1–112.
Gebo, D., and Simons, E. 1984. Puncture marks on early African anthropoids. *Am. J. Phys. Anthropol.* **65:**31–36.
Gebo, D., and Simons, E. 1987. Morphology and locomotor adaptations of the foot in early Oligocene anthropoids. *Am. J. Phys. Anthropol.* **74:**83–101.
Gengwu, L. 1987. Neogene climatic characteristics of northern China. Abstr. of pap. at The Paleoenvironment of East Asia from the mid-Tertiary. Centre of Asian Studies, Univ. Hong Kong, Jan. 9–13, 1987.

Gentry, A. 1970. The Bovidae of the Fort Ternan fossil fauna. In *Fossil vertebrates of Africa,* ed. Leakey and R. Savage, **2:**243–324. Academic Press, London.
Gentry, A. 1976. Bovidae of the Omo Group deposits. In *Earliest Man and environments in the Lake Rudolf Basin,* ed. Y. Coppens, F. Howell, G. Isaac, and R. Leakey, pp. 293–301. Univ. Chicago Press, Chicago.
Gervais, P. 1845. *Zoologie de la France.* Dubochet, Paris.
Gervais, P. 1872. Sur un singe fossile, d'espèce non enocre décrite, qui a été decouvert au Monte-Bamboli (Italie). *C. R. Acad. Sci. (Paris)* **74:**1217–1223.
Gingerich, P. 1973*a*. First record of the Paleocene primate *Chiromyoides* from North America. *Nature* **244:**517–518.
Gingerich, P. 1973*b*. Anatomy of the temporal bone in the Oligocene anthropoid *Apidium* and the origin of Anthropoidea. *Folia Primatol.* **19:**329–337.
Gingerich, P. 1974. Function of pointed premolars in *Phenacolemur* and other mammals. *J. Dent. Res.* **53:**497.
Gingerich, P. 1975*a*. Dentition of *Adapis parisiensis* and the evolution of lemuriform primates. In *Lemur biology,* ed. I. Tattersall and R. Sussman, pp. 65–80. Plenum, New York.
Gingerich, P. 1975*b*. New North American Plesiadapidae and a biostratigraphic zonation of the middle and upper Paleocene. *Contrib. Mus. Paleontol. Univ. Mich.* **24:**135–148.
Gingerich, P. 1975*c*. Systematic position of *Plesiadapis. Nature* **253:**111–113.
Gingerich, P. 1975*d*. A new genus of Adapidae from the late Eocene of southern France and its significance for the origin of higher primates. *Contrib. Mus. Paleontol. Univ. Mich.* **24:**163–170.
Gingerich, P. 1976. Cranial anatomy and evolution of early Tertiary Plesiadapidae. *Contrib. Mus. Paleontol. Univ. Mich.* **15:**1–141.
Gingerich, P. 1977*a*. Correlation of tooth size and body size in living hominoid primates, with a note on relative brain size in *Aegyptopithecus* and *Proconsul. Am. J. Phys. Anthropol.* **47:**395–398.
Gingerich, P. 1977*b*. New species of Eocene primates and the phylogeny of European Adapidae. *Folia Primatol.* **28:**60–80.
Gingerich, P. 1977*c*. Radiation of Eocene Adapidae in Europe. *Geobios (Lyon) Mem. Special* **1:**165–182.
Gingerich, P. 1979. Phylogeny of Middle Eocene Adapidae in North America: *Smilodectes and Notharctus. J. Paleontol.* **53:**153–163.
Gingerich, P. 1980*a*. Dental and cranial adaptations in Eocene Adapidae. *Z. Morph. Anthropol.* **71:**135–142.
Gingerich, P. 1980*b*. Eocene Adapidae, paleobiogeography, and the origin of South American Platyrrhini. In *Evolutionary biology of the New World monkeys and continental drift,* ed. R. Ciochon and A. Chiarelli, pp. 123–138. Plenum, New York.
Gingerich, P. 1981. Early Cenozoic Omomyidae and the evolutionary history of tarsiiform primates. *J. Hum. Evol.* **10:**345–374.
Gingerich, P. 1984. Paleobiology of tarsiiform primates. In *Biology of tarsiers,* ed. C. Niemitz, pp. 33–44. Gustav Fischer, Stuttgart.
Gingerich, P. 1985. South American mammals in the Paleocene of North America. In *The great American biotic interchange,* ed. F. Stehli and S. Webb, pp. 123–137. Plenum, New York.
Gingerich, P. 1986. Early Eocene *Cantius torresi,* oldest primate of modern aspect from North America. *Nature* **319:**319–321.
Gingerich, P., and Dorr, J. 1979. Mandible of *Chiromyoides minor* from the upper Paleocene Chappo member of the Wasatch Formation, Wyoming. *J. Paleontol.* **53:**550–552.

Gingerich, P., and Haskin, R. 1981. Dentition of early Eocene *Pelycodus jarrovii* and the generic attribution of species formerly referred to *Pelycodus*. *Contrib. Mus. Paleontol. Univ. Mich.* **25:**327–337.

Gingerich, P., and Martin, R. 1981. Cranial morphology and adaptations in Eocene Adapidae. II. The Cambridge skull of *Adapis parisiensis*. *Am. J. Phys. Anthropol.* **56:**235–257.

Gingerich, P., and Schoeniger, M. 1977. The fossil record and primate phylogeny. *J. Hum. Evol.* **6:**483–505.

Gingerich, P., and Simons, E. 1977. Systematics, phylogeny, and evolution of early Eocene Adapidae (Mammalia, Primates) in North America. *Contrib. Mus. Paleontol. Univ. Mich.* **24:**245–279.

Gingerich, P., and Smith, B. 1985. Allometric scaling in the dentition of primates and insectivores. In *Size and scaling in primate biology*, ed. W. Jungers, pp. 257–272. Plenum, New York.

Gingerich, P.; Smith, B.; and Rosenberger, K. 1982. Allometric scaling in the dentition of primates and prediction of body weight from tooth size in fossils. *Am. J. Phys. Anthropol.* **57:**81–100.

Gleadow, A. 1980. Fission track age of the KBS tuff and associated hominid remains in northern Kenya. *Nature* **284:**225–230.

Godinot, M., and Dagosto, M. 1983. The astragalus of *Necrolemur*. *J. Paleontol.* **57:**1321–1324.

Goodall, J. 1976. Continuities between chimpanzee and human behavior. In *Human origins: Louis Leakey and the East African evidence*, ed. G. Isaac and T. McCown, pp. 81–95. W. A. Benjamin, Menlo Park, Calif.

Goodman, M. 1975. Protein sequence and immunological specificity. In *Phylogeny of the primates*, ed. W. Luckett and F. Szalay, pp. 219–248. Plenum, New York.

Goodman, M. 1976. Protein sequences in phylogeny. In *Molecular evolution*, ed. F. Ayala, pp. 141–159. Sinauer Associates, Sunderland, Mass.

Goodman, M.; Baba, M.; and Darga, L. 1983. The bearing of molecular data on the cladogenesis and times of divergence of hominoid lineages. In *New interpretations of ape and human ancestry*, ed. R. Ciochon and R. Corruccini, pp. 67–86. Plenum, New York.

Gould, S. 1966. Allometry and size in ontogeny and phylogeny. *Biol. Rev. Camb. Philos. Soc.* **41:**587–640.

Gowlett, J.; Harris, J.; Walton, D.; and Wood, B. 1981. Early archaeological sites, hominid remains and traces of fire from Chesowanja, Kenya. *Nature* **294:**125–129.

Gray, J. 1968. *Animal locomotion*. Weidenfeld & Nicholson, London.

Greenfield, L. 1979. On the adaptive pattern of *"Ramapithecus"*. *Am. J. Phys. Anthropol.* **50:**527–548.

Greenfield, L. 1980. A late divergence hypothesis. *Am. J. Phys. Anthropol.* **52:**351–365.

Greenfield, L. 1983. Toward the resolution of discrepancies between phenetic and paleontological data bearing on the question of human origins. In *New interpretations of ape and human ancestry*, ed. R. Ciochon and R. Corruccini, pp. 695–704. Plenum, New York.

Gregory, J. 1896. *The great rift valley*. Seeley Service, London.

Gregory, W. 1916. Studies on the evolution of the primates. *Bull. Am. Mus. Nat. Hist.* **35:**239–355.

Gregory, W. 1920. On the structure and relations of *Notharctus*, an American Eocene primate. *Mem. Am. Mus. Nat. Hist.* **3:**45–243.

Gregory, W.; Hellman, M.; and Lewis, G. 1938. Fossil anthropoids of the Yale–

Cambridge India Expedition of 1935. *Carnegie Inst. Wash. Publ.* no. 495, pp. 1–27.

Grine, F. 1987. On the eruption pattern of the permanent incisors and first permanent molars in *Paranthropus. Am. J. Phys. Anthropol.* **72:**353–359.

Groves, C. 1972. Systematics and phylogeny of gibbons. In *Gibbon and Siamang,* ed. D. Rumbaugh, **1:**1–80. Karger, Basel.

Groves, C. 1974. New evidence of the evolution of the apes and man. *Vestn. Ustred. Ustavu Geol.* **49:**53–56.

Gunnell, G. 1985. Systematics of early Eocene Microsyopinae in the Clark's Fork Basin, Wyoming. *Contrib. Mus. Paleontol. Univ. Mich.* **27:**51–71.

Gurche, J. 1982. Early primate brain evolution. In *Primate brain evolution: methods and concepts,* ed. E. Armstrong and D. Falk, pp. 227–246. Plenum, New York.

Hallam, A. 1987. End-Cretaceous mass extinction event: argument for terrestrial causation. *Science* **238:**1237–1242.

Harle, E. 1898. Une machoire de Dryopithèque. *Bull. Soc. Geol. Fr.* **26:**377–383.

Harrison, T. 1981. New finds of small fossil apes from the Miocene locality at Koru in Kenya. *J. Hum. Evol.* **10:**129–137.

Harrison, T. 1985*a*. African oreopithecids and the origin of the family. *Am. J. Phys. Anthropol.* **66:**180.

Harrison, T. 1985*b*. Small bodied apes from the Miocene of East Africa. *Primate Eye* **25:**23.

Harrison, T. 1986. New fossil anthropoids from the middle Miocene of East Africa and their bearing on the origin of the Oreopithecidae. *Am. J. Phys. Anthropol.* **71:**265–284.

Harrison, T. 1987. The phylogenetic relationships of the early catarrhine primates: a review of the current evidence. *J. Hum. Evol.* **16:**41–80.

Hay, R. 1971. Geologic background of Beds I and II. In *Olduvai Gorge,* 4 vols., ed. M. Leakey, **3:**9–20. Cambridge Univ. Press.

Hay, R. 1973. Lithofacies and environments of Bed I, Olduvai Gorge, Tanzania. *Quat. Res. (N. Y.)* **3:**541–560.

Hay, R. 1981. Paleoenvironments of the Laetolil Beds, Northern Tanzania. In *Hominid sites: their geologic settings,* ed. G. Rapp and C. Vondra, pp. 7–23. Westview Press, Boulder, Colo.

Heintz, E.; Brunet, M.; and Battail, B. 1981. A cercopithecid primate from the late Miocene of Molayan, Afghanistan, with remarks on *Mesopithecus. Int. J. Primatol.* **2:**273–284.

Hershkovitz, P. 1970. Notes on tertiary platyrrhine monkeys and description of a new genus from the late Miocene of Colombia. *Folia Primatol.* **12:**1–37.

Hershkovitz, P. 1974*a*. The ectotympanic bone and origin of higher primates. *Folia Primatol.* **22:**237–242.

Hershkovitz, P. 1974*b*. A new genus of Late Oligocene monkey with notes on postorbital closure and platyrrhine evolution. *Folia Primatol.* **21:**1–35.

Hershkovitz, P. 1977. *Living New World monkeys.* Univ. Chicago Press, Chicago.

Hershkovitz, P. 1982. Supposed squirrel monkey affinities of the late Oligocene *Dolichocebus gaimanensis. Nature* **298:**201–202.

Hiiemae, K. 1981. Evolution of man (the Primates). In *Dental Anatomy and Embryology,* ed. J. Osborn, Blackwell, Oxford.

Hill, A., and Ward, S. 1988. Origin of the Hominidae: the record of African large hominoid evolution between 14 My and 4 My. *Yearb. Phys. Anthropol.* **31:**49–84.

Hillhouse, J.; Ndomoi, J.; Cox, A.; and Brock, A. 1977. Additional results on palaeomagnetic stratigraphy of the Koobi Fora Formation, east of Lake Turkan, Kenya. *Nature* **265:**411–415.

Hladik, C. 1977. Chimpanzees of Gabon and chimpanzees of Gombe: some comparative data on the diet. In *Primate ecology: studies of feeding and ranging behavior in lemurs, monkeys, and apes,* ed. T. Clutton-Brock, pp. 481–503. Academic Press, New York.

Hoffstetter, R. 1969. Un primate de l'Oligocène inferieur sud-americain: *Branisella boliviana* gen. et sp. nov. *C. R. Acad. Sci. (Paris) Ser. D* **269:**434–437.

Hoffstetter, R. 1974. Phylogeny and geographical deployment of the primates. *J. Hum. Evol.* **3:**327–350.

Hoffstetter, R. 1980. Origin and deployment of New World monkeys emphasizing the southern continents route. In *Evolutionary biology of the New World monkeys and continental drift,* ed. R. Ciochon and A. Chiarelli, pp. 103–122. Plenum, New York.

Holloway, R. 1984. The Taung endocast and the lunate sulcus: a rejection of the hypothesis of its anterior position. *Am. J. Phys. Anthropol.* **64:**285–287.

Holmes, A. 1965. *Principles of physical geology,* 2d ed. Thomas Nelson & Sons, London.

Hooijer, D. 1963. Miocene Mammalia of Congo. *Ann. Mus. Roy. Afr. Cent. Ser. Octavo. Sci. Geol.* **46:**1–77.

Hooton, E. 1931. *Up from the ape.* Macmillan, New York.

Hopwood, A. 1933. Miocene primates from Kenya. *J. Linn. Soc. London Zool.* **38:**437–464.

Howell, F. 1978. Hominidae. In *Evolution of African Mammals,* ed. V. Maglio and H. Cooke, pp. 154–248. Harvard University Press, Cambridge, Mass.

Howell, F., and Coppens, Y. 1974. Inventory of remains of Hominidae from Pliocene/Pleistocene formations of the lower Omo Basin, Ethiopia (1967–1972). *Am. J. Phys. Anthropol.* **40:**1–16.

Howell, F., and Coppens, Y. 1976. An overview of Hominidae from the Omo succession, Ethiopia. In *Earliest man and environments in the Lake Rudolf Basin,* ed. Y. Coppens, F. Howell, G. Isaac, R. Leakey, pp. 522–532. Univ. Chicago Press, Chicago.

Howell, F., and Wood, B. 1974. Early hominid ulna from the Omo Basin, Ethiopia. *Nature* **249:**174–176.

Howell, F.; Haesaerts, P.; and de Heinzelin, J. 1987. Depositional environments, archeological occurrences, and hominids from Members E and F of the Shungura Formation (Omo basin, Ethiopia). *J. Hum. Evol.* **16:**665–700.

Hsu, K.; Montadert, L.; Bernoulli, C.; Cita, M.; Erickson, A.; Garrison, R.; Kidd, R.; Melieres, F.; Muller, C.; and Wright, R. 1977. History of the Mediterranean salinity crises. *Nature* **267:**399–403.

Hubbard, R., and Boulter, M. 1983. Reconstruction of Palaeogene climate from palynological evidence. *Nature* **301:**147–150.

Hughes, A., and Tobias, P. 1977. A fossil skull probably of the genus *Homo* from Sterkfontein, Transvaal. *Nature* **265:**310–312.

Hunt, K., and Vitzthum, V. 1986. Dental metric assessment of the Omo fossils: implications for the phylogenetic position of *Australopithecus africanus. Am. J. Phys. Anthropol.* **71:**141–156.

Hurzeler, J. 1948. Zur stammesgeschichte der Necrolemuriden. *Schweiz. Palaeontol Abh.,* **66:**1–46.

Huxley, T. 1863. *Evidence as to man's place in nature.* Williams and Norgate, London.

Hylander, W. 1975. Incisor size and diet in anthropoids with special reference to Cercopithecidae. *Science* **198:**1095–1098.

Isaac, G. 1978. The Olorgesailie Formation: Stratigraphy, tectonics, and the palaeogeographic context of the middle Pleistocene archeological sites. In *Geological background to fossil man: recent research in the Gregory Rift Valley, East Africa,* ed. W. Bishop, pp. 173–206. Scottish Academic Press, Edinburgh.

Isaac, G.; Harris, J.; and Crader, D. 1976. Archeological evidence from the Koobi Fora Formation. In *Earliest man and environments in the Lake Rufolf Basin,* ed. Y. Coppens, F. Howell, G. Isaac, and R. Leakey, pp. 533–551. Univ. Chicago Press, Chicago.

Jarman, P. 1968. The effect of the creation of Lake Kariba upon the terrestrial ecology of the middle Zambezi Valley, with particular reference to the Mammalia. Ph.D. dissertation, Manchester University.

Jarman, P. 1974. The social organization of antelope in relation to their ecology. *Behaviour* **58:**215–267.

Jenkins, F. 1974. Tree shrew locomotion and the origins of primate arborealism. In *Primate locomotion,* ed. F. Jenkins, pp. 85–115. Academic Press, New York.

Jepsen, G. 1970. Bat origins and evolution. In *Biology of bats,* ed. W. Wimsatt, **1:**1–64. Academic Press, New York.

Jerison, H. 1973. *Evolution of the brain and intelligence.* Academic Press, New York.

Jerison, H. 1979. Brain, body, and encephalization in early Primates. *J. Hum. Evol.* **8:**615–635.

Johanson, D., and White, T. 1979. A systematic assessment of early African hominids. *Science* **202:**321–330.

Johanson, D.; Taieb, M.; and Coppens, Y. 1982. Pliocene hominids from the Hadar Formation, Ethiopia (1973–1977): Stratigraphic, chronologic, and paleoenvironmental contexts, with notes on hominid morphology and systematics. *Am. J. Phys. Anthropol.* **57:**373–402.

Johanson, D.; Masao, F.; Eck, G.; White, T.; Walter, R.; Kimbel, W.; Asfaw, B.; Manega, P.; Ndessokia, P.; and Suwa, G. 1987. New partial skeleton of *Homo habilis* from Olduvai Gorge, Tanzania. *Nature* **327:**205–209.

Johnson, G.; Zeitler, P.; Naseser, C.; Johnson, N.; Summers, D.; Frost, C.; Opdyke, N.; and Tahirkheli, R. 1982. The occurrence and fission track ages of late Neogene and Quaternary volcanic sediments, Siwalik Group, northern Pakistan. *Palaeogeogr., Palaeoclimatol., Palaeocol.* **37:**63–93.

Jolly, C. 1966. Introduction to the Cercopithecoidea, with notes on their use as laboratory animals. *Symp. Zool. Soc. Lond.* **17:**427–457.

Jolly, C. 1967. Evolution of baboons. In *The baboon in medical research,* ed. H. Vagtborg, **2:**427–457. Univ. Texas Press, Austin.

Jolly, C. 1970. The seed-eaters, a new model of hominid differentiation based on a baboon analogy. *Man* **5:**5–26.

Jolly, C. 1972. The classification and natural history of *Theropithecus,* baboons of the African Plio-Pleistocene. *Bull. Br. Mus. (Nat. Hist.) Geol.* **22:**1–122.

Jolly, C. and Plog, F. 1976. *Physical Anthropology and Archeology,* 1st ed. Knopf, New York.

Jolly, C., and Plog, F. 1986. *Physical Anthropology and Archeology,* 4th ed. McGraw–Hill, New York.

Jones, F. 1916. *Arboreal man.* E. Arnold, London.

Jungers, W. 1982. Lucy's limbs: skeletal allometry and locomotion in *Australopithecus afarensis, Nature* **297:**676–678.

Jungers, W. 1987. Body size and morphometric affinities of the appendicular skeleton in *Oreopithecus bambolii. J. Hum. Evol.* **16:**1987.

Jungers, W. 1988. New estimates of body size in australopithecines. In *Evolution-*

ary History of the Robust Australopithecines, ed. F. Grine, pp. 115–125. Aldine de Gruyter, New York.

Jungers, W., and Stern, J. 1983. Body proportions, skeletal allometry, and locomotion in the Hadar hominids: a reply to Wolpoff. *J. Hum. Evol.* **12:**673–684.

Kay, R. 1975. The functional adaptations of primate molar teeth. *Am. J. Phys. Anthropol.* **43:**195–216.

Kay, R. 1977. The evolution of molar occlusion in Cercopithecidae and early catarrhines. *Am. J. Phys. Anthropol.* **46:**327–352.

Kay, R. 1982. *Sivapithecus simonsi,* a new species of Miocene hominoid, with comments on the phylogenetic status of the Ramapithecinae. *Int. J. Primatol.* **3:**113–173.

Kay, R. 1984. On the use of anatomical features to infer foraging behavior in extinct primates. In *Adaptations for foraging in nonhuman primates,* ed. J. Cant and P. Rodman, pp. 21–53. Columbia Univ. Press, New York.

Kay, R., and Cartmill, M. 1977. Cranial morphology and adaptation of *Palaechthon nacimienti* and other Paromomyidae, with a description of a new genus and species. *J. Hum. Evol.* **6:**19–53.

Kay, R., and Covert, B. 1984. Anatomy and behavior of extinct primates. In *Food acquisition and processing in primates,* ed. D. Chivers, B. Wood, and A. Bilsborough, pp. 467–508. Plenum, New York.

Kay, R., and Hylander, W. 1978. The dental structure of mammalian folivores with special reference to primates and Phalangeroidea. In *The ecology of arboreal folivores,* ed. G. Montgomery, pp. 173–191. Smithsonian Institution Press, Washington, D.C.

Kay, R., and Sheine, W. 1979. On the relationship between chitin particle size and digestibility in the primate *Galago senegalensis. Am. J. Phys. Anthropol.* **50:**301–308.

Kay, R., and Simons, E. 1980. The ecology of Oligocene Anthropoidea. *Int. J. Primatol.* **1:**21–37.

Kay, R., and Simons, E. 1983*a*. Dental formulae and dental eruption patterns in Parapithecidae. *Am. J. Phys. Anthropol.* **62:**363–375.

Kay, R., and Simons, E. 1983*b*. A reassessment of the relationship between later Miocene and subsequent Hominoidea. In *New interpretations of ape and human ancestry,* ed. R. Ciochon and R. Corruccini, pp. 577–624. Plenum, New York.

Kay, R.; Fleagle, J.; Simons, E. 1981. A revision of the Oligocene apes of the Fayum Province, Egypt. *Am. J. Phys. Anthropol.* **55:**293–322.

Kay, R.; Madden, R.; Cifelli, R.; and Diaz, G. 1986. A new specimen of Miocene Colombian *Stirtonia. Am. J. Phys. Anthropol.* **69:**221.

Keast, A. 1972. Continental drift and the evolution of the biota on southern continents. In *Evolution, mammals, and southern continents,* ed. A. Keast, F. Erk, and B. Glass, pp. 23–87. S.U.N.Y. Albany Press, New York.

Keith, A. 1931. *New discoveries relating to the antiquity of man.* Williams and Norgate, London.

Keller, G.; D'Hondt, S.; and Vallier, T. 1983. Multiple microtektite horizons in upper Eocene marine sediments: no evidence for mass extinctions. *Science* **221:**150–152.

Kelley, J., and Pilbeam, D. 1986. The dryopithecines: taxonomy, comparative anatomy, and phylogeny of Miocene large hominoids. In *Comparative primate biology,* 4 vols. ed. D. Swindler, **1:**361–411. Alan Liss, New York.

Kennett, J. 1977. Cenozoic evolution of Antarctic glaciation, the circum-Antarctic Ocean, and their impact on global paleoceanography. *J. Geophys. Res.* **82:**3843–3860.

Kennett, J. 1978. The development of planktonic biogeography in the southern ocean during the Cenozoic. *Mar. Micropaleontol.* **3:**301–345.

Kennett, J., et al. 1985. Palaeotectonic implications of increased late Eocene–early Oligocene volcanism from South Pacific DSDP sites. *Nature* **316:**507–511.

Kimbel, W. 1984. Variation in the pattern of cranial venous sinuses and hominid phylogeny. *Am. J. Phys. Anthropol.* **63:**243–264.

Kimbel, W., and White, T. 1988. A revised reconstruction of the adult skull of *Australopithecus afarensis. J. Hum. Evol.* **17:**545–550.

Kimbel, W.; White, T.; and Johanson, D. 1984. Cranial morphology of *Australopithecus afarensis:* a comparative study based on a composite reconstruction of the adult skull. *Am. J. Phys. Anthropol.* **64:**337–388.

Kimbel, W.; White, T.; and Johanson, D. 1988. Implications of KNM-WT 17000 for the evolution of "robust" australopithecines. In *Evolutionary history of the "robust" australopithecines,* ed. F. Grine, pp. 259–268. Aldine de Gruyter, New York.

King, M., and Wilson, A. 1975. Evolution at two levels in humans and chimpanzees. *Science* **188:**107–116.

Kleiber, M. 1961. *The fire of life: an introduction to animal energetics.* Wiley, New York.

Kohne, D.; Chiscon, J.; and Hoyer, B. 1972. Evolution of primate DNA: a summary. In *Perspectives on human evolution,* ed. S. Washburn and P. Dolhinow, pp. 166–168. Holt, Rinehart & Winston, New York.

Kortland, A. 1980. The Fayum primate forest: did it exist? *J. Hum. Evol.* **9:**227–297.

Kortlandt, A. 1983. Facts and fallacies concerning Miocene ape habitats. In *New interpretations of ape and human ancestry,* ed. R. Ciochon and R. Corruccini, pp. 465–515. Plenum, New York.

Kramer, A. 1986. Distinctiveness in the Miocene-Pliocene fossil record: the Lothagam mandible. *Am. J. Phys. Anthropol.* **70:**457–474.

Krause, D. 1978. Paleocene primates from western Canada. *Can. J. Earth Sci.* **15:**1250–1271.

Krause, D. 1981. Extinction of multituberculates and plesiadapiform primates: examples of competitive exclusion in the mammalian fossil record. *Geol. Soc. Am. Abs.* **13:**491.

Krause, D. 1984. Mammalian evolution in the Paleocene: beginning of an era. *Stud. Geol. Univ. Tenn. Dept. Geol. Sci.* **8:**87–109.

Kretzoi, M. 1975. New ramapithecines and *Pliopithecus* from the lower Pliocene of Rudabanya in north-eastern Hungary. *Nature* **257:**578–581.

Krishtalka, L. 1978. Paleontology and geology of the Badwater Creek area, central Wyoming. Part 15. Review of the late Eocene primates from Wyoming and Utah, and the Plesitarsiiformes. *Ann. Carnegie Mus.* **47:**335–360.

Krishtalka, L., and Schwartz, J. 1978. Phylogenetic relationship of plesiadapiform–tarsiiform primates. *Ann. Carnegie Mus.* **47:**515–540.

Kurland, J., and Pearson, J. 1986. Ecological significance of hypometabolism in nonhuman primates: allometry, adaptation, and deviant diets. *Am. J. Phys. Anthropol.* **71:**445–458.

Kurten, B. 1972. *Not from the apes.* Vantage Books, New York.

Labeyrie, J. 1974. New approach to surface seawater paleotemperatures using $^{18}O/^{16}O$ ratios in silica of diatom frustules. *Nature* **248:**40–41.

LaBrecque, J.; Kent, D.; and Cande, S. 1977. Revised magnetic polarity time scale for late Cretaceous and Cenozoic time. *Geology* **5:**330–335.

Laitman, J. 1985. Evolution of the hominid upper respiratory tract: the fossil

evidence. In *Human evolution: past, present, and future,* ed. P. Tobias, pp. 281–286. Alan Liss, New York.
Laitman, J., and Heimbuch, R. 1982. The basicranium of Plio-Pleistocene hominids as an indicator of their upper respiratory systems. *Am. J. Phys. Anthropol.* **59:**323–343.
Langdon, J. 1985. Fossils and the origin of bipedalism. *J. Hum. Evol.* **14:**615–635.
Lartet, E. 1837. Note sur la découverte recente d'un machoire de singe fossile. *C. R. Acad. Sci. (Paris)* **4:**85–93.
Lartet, E. 1856. Note sur un grand singe fossile qui se rattache au groupe des singes superieurs. *C. R. Acad. Sci. (Paris)* **43:**219–223.
Lavocat, R. 1974. The interrelationships between the African and South American rodents and their bearing on the problem of the origin of South American monkeys. *J. Hum. Evol.* **3:**323–326.
Lavocat, R. 1980. The implications of rodent paleontology and biogeography to the geographical sources and origins of platyrrhine primates. In *Evolutionary biology of the New World monkeys and continental drift,* ed. R. Ciochon and A. Chiarelli, pp. 93–102. Plenum, New York.
Leakey, L. 1935. *Adam's ancestors.* Longmans, Green, New York.
Leakey, L. 1962. A new lower Pliocene fossil primate from Kenya. *Ann. Mag. Nat. Hist.* **4:**689–696.
Leakey, L. 1963. East African fossil Hominoidea and the classification within this super-family. In *Classification and human evolution,* ed. S. Washburn, pp. 32–49. Aldine, Chicago.
Leakey, L. 1968. Lower dentition of *Kenyapithecus africanus. Nature* **217:**827–830.
Leakey, L.; Tobias, P.; and Napier, J. 1964. A new species of the genus *Homo* from Olduvai Gorge. *Nature* **202:**7–9.
Leakey, M. 1985. Early Miocene cercopithecids from Buluk, northern Kenya. *Folia Primatol.* **44:**1–14.
Leakey, M. 1981. *Olduvai Gorge,* vol. 3. Cambridge Univ. Press, Cambridge.
Leakey, M., and Harris, J. 1987. *Laetoli: a pliocene site in northern Tanzania.* Oxford Univ. Press, Oxford.
Leakey, M. and Hay, R. 1979. Pliocene footprints in the Laetoli beds at Laetoli, northern Tanzania. *Nature* **278:**317–323.
Leakey, M.; Leakey, R.; Richtsmiere, J.; Simons, E.; and Walker, A. (in press) Morphological similarities in *Aegyptopithecus* and *Afropithecus. Folio Primatol.*
Leakey, R., and Leakey, M. 1986*a.* A new Miocene hominoid from Kenya. *Nature* **324:**143–146.
Leakey, R., and Leakey, M. 1986*b.* A second new Miocene hominoid from Kenya. *Nature* **324:**146–148.
Leakey, R., and Walker, A. 1985. New higher primates from the early Miocene of Buluk, Kenya. *Nature* **318:**173–175.
Leakey, R.; Leakey, M.; and Walker, A. 1988. Morphology of *Afropithecus turkanensis* from Kenya. *Am. J. Phys. Anthropol.* **76:**289–307.
Le Gros Clark, W. 1959. *The antecedents of man.* Edinburgh Univ. Press, Edinburgh.
Le Gros Clark, W. E., and Leakey, L. 1951. The Miocene Hominoidea of East Africa. *Br. Mus. (Nat. Hist.) Fossil Mamm. Afr.* **1:**1–117.
Le Gros Clark, W. E., and Thomas, D. 1951. Associated jaws and limb bones of *Limnopithecus macinnesi. Br. Mus. (Nat. Hist.) Fossil Mamm. Afr.* **3:**1–27.
Levinton, J., and Simon, C. 1980. A critique of the punctuated equilibria model and implications for the detection of speciation in the fossil record. *Syst. Zool.* **29:**130–142.

Lewin, R. 1984. *Human evolution: an illustrated introduction.* W. H. Freeman, New York.

Lewis, G. 1937. Taxonomic syllabus of Siwalik fossil anthropoids. *Am. J. Sci.* **34:**139–147.

Lewis, O. 1969. The hominoid wrist joint. *Am. J. Phys. Anthropol.* **30:**251–268.

Lewis, O. 1971. The contrasting morphology found in the wrist joints of semibrachiating monkeys and brachiating apes. *Folia Primatol.* **16:**248–256.

Lewis, O. 1972*a*. Osteological features characterizing the wrists of monkeys and apes, with a reconsideration of this region in *Dryopithecus africanus. Am. J. Phys. Anthropol.* **36:**45–58.

Lewis, O. 1972*b*. The evolution of the hallucal tarsometatarsal joint in the Anthropoidea. *Am. J. Phys. Anthropol.* **37:**13–34.

Li, C. 1978. A Miocene gibbon-like primate from Shihhung, Kiangsu Province. *Vertebr. Palasiat.* **16:**189–192.

Lieberman, S.; Gelvin, B.; Oxnard, C. 1985. Dental sexual dimorphism in some extant hominoids and ramapithecines from China: a quantitative approach. *Am. J. Primatol.* **9:**305–326.

Lillegraven, J. 1980. Primates from later Eocene rocks of southern California. *J. Mammal.* **61:**181–204.

Linnaeus, C. 1758. *Systema naturae per regna tria naturae, secundum classes, ordines genera, species cum characteribus, differentris, synonymis, locis.* Editis decima, reformata. Laurentii Salvii, Stockholm.

Lovejoy, O. 1975. Biomechanical perspectives on the lower limbs of early hominids. In *Primate functional morphology and evolution,* ed. R. Tuttle, pp. 291–326. Mouton, The Hague.

Lovejoy, O. 1981. The origin of man. *Science* **211:**341–350.

Lovejoy, O. 1988. Evolution of human walking. *Sci. Am.* **259:**118–125.

Lovejoy, O.; Heiple, K.; Burstein, A. 1973. The gait of *Australopithecus. Am. J. Phys. Anthropol.* **38:**757–779.

Lucas, S.; Rigby, K.; and Kues, B., eds. 1981. *Advances in San Juan Basin paleontology.* Univ. New Mexico Press, Albuquerque.

Luckett, W. 1975. Ontogeny of the fetal membranes and placenta. In *Phylogeny of the primates: a multidisciplinary approach,* ed. W. Luckett and F. Szalay, pp. 157–182. Plenum, New York.

Lydekker, R. 1879. Further notices of Siwalik Mammalia. *Rec. Geol. Surv. India* **12:**33–52.

McCown, T., and Kennedy, K., eds. 1972. *Climbing man's family tree: a collection of major writings on human phylogeny, 1699 to 1971.* Prentice-Hall, Englewood Cliffs, N.J.

MacDonald, J. 1963. The Miocene faunas from the Wounded Knee area of western South Dakota. *Bull. Am. Mus. Nat. Hist.* **125:**139–238.

McDougall, I. 1981. ^{40}Ar/^{39}Ar age spectra from the KBS tuff, Koobi Fora Formation. *Nature* **294:**120–124.

McDougall, I., and Watkins, R. 1985. Age of hominoid-bearing sequence at Buluk, northern Kenya. *Nature* **318:**175–178.

McDougall, I.; Maier, R.; Sutherland-Hawkes, P.; and Gleadow, A. 1980. K-Ar age estimate for the KBS tuff, East Turkana, Kenya. *Nature* **284:**230–234.

MacDougall, J. 1976. Fission-track dating. *Sci. Am.* **235:**114–122.

MacFadden, B.; Campbell, K.; Cifelli, R.; Siles, O.; Johnson, N.; Naeser, C.; and Zeitler, P. 1985. Magnetic polarity stratigraphy and mammalian fauna of the Deseadan Salla beds of northern Bolivia. *J. Geol.* **93:**223–250.

McHenry, H. 1974. How large were the australopithecines? *Am. J. Phys. Anthropol.* **40:**329–340.

McHenry, H. 1975*a*. Fossil hominid body weight and brain size. *Nature* **254:**686–688.

McHenry, H. 1975*b*. The ischium and hip extensor mechanism in human evolution. *Am. J. Phys. Anthropol.* **43:**39–46.

McHenry, H. 1975*c*. Fossils and the mosaic nature of human evolution. *Science* **190:**425–431.

McHenry, H. 1976. Early hominid body weight and encephalization. *Am. J. Phys. Anthropol.* **45:**77–84.

McHenry, H. 1982. The pattern of human evolution: studies on bipedalism, mastication, and encephalization. *Annu. Rev. Anthropol.* **11:**151–173.

McHenry, H. 1984. Relative cheek tooth size in *Australopithecus*. *Am. J. Phys. Anthropol.* **64:**297–306.

McHenry, H. 1985. Implications of postcanine megadontia for the origin of *Homo*. In *Ancestors: the hard evidence,* ed. E. Delson, pp. 178–183. Alan Liss, New York.

McHenry, H. 1988. New estimates of body weight in early hominids and their significance to encephalization and megadontia in robust australopithecines. In *Evolutionary history of the robust australopithecines,* ed. F. Grine, pp. 133–148. Aldine de Gruyter, New York.

McHenry, H., and Corruccini, R. 1975*a*. Distal humerus in hominoid evolution. *Folia Primatol.* **23:**227–244.

McHenry, H., and Corruccini, R. 1975*b*. Multivariate analysis of early hominid pelvic bones. *Am. J. Phys. Anthropol.* **43:**263–270.

McHenry, H., and Corruccini, R. 1983. The wrist of *Proconsul africanus* and the origin of hominoid postcranial adaptations. In *New interpretations of ape and human ancestry,* ed. R. Ciochon and R. Corruccini, pp. 353–368. Plenum, New York.

McHenry, H.; Andrews, P.; and Corruccini, R. 1980. Miocene hominoid palatofacial morphology. *Folia Primatol.* **33:**241–252.

McKenna, M. 1966. Paleontology and the origin of primates. *Folia Primatol.* **4:**1–25.

McKenna, M. 1975. Fossil mammals and early Eocene North Atlantic land continuity. *Ann. Mo. Bot. Gard.* **62:**335–353.

McKenna, M. 1980. Early history and biogeography of South America's extinct land mammals. In *Evolutionary biology of the New World monkeys and continental drift,* ed. R. Ciochon and A. Chiarelli, pp. 43–78. Plenum, New York.

McKenna, M. 1983*a*. Holarctic landmass rearrangement, cosmic events, and Cenozoic terrestrial organisms. *Ann. Mo. Bot. Gard.* **70:**459–489.

McKenna, M. 1983*b*. Cenozoic paleogeography of North Atlantic land bridges. In *Structure and development of the Greenlands-Scotland Ridge,* ed. M. Bott, S. Saxov, M. Talwani, and J. Thiede, pp. 351–399. Plenum, New York.

McLean, D. 1978. Terminal Mesozoic "greenhouse": lessons from the past. *Science* **201:**401–406.

McMahon, T., and Bonner, J. 1983. *On size and life.* W. H. Freeman, Scientific American Books, New York.

MacPhee, R., and Cartmill, M. 1986. Basicranial structures and primate systematics. In *Primate biology,* ed. D. Swindler, **1:**219–275. Alan Liss, New York.

MacPhee, R.; Cartmill, M.; and Gingerich, P. 1983. New Paleogene primate basicrania and the definition of the order Primates. *Nature* **301:**509–511.

Madden, C. 1975. *Pondaungia* needs restudy. Unpublished ms.

Maglio, V. 1972. Vertebrate faunas and chronology of hominid-bearing sediments east of Lake Rudolf, Kenya. *Nature* **239:**379–385.

Mahe, J. 1972. The Malagasy subfossils. In *Biogeography and ecology in Madagascar,* ed. R. Battistini and G. Richard-Vindard, pp. 339–365. W. Junk, The Hague.

Maier, W. 1984. Tooth morphology and diet. In *Food acquisition and processing in Primates,* ed. D. Chivers, B. Wood, and A. Bilsborough, pp. 303–330. Plenum, New York.

Mann, A. 1975. *Some paleodemographic aspects of the South African australopithecines.* Univ. Pennsylvania Publications in Anthropology, no. 1. Univ. Pennsylvania, Philadelphia.

Markl, R. 1974. Evidence for the breakup of eastern Gondwanaland by the early Cretaceous. *Nature* **251:**196–200.

Marshall, L.; Butler, R.; Drake, R.; Curtis, G.; and Tedford, R. 1979. Calibration of the Great American Interchange. *Science* **204:**272–279.

Martin, L. 1981. New specimens of *Proconsul* from Koru, Kenya. *J. Hum. Evol.* **10:**139–150.

Martin, L. 1985. Significance of enamel thickness in hominoid evolution. *Nature* **314:**260–263.

Martin, L., and Andrews, P. 1984. The phyletic position of *Graecopithecus freybergi. Cour. Forschungsinst. Senckenb.* **69:**25–40.

Martin, R. 1968. Towards a new definition of Primates. *Man* **3:**377–401.

Martin, R. 1982. Allometric approaches to the evolution of the primate nervous system. In *Primate brain evolution: methods and concepts,* ed. E. Armstrong and D. Falk, pp. 39–56. Plenum, New York.

Martin, R. 1986*a*. Primates: a definition. In *Major topics in primate and human evolution,* ed. B. Wood, L. Martin, and P. Andrews, pp. 1–31. Cambridge Univ. Press.

Martin, R. 1986*b*. Are fruit bats primates? *Nature* **320:**482–483.

Matthew, W. 1917. The dentition of *Nothodectes. Bull. Am. Mus. Nat. Hist.* **37:**831–839.

Matthews, R. 1974. *Dynamic stratigraphy.* Prentice-Hall, Englewood Cliffs, N.J.

Matthews, R., and Poore, R. 1980. Tertiary O^{18} record and glacio-eustatic sea-level fluctuations. *Geology* **8:**501–504.

Maw, B.; Ciochon, R.; and Savage, D. 1979. Late Eocene of Burma yields earliest anthropoid primate, *Pondaungia cotteri. Nature* **282:**65–67.

Maxwell, P.; Von Herzen, R.; Hsu, K.; Andrews, J.; Saito, T.; Percival, S.; Milow, E.; and Boyce, R. 1970. Deep sea drilling in the South Atlantic. *Science* **168:**1047–1059.

Mayr, E. 1981. Biological classification: toward a synthesis of opposing methodologies. *Science* **214:**510–516.

Miller, J. 1972. Dating Pliocene and Pleistocene strata using the potassium-argon and argon-40/argon-39 methods. In *Calibration of hominoid evolution: recent advances in isotopic and other dating methods applicable to the origin of man,* ed. W. Bishop and J. Miller, pp. 63–76. Scottish Academic Press, Edinburgh.

Mivart, St. G. 1873. On *Lepilemur* and *Cheirogaleus* and on the zoological rank of the Lemuroidea. *Proc. Zool. Soc. Lond.* **1873:**484–510.

Morbeck, M. 1975. *Dryopithecus africanus* forelimb. *J. Hum. Evol.* **4:**39–116.

Morner, N. 1978. Low sea-levels, droughts and mammalian extinctions. *Nature* **271:**738–739.

Napier, J. 1970. Paleoecology and catarrhine evolution. In *Old World monkeys,* ed. J. Napier and P. Napier, pp. 53–95. Academic Press, New York.

Napier, J., and Davis, P. 1959. The fore-limb skeleton and associated remains of *Proconsul africanus. Br. Mus. (Nat. Hist.) Fossil Mamm. Afr.* **16:**1–69.

Nash, L. 1986. Dietary, behavioral, and morphological aspects of gummivory in primates. *Yearb. Phys. Anthropol.* **29:**113–138.

Nelson, G., and Platnick, N. 1980. Multiple branching in cladograms: two interpretations. *Syst. Zool.* **29:**86–90.

Novacek, M. 1985. Evidence for echolocation in the oldest known bats. *Nature* **315:**140–141.

Novacek, M.; McKenna, M.; Neff, N.; and Cifelli, R. 1983. Evidence from earliest known erinaceomorph basicranium that insectivorans and primates are not closely related. *Nature* **306:**683–684.

O'Connor, B. 1975. The functional morphology of the cercopithecoid wrist and inferior radioulnar joints and their bearing on some problems in the evolution of the Hominoidea. *Am. J. Phys. Anthropol.* **43:**113–122.

Officer, C., and Drake, C. 1985. Terminal Cretaceous environmental events. *Science* **227:**1161–1167.

Oikonomides, A. 1982. *Hanno the Carthaginian: Periplus* (translation). Ares Publications, Chicago.

Olson, S., and Rasmussen, D. 1986. Paleoenvironment of the earliest hominoids: new evidence from the Oligocene avifauna of Egypt. *Science* **233:**1202–1204.

Olson, T. 1978. Hominid phylogenetics and the existence of *Homo* in Member I of the Swartkrans Formation, South Africa. *J. Hum. Evol.* **7:**159–178.

Olson, T. 1985*a*. Cranial morphology and systematics of the Hadar Formation hominids and *Australopithecus africanus*. In *Ancestors: the hard evidence*, ed. E. Delson, pp. 102–119. Alan Liss, New York.

Olson, T. 1985*b*. Taxonomic affinities of the immature hominid crania from Hadar and Taung. *Nature* **316:**539–540.

Opdyke, N.; Lindsay, E.; Johnson, G.; Johnson, N.; Tahirkheli, R.; and Mirza, M. 1979. Magnetic polarity stratigraphy and vertebrate paleontology of the upper Siwalik subgroup of northern Pakistan. *Palaeogeogr., Palaeoclimatol., Palaeoecol.* **27:**1–34.

Orlosky, F., and Swindler, D. 1975. Origins of New World monkeys. *J. Hum. Evol.* **4:**77–83.

Osborn, H. 1908. New fossil mammals from the Fayum Oligocene, Egypt. *Bull. Am. Mus. Nat. Hist.* **24:**265–272.

Owen, R. 1858. On the characters, principles of division, and primary groups of the class Mammalia. *J. Linn. Soc. London Zool.* **2:**1–37.

Ozansoy, F. 1957. Faunes des mammifères du Tartiare de Turquie et leurs revisions stratigraphique. *Bull. Miner. Res. Explor. Inst. Turkey* **49:**29–48.

Ozansoy, F. 1965. Etude des gisements continentaux et des mammifères du Cenozoique de Turquie. *Mem. Soc. Geol. Fr.* **44:**5–89.

Packer, D., and Sarmiento, E. 1984. External and middle ear characteristics of Primates, with reference to tarsier-anthropoid affinities. *Am. Mus. Novit.* **2787:**1–23.

Partridge, T. 1978. Re-appraisal of lithostratigraphy of Sterkfontein hominid site. *Nature* **275:**282–287.

Partridge, T. 1986. Palaeoecology of the Pliocene and lower Pleistocene hominids of Southern Africa: how good is the chronological and palaeoenvironmental evidence? *S. Afr. J. Sci.* **82:**80–83.

Patterson, C. 1987. Introduction. In *Molecules and morphology in evolution: conflict or compromise?*, ed. C. Patterson, pp. 1–22. Cambridge Univ. Press.

Patterson, B., and Howells, W. 1967. Hominid humeral fragment from early Pleistocene of northwest Kenya. *Science* **156:**64–66.

Patterson, B., and Pascual, R. 1972. The fossil mammal fauna of South America. In *Evolution, mammals, and southern continents,* ed. A. Keast, F. Erk, and B. Glass, pp. 247–309. S.U.N.Y. Albany Press, New York.

Patterson, B.; Behrensmeyer, A.; and Sill, W. 1970. Geology and fauna of a new Pliocene locality in northwestern Kenya. *Nature* **226:**918–921.

Pettigrew, J. 1986. Flying primates? Megabats have the advanced pathway from eye to midbrain. *Science* **231:**1304–1306.

Phillips, J. 1978. Dental variability in Ethiopian baboons. Ph.D. dissertation, New York University.

Phillips-Conroy, J., and Jolly, C. 1981. Sexual dimorphism in two subspecies of Ethiopian baboons and their hybrids. *Am. J. Phys. Anthropol.* **56:**115–129.

Pickford, M. 1981. Preliminary Miocene mammalian biostratigraphy for western Kenya. *J. Hum. Evol.* **10:**73–97.

Pickford, M. 1982. New higher primate fossils from the middle Miocene deposits at Majiwa and Kaloma, western Kenya. *Am. J. Phys. Anthropol.* **58:**1–19.

Pickford, M. 1983. Sequence and environments of the lower and middle Miocene hominoids of western Kenya. In *New interpretations of ape and human ancestry,* ed. R. Ciochon and R. Corruccini, pp. 421–440. Plenum, New York.

Pilbeam, D. 1967. Man's earliest ancestors. *Sci. J. (Lond.)* **3:**47–53.

Pilbeam, D. 1969. Tertiary Pongidae of East Africa. Evolutionary relationships and taxonomy. *Peabody Mus. Nat. Hist. Yale Univ. Bull.* **31:**1–185.

Pilbeam, D. 1979. Recent finds and interpretations of Miocene hominoids. *Annu. Rev. Anthropol.* **8:**333–352.

Pilbeam, D. 1982. New hominoid skull material from the Miocene of Pakistan. *Nature* **295:**232–234.

Pilbeam, D., and Gould, S. 1974. Size and scaling in human evolution. *Science* **186:**892–901.

Pilbeam, D., and Smith, R. 1981. New skull remains of *Sivapithecus* from Pakistan. *Mem. Geol. Surv. Pakistan* **11:**1–13.

Pilbeam, D., and Walker, A. 1968. Fossil monkeys from the Miocene of Napak, north-east Uganda. *Nature* **220:**657–660.

Pilbeam, D.; Meyer, G.; Badgley, C.; Rose, M.; Pickford, M.; Behrensmeyer, A.; and Shah, I. 1977*a*. New hominoid primates from the Siwaliks of Pakistan and their bearing on hominoid evolution. *Nature* **270:**689–695.

Pilbeam, D.; Barry, J.; Meyer, G.; Shah, I.; Pickford, M.; Bishop, W.; Thomas, H.; and Jacobs, L. 1977*b*. Geology and paleontology of Neogene strata of Pakistan. *Nature* **270:**684–689.

Pilbeam, D.; Rose, M.; Badgley, C.; and Lipschultz, B. 1980. Miocene hominoids from Pakistan. *Postilla* **181:**1–94.

Pilgrim, G. 1908. The Tertiary and post-Tertiary freshwater deposits of Baluchistan and Sind, with notices of new vertebrates. *Rec. Geol. Surv. India* **37:**139–166.

Pilgrim, G. 1910. Notice of new mammalian genera and species from the Tertiary of India. *Rec. Geol. Surv. India* **50:**63–71.

Pilgrim, G. 1913. The correlation of the Siwaliks with mammal horizons of Europe. *Rec. Geol. Surv. India* **43:**264–326.

Pilgrim, G. 1915. New Siwalik primates and their bearing on the question of the evolution of man and the Anthropoidea. *Rec. Geol. Surv. India* **45:**1–74.

Pilgrim, G. 1927. A new *Sivapithecus* palate, and other primate fossils from India. *Mem. Geol. Surv. India* **14:**1–26.

Pilgrim, G. 1934. Correlation of the ossiferous sections in the upper Cenozoic of India. *Am. Mus. Novit.* **704:**1–5.

Pocock, R. I. 1918. On the external characters of the lemurs and of *Tarsius*. *Proc. Zool. Soc. Lond.* **1918:**19–53.

Pollack, J.; Toon, O.; Ackerman, T.; McKay, C.; and Turco, R. 1983. Environmental effects of an impact-generated dust cloud: implications for the Cretaceous-Tertiary extinctions. *Science* **219:**287–289.

Prasad, K. 1962. Fossil primates from the Siwalik beds near Haritalyangar, Himachal Pradesh, India. *J. Geol. Soc. India* **3:**86–96.

Prasad, K. 1964. Upper Miocene anthropoids from the Siwalik beds of Haritalyangar, Himachal Pradesh, India. *Palaeontology* **7:**124–134.

Preuss, T. 1982. The face of *Sivapithecus indicus:* description of a new, relatively complete specimen from the Siwaliks of Pakistan. *Folia Primatol.* **38:**141–157.

Prothero, D. 1985. Mid-Oligocene extinction event in North American land mammals. *Science* **229:**550–551.

Rabinowitz, P.; Coffin, M.; and Falvey, D. 1983. The separation of Madagascar and Africa. *Science* **220:**67–69.

Radinsky, L. 1974. The fossil evidence of anthropoid brain evolution. *Am. J. Phys. Anthropol.* **41:**15–28.

Radinsky, L. 1975. Primate brain evolution. *Am. Sci.* **63:**656–663.

Radinsky, L. 1977. Early primate brains: facts and fiction. *J. Hum. Evol.* **6:**79–86.

Radinsky, L. 1982. Some cautionary notes on making inferences about relative brain size. In *Primate brain evolution: methods and concepts,* ed. E. Armstrong and D. Falk, pp. 29–37. Plenum, New York.

Rak, Y. 1978. The functional significance of the squamosal suture in *Australopithecus boisei. Am. J. Phys. Anthropol.* **49:**71–78.

Rak, Y. 1983. *The australopithecine face.* Academic Press, New York.

Rak, Y. 1985. Australopithecine taxonomy and phylogeny in light of facial morphology. *Am. J. Phys. Anthropol.* **66:**281–287.

Rak, Y., and Howell, F. 1978. Cranium of a juvenile *Australopithecus boisei* from the lower Omo Basin, Ethiopia. *Am. J. Phys. Anthropol.* **48:**345–366.

Rasmussen, D. T. 1986. Anthropoid origins: a possible solution to the Adapidae-Omomyidae paradox. *J. Hum. Evol.* **15:**1–12.

Raza, S.; Barry, J.; Pilbeam, D.; Rose, M.; Shah, I.; and Ward, S. 1983. New hominoid primates from the middle Miocene Chinji Formation, Potwar Plateau, Pakistan. *Nature* **306:**52–54.

Reid, E., and Chandler, M. 1933. *The flora of the London Clay.* British Museum of Natural History, London.

Remane, A. 1921. Beitrage zur morphologie des Anthropoidengebisses. *Weigmanns Arch. Naturgesch. Abt.* **87:**1–179.

Remane, A. 1965. Die Geschichte der Menschenaffen. In *Menschliche Abstammungslehre,* pp. 249–309. Gustav Fischer, Stuttgart.

Richard, A. 1985. *Primates in nature.* W. H. Freeman, New York.

Ridley, M. 1986. *Evolution and classification: the reformation of cladism.* Longman, London.

Robinson, J. 1956. The dentition of the Australopithecinae. *Mem. Trans. Mus.* no. 9, pp. 1–179.

Robinson, J. 1963. Adaptive radiation in the australopithecines and the origin of man. In *African ecology and human evolution,* ed. C. F. Howell and F. Bourliere, pp. 385–416. Aldine, Chicago.

Robinson, J. 1972. *Early hominid posture and locomotion.* Univ. Chicago Press, Chicago.

Rodman, P., and McHenry, H. 1980. Bioenergetics and the origin of hominid bipedalism. *Am. J. Phys. Anthropol.* **52:**103–106.

Rose, K. 1975. The Carpolestidae, early Tertiary primates from North America. *Bull. Mus. Comp. Zool.* **147:**1–74.

Rose, K. 1984. Evolution and radiation of mammals in the Eocene, and the diversification of modern orders. *Stud. Geol. Univ. Tenn. Dep. Geol. Sci.* **8:**110–127.

Rose, K., and Fleagle, J. 1981. The fossil history of non-human primates in the Americas. In *Ecology and behavior of neotropical primates,* ed. A. Coimbra-filho and R. Mittermeier, **1:**111–167. Academia Brasiliera de Ciencias, Rio de Janeiro.

Rose, K., and Gingerich, P. 1977. Partial skull of the plesiadapiform primate *Ignacius* from the early Eocene of Wyoming. *Contrib. Mus. Paleontol. Univ. Mich.* **24:**181–189.

Rose, K., and Krause, D. 1984. Affinities of the primate *Altanius* from the early Tertiary of Mongolia. *J. Mammal.* **65:**721–726.

Rose, K., and Rensberger, J. 1983. Upper dentition of *Ekgmowechashala* from the John Day Formation, Oligo-Miocene of Oregon. *Folia Primatol.* **41:**102–111.

Rose, K., and Walker, A. 1985. The skeleton of early Eocene *Cantius,* oldest lemuriform primate. *Am. J. Phys. Anthropol.* **66:**73–89.

Rose, M. 1983. Miocene hominoid postcranial morphology: monkey-like, ape-like, neither, or both? In *New interpretations of ape and human ancestry,* ed. R. Ciochon and R. Corruccini, pp. 405–420. Plenum, New York.

Rose, M. 1986. Further hominoid postcranial specimens from the late Miocene Nagri Formation of Pakistan. *J. Hum. Evol.* **15:**333–368.

Rosenberger, A. 1977. *Xenothrix* and ceboid phylogeny. *J. Hum. Evol.* **6:**461–481.

Rosenberger, A. 1979. Cranial anatomy and implications of *Dolichocebus,* a late Oligocene ceboid primate. *Nature* **279:**416–418.

Rosenberger, A. 1980. Gradistic views and adaptative radiation of platyrrhine primates. *Z. Morph. Anthropol.* **71:**157–163.

Rosenberger, A. 1981. A mandible of *Branisella boliviana* from the Oligocene of South America. *Int. J. Primatol.* **2:**1–7.

Rosenberger, A. 1984. Fossil New World monkeys dispute the molecular clock. *J. Hum. Evol.* **13:**737–742.

Rosenberger, A. 1986. Platyrrhines, catarrhines, and the anthropoid transition. In *Major topics in primate and human evolution,* ed. B. Wood, L. Martin, and P. Andrews. Cambridge Univ. Press.

Rosenberger, A., and Szalay, F. 1980. On the tarsiiform origins of Anthropoidea. In *Evolutionary biology of the New World monkeys and continental drift,* ed. R. Ciochon and A. Chiarelli, pp. 139–157. Plenum, New York.

Russell, D. 1964. Les mammifères paleocènes d'Europe. *Mem. Mus. Natl. Hist. Nat. Ser. C Geol.* **13:**1–324.

Russell, D. 1975. Paleoecology of the Paleocene-Eocene transition in Europe. *Contrib. Primatol.* **5:**28–61.

Rutimeyer, L. 1862. Eocaene saugethiere aus dem gebiet des Schweizerischen Jura. *Allg. Schweizerische Gesell., neue Denkschrifte* **19:**1–98.

Sailer, L., and Gaulin, S. 1985. Measuring the relationship between dietary quality and body size in primates. *Primates* **26:**14–27.

Sarich, V. 1971. A molecular approach to the question of human origins. In *Background for man,* ed. P. Dolhinow and V. Sarich, pp. 60–81. Little, Brown, Boston.

Sarich, V., and Cronin, J. 1976. Molecular systematics of the primates. In *Molecular anthropology: genes and proteins in the evolutionary ascent of the primates,* ed. M. Goodman and R. Tashian, pp. 141–171. Plenum, New York.

Sarich, V., and Cronin, J. 1977. Generation length and rates of hominoid molecular evolution. *Nature* **269:**354–355.
Sarich, V., and Wilson, A. 1967. Immunological time scale for hominid evolution. *Science* **158:**1200–1203.
Savage, D., and Russell, D. 1983. *Mammalian paleofaunas of the world.* Addison-Wesley, Reading, Mass.
Schlosser, M. 1901. Dei menschenahnlichen Zahne aus dem Bohnerz der Schwabischen Alb. *Zool. Anz.* **24:**261–271.
Schlosser, M. 1910. Uber einige fossile Saugetierre aus dem Oligocan von Agypten. *Zool. Anz.* **34:**500–508.
Schlosser, M. 1911. Beitrage zur kenntnis der Oligozanen Landsaugetiere aus dem Fayum: Aegypten. *Beitr. Palaeontol. Oesterreich-Ungarns Orients* **6:**1–227.
Schmidt-Nielsen, K. 1979. *Animal physiology.* Cambridge Univ. Press.
Schoeninger, M. 1976. Functional significance of the development of a mesostyle in the Eocene primates *Pelycodus* and *Notharctus. Am. J. Phys. Anthropol.* **44:**204.
Schon, M., and Ziemer, L. 1973. Wrist mechanism and locomotor behavior of *Dryopithecus africanus. Folia Primatol.* **20:**1–11.
Schon-Ybarra, M., and Conroy, G. 1978. Nonmetric features in the ulna of *Aegyptopithecus, Alouatta, Lagothrix. Folia Primatol.* **29:**178–195.
Schultz, A. 1968. The recent hominoid primates. In *Perspective on human evolution,* ed. S. Washburn and P. Jay, pp. 122–195. Holt, Rinehart & Winston, New York.
Schultz, A. 1969. *The life of primates.* Weidenfeld & Nicholson, London.
Schwartz, J. 1984. The evolutionary relationships of man and orang-utans. *Nature* **308:**501–505.
Schwartz, J. 1986. Primate systematics and a classification of the order. In *Comparative Primate Biology,* 4 vols., ed. D. Swindler, **1:**1–41. Alan Liss, New York.
Schwartz, J., and Krishtalka, L. 1977. Revision of Picrodontidae: dental homologies and relationship. *Ann. Carnegie Mus.* **46:**55–70.
Schwartz, J.; Tattersall, I.; and Eldredge, N. 1978. Phylogeny and classification of the Primates revisited. *Yearb. Phys. Anthropol.* **21:**95–133.
Shipman, P. 1987. An age old question: why did the human lineage survive? *Discover,* April, pp. 60–64.
Shipman, P.; Walker, A.; Van Couvering, J.; Hooker, P.; and Miller, J. 1981. The Fort Ternan hominoid site, Kenya: geology, age, taphonomy and paleoecology. *J. Hum. Evol.* **10:**49–72.
Sibley, C., and Ahlquist, J. 1984. The phylogeny of the hominoid primates, as indicated by DNA-DNA hybridization. *J. Mol. Evol.* **20:**1–15.
Simons, E. 1960. *Apidium* and *Oreopithecus. Nature* **186:**824–826.
Simons, E. 1961. Notes on Eocene tarsioids and a revision of some Necrolemurinae. *Bull. Brit. Mus. Natur. Hist.* **5:**45–75.
Simons, E. 1962. Two new primate species from the African Oligocene. *Postilla* **64:**1–12.
Simons, E. 1965. New fossil apes from Egypt and the initial differentiation of the Hominoidea. *Nature* **205:**135–139.
Simons, E. 1969. Miocene monkey *Prohylobates* from northern Egypt. *Nature* **223:**687–689.
Simons, E. 1970. The deployment and history of Old World monkeys. In *Old World monkeys,* ed. J. Napier and P. Napier, pp. 97–137. Academic Press, New York.
Simons, E. 1971. Relationships of *Amphipithecus* and *Oligopithecus. Nature* **232:**489–491.

Simons, E. 1972. *Primate evolution.* Macmillan, New York.
Simons, E. 1974. Notes on early Tertiary prosimians. In *Prosimian biology,* ed. R. Martin, G. Doyle, and A. Walker, pp. 415–433. Duckworth, London.
Simons, E. 1976. The fossil record of primate phylogeny. In *Molecular Anthropology,* ed. M. Goodman, E. Tashian, and J. Tashian, pp. 35–62. Plenum, New York.
Simons, E. 1985*a*. Origins and characteristics of the first hominoids.In *Ancestors: the hard evidence,* ed. E. Delson, pp. 37–41. Alan Liss, New York.
Simons, E. 1985*b*. African origin, characteristics, and context of earliest higher primates. In *Hominid evolution: past, present and future,* ed. P. Tobias, pp. 101–106. Alan Liss, New York.
Simons, E. 1986. *Parapithecus grangeri* of the African Oligocene: an archaic catarrhine without lower incisors. *J. Hum. Evol.* **15:**205–213.
Simons, E. 1987. New faces of *Aegyptopithecus* from the Oligocene of Egypt. *J. Hum. Evol.* **16:**273–290.
Simons, E., and Bown, T. 1985. *Afrotarsius chatrahi,* 1st tarsiiform primate from Africa. *Nature* **313:**475–477.
Simons, E., and Chopra, S. 1969. *Gigantopithecus,* a new species from north India. *Postilla* **138:**1–18.
Simons, E., and Fleagle, J. 1973. The history of extinct gibbon-like primates. In *Gibbon and Siamang,* ed. D. Rumbaugh, **2:**121–148. Karger, Basel.
Simons, E., and Kay, R. 1983. *Qatrania,* new basal anthropoid primate from the Fayum, Oligocene of Egypt. *Nature* **304:**624–626.
Simons, E., and Pilbeam, D. 1965. Preliminary revision of the Dryopithecinae. *Folia Primatol.* **3:**81–152.
Simons, E., and Russell, D. 1960. Notes on the cranial anatomy of *Necrolemur. Breviora* **127:**1–14.
Simons, E., and Wood, A. 1968. Early Cenozoic mammalian faunas, Fayum Province, Egypt. *Peabody Mus. Nat. Hist. Yale Univ. Bull.* **28:**1–105.
Simons, E.; Andrews, P.; and Pilbeam, D. 1978. Cenozoic apes. In *Evolution of African mammals,* ed. V. Maglio and H. Cooke, pp. 120–146. Harvard Univ. Press, Cambridge, Mass.
Simons, E.; Bown, T.; and Rasmussen, D. T. 1986. Discovery of two additional prosimian primate families (Omomyidae, Lorisidae) in the African Oligocene. *J. Hum. Evol.* **15:**431–437.
Simons, E.; Rasmussen, D. T.; and Gebo, D. 1987. A new species of *Propliopithecus* from the Fayum, Egypt. *Am. J. Phys. Anthropol.* **73:**139–148.
Simpson, G. 1933. The plagiaulacoid type of mammalian dentition. *J. Mammal.* **14:**97–107.
Simpson, G. 1935. The Tiffany fauna. Upper Paleocene. 2 Structure of relationships of *Plesiadapis. Am. Mus. Novit.* no. 816, pp. 1–30.
Simpson, G. 1937. The beginning of the Age of Mammals. *Biol. Rev.* **7:**1–47.
Simpson, G. 1940*a*. Mammals and land bridges. *J. Wash. Acad. Sci.* **30:**137–163.
Simpson, G. 1940*b*. Studies on the earliest primates. *Bull. Am. Mus. Nat. Hist.* **77:**185–212.
Simpson, G. 1953. *The major features of evolution.* Columbia Univ. Press, New York.
Simpson, G. 1955. The Phenacolemuridae, new family of early primates. *Bull. Am. Mus. Nat. Hist.* **105:**415–441.
Simpson, G. 1961. *Principles of animal taxonomy.* Columbia Univ. Press, New York.
Simpson, G 1963. The meaning of taxonomic statements. In *Classification and Human Evolution,* ed. S. Washburn, pp. 1–31. Aldine, Chicago.
Simpson, G. 1975. Recent advances in methods of phylogenetic inference. In

Phylogeny of the Primates: a multidisciplinary approach, ed. W. Luckett and F. Szalay, pp. 3–19. Plenum, New York.

Simpson, G. 1981. History of vertebrate paleontology in the San Juan Basin. In *Advances in San Juan Basin paleontology,* ed. S. Lucas, K. Rigby, and B. Kues, pp. 3–28. Univ. New Mexico Press, Albuquerque.

Skelton, R.; McHenry, H.; and Drawhorn, G. 1986. Phylogenetic analysis of early hominids. *Curr. Anthropol.* **27:**21–43.

Sloan, R.; Rigby, J.; Van Valen, L.; and Gabriel, D. 1986. Gradual dinosaur extinction and simultaneous ungulate radiation in the Hell Creek Formation. *Science* **232:**629–633.

Smith, B. 1986. Dental development in *Australopithecus* and early *Homo. Nature* **323:**327–330.

Smith, G. 1912. The origin of man. *Smithson. Inst. Annu. Rep.* **1912:**553–572.

Smith, J., and Savage, R. 1956. Some locomotory adaptations in mammals. *J. Linn. Soc. A (Zool.)* **42:**603–622.

Smith, R. 1980. Rethinking allometry. *J. Theor. Biol.* **87:**97–111.

Smith, R. 1985. The present as a key to the past: body weight of Miocene hominoids as a test of allometric methods for paleontological inference. In *Size and scaling in primate biology,* ed. W. Jungers, pp. 437–448. Plenum, New York.

Sokal, R. 1974. Classification purposes, principles, progress, prospects. *Science* **105:**1115–1123.

Spuhler, J. 1988. Evolution of mitochondrial DNA in monkeys, apes and humans. *Yearb. Phys. Anthropol.* **31:**15–48.

Stains, H. 1959. Use of the calcaneum in studies of taxonomy and food habits. *J. Mamm.* **40:**392–401.

Stehlin, H. 1909. Remarques sur les faunules de Mammifères des couches eocènes et oligocènes du Bassin de Paris. *Bull. Soc. Geol. Fr.* **9:**488–520.

Steininger, F. von; Rogl, F.; and Martini, E. 1976. Current Oligocene / Miocene biostratigraphic concept of the Central Paratethys (Middle Europe). *Newsl. Stratigr.* **4:**174–202.

Stephan, H.; Bauchot, R.; and Andy, O. 1970. Data on the size of the brain and of various parts in insectivores and primates. In *The primate brain,* ed. C. Noback and W. Montagna, pp. 289–297. Appleton-Century-Crofts, New York.

Stern, J., and Susman, R. 1983. The locomotor anatomy of *Australopithecus afarensis. Am. J. Phys. Anthropol.* **60:**279–317.

Stirton, R. 1951. Ceboid monkeys from the Miocene of Colombia. *Univ. Calif. Publ. Geol. Sci.* **28:**315–356.

Stringer, C., and Andrews, P. 1988. Genetic and fossil evidence for the origin of modern humans. *Science* **239:**1263–1268.

Stringer, C.; Grun, R.; Schwarcz, H.; and Goldberg, P. 1989. ESR dates for the hominid burial site of Es Skuhl in Israel. *Nature* **338:**756–758.

Sudre, J. 1975. Un prosimien du palaeogène ancient du Sahara nord-occidental: *Azibius trerki. C. R. Acad. Sci. (Paris)* **280:**1539–1542.

Sullivan, W. 1974. *Continents in motion.* McGraw Hill, New York.

Susman, R. 1988. Evidence for tool behavior in *Paranthropus robustus* from Member I, Swartkrans. *Science* **240:**781–784.

Susman, R., and Stern, J. 1982. Functional morphology of *Homo habilis. Science* **217:**931–933.

Sussman, R. (in press). *The ecology and behavior of free-ranging primates.* Macmillan, New York.

Sussman, R., and Kinzey, W. 1984. The ecological role of the Callitrichidae: a review. *Am. J. Phys. Anthropol.* **64:**419–449.

Sussman, R., and Raven, P. 1978. Pollination by lemurs and marsupials: an archaic coevolutionary system. *Science* **200:**731–736.

Swindler, D., and Wood, C. 1973. *An atlas of primate gross anatomy.* Univ. Washington Press, Seattle.

Szalay, F. 1968*a*. The beginnings of primates. *Evolution* **22:**19–36.

Szalay, F. 1968*b*. The Picrodontidae, a family of early primates. *Am. Mus. Novit.* no. 2329, pp. 1–55.

Szalay, F. 1969*a*. Mixodectidae, Microsyopidae, and the insectivore-primate boundary. *Bull. Am. Mus. Nat. Hist.* **140:**195–330.

Szalay, F. 1969*b*. Uintasoricinae, a new subfamily of early Tertiary mammals. *Am. Mus. Novit.* no. 2363, pp. 1–36.

Szalay, F. 1970. Late Eocene *Amphipithecus* and the origins of catarrhine primates. *Nature* **227:**355–357.

Szalay, F. 1972*a*. *Amphipithecus* revisited. *Nature* **236:**179–180.

Szalay, F. 1972*b*. Paleobiology of the earliest primates. In *The functional and evolutionary biology of primates,* ed. R. Tuttle, pp. 3–35. Aldine, Chicago.

Szalay, F. 1973. New Paleocene primates and a diagnosis of the new suborder Paromomyiformes. *Folia Primatol.* **19:**73–87.

Szalay, F. 1975*a*. Where to draw the nonprimate-primate taxonomic boundary. *Folia Primatol.* **23:**158–163.

Szalay, F. 1975*b*. Phylogeny of primate higher taxa: the basicranial evidence. In *Phylogeny of the primates: a multidisciplinary approach,* ed. W. Luckett and F. Szalay, pp. 91–125. Plenum, New York.

Szalay, F. 1975*c*. Phylogeny, adaptations and dispersal of tarsiiform primates. In *Phylogeny of the primates: a multidisciplinary approach,* ed. W. Luckett and F. Szalay, pp. 357–404. Plenum, New York.

Szalay, F. 1976. Systematics of the Omomyidae: taxonomy, phylogeny and adaptations. *Bull. Am. Mus. Nat. Hist.* **156:**157–450.

Szalay, F. 1977. Ancestors, descendants, sister groups and testing of phylogenetic inferences. *Syst. Zool.* **26:**12–18.

Szalay, F., and Dagosto, M. 1980. Locomotor adaptations as reflected in the humerus of Paleogene primates. *Folia Primatol.* **34:**1–45.

Szalay, F., and Decker, R. 1974. Origins, evolution and function of the tarsus in late Cretaceous eutherians and Paleocene primates. In *Primate locomotion,* ed. F. Jenkins, pp. 239–259. Academic Press, New York.

Szalay, F., and Delson, E. 1979. *Evolutionary history of the primates.* Academic Press, New York.

Szalay, F., and Li, C. 1986. Middle Paleocene euprimate from southern China and the distribution of primates in the Paleogene. *J. Hum. Evol.* **15:**387–398.

Szalay, F., and McKenna, M. 1971. Beginning of the Age of Mammals in Asia: the late Paleocene Gashato fauna, Mongolia. *Bull. Am. Mus. Nat. Hist.* **144:**273–316.

Szalay, F., and Wilson, J. 1976. Basicranial morphology of the early Tertiary tarsiiform *Rooneyia* from Texas. *Folia Primatol.* **25:**288–293.

Szalay, F.; Tattersall, I.; and Decker, R. 1975. Phylogenetic relationships of *Plesiadapis*—postcranial evidence. In *Approaches to primate paleobiology. Contrib. Primat.,* ed. F. Szalay, **5:**136–166. S. Karger AG, Basel.

Szalay, F.; Li, C.; and Wang, B. 1986. Middle Paleocene omomyid primate from Anhui Province, China. *Am. J. Phys. Anthropol.* **69:**269.

Tarling, D. 1980. The geologic evolution of South America with special reference to the last 200 million years. In *Evolutionary biology of the New World monkeys and continental drift,* ed. R. Ciochon and A. Chiarelli, pp. 1–41. Plenum, New York.

Tarling, D., and Tarling, M. 1971. *Continental drift.* Doubleday, New York.

Tattersall, I. 1982. *The primates of Madagascar.* Columbia Univ. Press, New York.

Tattersall, I.; Delson, E.; and Van Couvering, J., eds. 1988. *Encyclopedia of human evolution and prehistory.* Garland Publishing, New York.

Tattersall, I., and Eldredge, N. 1977. Fact, theory, and fantasy in human paleontology. *Am. Sci.* **65:**204–211.

Taylor, R., and Rowntree, V. 1973. Running on two or on four legs: which consumes more energy? *Science* **179:**186–187.

Taylor, R.; Schmidt-Nielsen, K.; and Raab, J. 1970. Scaling of energetic cost of running to body size in mammals. *Am. J. Physiol.* **219:**1104–1107.

Teaford, M., and Walker, A. 1984. Quantitative differences in dental microwear between primate species with different diets and a comment on the presumed diet of *Sivapithecus. Am. J. Phys. Anthropol.* **64:**191–200.

Teaford, M.; Beard, K.; Leakey, R.; and Walker, A. 1988. New hominoid facial skeleton from the early Miocene of Rusinga Island, Kenya, and its bearing on the relationship between *Proconsul nyanzae* and *Proconsul africanus. J. Hum. Evol.* **17:**461–478.

Teilhard de Chardin, P. 1921. Les mammifères de l'Eocène inferieur française et leurs gisements. *Ann. Paleontol.* **10:**171–176.

Tekkaya, I. 1974. A new species of Tortonian anthropoid from Anatolia. *Bull. Miner. Res. Explor. Inst. Turk. Foreign Ed.* **83:**148–165.

Templeton, A. 1984. Phylogenetic inference from restriction endonuclease site maps with particular reference to the evolution of humans and apes. *Evolution* **37:**221–244.

Thomas, H. 1985. The early and middle Miocene land connection of the Afro-Arabian plate and Asia: a major event for hominoid dispersal? In *Ancestors: the hard evidence,* ed. E. Delson, pp. 42–50. Alan Liss, New York.

Thomas, O. 1913. On some rare Amazonian mammals from the collection of the Para Museum. *Ann. Mag. Nat. Hist.* **7:**188–196.

Tobias, P. 1967. *Olduvai Gorge,* vol. 2. Cambridge Univ. Press.

Tobias, P. 1980. *"Australopithecus afarensis"* and *A. africanus*: a critique and an alternative hypothesis. *Palaeontol. Afr.* **23:**1–17.

Tobias, P. 1984*a*. The child from Taung. *Science 84* **5**(9)99–100.

Tobias, P. 1984*b*. *Dart, Taung, and the "missing link."* Witwatersrand Univ. Press., Johannesburg.

Tobias, P. 1985. Taung. In *Excursions for the Taung Diamond Jubilee.* Univ. Witwatersrand.

Tobias, P. 1987. The brain of *Homo habilis:* a new level of organization in cerebral evolution. *J. Hum. Evol.* **16:**741–762.

Tong, Y. 1979. A late Paleocene primate from S. China. *Vertebr. Palasiat.* **17:**65–70.

Toth, N. 1987. Behavioral inferences from Early Stone artifact assemblages: an experimental model. *J. Hum. Evol.* **16:**763–788.

Tracy, C. 1977. Minimum size of mammalian homeotherms: role of the thermal environment. *Science* **198:**1034–1035.

Tschudy, R.; Pillmore, C.; Orth, C.; Gilmore, J.; and Knight, J. 1984. Disruption of the terrestrial plant ecosystem at the Cretaceous-Tertiary boundary, Western Interior. *Science* **225:**1030–1032.

Tuttle, R. 1972. Functional and evolutionary boundary of hylobatid hands and feet. In *Gibbon and Siamang,* ed. D. Rumbaugh, **1:**136–206.

Tuttle, R. 1974. Darwin's apes, dental apes, and the descent of man: normal science in evolutionary anthropology. *Curr. Anthropol.* **15:**389–398.

Tuttle, R. 1985. Ape footprints and Laetoli impressions: a response to the SUNY claims. In *Hominid evolution: past, present, and future,* ed. P. Tobias, pp. 129–134. Alan Liss, New York.

Vail, P.; Mitchum, R.; Todd, R.; Wedmier, J.; Thomson, S.; Sangree, J., Bubb, J.; and Hatelid, W. 1977. *Seismic-stratigraphy applications to hydrocarbon explorations,* ed. C. E. Payton, pp. 49–212. *A. A. P. G. Memoir* no. 26.

Van Couvering, J. 1972. Radiometric calibration of the European Neogene. In *Calibration of hominoid evolution: recent advances in isotopic and other dating methods applicable to the origin of man,* ed. W. Bishop and J. Miller, pp. 247–272. Scottish Academic Press, Edinburgh.

Van Couvering, J., and Miller, J. 1969. Miocene stratigraphy and age determinations, Rusinga Island, Kenya. *Nature* **221:**628–632.

Van Couvering, J., and Van Couvering, J. 1976. Early Miocene mammal fossils from East Africa: aspects of geology, faunistics and paleoecology. In *Human origins: Louis Leakey and the East African evidence,* ed. G. Isaac and E. McCown, pp. 155–207. Univ. of Calif. Press, Berkeley.

Van Valen, L., and Sloan, R. 1965. The earliest primates. *Science* **150:**743–745.

Villalta, J., and Crusafont, M. 1944. Dos nuevos antropomorfos del Mioceno español y su situacion dentro de la moderna sistematica de los simidos. *Notas. Comun. Inst. Geol. Min. Esp.* **13:**91–139.

Vogel, J. 1985. Further attempts at dating the Taung tufas. In *Hominid evolution: past, present, and future,* ed. P. Tobias, pp. 189–194. Alan Liss, New York.

von Koenigswald, G. 1965. Critical observations upon the so-called higher primates from the upper Eocene of Burma. *Proc. K. Ned. Akad. Wet.* **68:**165–167.

von Koenigswald, G. 1969. Miocene Cercopithecoidea and Oreopithecoidea from the Miocene of East Africa. In *Fossil Vertebrates of Africa,* ed. L. Leakey, **1:**39–51. Academic Press, London.

von Koenigswald, G. 1972. Ein unterkiefer eines fossilen hominoiden aus dem Terpliozan Griechenlands. *Koninkl. Nederlanse Akad. Wetensch. Proc.* **75:**385–394.

Vrba, E. 1974. Chronological and ecological implications of the fossil Bovidae at the Sterkfontein australopithecine site. *Nature* **250:**19–23.

Vrba, E. 1975. Some evidence of chronology and palaeoecology of Sterkfontein, Swartkrans and Kromdraai from the fossil record. *Nature* **254:**301–304.

Vrba, E. 1980*a*. Evolution, species and fossils: how does life evolve? *S. Afr. J. Sci.* **76:**61–84.

Vrba, E. 1980*b*. The significance of bovid remains as indicators of environment and predation patterns. In *Fossils in the making: vertebrate taphonomy and paleoecology,* ed. A. Behrensmeyer and A. Hill, pp. 247–271. Univ. Chicago Press, Chicago.

Vrba, E. 1981. The Kromdraai australopithecine site revisted in 1980; recent investigations and results. *Ann. Transvaal Mus.* **33:**17–60.

Vrba, E. 1982. Biostratigraphy and chronology, based particularly on Bovidae of southern African hominid-associated assemblages: Makapansgat, Sterkfontein, Taung, Kromdraai, Swartkrans; also Elandsfontein, Broken Hill and Cave of Hearths. In *Proc. Congress International de Paleontologie Humaine,* ed. H. de Lumley and M. de Lumley, **2:**707–752. Union Internationale des Sciences Prehistoriques et Protohistorique, Nice.

Vrba, E. 1985*a*. Ecological and adaptive changes associated with early hominid evolution. In *Ancestors: the hard evidence,* ed. E. Delson, pp. 63–71. Alan Liss, New York.

Vrba, E. 1985*b*. The Kromdraai australopithecine site. In *Excursions for the Taung Diamond Jubilee*. Univ. Witwatersrand.

Walker, A. 1969. Lower Miocene fossils from Mount Elgon, Uganda. *Nature* **223:**591–593.

Walker, A., and Pickford, M. 1983. New postcranial fossils of *Proconsul africanus* and *Proconsul nyanzae*. In *New interpretations of ape and human ancestry,* ed. R. Ciochon and R. Corruccini, pp. 325–352. Plenum, New York.

Walker, A., and Rose, M. 1968. Fossil hominoid vertebra from the Miocene of Uganda. *Nature* **217:**980–981.

Walker, A.; Zimmerman, M.; and Leakey, R. 1982. A possible case of hypervitaminosis A in *Homo erectus. Nature* **296:**248–250.

Walker, A.; Falk, D.; Smith, R.; and Pickford, M. 1983. The skull of *Proconsul africanus:* reconstruction and cranial capacity. *Nature* **305:**525–527.

Walker, A.; Teaford, M.; and Leakey, R. 1985. New *Proconsul* fossils from the early Miocene of Kenya. *Am. J. Phys. Anthropol.* **66:**239–240.

Walker, A.; Leakey, R.; Harris, J.; and Brown, F. 1986. 2.5 Myr *Australopithecus boisei* from west of Lake Turkana, Kenya. *Nature* **322:**517–522.

Wallace, J. 1972. The dentition of the South African early hominids: a study of form and function. Ph.D. dissertation, Univ. Witwatersrand.

Walter, R., and Aronson, J. 1982. Revisions of K / Ar ages for the Hadar hominid site, Ethiopia. *Nature* **296:**122–127.

Wang, C.; Shi, Y.; and Zhou, W. 1982. Dynamic uplift of the Himalaya. *Nature* **298:**553–556.

Wanyong, C.; Yufen, L.; and Qianli, Y. 1986. On the paleoclimate during the period of *Ramapithecus* in Lufeng County, Yunnan Province. *Acta Anthropol. Sin.* **4:**88.

Ward, J.; Seely, M.; and Lancaster, N. 1983. On the antiquity of the Namib. *S. Afr. J. Sci.* **79:**175–183.

Ward, S., and Hill, A. 1987. Pliocene hominid partial mandible from Tabarin, Baringo, Kenya. *Am. J. Phys. Anthropol.* **72:**21–38.

Ward, S., and Kimbel, W. 1983. Subnasal alveolar morphology and the systematic position of *Sivapithecus. Am. J. Phys. Anthropol.* **61:**157–171.

Ward, S., and Pilbeam, D. 1983. Maxillofacial morphology of Miocene hominoids from Africa and Indo-Pakistan. In *New interpretations of ape and human ancestry,* ed. R. Ciochon and R. Corruccini, pp. 211–238. Plenum, New York.

Warwick, R., and Williams, P., eds. 1973. *Gray's anatomy,* 35th ed. W. B. Saunders, Philadelphia.

Waterman, P. 1984. Food acquisition and processing as a function of plant chemistry. In *Food acquisition and processing in primates,* ed. D. Chivers, B. Wood, and A. Bilsborough, pp. 177–211. Plenum, New York.

Watters, J., and Krause, D. 1986. Plesiadapid primates and biostratigraphy of the North American late Paleocene. *Am. J. Phys. Anthropol.* **69:**277.

Wells, K. 1971. *Kinesiology: the scientific basis of human motion.* W. B. Saunders, Philadelphia.

White, T. 1977*a*. The anterior mandibular corpus of early African Hominidae. Ph.D. dissertation, Univ. Michigan.

White, T. 1977*b*. New fossil hominids from Laetolil, Tanzania. *Am. J. Phys. Anthropol.* **46:**197–230.

White, T. 1981. Primitive hominid canine from Tanzania. *Science* **213:**348–349.

White, T. 1984. Pliocene hominids from the Middle Awash, Ethiopia. *Cour. Forschungsinst. Senckenb.* **69:**57–68.

White, T. 1988. The comparative biology of "robust" *Australopithecus:* clues from context. In *Evolutionary history of the "robust" australopithecines,* ed. F. Grine, pp. 449–484. Aldine de Gruyter, New York.

White, T., and Harris, J. 1977. Suid evolution and correlation of African hominid localities. *Science* **198:**13–21.

White, T., and Suwa, G. 1987. Hominid footprints at Laetoli: facts and interpretations. *Am. J. Phys. Anthropol.* **72:**485–514.

White, T.; Johanson, D.; and Kimbel, W. 1981. *Australopithecus africanus:* its phyletic position reconsidered. *S. Afr. J. Sci.* **77:**445–470.

Wible, J. 1983. The internal carotid artery in early eutherians. *Acta Palaeontol.* **28:**281–293.

Wikander, R.; Covert, H.; and Deblieux, D. 1986. Ontogenetic, intraspecific, and interspecific variation of the prehallux in primates: implications for its utility in the assessment of phylogeny. *Am. J. Phys. Anthropol.* **70:**513–524.

Williams, J. 1985. Morphology and variation in the posterior dentition of *Picrodus silberlingi. Folia Primatol.* **45:**48–58.

Williamson, P. 1985. Evidence for an early Plio-Pleistocene rain forest expansion in East Africa. *Nature* **315:**487–489.

Wilson, J., and Szalay, F. 1976. New adapid primate of European affinities from Texas. *Folia Primatol.* **25:**294–312.

Wolfe, J. 1971. Tertiary climatic fluctuations and methods of analysis of Tertiary floras. *Palaeogeogr., Palaeoclimatol., Palaeoecol.* **9:**27–57.

Wolfe, J. 1978. A palaeobotanical interpretation of Tertiary climates in the Northern Hemisphere. *Am. Sci.* **66:**694–703.

Wolfe, J. 1980. Tertiary climates and floristic relationships at high latitudes in the Northern Hemisphere. *Palaeogeogr., Palaeoclimatol., Palaeoecol.* **30:**313–323.

Wolfe, J., and Hopkins, D. 1967. Climatic changes recorded by Tertiary land floras in northwestern North America. In *Tertiary correlations and climatic changes in the Pacific,* ed. E. Hatai, pp. 67–76. 11th Symp. Pacific Sci. Cong., Tokyo.

Wolfe, J., and Upchurch, G. 1986. Vegetation, climatic and floral changes at the Cretaceous-Tertiary boundary. *Nature* **324:**148–152.

Wolpoff, M. 1971. Competitive exclusion among lower Pleistocene hominids: the single species hypothesis. *Man* **6:**601–614.

Wolpoff, M. 1980. *Paleoanthropology.* Knopf, New York.

Wolpoff, M. 1982. *Ramapithecus* and hominid origins. *Curr. Anthropol.* **23:**501–510.

Wolpoff, M. 1983. *Ramapithecus* and human origins: an anthropologist's perspective of changing interpretations. In *New interpretations of ape and human ancestry,* ed. R. Ciochon and R. Corruccini, pp. 651–676. Plenum, New York.

Wolpoff, M. (in press). Multiregional evolution: the fossil alternative to Eden. In *The origins and dispersal of modern humans: behavioral and biological perspectives,* ed. P. Mellars and C. Stringer. Edinburgh Univ. Press, Edinburgh.

Woo, J. 1957. *Dryopithecus* teeth from Keiyuan, Yunnan Province. *Vertebr. Palasiat.* **1:**25–32.

Wood, A. 1972. An Eocene hystricognathous rodent from Texas: its significance in interpretations of continental drift. *Science* **175:**1250–1251.

Wood, A. 1973. Eocene rodents, Pruett Formation, southwest Texas. Pearce-Sellards Series, *Tex. Mem. Mus.* **20:**1–41.

Wood, A. 1980. The origin of the caviomorph rodents from a source in middle America: a clue to the area of origin of the platyrrhine primates. In *Evolutionary biology of the New World monkeys and continental drift,* ed. R. Ciochon and A. Chiarelli, pp. 79–92. Plenum, New York.

Wood, B. A., and Chamberlain, A. T. 1987. The nature and affinities of the "robust" australopithecines: a review. *J. Hum. Evol.* **16:**625–642.

Wood, C.; Chaney, R.; Clark, J.; Colbert, E.; Jepsen, G.; Reeside, J.; and Stock, C. 1941. Nomenclature and correlation of the North American continental Tertiary. *Geol. Soc. Am. Bull.* **52:**1–48.

Woodburne, M., ed. 1987. *Cenozoic mammals of North America.* Univ. Calif. Press, Berkeley.

Woodward, A. 1914. On the lower jaw of an anthropoid ape from the upper Miocene of Lerida (Spain). *Q. J. Geol. Soc. Lond.* **70:**316–320.

Wu, R., and Yuerong, P. 1984. A late Miocene gibbon-like primate from Lufeng, Yunnan Province. *Acta Anthropolog. Sin.* **3:**185–195.

Wu, R., and Yuerong, P. 1985. Preliminary observations on the cranium of *Laccopithecus robustus* from Lufeng, Yunnan with reference to its phylogenetic relationship. *Acta Anthropolog. Sin.* **4:**12.

Wu, R.; Han, R.; Xu, R.; Lu, R.; Pan, Y.; Zhang, X.; Zheng, L.; and Xiao, M. 1981. *Ramapithecus* skulls found first time in the world. *Kexue Tongbao (Foreign Lang. Ed.)* **26:**1018–1021.

Wu, R.; Qinghua, X.; and Qingwu, L. 1983. Morphological features of *Ramapithecus* and *Sivapithecus* and their phylogenetic relationships—morphology and comparison of the crania. *Acta Anthropolog. Sin.* **2:**9.

Wu, R.; Qinghua, X.; and Qingwu, L. 1984. Morphological features of *Ramapithecus* and *Sivapithecus* and their phylogenetic relationships—morphology and comparison of the mandibles. *Acta Anthropolog. Sin.* **3:**9.

Wu, R.; Qinghua, X.; and Qingwu, L. 1986. Relationship between Lufeng *Sivapithecus* and *Ramapithecus* and their phylogenetic position. *Acta Anthropolog. Sin.* **4:**1–31.

Xu, Q., and Lu, Q. 1979. The mandibles of *Ramapithecus* and *Sivapithecus,* Lufeng, Yunnan. *Vertebr. Palasiat.* **17:**1–13.

Yuerong, P., and Jablonski, N. 1987. The evolution and paleobiogeography of monkeys in China. Abstr. of pap. at The Palaeoenvironment of East Asia from the mid-Tertiary. Centre of Asian Studies, Univ. Hong Kong, Jan. 9–13, 1987.

Zapfe, H. 1958. The skeleton of *Pliopithecus vindobonensis. Am. J. Phys. Anthropol.* **16:**441–457.

Zapfe, H. 1960. Die primatenfunde aus dem Miozan von Klein-Hadersdorf in Niederosterreich. *Schweiz. Palaeontol. Abh.* **78:**1–293.

Zihlman, A., and Brunker, J. 1979. Hominid bipedalism: then and now. *Yearb. Phys. Anthropol.* **22:**132–162.

Zihlman, A.; Cronin, J.; Cramer, D.; and Sarich, V. 1978. Pygmy chimpanzee as a possible prototype for the common ancestor of humans, chimpanzees and gorillas. *Nature* **275:**744–746.

Zingeser, M. 1973. Dentition of *Brachyteles arachnoides* with reference to alouattine and atelenine affinities. *Folia Primatol.* **20:**351–390.

Zwell, M. 1972. On the supposed *Kenyapithecus africanus* mandible. *Nature* **240:**236–239.

Zwell, M., and Conroy, G. 1973. Multivariate analysis of the *Dryopithecus africanus* forelimb. *Nature* **244:**373–375.

Credits

Boldface numbers refer to figures and tables, with the latter preceded by a **T**.

PREFACE

Opening quotation from the composition "A Little Priest" by Stephen Sondheim. From the Broadway show *Sweeney Todd.* Copyright © 1979 Revelation Music Publicity Corp. and Rilting Music, Inc. A Tommy Valando Publication.

From "Your Song" written by Elton John and Bernie Taupin. Copyright © 1969 Dick James Music Limited. All Rights for the United States and Canada controlled by Dick James Music, Inc. International copyright secured. All Rights Reserved. Used By Permission.

CHAPTER ONE

Quotations on pp. 14 and 19 from Dr. Ernest Mayr, "Biological Classification: Toward a Synthesis of Opposing Methodologies," *Science,* Vol. 214, pp. 510–16, 30 October 1981. Copyright © 1981 by the AAAS. **1.1a–c** C. Jolly and F. Plog, *Physical Anthropology and Archaeology,* 4th ed., McGraw-Hill, New York, 1986; **d,e** M. Cartmill, 1978, "The orbital mosaic in prosimians and the use of variable traits in systematics," *Folia Primatologica,* 30:39–114. Copyright © S. Karger AG. **1.2** C. Patterson, "Introduction," *Molecules and Morphology in Evolution: Conflict or Compromise* (C. Patterson, ed.), Cambridge University Press, Cambridge, 1987. **1.3** Courtesy of J. Schwartz, University of Pittsburgh. **1.4** From G. G. Simpson, reprinted from *Classification and Human Evolution,* edited by Sherwood L. Washburn, Viking Fund Publications in Anthropology, No. 37. Copyright 1963 by the Wenner-Gren Foundation for Anthropological Research, Inc., New York. **1.5** F. Szalay, 1977, "Ancestors, descendants, sister groups and testing of phylogenetic inferences," *Systematic Zoology,* 26:12–18. Copyright © The Society of Systematic Zoology. **1.6, T1.1** E. Vrba, 1980, "Evolution, species and fossils: how does life evolve?" *South African Journal of Science,* 76:61–84. **T1.2** A. Richard, *Primates in Nature,* W. H. Freemann, New York, 1985. **1.8** P. Gingerich and B. Smith, "Allometric scaling in the dentition of primates and insectivores," *Size and Scaling in Primate Evolution* (W. Jungers, ed.), Plenum, New York, 1985, pp. 257–72. **1.9** A. Richard, *Primates in Nature,* W. H. Freeman, New York,

1985. **1.10** C. Jolly and F. Plog, *Physical Anthropology and Archaeology,* 4th ed., McGraw-Hill, New York, 1986. **1.11** W. E. Le Gros Clark, *The Antecedents of Man,* Edinburgh University Press, Edinburgh, 1959. **T1.3** C. Jolly and F. Plog, *Physical Anthropology and Archaeology,* 4th ed., McGraw-Hill, New York, 1986. **1.12** MacPhee et al. 1983. Reprinted by permission from *Nature,* Vol. 301, pp. 509–11. Copyright © 1983 Macmillan Magazines Ltd.

CHAPTER TWO

2.2a P. Vail et al., "Seismic-stratigraphy applications to hydrocarbon explorations" (C. E. Payton, ed.), *AAPG Memoir,* no. 26, 1977, pp. 49–212. Reprinted by permission; **b** R. Lewin, *Human Evolution: An Illustrated Introduction,* W. H. Freeman, New York, 1984. **T2.1** K. Rose and J. Fleagle, "The fossil history of nonhuman primates in the Americas," *Ecology and Behavior of Neotropical Primates* (A. Coimbra-filho and R. Mittermeier, eds.), Vol. 1, Academia Brasiliera de Ciencias, Rio de Janeiro, 1981, pp. 111–67. **2.4** P. Gingerich, 1976, "Cranial anatomy and evolution of early Tertiary Plesiadapidae," *Papers on Paleontology,* 15:1–141, The Museum of Paleontology, University of Michigan. **2.5** P. Hershkovitz, *Living New World Monkeys,* University of Chicago Press, Chicago, 1977. Copyright © 1977 by the University of Chicago. **2.6a,b** R. MacPhee and M. Cartmill, "Basicranial structures and primate systematics," *Primate Biology,* Vol. 1, Alan R. Liss, New York, 1986; **c** P. Hershkovitz, *Living New World Monkeys,* University of Chicago Press, Chicago, 1977. Copyright © 1977 by the University of Chicago. **2.7** P. Gingerich, 1976, "Cranial anatomy and evolution of early Tertiary Plesiadapidae," *Papers on Paleontology,* 15:1–141, The Museum of Paleontology, University of Michigan. **2.8, 2.9** W. E. Le Gros Clark, *The Antecedents of Man,* Edinburgh University Press, Edinburgh, 1959. **2.10** Photos from W. Matthew (1917). Courtesy, American Museum of Natural History. **2.11, 2.12** P. Gingerich, 1976, "Cranial anatomy and evolution of early Tertiary Plesiadapidae," *Papers on Paleontology,* 15:1–141, The Museum of Paleontology, University of Michigan. **2.13** Photos from W. Matthew (1917). Courtesy, American Museum of Natural History, **2.15** J. Fleagle, *Primate Adaptation and Evolution,* Academic Press, Orlando, FL, 1988. **2.16** K. Rose, 1975, "The Carpolestidae, early Tertiary Primates from North America, " Bulletin of the Museum of Comparative Zoology, 147:1–74. Courtesy, Museum of Comparative Zoology, Harvard University. **2.17** F. Szalay, *The Functional and Evolutionary Biology of Primates* (R. Tuttle, ed.), Aldine-Atherton, Chicago, 1972, pp. 3–35. **2.18a–c** G. Simpson, 1955, "The Phenacolemuridae, new family of early primates," *Bulletin of the American Museum of Natural History,* 105:415–41. Courtesy, American Museum of Natural History; **d–f** F. Szalay, 1973, "New Paleocene Primates and a diagnosis of the new Suborder Paromomyiformes," *Folia Primatologica,* 19:73–87. Copyright © S. Karger AG. **219a** K. Rose and P. Gingerich, 1976, "Partial skull of plesiadapiforms primate *Ignacius* from the early Eocene of Wyoming," *Contributions from the Museum of Paleontology,* 24:181–89, The Museum of Paleontology, University of Michigan; **b,c** G. Simpson, "The Phenacolemuridae, new family of early primates," *Bulletin of the American Museum of Natural History,* 105:415–41. Courtesy, American Museum of Natural History. **2.20a** Courtesy of J. Schwartz, University of Pittsburgh; **b** F. Szalay, "The Picrodontidae, a family of early primates," *American Museum Novitates,* No. 2329:1–55. Courtesy, American Museum of Natural History. **2.21** P. Gingerich, 1976, "Cranial anatomy and evolution of early Tertiary Plesiadapidae," *Papers on Paleontology,* 15:1–141, The Museum of Paleontology, University of Michigan.

CHAPTER THREE

3.1 D. Savage and D. Russell, *Mammalian Paleofaunas of the World,* Addison-Wesley, Reading, MA, 1983. **T3.1** K. Rose and J. Fleagle, "The fossil history of non-human primates in the Americas," *Ecology and Behavior of Neotropical Primates* (A. Coimbra-filho and R. Mittermeier, eds.), Vol. 1, Academia Brasiliera de Ciencias, Rio de Janeiro, 1981, pp. 111–67. **3.2a** G. Simpson, 1940, "Studies on the earliest primates," *Bulletin of the American Museum of Natural History,* 77:185–212. Courtesy, American Museum of Natural History; **b** courtesy of J. Schwartz, University of Pittsburgh. **T3.2** P. Gingerich, "Eocene Adapidae, paleobiogeography, and the origin of South American Platyrrhini," *Evolutionary Biology of the New World Monkeys and Continental Drift* (R. Ciochon and A. Chiarelli, eds.), Plenum, New York, 1980, pp. 123–38. **3.3** R. Kay and M. Cartmill, 1977, "Cranial morphology and adaptation of *Palaechthon nacimienti* and other Paromomyidae, with a description of a new genus and species," *Journal of Human Evolution,* 6:19–53. Copyright © Academic Press (London). **T.3.3** From J. Gurche, "Early primate brain evolution," *Primate Brain Evolution* (E. Armstrong and D. Falk, eds.), Plenum, New York, 1982, pp. 227–46. **T3.4** E. Delson and A. Rosenberger, "Phyletic perspectives on platyrrhine origins and anthropoid relationships," *Evolutionary Biology of the New World Monkeys and Continental Drift* (R. Ciochon and A. Chiarelli, eds.), Plenum, New York, 1980, pp. 445–58. **3.4** P. Gingerich and E. Simons, 1977, "Systematics, phylogeny, and evolution of early Eocene Adapidae (Mammalia, Primates) in North America," *Contributions from the Museum of Paleontology,* 24:245–79, The Museum of Paleontology, University of Michigan. **3.5** W. Gregory (1920). Courtesy, American Museum of Natural History. **3.6** *Plesiadapis* bones: F. Szalay, I. Tattersall, and R. Decker, "Phylogenetic relationships of *Plesiadapis*—postcranial evidence," *Approaches to Primate Paleobiology Contrib. Primat.* (F. Szalay, ed.), Vol. 5, S. Karger AG, Basel, Switzerland, 1975, pp. 136–66; all others: W. Gregory (1920). Courtesy, American Museum of Natural History. **3.7a, 3.8a** F. Szalay, I. Tattersall, and R. Decker, "Phylogenetic relationships of *Plesiadapis*—postcranial evidence," *Approaches to Primate Paleobiology Contrib. Primat.* (F. Szalay, ed.), Vol. 5, S. Karger AG, Basel, Switzerland, 1975, pp. 136–66; **3.7b,c, 3.8b,c** W. Gregory (1920). Courtesy, American Museum of Natural History. **3.9** P. Gingerich and R. Martin, 1981, "Cranial morphology and adaptations in Eocene Adapidae. II. The Cambridge skull of *Adapis parisiensis,*" *American Journal of Physical Anthropology,* 56:235–57. Copyright © Alan R. Liss. **3.10** W. Gregory (1920). Courtesy, American Museum of Natural History. **3.11** T. Bown and K. Rose, 1987, "Patterns of dental evolution in early Eocene anaptomorphine primates (Omomyidae) from the Bighorn basin, Wyoming," *The Paleontological Society Memoir,* 23. Copyright © The Paleontological Society, Inc. **3.12** Courtesy of K. Rose, Johns Hopkins University. **3.13a,b** E. Simons and D. Russell, 1960, "Notes on the cranial anatomy of *Necrolemur,*" *Breviora,* 127:1–14, Museum of Comparative Zoology, Harvard University; **c** courtesy of J. Schwartz, University of Pittsburgh. **3.14** G. Simpson, 1940, "Studies on the earliest primates," *Bulletin of the American Museum of Natural History,* 77:185–212.

Courtesy, American Museum of Natural History. **T3.5** A. Rosenberger and F. Szalay, "On the tarsiiform origins of Anthropoidea," *Evolutionary Biology of the New World Monkeys and Continental Drift* (R. Ciochon and A. Chiarelli, eds.), Plenum, New York, 1980, pp. 139–57. **3.15** P. Gingerich, "Dental and cranial adaptations in Eocene Adapidae," *Zeitschrift für Morphologie und Anthropologie,* Vol. 71, Schweizerbart'sche Verlagsbuchhandlung, Stuttgart, 1980, pp. 135–42. **3.16** P. Gingerich, "Eocene Adapidae, paleobiogeography, and the origin of South American platyrrhines," *Evolutionary Biology of the New World Monkeys and Continental Drift* (R. Ciochon and A. Chiarelli, eds.), Plenum, New York, 1980, pp. 123–38. **3.17** R. MacPhee and M. Cartmill, "Basicranial structures and primate systematics," *Primate Biology,* Vol. 1, Alan R. Liss, New York, 1986. **3.18** A. Schultz, *The Life of Primates,* Weidenfeld & Nicolson, London, 1969. **3.19** L. Radinsky, 1975, "Primate brain evolution," *American Scientist,* 63:656–63. Copyright © Sigma Xi, The Scientific Research Society of North America, Inc. **3.20** D. Rasmussen, 1986, "Anthropoid origins: a possible solution to the Adapidae–Omomyidae paradox," *Journal of Human Evolution,* 15:1–12. Copyright © Academic Press (London).

CHAPTER FOUR

4.1 D. Savage and D. Russell, *Mammalian Paleofaunas of the World,* Addison-Wesley, Reading, MA, 1983. **4.2** D. Tarling, "The geologic evolution of South America with special reference to the last 200 million years," *Evolutionary Biology of the New World Monkeys and Continental Drift* (R. Ciochon and A. Chiarelli, eds.), Plenum, New York, 1980, pp. 1–41; current arrows: R. Ciochon and A. Chiarelli, "Paleographic perspectives on the origin of the Platyrrhine," *Evolutionary Biology of the New World Monkeys and Continental Drift* (R. Ciochon and A. Chiarelli, eds.), Plenum, New York, 1980. **4.3** J. Fleagle and R. Kay, 1987, "The phyletic position of the Parapithecidae," *Journal of Human Evolution,* 16:483–532. Copyright © Academic Press (London). **4.4** C. Jolly and F. Plog, *Physical Anthropology and Archaeology,* 4th ed., McGraw-Hill, New York, 1986; P. Hershkovitz, *Living New World Monkeys,* University of Chicago Press, Chicago, 1977. Copyright © 1977 by the University of Chicago; D. Swindler and C. Wood, *An Atlas of Primate Gross Anatomy,* Robert E. Krieger, Melbourne, FL, 1973. **T4.2** J. Fleagle and R. Kay, 1987, "The phyletic position of the Parapithecidae," *Journal of Human Evolution,* 16:483–532. Copyright © Academic Press (London). **4.5** Courtesy of and © Eric Delson, American Museum of Natural History. **4.6a** F. Szalay and E. Delson, *Evolutionary History of the Primates,* Academic Press, Orlando, FL, 1979. **4.6b, 4.7** J. Fleagle and R. Kay, "The phyletic position of the Parapithecidae," *Journal of Human Evolution,* 16:483–532. Copyright © Academic Press (London). **4.8** Courtesy of and © Eric Delson, American Museum of Natural History. **4.9** Courtesy of E. Simons, Duke University. **4.11** E. Simons, "Two new primate species from the African Oligocene," *Postilla,* 64:1–12. Copyright © Peabody Museum, Yale University. **4.12** Graph: R. Kay and E. Simons, "The ecology of Oligocene Anthropoidea," *International Journal of Primatology,* 1:21–37. Copyright © 1980, Plenum Publishing Company. **4.13, 4.14** P. Hershkovitz, 1974, "A new genus of late Oligocene monkey with notes on postorbital closure and platyrrhine evolution," *Folia Primatologica,* 21:1–35. Copyright © S. Karger AG. **4.15** E. Colbert, "A new primate from the Upper Eocene Pondaung formations of Burma," *American Museum Novitates,* No. 951, 1937, pp. 1–18. Courtesy, American Museum of Natural History. **4.16** J. Fleagle and R. Kay, 1987, "The phyletic position of the

Parapithecidae," *Journal of Human Evolution,* 16:483–532. Copyright © Academic Press (London).

CHAPTER FIVE

Quotations on p. 267 from R. Leakey and A. Walker, "New higher primates from the early Miocene of Buluk, Kenya." Reprinted by permission from *Nature,* Vol. 318, pp. 173–75. Copyright © 1985 Macmillan Magazines Ltd. **5.1** W. Bishop, *Geological Background to Fossil Man,* Scottish American Press, Edinburgh, 1978. **5.2** P. Andrews and J. Van Couvering, "Paleoenvironments in the East African Miocene," *Approaches to Primate Paleobiology* (F. Szalay, ed.), S. Karger AG, Basel, Switzerland, 1975, pp. 62–103. **5.3** R. Bernor, "Geochronology and zoogeographic relationships of Miocene Hominoidea," *New Interpretations of Ape and Human Ancestry* (R. Ciochon and R. Corruccini, eds.), Plenum, New York, 1983, pp. 325–52. **5.4a** D. Savage and D. Russell, *Mammalian Paleofaunas of the World,* Addison-Wesley, Reading, MA, 1983; **b** P. Andrews and J. Van Couvering, "Paleoenvironments in the East African Miocene," *Approaches to Primate Paleobiology* (F. Szalay, ed.), S. Karger AG, Basel, Switzerland, 1975, pp. 62–103; **c** M. Pickford, "Sequence and environments of the lower and middle Miocene hominoids of western Kenya," *New Interpretations of Ape and Human Ancestry* (R. Ciochon and R. Corruccini, eds.), Plenum, New York, 1983, pp. 421–40. **5.5** D. Savage and D. Russell, *Mammalian Paleofaunas of the World,* Addison-Wesley, Reading, MA, 1983. **T5.2** M. Pickford, "Sequence and environments of the lower and middle Miocene hominoids of western Kenya," *New Interpretations of Ape and Human Ancestry* (R. Ciochon and R. Corruccini, eds.), Plenum, New York, 1983, pp. 421–40. **5.7** R. Bernor, "Geochronology and zoogeographic relationships of Miocene Hominoidea," *New Interpretations of Ape and Human Ancestry* (R. Ciochon and R. Corruccini, eds.), Plenum, New York, 1983, pp. 325–52. **5.8–5.10a,b** Courtesy of P. Andrews, British Museum of Natural History. **5.10c** Courtesy of and © Eric Delson, American Museum of Natural History. **5.11–5.13** Courtesy of P. Andrews, British Museum of Natural History. **5.14** J. Fleagle and E. Simons, 1978, "*Micropithecus clarki,* a small ape from the Miocene of Uganda," *American Journal of Physical Anthropology,* 49:427–40. Copyright © Alan R. Liss (New York.). **5.15a** A. Cave, "Frontal sinus of the gorilla," *Proceedings of the Zoological Society of London,* Vol. 136, pp. 359–73, 1961. Courtesy of the Zoological Society of London. **5.16** A. Walker and M. Pickford, "New postcranial fossils of *Proconsul africanus* and *Proconsul nyanzae,*" *New Interpretations of Ape and Human Ancestry* (R. Ciochon and R. Corruccini), Plenum, New York, 1983, pp. 325–52. **5.17** A. Walker and M. Pickford, "New postcranial fossils of *Proconsul africanus* and *Proconsul nyanzae,*" *New Interpretations of Ape and Human Ancestry* (R. Ciochon and R. Corruccini), Plenum, New York, 1983, pp. 325–52. **5.18** D. Swindler and C. Wood, *An Atlas of Primate Gross Anatomy,* Robert E. Krieger, Melbourne, FL, 1973. **5.19** A. Walker and M. Pickford, "New postcranial fossils of *Proconsul africanus* and *Proconsul nyanzae,*" *New Interpretations of Ape and Human Ancestry* (R. Ciochon and R. Corruccini), Plenum, New York, 1983, pp. 325–52. **5.20–5.22a** Courtesy of and © Eric Delson, American Museum of Natural History. **5.22b** Courtesy of D. Pilbeam, Harvard University. **5.23** R. Kay, 1982, "*Sivapithecus simonsi,* a new species of Miocene hominoid, with comments on the phylogenetic status of the Ramapithecinae," *International Journal of Primatology,* 3:13–73. Copyright © Plenum Publishing Company. **5.24** Courtesy of R. Wu, Academia Sinica (Beijing). **5.25** S. Ward and W. Kimbel, 1983, "Subnasal alveolar morphology and the systematic position of *Sivapithecus," American Journal of Physical Anthropology,* 61:157–71. Copyright © Alan R. Liss, Inc. **5.26** Courtesy of and © Eric Delson, American Museum of Natural History. **T5.5** J. Fleagle, *Primate Adaptation and Evolution,* Academic Press, Orlando, FL, 1988. **5.27** M. Leakey et al., "Morphological similarities in *Aegyptopithecus* and *Afropithecus," Folia Primatologica,* 1988. Copyright © S. Karger AG. **5.28, 5.29** Courtesy of and © Eric Delson, American Museum of Natural History. **5.30, 5.31a** Courtesy of D. Pilbeam; reproduced, with permission, from the *Annual Review of Anthropology,* Vol. 8. © 1979 by Annual Reviews, Inc. **5.30b** L. Greenfield, 1980, "A late divergence hypothesis," *American Journal of Physical Anthropology,* 52:351–65. Copyright © Alan R. Liss, Inc.; **c** P. Andrews, 1978, "A revision of the Miocene Hominoidea of East Africa," *Bulletin of the British Museum Geology Series,* Vol. 30, no. 2. By courtesy of the British Museum (Natural History). **5.32** J. Spuhler, 1988, "Evolution of mitochondrial DNA in monkeys, apes and humans," *Yearbook of Physical Anthropology,* 31:15–48. Copyright © Alan R. Liss, Inc. **5.33** Courtesy of D. Pilbeam, Harvard University. **5.34** P. Andrews and L. Aiello, "An evolutionary model for feeding and positional behaviour," *Food Acquisition and Processing in Primates* (D. Chivers, B. Wood, and A. Bilsborough, eds.), Plenum, New York, 1984.

CHAPTER SIX

6.1 R. Flint, *Glacial and Quaternary Geology,* Wiley, New York, 1971. **6.2a** F. Howell, "Hominidae," *Evolution of African Mammals* (V. Maglio and H. Cooke, eds.), Harvard University Press, 1978. Reproduced by permission. Copyright © 1978 by the President and Fellows of Harvard College; **b** T. White, D. Johanson, and W. Kimbel, 1981, "*Australopithecus africanus:* its phyletic position reconsidered," *South African Journal of Science,* 77:445–70. **6.3** H. B. Cooke, 1983, "Human evolution: the geological framework," *Canadian Journal of Anthropology,* 3:143–61. **6.4** P. Tobias, 1980, "*Australopithecus afarensis* and *A. africanus:* a critique and an alternative hypothesis," *Paleontologia Africana,* 23:1–17. **6.5** E. Vrba, "The Kromdraai australopithecine site," *Excursions for the Taung Diamond Jubilee,* University of Witwatersrand Press, Johannesburg, 1985. **6.6a** P. Tobias, 1980, "*Australopithecus afarensis* and *A. africanus:* a critique and an alternative hypothesis," *Paleontologia Africana* 23:1–17; **b** Cooke, H. B. S. 1986. Changing Perspectives on the Age of Man. Raymond Dart Lecture No. 21. Reproduced by permission. © Witwatersrand University Press, Johannesberg, for the Institute for the Study of Man in Africa. **6.7** Reprinted with permission from: Charles K. Brain, "New Information from the Swartkrans Cave of Relevance to Robust Australopithecines," in: Frederick E. Grine, Editor, *Evolutionary History of the Robust Australopithecines* (New York: Aldine de Gruyter). Copyright 1988 Aldine de Gruyter. **6.8** E. Vrba, "Ecological and adaptive changes associated with early hominid evolution," *Ancestors: The Hard Evidence* (E. Delson, ed.), Alan R. Liss, New York, 1985. **6.9** T. White, "Ethiopia," *Encyclopedia of Human Evolution and Prehistory* (I. Tattersall, E. Delson, and J. Van Couvering, eds.), Garland, New York, 1988, p. 182. Courtesy of T. White. **6.10** F. Brown, "Omo," *Encyclopedia of Human Evolution and Prehistory* (I. Tattersall, E. Delson, and J. Van Couvering, eds.), Garland, New York, 1988, p. 394–96. Courtesy of F. Brown. **6.11** F. Brown, "East Turkana," *Encyclopedia of Human Evolution and Prehistory* (I. Tattersall, E. Delson, and J. Van Couvering, eds.), Garland, New York, 1988, pp. 174–76. Courtesy of F. Brown. **6.12** R. Hay, "Lithofacies and environments of Bed I, Olduvai

Gorge, Tanzania," *Quaternary Research,* 3:541–60. Copyright © Academic Press (Orlando, FL). **6.13** Walker et al. 1983. Reprinted by permission from *Nature,* Vol. 305, pp. 525–27. Copyright © 1983 Macmillan Magazine Ltd.; W. Kimbel and T. White, 1988, "A revised reconstruction of the adult skull of *Australopithecus afarensis,*" *Journal of Human Evolution,* 17:545–50. Copyright © Academic Press (London); F. Howell, "Hominidae," *Evolution of African Mammals* (V. Maglio and H. Cooke, eds.), 1978, pp. 154–248. Reproduced by permission of Harvard University Press. Copyright © 1978 by the President and Fellows of Harvard College; Y. Rak. *The Australopithecine Face,* Academic Press, Orlando, FL, 1983; C. Jolly and F. Plog, *Physical Anthropology and Archaeology,* McGraw-Hill, New York, 1986. **6.14a** Photos courtesy of the Institute of Human Origins: **b** Y. Rak, *The Australopithecine Face,* Academic Press, Orlando, FL, 1983. **6.15a** T. White, D. Johanson, and W. Kimbel, 1981, "*Australopithecus africanus:* its phyletic position reconsidered," *South African Journal of Science,* 77:445–470; **b** K. Hiiemae, *Dental Anatomy and Embryology* (J. Osborne, ed.), Blackwell Scientific Press, Oxford, 1981; **c** T. White, 1977, "New fossil hominids from Laetolil, Tanzania," *American Journal of Physical Anthropology,* 46:197–230. **6.17** R. Holloway, "Brain," *Encyclopedia of Human Evolution and Prehistory* (I. Tattersall, E. Delson, and J. Van Couvering, eds.), Garland, New York, 1988, pp. 98–105. Courtesy of R. Holloway. **6.21a** Photo by R. G. Klomfass, by permission of P. V. Tobias; **b** F. Howell, "Hominidae," *Evolution of African Mammals* (V. Maglio and H. Cooke, eds.), 1978, pp. 154–248. Reproduced by permission of Harvard University Press. Copyright © 1978 by the President and Fellows of Harvard College. **6.22** C. Jolly and F. Plog, *Physical Anthropology and Archaeology,* 4th ed., McGraw-Hill, New York, 1986. **6.23** Courtesy of A. Walker, copyright © National Museums of Kenya. **T6.2** From P. Tobias, 1987, "The brain of *Homo habilis:* a new level of organization in cerebral evolution," *Journal of Human Evolution,* 16:741–62. Copyright © Academic Press (London). **6.24a, 6.25a** Courtesy of A. Walker, copyright © National Museums of Kenya. **6.24b,c, 6.25b** F. Howell, "Hominidae," *Evolution of African Mammals* (V. Maglio and H. Cooke, eds.), 1978, pp. 154–248. Reproduced by permission of Harvard University Press. Copyright © 1978 by the President and Fellows of Harvard College. **6.26** Photo by and courtesy of Tim White. **6.27** Photo by Tim White. **6.28a** O. Lovejoy, 1988, "The Evolution of Human Walking," *Scientific American,* 259:118–25; (b) K. Wells, *Kinesiology,* 5th ed., W. B. Saunders, Philadelphia, 1971. **6.29** All except *Pan* in (b) from O. Lovejoy, 1988, "The Evolution of Human Walking," *Scientific American,* 259:118–25. **6.31a** A. Zihlman and L. Brunker, "Hominid bipedalism: then and now," *Yearbook of Physical Anthropology,* 22:132–62. Copyright © Alan R. Liss, Inc.; **b** *Gray's Anatomy,* 35th ed. (R. Warwick and P. Williams, eds.), Churchill Livingstone, New York, 1973. **6.32** A. Zihlman and L. Brunker, "Hominid bipedalism: then and now," *Yearbook of Physical Anthropology,* 22:132–62. Copyright © Alan R. Liss, Inc. **6.33** Courtesy of the Institute of Human Origins. **6.34a,c** O. Lovejoy, 1988, "The Evolution of Human Walking," *Scientific American,* 259:118–25; **b** K. Wells, *Kinesiology,* 5th ed., W. B. Saunders, Philadelphia, 1971. **6.35b** J. T. Robinson, *Early Hominid Posture and Locomotion,* University of Chicago Press, Chicago, 1972. Copyright © 1972 by the University of Chicago. **6.36a–c** O. Lovejoy, 1988, "The Evolution of Human Walking," *Scientific American,* 259:118–25; **d** *Gray's Anatomy,* 35th ed. (R. Warwick and P. Williams, eds.), Churchill Livingstone, New York, 1973. **6.37** J. D. Clark, "Leaving no stone unturned: archaelogical advances and behavioral adaptation," *Hominid*

Evolution: Past, Present, and Future (P. Tobias, ed.), Alan R. Liss: New York, 1985, pp. 65–88. **6.38** R. Potts, "Oldowan," *Encyclopedia of Human Evolution and Prehistory* (I. Tattersall, E. Delson, and J. Van Couvering, eds.), Garland, New York, 1988, pp. 387–90. Courtesy of R. Potts. **6.39** J. Clark, *The Prehistory of Africa,* Praeger, New York, 1970. **6.40** F. Grine, "Australopithecine," *Encyclopedia of Human Evolution and Prehistory* (I. Tattersall, E. Delson, and J. Van Couvering, eds.), Garland, New York, 1988, pp. 67–74. Courtesy of F. Grine. **6.41a** Y. Rak, *The Australopithecine Face,* Academic Press, Orlando, FL, 1983; **b** Y. Rak, 1985, "Australopithecine taxonomy and phylogeny in light of facial morphology," *American Journal of Physical Anthropology,* 66:281–87. Copyright © Academic Press (Orlando, FL). **T6.3** C. Stringer and P. Andrews, 1988, "Genetic and fossil evidence for the origin of modern humans," *Science,* 239:1263–68.

APPENDICES

A.1 J. LaBrecque, D. Kent, and S. Cande, 1977, "Revised magnetic polarity time scale for Late Cretaceous and Cenozoic time," originally published in *Geology,* 5:330–35; and from J. P. Kennett, "The development of planktonic biogeography in the Southern Ocean during the Cenozoic," *Marine Micropaleontology,* Elsevier Science Publishers (Physical Science and Engineering Division): Amsterdam, 1978. **A.2** F. Szalay and E. Delson, *Evolutionary History of the Primates,* Academic Press, Orlando, FL, 1979.

B.1a W. Gregory (1920). Courtesy, American Museum of Natural History; **b** *Gray's Anatomy,* 35th ed. (R. Warwick and P. Williams, eds.), Churchill Livingstone, New York, 1973; **c** W. E. Le Gros Clark, *The Antecedents of Man,* Edinburgh University Press, Edinburgh, 1959, **B.2** *Lemur* from Elliot (1913); *Pan* and *Homo* from D. Swindler and C. Wood, *An Atlas of Primate Gross Anatomy,* Robert E. Krieger, Melbourne, FL, 1973. **B.3a** W. E. Le Gros Clark, *The Antecedents of Man,* Edinburgh University Press: Edinburgh, 1959; **b,c** G. Simpson, "The beginning of the Age of Mammals," *Biological Reviews* XII, January, 1937, pp. 1–47. Copyright © Cambridge University Press. **TB.1** K. Hiiemae, "Evolution of Man (The Primates)," *Dental Anatomy and Embryology* (J. Osborne, ed.), Blackwell Scientific Press, Oxford, 1981. **B.4** K. Hiiemae, "Evolution of Man (The Primates)," *Dental Anatomy and Embryology* (J. Osborne, ed.) Blackwell Scientific Press: Oxford, 1981. **B.5** D. Swindler and C. Wood, *An Atlas of Primate Gross Anatomy,* Robert E. Krieger, Melbourne, FL, 1973.

C.1 D. J. Chivers et al., "Food acquisition and processing in primates: concluding discussion," *Food Acquisition and Processing in Primates* (D. Chivers, B. Wood, and A. Bilsborough, eds.), Plenum, New York, 1984, pp. 545–56.

TD.1 Reprinted with permission of Macmillan Publishing Company from *Primate Evolution* by Elwyn L. Simons. Copyright © 1972 by Elwyn L. Simons. **TD.2** J. Schwartz, I. Tattersall, and N. Eldredge, 1978, "Phylogeny and classification of the Primates revisited," *Yearbook of Physical Anthropology,* 21:95–133. Copyright © Alan R. Liss. **TD.3** F. Szalay and E. Delson, *Evolutionary History of the Primates,* Academic Press, Orlando, FL, 1979. **TD.4** J. Fleagle, *Primate Adaptation and Evolution,* Academic Press, Orlando, FL, 1988.

Glossary

abductors muscle that moves a body part away from the midline

acetabulum depression in the hip bone into which the head of the femur fits: the "socket" in the ball-and-socket hip joint. In hominids the three hip bones (ilium, ischium, pubis) all meet in the acetabulum

adapid member of the family Adapidae, one of the two families of Eocene euprimates (the other being the Omomyidae), both of which became extinct in the mid-Tertiary

adapine member of the Adapinae, one of two extinct subfamilies of Eocene adapids (the other being the Notharctinae); most species occurred in Europe

Age of Mammals Cenozoic Era, covering the last 65 million years of earth's history

alcelaphine member of the tribe Alcelaphini, medium to large-sized antelopes (hartebeests, topis, wildebeests)

alisphenoid bone portion of the sphenoid bone forming part of the lateral wall of the skull

allometry study of the relative growth relationships between different parts of an organism; mathematically expressed by the **allometric equation** $Y = aX^b$ where Y and X are the two variables under consideration

allopatric speciation emergence of new species from populations that are geographically separated from each other

alpha decay radioactive decay in which the nucleus of the parent atom loses an alpha particle (2 protons and 2 neutrons)

alveolar processes the tooth-bearing portion of the jaws

alveolar prognathism *see* prognathous

alveoli tooth sockets

analogous describes the similarity of form or structure between two species that they do not share with their nearest common ancestor; that is, the similar structure has evolved independently in the two species and is due to convergent evolution

anaptomorphine member of the Anaptomorphinae, one of the three extinct subfamilies of the Eocene Omomyidae

angular process in the skull, the portion of the inferior aspect of the mandible where the ascending mandibular ramus and mandibular body meet; the masseter muscle attaches to its lateral side and the medial pterygoid muscle to its medial side. Also called the mandibular angle

antemolar dentition all the teeth mesial to (forward of) the molars; incisors, canines, and premolars

anterior iliac spines in hominids, two bony projections (superior and inferior) for muscle attachments that protrude from the anterior edge of the ilium

anterior pillars two vertical columns of bone on either side of the nasal aperture in the faces of some australopithecines; developed to offset the large chewing stresses of massive jaws and teeth

anterocone small cusp on the front of the incisors in some plesiadapiforms

anthracothere member of the family Anthracotheriidae, an extinct group of large, hippolike ungulates

anthropoid member of the suborder Anthropoidea (higher primates), which includes Old and New World monkeys, apes, and humans; sometimes the term is used to describe just the great apes, but this text uses the former definition. For contrast, *see also* prosimian

anthropometry science dealing with measurements of the size, weight, and proportions of the human body

antilopine member of the tribe Antilopini, small to medium-sized antelopes (gazelles and springbok)

apatemyid member of the family Apatemyidae, primitive Paleogene insectivorans; one of several groups once proposed as candidates for primate origins

apomorphy *see* derived character state

arboreal living mainly in trees; **arboreal quadrupeds** are animals that use all four limbs in walking and running on tree limbs

arboreal theory a theory that holds that most of the cranial and postcranial trends found in primates initially evolved as adaptations for life in the trees

Arctic Ocean spillover model theory that the mass extinction events at the Cretaceous–Tertiary boundary (about 65 MYA) occurred because of a chain of ecological events triggered by passage of colder, less salty water from the Arctic out into the North Atlantic; one of several models proposed to explain the K/T extinctions

artiodactyl member of the order Artiodactyla (even-toed ungulates, or hoofed animals)

ascending mandibular ramus (pl. **rami**) vertical portion of the mandible; also called the vertical ramus of the mandible

asteroid-impact model theory that the mass extinctions evident in the fossil record at the Cretaceous–Tertiary boundary resulted from the sudden climatic changes induced by collision of the earth with an asteroid or meteorite; one of several models proposed to explain the K/T mass extinctions.

astragalus ankle bone; also called the talus

auditory bulla bone surrounding the middle-ear cavity; formed by the petrosal bone in primates

auditory meatus external opening of the ear canal. Also called the external auditory tube

australopithecine member of the subfamily Australopithecinae, a group of early hominids living in East and South Africa during the Plio-Pleistocene

autapomorphic describes a derived character that arose for the first time in a particular group; autapomorphic characters of a taxon are those not shared by its sister groups or by their most recent common ancester

basicranium base of the skull, formed mainly by the occipital and sphenoid bones

bed in geology, a small, distinct rock unit that is identifiable in the field

Bergmann's rule general rule in zoology that animals living in cold climates tend to be larger than closely related species living in warm climates in order to lower their surface-area-to-volume ratio and thus reduce heat loss

Beringia land bridge across the Bering Strait between what is now Alaska and Siberia; formed during the Cretaceous and persisted into the early Tertiary

beta decay radioactive decay in which the nucleus emits an electron and one of its neutrons turns into a proton

bilophodont refers to a type of molar construction in which there are two parallel enamel ridges running from the protocone and paracone anteriorly and the hypocone and metacone posteriorly on the upper molars and from the protoconid and metaconid anteriorly and the hypoconid and entoconid posteriorly on the lower molars; typical in Old World monkeys

binocular vision vision in which both visual axes can focus on a distant object to produce a stereoscopic (three-dimensional) image

bovid member of the family Bovidae, cloven-hooved ungulates (bison, antelopes, deer, goats, sheep, etc.)

brachial index ratio of the length of the forearm (radius) divided by the length of the upper arm (humerus) × 100

brachiation arboreal locomotion in which the animal progresses below branches by using only the forelimbs

breccia poorly sorted, cemented material generally consisting of angular rock fragments

buccal toward the lateral (cheek) side of a tooth; in a **buccal view** the jaw and/or teeth are seen from the cheek side. For constrast, *see also* lingual

bunodont refers to teeth with low, rounded cusps

calcaneus heel bone

calcarine fissure characteristic groove found on the medial surface of each occipital lobe (visual cortex) of the primate brain

callitrichid member of the New World monkey family Callitrichidae, which consists of the tamarins and marmosets

calvaria top of the skull; the skullcap

canine fossa (pl. **fossae**) depression or slight concavity in the alveolar process of the maxilla just behind the canine jugum

canine jugum (pl. **juga**) vertical ridge in the alveolar process of the maxilla caused by an enlarged canine root

caniniform shaped like a canine

capitulum small, rounded head on the lateral articular surface of the distal humerus that articulates with the radius

carnivoran member of the order Carnivora, true carnivores including cats, dogs, seals, bears, etc. For contrast, *see also* carnivore

carnivorous, carnivory meat eating; **carnivores** are animals whose primary or sole source of food is meat. Not a taxonomic term; for contrast, *see also* carnivoran

carpal flexors muscles that flex the wrist joint (carpus)

carpolestid member of the Carpolestidae, a small family of plesiadapids that existed in North America from the middle Paleocene through early Eocene

catarrhine member of the infraorder Catarrhini, Old World anthropoids (monkeys, apes, and humans)

caviomorph member of the suborder Caviomorpha, which includes South American rodents such as guinea pigs and porcupines

cebid member of the New World monkey family Cebidae, which consists of all the New World monkeys except the callitrichids (tamarins and marmosets)

ceboid member of the superfamily Ceboidea (living New World primates)

cecum (caecum) a blind pouch at the junction of the small and large intestine

central sulcus transverse groove across the top of the cerebral cortex separating the somatic sensory area from the somatic motor area

cercopithecid member of the Cercopithecidae, the family of Old World monkeys from the late Miocene through the present

cercopithecine member of the Cercopithecinae, the subfamily of Old World monkeys characterized by cheek pouches

cercopithecoid member of the superfamily Cercopithecoidea (Old World monkeys)

character morphological feature or trait chosen for study

character state assessment of a morphological feature (character) as either primitive or derived

character weighting assigning more importance to certain characters over others when using them to elucidate evolutionary relationships

chiropteran member of the order Chiroptera (bats)

cingulum (pl. **cingula**) shelf of enamel running around the periphery of a tooth

clade group composed of all the species descended from a single common ancestor

cladism classification by means of shared derived characters; also called phylogenetic systematics

cladogram branching tree diagram used to represent phyletic relationships; also called a cladistic tree

clavicle collarbone

claviculate possessing clavicles

close-packed joint position of a joint when all the articular surfaces are as close together as possible

coefficient of variation statistical measure of variability that is independent of size

colobine member of the Colobinae, the subfamily of Old World monkeys that are leaf eaters

condylarth member of the order Condylarthra (early Paleogene ancestors of the ungulates)

conspecific of the same species

continental drift slow movements of continents (and the crustal plates to which they are attached) over the surface of the earth; *see also* plate tectonics

convergent *see* analogous

coronoid process part of the ascending mandibular ramus of the mandible where the temporalis muscle attaches

coronolateral sulcus in the brain, a groove running longitudinally along the lateral side of the cortex; typically found in many prosimians

crenulation wrinkled surface of the tooth enamel of some primates

creodont member of the order Creodonta (primitive early Tertiary carnivores)

crepuscular active primarily around the hours of dawn and dusk

Cretaceous–Tertiary (K/T) boundary in geology, the boundary between the strata of the Mesozoic era (Age of Reptiles) and those of the Cenozoic era (Age of Mammals), approximately 65 MYA

cricetid member of the Cricetidae, a family of rodents (including true hamsters, gerbils, and mole rats)

cristid obliqua ridge of enamel on the lower molars running obliquely from the hypoconid to the back of the trigonid

crural index ratio of the length of the leg (tibia) divided by the length of the thigh (femur) × 100

cursorial refers to a type of terrestrial locomotion characterized by fast running

cusp pointed or rounded protuberance on the occlusal surface of a tooth

cuspules small accessory cusps on teeth

decay constant constant rate by which radioactive elements change spontaneously to lower energy states

deciduous teeth teeth that are shed and replaced by other teeth during the normal course of a mammal's lifetime

dental alveoli (s. **alveolus**) sockets in the jaws that house the tooth roots

dental formula shorthand notation denoting the number of teeth in each quadrant of the upper and lower jaws; for example, 2.1.3.3./1.0.2.3. denotes two incisors, one canine, three premolars, and three molars on each side of the upper jaw and one incisor, no canines, two premolars, and three molars on each side of the lower jaw

derived character state the evolutionary later state of a character, relative to its ancestral state; also called an apomorphy

dermatoglyphs patterns of ridges on the skin of the fingers, palms, toes, and soles

dermopteran member of the order Dermoptera (the "flying lemur")

diastema (pl. **diastemata**) space or gap between adjacent teeth in the dental row

digital flexors muscles that flex the fingers or toes

digitigrade refers to a type of quadrupedal locomotion in which animals support their body weight on their phalanges; for contrast, *see also* palmigrade

diphyletic taxon in phylogeny, a group in which all the members evolve from one of two separate groups

distal further away from any line of reference; in dentition, refers to the side of the tooth or portion of the jaw toward the back of the mouth. For contrast, *see also* proximal, mesial

dolichocephalic having a long, narrow skull

dorsiflexion bending the dorsal surface of the hand or foot toward the arm or leg

dryomorph vernacular term to describe a group of primitive catarrhines found mainly in the Miocene of Africa and Europe

ectotympanic bone bone in the middle ear that supports the tympanic membrane or eardrum

edentate member of the order Edentata, a group of South American toothless mammals including anteaters, armadillos, and sloths

electron capture radioactive decay in which a proton in the nucleus picks up an orbital electron and turns into a neutron

encephalization quotient (EQ) measure of relative brain size in which the brain weight is compared with that of the average living mammal of equal body weight

endocast naturally occurring (fossilized) or artificially made mold of the external surface of the brain

endothermic describes "warm-blooded" animals that maintain a fairly constant internal temperature through metabolic processes

entepicondylar foramen hole or notch in the distal end of the humerus for passage of the brachial artery and median nerve

entoconid cusp on the lingual rim of the talonid of lower molars

entocuneiform one of the small bones of the mammalian foot that articulates with the first metatarsal, the mesocuneiform, and the navicular

epitrochlear fossa depression on the posterior surface of the medial epicondyle of the humerus for attachment of a strong ligament connecting the humerus and ulna; found in several Oligocene primates

erector spinae deep back muscles

erinaceomorph member of the family Erinaceidae, hedgehoglike insectivorans; erinaceomorphs are one of several groups proposed as candidates for primate origins

ethmoid bone small bone in the skull that contributes to the medial orbital wall in some primates and also forms a small portion of the floor of the braincase under the frontal lobes

ethmoidal sinuses air spaces within the ethmoid bone

euprimates vernacular term for the "modern" primates, that is, all the nonplesiadapiform primates including strepsirhines and haplorhines

eutherian placental mammal

eversion turning outward, for example, of the bottom of the foot

evolutionary taxonomy system of taxonomy that takes into account both the branching pattern of evolution and the amount of evolutionary change occurring along each lineage

extant currently existing; not extinct

extensor any muscle that acts to increase the angle between two bones in a joint

external auditory tube outer-ear passage

external oblique most superficial of three large flat muscles forming the lower body wall (the other two being the internal oblique and the transversus abdominis)

faunal assemblage group of living or fossil animals found in a particular geographic or geological context

fenestra cochleae the "round" window on the medial surface of the middle ear covered by the secondary tympanic membrane; also called the fenestra rotunda

fission track dating method of dating rocks by counting the number of spontaneously produced fission tracks caused by the radioactive decay of uranium-238 in the rocks

flexor any muscle that acts to decrease the angle between two bones in a joint

floral assemblage group of living or fossil plants found in a particular geographic or geological context

folivorous, folivory leaf eating; **folivores** are animals whose primary source of food is foliage

foramen magnum large hole in the base of the skull through which the spinal cord passes, joining the base of the brain

formation in geology, a fundamental rock unit of a stratigraphic section at a given locality

frenulum small fold of skin immobilizing the upper lip; present in all extant strepsirhines but reduced or absent in extant haplorhines

frontal sinus air space found in the frontal bone

frugivorous, frugivory fruit eating; **frugivores** are animals whose primary source of food is fruit

genioglossal pit notch on the inside border of the mandibular symphysis marking the origin of the genioglossus (tongue) muscle

geological time units abstract units of time that exist whether or not any rock units actually record the passage of that particular interval of time; the geological time units of era, period, epoch, and age correspond to the time-stratigraphic units of erathem, system, series, and standard stage

geometric scaling *see* isometry

glabella area on the frontal bone between the brow ridges

gluteus minimus muscle running from the lateral surface of the ilium to the greater trochanter of the femur; important in stabilizing the hip joint in bipedal locomotion

gluteus medius *see* gluteus minimus

Gondwanaland name given to one of the two supercontinents resulting from the breakup of Pangaea around 200 MYA; consisted of South America, Africa, Madagascar, India, Arabia, Malaya, East Indies, New Guinea, Australia, and Antarctica and broke up in the Mesozoic due to continental drift

graben in geology, a fault trough

gracile describes any slender, lightly built body or body part; often used to categorize the less robust australopithecine species *A. africanus*

Grand Coupure refers to the time near the end of the Eocene when many faunal groups, including primates, became extinct in the Northern Hemisphere

granivorous, granivory grain eating; **granivores** are animals that eat primarily grains, although the term is sometimes used to describe animals that eat seeds as well

greater palatine foramina (s. **foramen**) small holes in the posterior part of the hard palate for transmission of blood vessels and nerves to the palate

greater trochanter that part of the proximal femur where the gluteus medius and gluteus minimus attach

greenhouse model theory that the mass extinction events occurring at the Cretaceous–Tertiary boundary resulted from increased temperatures worldwide; one of several models proposed to explain K/T extinction events

group in geology, a rock unit consisting of two or more formations

gummivorous, gummivory gum eating; **gummivores** are animals that eat primarily gums, saps, and other tree exudates

half-life the amount of time in which one-half of the atoms in a radioactive element decays

hallux first digit on the hindlimb foot ("big toe")

haplorhine member of the suborder Haplorhini, which includes the Tarsiiformes (tarsiers and omomyids) and the Anthropoidea (monkeys, apes, and humans)

hard palate bony separation between the oral and nasal cavities

heteromorphic of different size and shape

higher primate *see* anthropoid

hominid member of the family Hominidae (australopithecines and humans)

hominoid member of the superfamily Hominoidea (apes and humans)

homologous describes a similarity of form or structure between two species that they share with their nearest common ancestor

homomorphic of similar size and shape

honing facet surface on the anterior portion of the anterior lower premolar (a sectorial tooth) used to sharpen the posterior edge of the upper canine when the two teeth come into contact; *see also* sectorial tooth

horizontal ramus *see* mandibular body

humerofemoral index ratio of the length of the humerus divided by the length of the femur × 100

hylobatid member of the family Hylobatidae (gibbons and siamangs)

hypocone cusp on the posterior lingual surface of the upper molars

hypoconid cusp on the lateral rim of the talonid ("heel" portion) of lower molars

hyracoid member of the order Hyracoidea (rock hyraxes or conies)

iliac pillar a bony buttress running down the iliac blade in hominids for resisting the powerful muscular forces generated by the hip abductor muscles

iliofemoral ligament ligament that runs from the ilium to the femur and helps prevent the trunk from extending posteriorly at the hip joint

incertae sedis of uncertain taxonomic position; placed after the name of a taxon at any level of a classification hierarchy to indicate that the affinities of that taxon are not precisely determinable

incisiform shaped like an incisor

incisive canal midline canal at the anterior portion of the hard palate for passage of nerves and vessels running between the nasal and oral cavities

inferior transverse torus *see* simian shelf

infraorbital foramen (pl. **foramina**) external opening of the infraorbital canal on the anterior surface of the maxilla for passage of blood vessels and nerves to the lower face

insectivoran member of the order Insectivora, a group of insect-eating mammals including true shrews, moles, hedgehogs, tree shrews, and elephant shrews; for contrast, *see also* insectivore

insectivorous, insectivory insect eating; an **insectivore** is any mammal whose primary source of food is insects. Not a taxonomic term; for contrast, *see also* insectivoran

intermembral index ratio of the length of the forelimb (humerus + radius) divided by the length of the hindlimb (femur + tibia) × 100

internal carotid artery one of the main arteries supplying blood to the brain

internal oblique *see* external oblique

interorbital distance distance between the orbits measured at their medial margins

inversion turning inward, for example, of the sole of the foot

ischial callosities well-developed sitting pads on the ischium of all Old World monkeys and gibbons

ischial tuberosities bony expansions of the ischium that support ischial callosities

isometry change in overall size of an object or organism that maintains the same shape

Jarmin-Bell principle rule relating body size of mammals to the nutritional quality of the foods they eat: small mammals eat high-quality (high-energy), quickly digested foods such as insects, while larger mammals, needing relatively fewer calories per unit of body weight, can subsist on more low-quality, harder-to-digest foods processed in bulk, such as leaves

Kay's threshold approximate body weight (around 500 g) that separates primarily insectivorous from noninsectivorous primates

knuckle walking type of quadrupedal locomotion in which the upper body is supported by the dorsal surfaces of the middle phalanges of the hand; practiced by chimpanzees and gorillas

lacrimal bone in the skull, a small bone forming part of the medial orbital wall

lacrimal foramen (pl. **foramina**) opening of the tear duct connecting the orbit with the nasal cavity

lacrimal fossa (pl. **fossae**) the depression in which the lacrimal foramen sits

land-mammal ages segments of geological time characterized by recognizable suites of mammal fossils

lateral sulcus on the brain, the groove on the lateral side of each cerebral hemisphere that separates the temporal lobe from the frontal and parietal lobes; also called the sylvian sulcus

laterocone small pointed cusp on the lateral side of the incisors of some plesiadapiforms

Laurasia name given to one of the two supercontinents that resulted from the breakup of Pangaea in the Mesozoic around 200 MYA (the other being Gondwanaland); consisted of the future continents of North America, Europe, and Asia prior to its breakup later in the Mesozoic due to continental drift

lemuriform member of the infraorder Lemuriformes (lemurs and lorises); in some classifications discussed in the text, lorises are accorded separate, equal taxonomic rank with lemurs. *See also* lemurophile hypothesis

lemurophile hypothesis hypothesis of primate phylogeny that considers the plesiadapiforms to be the sister group of the omomyids and tarsiers (all three groups comprising the Plesitarsiiformes) and lemurs, lorises, and adapids to form a sister group of the anthropoids (together comprising the Simiolemuriformes)

leptictid member of the Leptictidae, a family of primitive early Tertiary insectivorans; one of several groups proposed as candidates for primate origins

lesser trochanter protuberance on the femur for attachment of the iliopsoas muscle, a flexor of the thigh

lingual toward the tongue; in a **lingual view** the jaw and/or teeth are seen from the side adjacent to the tongue. For contrast, *see also* buccal

lophs transverse crests on molars

lorisid member of the family Lorisidae (lorises and galagos)

lorisiform in several primate taxonomies, refers to a member of the infraorder Lorisiformes, lorises and galagos; *see also* lemuriform

mandibular body horizontal portion of the mandible; also called the horizontal ramus of the mandible

mandibular fossae (s. **fossa**) shallow depressions, one on either side of the base of the skull, into which the mandibular condyles fit

mandibular symphysis the midline joint connecting the two sides of the mandible

mandibular torus (pl. **tori**) shelflike thickening of bone on the inside of the mandibular symphysis; superior and inferior transverse tori are two transverse bony supports on the posterior side of the mandibular symphysis. The inferior transverse torus is often referred to as the simian shelf

manus hand

masseter one of the muscles of mastication; runs from the zygomatic arch to the lateral surface of the ascending ramus of the mandible

maxillary sinus air space within the maxillary bone

medial epicondyle bony projection on the medial side of the distal humerus for attachment of many of the digital and carpal flexors

medial torsion an inward rotation along a longitudinal axis

mediocone small cusp on the front of the incisors of some plesiadapiforms

Mediterranean crisis refers to the "drying up" of the Mediterreanean Sea about 6 MYA

megadont having large teeth relative to body size

member in geology, a rock unit that is a subdivision of a formation

meniscus disk of fibrocartilage found in certain joints (such as the knee and ulnar–carpal joints)

mesial toward the anterior side of a molar or premolar tooth or to the side of an incisor tooth nearest the midline of the jaw; for contrast, *see also* distal

metacone one of three primary cusps on the crown of upper molars; found on the posterior buccal (outer back) corner of the tooth

metaconid one of three main cusps on the trigonid (anterior triangular portion) of lower molars; found on the distal lingual (inner back) corner of the trigonid

metacrista ridge of enamel extending from the metacone in the upper molars of primitive eutherians and early primates

metopic suture midline joint between the two frontal bones in the skull

microchoerine member of the Microchoerinae, one of three omomyid subfamilies from the middle and late Eocene of Europe

microsyopid member of the Microsyopidae, a successful plesiadapiform family that flourished in Europe and North America from the middle Paleocene to the late Eocene

molariform molarlike in form and function

molecular clock means of determining dates of evolutionary divergence using biomolecular similarities and differences in extant species; assumes that molecular evolution proceeds at a fairly constant rate

moment of inertia distribution of the mass of an object that affects its resistance to movement

monophyletic group refers to a group containing all the known descendants of an ancestral species; also, a group of species that share a common ancestor that would be classified as a member of the group

morphocline arrangement of the morphological variations of a homologous character into a continuum of primitive to derived states

morphotypes list of the known character states likely to be diagnostic of the ancestor of a species

multituberculate member of the order Multituberculata, a very successful group of primitive herbivorous mammals that flourished during the Mesozoic

***Nannopithex* fold** *see* postprotocingulum

nasal incisive foramen (pl. **foramina**) opening of the incisive canal into the nasal cavity; in some primates the foramen widens into a fossa at the opening

nasoalveolar clivus portion of the premaxilla extending from the nasal cavity to the incisor root sockets

nasolacrimal duct tear duct connecting the orbit with the nasal cavity

navicular small bone that articulates with the talus in the mammalian foot

negative allometry an allometric relationship in which the slope of the line comparing two variables is less than unity

neocortex cerebral hemispheres

Neogene system in geology, the time-stratigraphic unit composed of the Miocene, Pliocene, and Pleistocene epochs

neurocranium portion of the skull enclosing the brain, as distinct from the facial bones and the basicranium

normal polarity epoch period of geological time in which the magnetic field pointed north, as at present; for contrast, *see also* reversed polarity epoch

notharctine member of the subfamily Notharctinae, one of two Eocene adapid subfamilies, the other being the Adapinae

notoungulate member of the order Notoungulata, one of several extinct orders of archaic South American ungulates

nuchal crest bony shelf on the back of the skull for the attachment of powerful neck muscles

obturator foramen (pl. **foramina**) space in the pelvis enclosed by the pubis, ischium, and ilium

occipital bone bone forming the posterior and much of the basicranium of the mammalian skull

occlusal refers to the surfaces of the teeth that meet during occlusion (chewing); in an **occlusal view** the crowns of the teeth as seen from above

olecranon fossa (pl. **fossae**) depression at the posterior side of the distal humerus for accommodating the olecranon process of the ulna when the elbow is extended

olecranon process projection on the proximal end of the ulna for attachment of the triceps muscle; the elbow

omomyid member of the tarsierlike family Omomyidae, one of the two Eocene euprimate families (the other being the Adapidae), both of which are now extinct

omomyine member of the Omomyinae, one of three subfamilies of the Eocene Omomyidae

omomyophile hypothesis hypothesis of primate phylogeny that considers the omomyids to be the sister group of the tarsiers

ontogenetic relating to the growth and development of an individual organism

oral incisive foramen (pl. **foramina**) opening of the incisive canal into the oral cavity of the skull; in some primates the foramen widens into a fossa at the opening

orbit bony socket for the eye

orbital convergence realignment of the visual axes of the orbits during primate evolution from laterally facing to anteriorly facing lines of sight

orbitosphenoid bone portion of the sphenoid bone that forms part of the posterior wall of the orbit

orthognathous having a relatively vertical, nonprotruding face; for contrast, *see also* prognathous

otic relating to the ear

palatine process horizontal portion of the maxilla forming the hard palate

Paleogene system in geology, the time-stratigraphic unit composed of the Paleocene, Eocene, and Oligocene epochs

paleomagnetic column depiction of the alternating normal and reversed polarity epochs through geological time

palmigrade refers to a type of quadrupedal locomotion characterized by weight bearing on the palms of the hands rather than on the digits or knuckles; for contrast, *see also* digitigrade

palynology study of plant pollens

Pangaea single supercontinent containing all landmasses prior to about 200 MYA

paracone one of three primary cusps on the crowns of primate upper molars; found on the mesial buccal (outer front) corner of the tooth

paraconid cusp on the mesial lingual surface of the lower molars

paracrista enamel ridge running from the paracone in the upper molars of primitive mammals and early primates

paraphyletic refers to a group containing some but not all the known descendants of the common ancestor of the group

parapithecid member of the family Parapithecidae, a group of early Oligocene anthropoids known from the Fayum area of Egypt

parasagittal refers to a plane through the body parallel to the sagittal plane, which divides the body into left and right sides

parietal bone flat bone forming part of the lateral wall of the skull

parietal lobe portion of the cerebral cortex that lies beneath the parietal bone; bounded posteriorly by the occipital lobe, anteriorly by the frontal lobe, and inferiorly by the temporal lobe

paramomyid member of the Paromomyidae, a small plesiadapid family that existed from the middle Paleocene to the late Eocene

patella kneecap

patellar groove depression on the distal femur in which the kneecap (patella) moves

periods in geology, the abstract geological time unit that corresponds to a geological system

perissodactyl member of the order Perissodactyla (odd-toed ungulates, or hooved mammals)

peroneal tubercle bony projection on the hallucal metatarsal bone for insertion of the peroneus longus tendon

peroneus longus muscle on the lateral side of the leg that adducts the big toe (hallux)

pes foot

petrosal bone portion of the temporal bone that forms the auditory bulla in primates

phalanges (s. **phalanx**) finger and toe bones

phenetic taxonomy classification system based solely on the similarity of phenotypes without regard for primitive or derived characters

phenotype appearance of an organism; its observed set of characteristics

philtrum vertical cleft in the rhinarium of extant strepsirhines; not present in haplorhines

phyletic gradualism model of evolution in which change takes place slowly and in small steps; for contrast *see also* punctuated equilibrium

phylogenetic systematics *see* cladism

phylogenetic tree *see* phylogram

phylogeny evolutionary or geneological relationships among a group of organisms

phylogram branching diagram for a set of species that shows their ancestral relationships; one direction, usually the vertical axis, represents time. Also called a phylogenetic tree

picrodontid member of the Picrodontidae, a family of small plesiadapids from the middle to late Paleocene with morphological and dietary similarities to bats

plagiaulacoid dentition dentition in which the molars and premolars are arranged like a cutting board and cleaver; characteristic of carpolestids and saxonellids

plantarflexion flexion at the ankle joint so that the toes point downward

plate tectonics in geology, the movement of segments of the lithosphere (including the rigid upper mantle, plus oceanic and continental crust), which floats on the underlying, more gelatinous asthenosphere

platyrrhine member of the infraorder Platyrrhini (New World anthropoids)

plesiadapid member of the Plesiadapidae, one of the most diverse and abundant of the plesiadapiform families, occurring from the middle Paleocene to early Eocene in Europe and North America

plesiadapiform member of the suborder Plesiadapiformes, an archaic group of mostly Paleocene primates

plesiomorphic describes an ancestral character

plesitarsiiform member of the taxon Plesitarsiiformes; *see also* lemurophile hypothesis

pliomorphs vernacular term to describe a group of primitive catarrhines mostly from the Miocene of Europe and Asia

pneumatized filled with air spaces, as in pneumatized portions of the skull

polarity epoch a sustained period of time characterized by either normal or reversed magnetic polarity; *see also* paleomagnetic column

polarity event short-lived polarity reversal occurring within a polarity epoch

pollex thumb

polymorphic showing a variety of forms

pongid member of the family Pongidae, the great apes (chimpanzees, gorillas, and orangutans)

positive allometry allometric relationship in which the slope of the line comparing two variables is greater than unity

postcanine dentition all the teeth distal to (in back of) the canines: premolars and molars

postcranial skeleton all of the skeleton, except the skull, that is, the limbs and vertebral column

posterocone small, pointed cusp at the back of the incisors of some plesiadapiforms

postorbital bar bony ring surrounding the lateral side of the orbit in lower primates and many other mammals

postorbital closure "walling off" of the orbit posteriorly by means of a bony partition so that the orbit forms a cup-shaped structure

postorbital constriction narrowing of the skull behind the orbits

postprotocingulum structure occurring on the upper molars of some early primates in which a ridge of enamel runs from the distal lingual corner of the protocone to the cingulum; also called the Nannopithex fold (for the Eocene omomyid *Nannopithex*)

potassium–argon (K/A) dating radiometric technique for dating rocks by measuring the amount of decay of potassium into argon

prehallux small sesamoid (accessory) bone sometimes occurring in the tarsometatarsal joint of the hallux

prehension ability to grasp objects with hands or feet

premaxilla bony part of the upper jaw that houses the incisors

premolariform shaped like a premolar

procumbent refers to incisors that project more in the horizontal than in the vertical plane

prognathous having a protruding jaw in which the bone housing the incisors projects anteriorly; for contrasts, *see* orthognathous

promontory artery one of the two main branches of the internal carotid artery in the middle ear of primates, the other being the stapedial artery

pronation medial rotation of the forearm to turn the palm backward or downward; for contrast, *see also* supination

propliopithecid member of the family Propliopithecidae (sometimes called the Pliopithecidae), a group of Oligocene Old World anthropoids from the Fayum area of Egypt

prosimian member of the suborder Prosimii (lower, or primitive primates), one of the two primate suborders (the other being the Anthropoidea, or higher primates); includes lemurs, lorises, and tarsiers, as well as adapids and omomyids

protocone one of three primary cusps on the crowns of primate upper molars; found on the lingual side of the tooth

protoconid one of three main cusps on the trigonid (anterior triangular portion) of lower molars; found on the buccal side of the tooth

proximal closer to any particular point of reference; for contrast, *see also* distal

pseudohypocone type of cusp that develops through cleavage of the protocone, unlike the true hypocone, which arises from the cingulum; found in Eocene notharctines

pterion temple region of the skull

pterygoid plates bony plates on the inferior surface of the sphenoid bone for attachment of two muscles of mastication, the lateral and medial pterygoid muscles

punctuated equilibrium model of evolution in which the changes leading to new species occur quickly, by abrupt genetic shifts rather than slowly and gradually; for contrast, *see also* phyletic gradualism

pyroclastic rocks volcanic rocks that have been blown into the atmosphere

quadritubercular having four cusps; refers to a four-sided (squarish) molar tooth

radiometric dating techniques that make use of the fact that many kinds of atoms are unstable and change spontaneously into a lower energy state by radioactive emission; each radioactive element has one particular mode of decay and its own unique rate of decay, which is a constant. Measuring ratios of undecayed atoms to the products of decay allows one to extrapolate back to the ages of many fossil-bearing rocks

ramamorph vernacular term to describe a group of Miocene hominoids from Africa and Eurasia characterized by thick-enameled dentitions

reversed polarity epoch periods of geological time in which the magnetic field points to the south; for contrast, *see also* normal polarity epoch

rhinarium hairless patch of skin between the nose and upper lip; present in extant strepsirhines

rift system elongated trough in the earth's crust bounded on either side by faults

robust describes any large, or heavily built body of body part; often used to categorize the robust australopithecines, *A. robustus* and *A. boisei*

rock units body of distinct rock types that are distinguishable in the field (group, formation, member, bed, etc.)

sacroiliac joint joint between the sacrum and the ilium

sagittal crest bony crest running along the midline of the skull for attachment of enlarged temporalis muscles

savanna plain characterized by coarse grasses and scattered tree growth and in which rainfall is seasonal

saxonellid member of the Saxonellidae, a plesiadapid family (with only one genus) from the late Paleocene of Europe and North America

scapula (pl. **scapulae**) shoulder blade

sciatic notch broad notch on the posterior surface of the pelvis separating the ilium and ischium; prominent in humans and later australopithecines but not in earlier primates

sclerophyllous vegetation vegetation adapted to dry conditions

sectorial tooth tooth in the lower jaw (usually the anterior premolar) with a honing facet for sharpening the upper canine during occlusion

semitendinosus one of the hamstring muscles, which extend the thigh and flex the leg

sexual dimorphism phenomenon in which homologous nonreproductive structures are of greatly different size and/or shape in males and females of the same species

sigmoid notch concave surface at the proximal end of the ulna that articulates with the trochlea of the humerus

simian shelf distinct mandibular torus that projects posteriorly from the inferior surface of the mandibular symphysis; so-called because it is found in some fossil and many extant apes. Also called the inferior transverse torus

simiolemuriform member of the taxon Simiolemuriformes; *see also* lemurophile hypothesis

sister groups two groups that result from a single split in a cladogram; that is, they (and only they) share the same parent taxon. Also called daughter taxa

Siwaliks extensive Miocene-through-Present sediments deposited along the base of the rising Himalayas

spatulate spatula-shaped; refers to incisors that are broad

speciation emergence of new species

sphenoidal sinus air cavities within the sphenoid bone

sphenoid bone irregularly shaped bone forming part of the base of the skull

squamosal suture suture between the parietal and temporal bones

stapedial artery one of the two branches of the internal carotid artery in the middle ear of primates (the other being the promontory artery)

strepsirhine member of the suborder Strepsirhini, which includes the Adapidae and the Lemuriformes

stylar shelf shelflike extension of enamel on some cheek teeth

subnasal alveolar process part of the premaxilla housing the roots of the upper incisors

subtalar joint joint between the talus and the underlying calcaneus

supination rotation of the forearm so that the palm faces forward; for contrast, *see also* pronation

supraglabella forehead region

supraorbital tori (s. **torus**) brow ridges

supratoral sulcus groove in the skull between the brow ridges and the frontal bone

suspensory posture behavior characterized by suspension of the body below branches

sylvian sulcus (fissure) *see* lateral sulcus

symphysis (pl. **symphyses**) one of several fibrocartilaginous joints found in the midline of the body, such as the mandibular symphysis and the pubic symphysis

symplesiomorphic describes a shared, primitive character (symplesiomorph)

synapomorphic describes a shared, derived character (synapomorph)

syndesmosis fibrous connection between adjacent but unfused bones; for contrast, *see also* synostosis

synostosis union of adjacent bones formed by an osseous material

talocrural joint ankle joint

talonid posterior heellike portion of lower molars

talus ankle bone; also called the astragalus

taphonomy study of the processes by which animal bones become fossilized

tarsiid member of the family Tarsiidae, which consists of only the genus *Tarsius*

tarsiiform member of the infraorder Tarsiiformes, which includes the tarsiers and in some classifications the Omomyidae as well

tarsiphile hypothesis hypothesis in primate phylogeny that considers tarsiers to be the sister group of the platyrrhines and catarrhines

taxon (pl. **taxa**) a group of organisms that are classified together

taxonomic rank position of a category in the Linnaean taxonomic hierarchy: phylum, class, order, family, etc.

temporal bone bone of the skull that forms part of the lateral wall of the skull and covers the various structures of the ear

temporal fossa (pl. **fossae**) space enclosed by the side of the skull and the zygomatic arch which is occupied by the temporalis muscle

temporalis one of the major muscles of mastication arising from the side of the skull and inserting onto the coronoid process of the mandible

temporal lobe *see* parietal lobe

temporonuchal crest compound bony crest on the skull formed by convergence of the temporal lines (or sagittal crest) and the nuchal crest

terrestrial quadruped ground-living animal that moves about primarily on all four limbs

Tethys Sea large prehistoric seaway that separated Eurasia from Africa and Arabia during the early Tertiary; the remnant of it today is the Mediterranean

thermoluminescence dating technique that measures the number of alpha particles produced by radiation trapped in crystal lattices between the present and the last heating of the quartz crystal

tibial malleolus medial portion of the distal tibia forming the medial part of the ankle joint

time-stratigraphic units body of rock strata unified by having been formed during a specific interval of geological time; represents all the rocks (and only those rocks) formed during a certain time span of earth's history; the units are erathem, system, series, and stage

tragulid member of the family Tragulidae (mouse deer)

transverse torus (pl. **tori**), **superior or inferior** *see* mandibular torus

triceps muscle of the arm that extends the elbow joint

trigone portion of the upper molars formed from the three main cusps: protocone, metacone, and paracone

trigonid anterior triangular portion of the lower molars; contains the paraconid (when present), protoconid, and metaconid

tritubercular having three cusps; refers to a triangular-shaped tooth

trochlea any smooth, saddle-shaped bony surface that resembles a pulley as it articulates with other bones; specifically, in the elbow joint the surface on the humerus that articulates with the ulna; in the knee joint the surface on the femur that articulates with the patella; and in the ankle joint the surface on the talus that articulates with the tibia. The raised margins of the trochlea are referred to as **trochlear ridges**

tuff rock composed of volcanic ash

tupaiid member of the family Tupaiidae (tree shrews), one of several groups proposed as candidates for primate origins

tympanic membrane eardrum

type locality site from which the type specimen for a particular species or rock unit was taken

type species species for which a genus or higher-level taxon was first named and described

type specimen individual specimen—usually the first to be found for a particular species—that serves as the basis for identifying all other individuals of the same species. Also called a holotype

ungulate hooved mammal

uranium–lead dating radiometric dating technique that measures the decay of uranium into lead

vertical ramus (pl. **rami**) *see* ascending mandibular ramus

victoriapithecine member of the Victoriapithecinae, the only subfamily of the Victoriapithecidae, a group of extinct East African colobines that may be ancestral to cercopithecines and colobines

zygomatic arch bony arch on the lateral part of the skull formed by projections of the zygomatic bone and the temporal bone for attachment of the masseter muscle

INDEX

Page numbers in *italics* refer to figures. Tables are indicated by **bold** type.

acetabulum, 334
Acheulian industrial complex, 354–56, *355*, 357
Adapidae, 103–17
  Adapinae subfamily of, 115–17, *115*
  Aegyptopithecus compared to, 160
  body weight range in, 125–28, *126*
  canines of, 103, 105–6, 130
  characters common to Anthropoidea and, **109,** 130
  comparison of Omomyidae and, 101–3
  in comparison with later primates, 106–9
  dentition of, 106, 116–17, *116*
  ectotympanic ring of, 103, 105, 108
  from Europe, 103–5, *126,* 130
  genera of, 99
  general morphology of, 105–6
  hindlimbs of, *114,* 128
  incisors of, 103, 105–6, 130
  jaws of, 103
  mandibular symphysis of, 103, 105, 108, 130
  molars of, 103, 106
  Notharctinae subfamily of, *see* Notharctinae
  origins and distribution of, 103–5, 130
  premolars of, 103, 105, 108, 130
  skulls of, *102,* 105–6, 128
  snout of, 103
  see also euprimates
Adapinae, 105, 115–17, *115*
  dental distinctions between Notharctinae and, 116–17, *116*
  genera of, 115–16
  reasons for interest in, 115
Adapis, 106–7
Aegyptopithecus, 108, 160–64, *161, 162,* 166–68, 183, 219, 245–46
Africa:
  dinosaurs of, 141
  dryomorphs from, 204
  early Miocene fossil record of, 194–96, *195*
  during Eocene epoch, 96
  middle and late Miocene fossil record of, 196–97
  migration of New World primates from, 181–82
  Miocene fossil record of, 194–97
  Miocene geological events in, 191
  Miocene mammalian migration routes between Eurasia and, 191, *193,* 197–98
  Oligocene fossil record of, 142–43
  Oligocene geographical separation between South America and, 139
  Oligocene migration route between South America and, 140
  Paleocene fossil record of, 56
  Plio-Pleistocene primates from, 273–75, **274**
  ramamorphs from, 240–42
  Stone Age use of tools in, *352*
  Tertiary deposits in, 18
African Genesis (Ardrey), 324*n*
Afropithecus, 247, 267
Afropithecus turkanensis, 247, *248*
Afrotarsius, 165–67
Agenian age, 197
age ranges of Miocene hominoids, *205*

Albany, State University of New York (SUNY) at, 362
alcelaphines, association of antilopines and early hominids in South Africa with, **289**
alisphenoid bone, 161
allometric equation, 29–30
allometric relationships, 30, 33
allometry, 28–32
basic principles of, 28–29
correlation between area of lower first molar and body weight in, 30
correlation between metabolic rate and body weight in, 29–32
definition of, 28
allopatric speciation, 15–16
alluvial fans, 303
alluvial plains, 303
Alouatta, 128
alpha decay, 21
Alvarez, Luis, 51
Alvarez, Walter, 51
alveolar processes, 217
alveoli, 130, 158
American Museum of Natural History, 15, 64, 145, 174, 182, 233
Amphipithecus, 104, 174–77, *176*
analogous, 10
Anaptomorphinae, 118, 120, 125–26
Anchomomys, 128
Andrews, Peter, 189, 206, 230, 231–33, 255, 264
angiosperm radiation theory, 41–43
angular process, 179
Antarctica, 96
Antarctic ice sheet, formation of, 96
antemolar dentition, 68, 82
antemolars, 85
anterior iliac spines, 334
anterior inferior iliac spine, 113–14
anterior pillars, 308
Anthropoidea, 5, 45, 182–83
Amphipithecus genus of, 104, 174–77, *176*
differences in cortical folding of Prosimii and, *133*
different views on origins of, *135*
as diphyletic taxon, 181
distinctions between Prosimii and, *133, 148,* 149
ear region of, 134
from Fayum Depression, 142
features common to Adapidae and, **109,** 130
features common to Tarsiiformes and, **109,** 124, **125**
features in ancestors of, **180**
morphology of, **104**
origins of, 172–77
phyletic link between Adapidae and, 130
Pondaungia genus of, 173–77
relationship between Parapithecidae and, 177, 179, 181
skulls of, 132–33
anthropometric, 4
antilopines, association of alcelaphines and early hominids in South Africa with, **289**
Apatemyidae, ties between primates and, 59
apes:
anthropometric criteria distinguishing humans from, 4
Barbary, 3
European disappearance of, 198
evolutionary link between humans and, 4–5
apes, modern:
phylogenetic relationships among Miocene hominoids and, 259–64, *262*
ramamorph affinities with, 264–66
sinus system of, *218*
Apidium, 154–57, *157,* 166, 168
apomorphic state, 10
Arambourg, Camille, 294
arboreal, 6
arboreal quadrupeds, postcranial adaptations in, 76–78
arboreal theory, 39–41
Arctic Circle, 97
Arctic Ocean, 55
Arctic Ocean spillover model, 51
Arctostylops, 141
Ardrey, Robert, 324*n*
Argentina, New World monkeys from, 253–54
Aristotle, 3
articular eminence, 360
Asia:
Adapidae from, 103–5
anthropoid origins in, 177
dryomorphs compared with ramamorphs from, 239–40
fossils of Paleocene mammals from, 56
during Miocene epoch, 186
Miocene geological events in, 192–93
Miocene mammalian migrations to, 193–94
Omomyidae from, 117–18
origins of *Afrotarsius* in, 165–66
Asia Minor, ramamorphs from, 230–33
Astaracian age, 197–98, 203–4
asteroid-impact model, 51–52
astragalus, 163

auditory bulla, 61–63
 im mammals, *61*
 of Plesiadapidae, 74
 in Plesiadapiformes, 63–66
auditory region, *62*
auditory tube:
 in Omomyidae and Anthropoidea, 133
 in Omomyidae and *Tarsius*, 124
Australia, 96–97
Australopithecus, 305–25
 behavioral and cultural trends in, 346–57
 bipedalism in, 337–43, 347–48
 body size and EQs of, 324–25
 body types of, 274–75
 craniodental skeleton of, 306–24, *307*, 326
 from East Africa, 289–92, 294–98, 301–2, 304
 fossil record of, 274
 gracile, 305, 308–15, 322–24, 362–63
 gracile vs. robust, 322–24
 Homo compared with, 271–72
 hypotheses of evolutionary relationships among, *358*
 from Kromdraai, 283
 from Makapansgat, 284
 maturation patterns in, 349–51
 opportunistic tools used by, 353–54
 pelvis in, *343*, 345–46
 phylogenetic relationships within, 357–63, *361*
 Plio-Pleistocene appearance of, 271
 robust, 305, 315–24, 362–63
 species of, 274–75, 305
 from Sterkfontein, 281
 from Swartkrans, 286
 from Taung, 277–80
Australopithecus aethiopicus, 297–98
Australopithecus afarensis, 305, *312*, 350, 356
 bipedalism in, 337–40, 348
 body size of, 324*n*
 brain of, 325
 craniodental skeleton of, 307–11, *311*, 325
 from East Africa, 289–92, 296–97, 304
 partial skeleton of, 337, *339*
 pelvis in, 338–42, *341*
 phylogenetic relationships of, 357–62
Australopithecus africanus, 305, 350, 356
 bipedalism in, 342
 body size of, 324–25
 brain of, 315, 326
 craniodental skeleton of, 307–8, 311–15, *313*, *316*, 324
 from East Africa, 290, 295–97, 302
 masticatory system of, 360
 phylogenetic relationships of, 357–60, 362–63
 from South Africa, 280–82, 284, 286, 288
Australopithecus boisei, 305
 bipedalism in, 343
 brain of, 326
 craniodental skeleton of, 307–8, 320–23, *320*, *321*, *322*
 from East Africa, 289, 296–97, 301–2, 304
 opportunistic tools used by, 353–54, 357
 phylogenetic relationships of, 357–63
Australopithecus robustus, 305, *318*
 bipedalism in, 342–43
 body size of, 324–25
 brain of, 317, 326
 craniodental skeleton of, 307–8, 315–20, *319*
 from East Africa, 302
 life span of, 349–51
 phylogenetic relationships of, 357–60, 362–63
 from South Africa, 282, 286, 288
 tool use in, 354–57
autapomorphic characters, 10
 cladists vs. evolutionary taxonomists on, 13–14
Awash valley, Plio-Pleistocene primates from, 290, 292, *293*, 304
aye-ayes, 72
Azibius, 104

baboons, 3, 76, 281
Baldwin, David, 56
Barbary apes, 3
Barlow, Mr., 282
Basal Member, 292, 294
basicranium, 308
bats, 4–5, 10
 in Eocene Europe, 98
 similarities between Paleocene primates and, 86
Battell, Andrew, 4
Becquerel, Henri, 20
beds, 367
behavior:
 of *Australopithecus* and archaic humans, 346–57
 of euprimates, 125–28, *126*, *127*
 of Fayum primates, 166–68
 of Miocene hominoids, 255–56
 predatory, 88–89
Behrensmeyer, A. Kay, 298, 301
Bergmann's rule, 32
Beringia, 55

Berkeley, University of California at, 51, 164, 302
Bernor, Raymond, 197
beta decay, 21
Beyrich, August von, 371–72
bicuspids, 380
bilophodont, 36, 381
binocular vision, 6
biogeography:
  of Eocene epoch, 52–55, 93–97
  of Miocene epoch, 185–94
  of Oligocene epoch, 137–41
  of Paleocene epoch, 49–55
  of Plio-Pleistocene, 272–73
biology:
  allometric analysis and, 28–32
  effects of diet on dentition in, 34–36, *35*, *37*, *38*
  of extinct primates, 27–38
  relationship between size and diet in, 32–34
biomolecular data, phylogeny of Hominoidea and, 258–64
biorbital breadth, **239**
bipedalism:
  in *Australopithecus*, 337–43, 347–48
  behavioral theories on evolution of, 347–51
  biomechanical principles of, 332–37, *335*
  center of gravity in, 335–37
  efficiency of, 347–48
  in *Homo*, 344
  in modern humans, 349
  muscles in, 332–33
  quadrupedalism vs., 347
  seed-eating hypothesis of, 348
  sexual-and-reproductive-strategy model of, 348–49
  and trends in hominid pelvic morphology, 344–46, *345*
birds, 10
  cladists vs. evolutionary taxonomists on, 11–12
  from Fayum Depression, 142–43
body sizes:
  australopithecine phylogenetic relationships and, 363
  of *Australopithecus*, 324–25
  of euprimates, 125–28, *126*, *127*
  of Fayum primates, 166–68
  feeding behavior in plesiadapids and, *89*
  relationship between diet and, 32–34, *167*
body weights:
  of adapids, 125–28, *126*
  of African Miocene dryomorphs, 226–27
  of *Australopithecus robustus*, 320
  correlation between area of lower first molar and, 30
  correlation between mean endocranial capacity, EQs and, **325**
  correlation between metabolic rate and, 29–32
  distribution of, 337, *337*
  frequency distribution of primates according to, *35*
  as function of tooth size, *30*
  of *Proconsul*, 207–9
Boise, Charles, 302
Boltwood, B., 21
Bondt, Jakob de, 4
Boulder Conglomerate, 198
Bowen, Bruce, 298
brachial index, 76
  of African Miocene dryomorphs, 227
  of Notharctinae, 113–15, *114*
brachiation, 76
Brain, C. K., 275–77, 283, 26, 317
brains, 6
  of Adapidae, 107
  of *Aegyptopithecus*, 160–61
  in *Apidium*, 155
  australopithecine phylogenetic relationships and, 363
  of *Australopithecus afarensis*, 325
  of *Australopithecus africanus*, 315, 326
  of *Australopithecus robustus*, 317, 326
  of hominoids, *314*
  of *Homo*, 326
  of *Homo sapiens*, 326, 328–30
  of Omomyidae, 121
  in Plesiadapidae, 74
  in Strepsirhini, 133
Branisella, 169–70
  taxonomic relationships of, 182–83
breccia, 275, 277
Bridgerian land-mammal age, 103
Bristol, University of, 89
British Empire Exhibition, 280
British-Kenya Miocene Expeditions, 206
Broca's area, 326
Bromage, Timothy, 350–51
Broom, Robert, 280–82, 286, 315, 350
Brown, Barnum, 82–84, 174
buccal, 378
bunodont, 174
Burma, Pondaung Formation of, 173

calcaneus, 78
callitrichids, 126

calvaria, 318, 378
Candir Formation, 232
canine fossae, 240
canine juga, 309
canines, 378, 380
of Adapidae, 103, 105–6, 130
of *Aegyptopithecus,* 162, 166
of African dryomorphs vs. modern African hominoids, 217–19
of *Amphipithecus,* 175, *176*
of Anthropoidea, 130
of *Australopithecus afarensis,* 309–11
of *Australopithecus africanus,* 315
of *Australopithecus robustus,* 319
of *Dendropithecus,* 215
of Fayum primates, 149
of *Micropithecus,* 216
of New World monkeys, 254–55
of Old World monkeys, 249
of *Oligopithecus,* 164
of Omomyidae, 103, 119–20, 124
of Paleocene primates, 88–89
of Parapithecidae, 179
of *Parapithecus,* 157
of Paromomyidae, 85
of Plesiadapidae, 72–73
of *Pliopithecus,* 243–45
premolariform, 103, 120
of *Proconsul,* 209–10
of Propliopithecidae, 158–59
of ramamorphs, 231, 234–35, 240–41
of ramamorphs vs. dryomorphs, 239
of *Rangwapithecus,* 212
sectorial, 103
sexually dimorphic, 103, 128, 166
Cannonball Sea, 55
Cantius, 105, 109–10
dental trends in, *110*
capitulum, 223
carbon-14 dating, 24
Caribbean Sea, 54, 182
carnivorans, 40, 44*n*
carnivores, 44*n*
carotid circulation, 123
carpal flexors, 156
Carpodaptes, 79, *80*
Carpolestidae, 57, 78–81, 98–99
craniodental remains of, *80*
dentition of, 79–81
carpometacarpal joint, 344
Cartmill, Matt, 39–41, 89, 134–35
Catarrhini, 45, 155
dryomorphs and, 245
features in ancestors of, **180**
relationship between Parapithecidae and, 177–79, 181
relationship between *Pliopithecus* and, 243–46, *244*
relationship between *Proconsul* and, 207
relationship between Propliopithecidae and, 180–81, **180**
cats, 41
cave breccia, 277
Cebidae, 154
Ceboidea, *see* New World monkeys
Cebupithecia, 254–55
Cenozoic era, 20, 49–50
climatic cooling through, *272*
center of gravity, 335–37, *336, 337*
central sulcus, 149
cephalic circulation, *62*
Cercopithecinae, 161, 249
Cercopithecoidea, *see* Old World monkeys
cercopithecoid vicars, 178
characters, 8
autapomorphic, 10, 13–14
cladistic analysis of, 8–15
common to tarsiiformes and anthropoids, **109**
convergent, 10
evolutionary taxonomic analysis of, 8, 11–15
of Fayum primates and other primates, **150**
morphological, 11
polymorphic, 11
shared derived, 9
symplesiomorphic, 10
synapomorphic, 10–11, 14
character states, 8
character weighting, 8
cheek teeth, 232
australopithecine phylogenetic relationships and, 362–63
of *Australopithecus,* 307–8
of *Micropithecus,* 216
of New World monkeys, 254
of Old World monkeys, 249–51
of Parapithecidae, 152
of Plesiadapidae, 72–73
of ramamorphs, 236
of *Tremacebus,* 170
Chemeron Formation, 290
Chicago, University of, 294
chimpanzees, 5, 45
center of gravity in, 336–37, *336, 337*
cladists vs. evolutionary taxonomists on, 12–14
comparison of intraspecific dental variability in Siwalik ramamorphs and, *235*
early descriptions of, 3–4

chimpanzees (*continued*)
hindlimbs of, *226*
hip bone of, *335*
pentadactyl limbs of, 10
phylogenetic relationships among Miocene hominoids and, 259–63, *262*
skeleton of, 373, *375*
upper dentition of, *321*
China:
during Miocene epoch, 16
Miocene mammalian migrations to, 194
pliomorphs from, 246
ramamorphs from, 236–39, *237*, **238, 239**
Chinji Formation, 198
Chiromyoides, 73, 90
chitin, 88
chronostratigraphic units, *369,* 370
chrons, 285–86, 372
cingula, 116, 380–81
Ciochon, Russell, 254, 263
clades, 9, 18
cladism:
character analysis in, 8–15
comparison of evolutionary taxonomy and, 11–15, *13*
definition of, 8
cladograms, 9–10, *9*
phylograms vs., *14,* 15
Clarke, R., 282
Clarkforkian-Wasatchian mammal-age boundary, 57
classification, 393–403
approaches to, 8–15
cladistic, 8–15, *13*
of early hominids, 357–65
of euprimates, 99–101, **100,** 125–35
evolutionary taxnomic, 8, 11–15, *13*
of Fayum primates, 177–81, **178**
of food types, 34
impact of biomolecular data and recent fossil evidence on, 258–64
Linnaean hierarchy in, 44, **44**
of Microsyopidae, 82–84
of Miocene hominoids, **202,** 256–68
of New World monkeys, **250,** 267–68
of New World Oligocene Anthropoidea, 182–83
of Old World monkeys, **250,** 267–68, *268*
of Oligocene primates, **144,** 172–83
of Paleocene primates, 90–91
phenetic taxonomic, 8
of Plesiadapiformes, **58**
plesitarsiiform-simiolemuriform, 128–31
of ramamorphs, 264–67
strepsirhine-haplorhine, 131–35
traditional, **42**
clavicles, 4, 382–83
cleavers, 356
Clemens, W., 82
climates, *see* paleoclimates
close packed, 225
coefficient of variation, 234
Colbert, Edwin, 174–75
Colobinae, 161, 249
Colombia, New World monkeys from, 253–55
colugos, 5
conspecific, 231
continental drift, 27, *54*
convergent characters, 10
Copelemur, 109–10
Coppens, Yves, 292, 294–96
coronoid process, 179
coronolateral sulcus, 133
Corruccini, Robert, 223–25, 254, 290
cranial morphology, 373–78
craniodental skeletons:
of African dryomorphs vs. modern African hominoids, 217–20, *218*
of ancestors of Catarrhini and Anthopoidea, **180**
of *Australopithecus,* 306–24, *307,* 326
of *Australopithecus afarensis,* 307–11, *311,* 325
of *Australopithecus africanus,* 307–8, 311–15, *313, 316,* 324
of *Australopithecus boisei,* 307–8, 320–23, *320, 321, 322*
of *Australopithecus robustus,* 307–8, 315–20, *319*
of Carpolestidae, *80*
of gracile vs. robust *Australopithecus,* 323–24
of *Homo,* 326–31, *327, 329, 330*
of Paromomyidae, *85*
of ramamorphs from Lufeng, **238**
craniofacial skeletons:
of humans, chimpanzess, and gorillas, 263
of *Sivapithecus* and orangutans, 264–66, *265*
crenulations, 72–73
crepuscular, 106
Cretaceous period:
fossil record of, 57
geography during, 53–55
sea-level changes during, *55*
Cretaceous-Tertiary (K/T) boundary, extinctions at, 49–52
Arctic Ocean spillover model explanation of, 51

asteroid-impact model explanation of, 51–52
greenhouse model explanation of, 50–51
cristid obliqua, 164
crocodiles, cladists vs. evolutionary taxonomists on, 11–12
Crook, John, 89
crural index, 76
of African Miocene dryomorphs, 227
of Notharctinae, 113
culture:
in *Australopithecus* and archaic humans, 346–57
beginnings of, 351–57
definitions of, 351
Curie, Marie, 20
cursorial animals, 39
cuspules, 380
Cynodontomys, 64

Dagosto, M., 78
Dart, Raymond, 277–81, 284, 312–15
Dartmouth College, 52
Darwin, Charles, 5, 19, 201
Davis, Peter, 220
Davis, University of California at, 347
Dean, M. C., 350–51
de Bonis, L., 230–31
decay constant, 24
Decker, R., 78
Delson, E., 164, 174, 178–79
Dendropithecus, 214–16, 219–20
oral incisive fossae of, 239–40
Dendropithecus macinnesi, 214–15, *215,* 217
Denen Dora Member, 292
dental alveoli, 130
dental combs:
of lemurs, 108
of Prosimii, *12*
taxonomic significance of, 11
dental formulas, 36
of *Amphipithecus,* 174
of *Apidium,* 154
of *Dolichocebus,* 172
of Fayum primates, 149
of Microsyopidae, 82
of New World monkeys, 253–54
of North American plesiadapids, 69–71, *71*
of *Oligopithecus,* 164, 179
of Omomyidae, 119, 124
of Parapithecidae, 177
of *Parapithecus,* 158
of Propliopithecidae, 159
of *Tremacebus,* 170
dental variability, studies on, 18
dentition, 378–82
of Adapidae, 106, 116–17, *116*
of *Aegyptopithecus,* 162
of *Afrotarsius,* 165, 167
antemolar, 68
of *Apidium,* 154
australopithecine phylogenetic relationships and, 358
of *Australopithecus afarensis,* 309–11, *312*
of *Australopithecus africanus,* 315
of *Australopithecus boisei,* 321, *321*
of *Australopithecus robustus,* 318–19, *319*
of *Branisella,* 169–70
of Carpolestidae, 79–81
of *Dendropithecus,* 215–16, *215*
effect of diet on, 34–36, *35, 37, 38*
euprimate dietary adaptation and, 125–28
of Fayum primates, 149
of *Gigantopithecus,* 229
intraspecific variability in, 234, *235*
of *Limnopithecus,* 213–14, *214*
of Microsyopidae, 82, *83*
of Notharctinae, 110, *110,* 116–17, *116*
of Old World monkeys, 249
of *Oligopithecus,* 164
of Omomyidae, *118,* 119–20, *120,* 124
of Omomyidae and Plesiadapidae, 130
Paleocene primate feeding adaptations and, 86–90
of Parapithecidae, 152, 154, *154,* 177, 179
of Paromomyidae, 84–85
of Picrodontidae, 86–87, *87*
plagiaulacoid, 79–81
of Plesiadapidae, 69–73, *71, 73,* 79
of Plesiadapiformes, 68, 90–91
of *Pliopithecus,* 243–45
of *Pondaungia,* 173–74
of *Proconsul,* 210, *210, 211*
of Propliopithecidae, 158–59, *159,* 162, 166
of *Rangwapithecus,* 212–13, *212*
Dermoptera, 98
Descent of Man, The (Darwin), 5
Developed Oldowan industry, 353, *355,* 356–57
Dhok Pathan Formation, 198, 252
diachronous, 367
Diacodon, 63
diastemata, 73
diet:
of Carpolestidae, 79
dentition and, 34–36, *35, 37, 38*
of euprimates, 125–28, *126, 127*
of Fayum primates, 166–68

diet (*continued*)
frequency distribution of weights of primates according to, *35*
of Miocene hominoids, 255–56
of Paleocene primates, 86–90
plesiadapid body size and, *89*
of *Pliopithecus*, 243
relationship between body size and, 32–34, *167*
digital flexors, 156
digitigrade, 222
dinosaurs:
of Africa and Eurasia, 141
disappearance of, 50–51
dominance of, 19
in South and North America, 141
Dinosaurs, Age of, 366
Dionysopithecus shuangouensis, 217
diphyletic taxa, 181
distal, 378
Dobzhansky, Theodosius, 349
Dolichocebus, *171*, 172, 183
dolichocephalic, 313
dorsiflexion, 225
Drake, Charles, 52
dryomorphs, 204–29, *226*
ancestral catarrhine features in, 245
comparison of Asian ramamorphs and Asian, 239–40
craniodental comparison of modern African hominoids and, 217–20, *218*
Dendropithecus genus of, 214–16, 219–20, 239–40
Dryopithecus genus of, *see Dryopithecus*
from East Africa, 201, **203,** 204–6, 211, 213, 216–17, 219–20
in Eurasia, 203–4
genera of, 201
Limnopithecus genus of, 213–14, *214*, 243, 258
Micropithecus genus of, 216–17, *216*
postcranial adaptations of, 220–27
Proconsul genus of, *see Proconsul*
Rangwapithecus genus of, 211–13, *212*, 217–19, 246
species of, 201
Dryopithecus, 201–4, 227–29, 243
first description of, 201
in Simons and Pilbeam's taxonomy, 256–58
Dryopithecus fontani, 227–29, *228*, 231, 256–57
Dryopithecus laietanus, 228, 256–57
Duke University, 34, 39, 99

Early Stone Age, 302, 356
ear region:
of Fayum primates, 155
of Omomyidae vs. Anthropoidea, 134
earth:
estimated age of, 19
magnetic field of, 26–27
East Africa:
dryomorphs from, 201, **203,** 204–6, 211, 213, 216–20
early Miocene environments of, 189–90
early Miocene primate sites in, 194–96, *195*
Hadar Formation in, 292–94
Laetoli in, 290–92, 304
Lake Turkana in, 297–301, *299*
late Miocene faunas from, 191
middle Miocene faunas from, 190–91
during Miocene epoch, 186–91, *188*, *189*
Miocene hominoids from, 201–3, **203**
Olduvai Gorge in, 301–4
Old World monkeys from, 251
Omo Basin in, 294–97, *295*
ramamorph appearance in, 202
sites older than 4 million years in, 289–90
East African Rift System, 186–91, *188*, *189*
Plio-Pleistocene primates from, 275, *276*, *278*, 288–305
East Turkana Basin, 298
ectotympanic bones, 64–65
in Adapidae, 103, 105, 108
in *Aegyptopithecus*, 161
in *Apidium*, 155
in Omomyidae, 103, 108, 124, 130
in Parapithecidae, 177, 179
in Plesiadapidae, 74, 130
in Propliopithecidae, 180
in *Tarsius*, 124
Edentata, 141
Egyptian Geological Survey, 145
Eitzman, W. I., 284
elbows, 223
Eldredge, Niles, 15
electron capture, 21
Elphidotarsius, 79, *80*, 89
encephalization, pelvic dimensions and parturition related to, *345*, 346
encephalization quotients (EQs): of *Aegyptopithecus*, 161–62
of *Australopithecus*, 324–25
correlation between mean endocranial capacity, body mass, and, **325**
of early and modern prosimians, **107**
of Omomyidae, 121
of Plesiadapidae, 74
of *Proconsul*, 208–9
endocasts, 121
endocrancial capacity, correlation between estimated body mass, EQs, and, **325**

endothermic, 28
entepicondylar foramen, 179
entoconid, 382
entocuneiform bone, 123
Eocene epoch, 20, 49–50, 93–135
  fossil record of, 56–57, 63, 97–99, 105
  land movements and mammal migration routes during, 94–97, *95*
  mammals of Paleocene epoch vs. mammals of, 98
  paleoclimates and biogeography of, 52–55, 93–97
  Plesiadapiformes in, 59
Eocene-Oligocene boundary, 96
epitrochlear fossae, 179
epochs, 371–72
Equatorial Africa, vegetation of, *189*
erector spinae, 332
Erinaceomorpha, ties between primates and, 59–63
ethmoidal sinuses, 217
ethmoid bone, 6, 373
Etler, D., 236
euprimates, 93
  body size, feeding adaptations, and behavior of, 125–28, *126, 127*
  classification of, 99–101, **100,** 125–35
  in Eocene Europe, 98
  first appearance of, 96
  morphology of, **104**
  orbital diameter and skull length in, 106, *106*
  plesitarsiiform-simiolemuriform classification of, 128–31
  rise of, 99–125
  strepsirhine-haplorhine classifications of, 131–35
  see also Adapidae; Omomyidae
Eurasia:
  dinosaurs of, 141
  first appearance of hominoids in, 191, 197
  first appearance of monkeys in, 198
  migration of New World primates from, 181
  Miocene fossil record of, 197–99, *199*
  Miocene geological events in, 191–93
  Miocene hominoids in, 203–4
  Miocene mammalian migration routes between Africa and, 191, *193,* 197–98
  pliomorphs from, 242
  Plio-Pleistocene primates from, 273–74
  ramamorphs from, 204, 229
Europe:
  Adapidae from, 103–5, *126,* 130
  disappearance of primitive apes from, 198
  dryomorphs from, 227–28
  Eocene fossil-bearing sites in, 97–98
  fossils of Paleocene mammals from, 56
  land bridge between North America and, 95, 99
  during Miocene epoch, 186
  Old World monkeys from, 252
  during Oligocene epoch, 138
  Omomyidae from, 117–18, 130
  plesiadapid fossils from, *60*
  ramamorphs from, 230–33
  stratigraphic distribution of Omomyidae in, *127*
eutherians, 64
eversion, 225
evolutionary taxonomy:
  comparison of cladism and, 11–15, *13*
  definition of, 8
extant, 18
external cingulum, 381
external obliques, 332
eyes, 373

faces:
  australopithecine phylogenetic relationships and, 359–61, *361*
  of *Australopithecus afarensis,* 309
  of *Australopithecus africanus,* 312, 315
  of *Australopithecus robustus,* 318
  of *Homo sapiens,* 330
  of Miocene and extant hominoids, *241*
  of Old World monkeys, 249
  in Omomyidae and Anthropoidea, 133
  orthognathous, 239
  of Paleocene primates, 73
  of ramamorphs vs. dryomorphs, 239–40
facial buttressing, 360, *361*
Falk, Dean, 362
Faunal Succession, Law of, 367
faunas:
  from East African late Miocene, 191
  from East African middle Miocene, 190–91
  mid-Oligocene extinction of, 138
faunivorous primates, 390–92
Fayum Depression, 138–39
  history and stratigraphy of, 145–47
  Jebel Qatrani Formation of, *146,* 147, 154, 164
  Oligocene fossil record in, 142–43
  Widan el Faras Basalt of, 147
Fayum primates, 143–68
  Afrotarsius genus of, 165–67
  Amphipithecus compared with, 175
  body size, feeding adaptations, and behavior in, 166–68
  classification of, 177–81, **178**
  dentition of, 149

Fayum primates (*continued*)
ear region of, 155
families of, 144
general morphology of, 149
morphological characters of other primates compared with, **150**
Oligopithecus genus of, 104, 164, 179
species of, 144
see also Oligocene primates; Parapithecidae; Propliopithecidae
feeding behavior, correlates of, 390–92
feet, 6
of *Aegyptopithecus,* 163
grasping, 39–41
of Omomyidae, 123–24, *123*
prehensile, 40
see also hindlimbs
fenestra cochleae, 63
fibula, 383
fire:
in cultural behavior, 356
earliest use of, 286
fission, 26
fission track dating, 26
Fitch, Frank, 298
floras, 93–94, 143, 186, *189*
folivorous primates, 30, 390
dentition of, 88–90
effects of diet on dentition in, 36, *37*
relationship between body size and diet in, 32–34
tooth size as function of body weight in, *30*
food types, classification of, 34
footprints, 331, *331*
foramen magnum, 378
Foraminifera, disappearance of, 50–51
forelimbs:
of *Australopithecus afarensis,* 337–38
of hominoids vs. cercopithecoids, 223–25, *224*
of Notharctinae, Plesiadapiformes, and Lemuriformes compared, *112*
of *Pliopithecus,* 245
of *Proconsul,* 220–25, *221*
of ramamorphs, 236
formations, 367
formen lacerum, 108
Fort Ternan:
hominoid fossils associated with, 194–97
ramamorphs from, 240
fossil ages, 19–27
carbon-14 dating in, 24
fission track dating in, 26
geologic time scale and, 19–20
K/A dating in, 25–26
paleomagnetic measurement of, 26–27
radiometric dating in, 20–27
uranium-lead dating in, 26
fossil lineages, documentation of, 18–19
fossils:
during early Miocene in Africa, 194–96, *195*
of Eocene epoch, 56–57, 63, 97–99
of Eocene primates in San Diego, 105
of Microsyopidae, 82
during middle and late Miocene in Africa, 196–97
of Miocene epoch, 194–200, *195, 199, 200*
during Miocene epoch in Eurasia, 197–99, *199*
of Old World monkeys, 251–53
of Oligocene epoch, 142–43
of Paleocene primates, 56–59
and phylogeny of Miocene hominoids, 258–64
of plesiadapids from North America and Europe, *60*
of Plesiadapiformes, 57–64
of Plio-Pleistocene,273–305
frenulum, 131
frontal bones, 154–55, 373
frontal sinuses, 217
frugivorous primates, 30, 390
body-size distribution of, 166–67
dentition of, 88, 90
effects of diet on dentition in, 36, *37*
tooth size as function of body weight in, *30*

galagos, orbital convergence in, 40
Galen, 3
Gartlan, J. S., 89
Gauss normal polarity epoch, 286, 296, 372
Gazin, C. Lewis, 105
Gebel Zelten, 251
genioglossal pit:
of *Amphipithecus,* 175
in Propliopithecidae, 159
Geological Survey, U.S., 52
Geological Survey of Pakistan, 233
geological systems, 367–70
geological time scale, 367–72
fossil ages and, 19–20
Gervais, P., 212
gestation periods, 6
gibbons, 12
phyletogenetic relationships among Miocene hominoids and, 259–62, *262*
Gibraltar, 2, 351
Gigantopithecus, 229, 231
phyletic affinities of, 264

in Simons and Pilbeam's taxonomy, 257
from Siwalik Group, 233, 235
Gigantopithecus bilaspurensis, 257
Gigantopithecus blacki, 257
Gigantopithecus giganteus, 230–31, 233, 235
Gilbert reversed epoch, 286
Gingerich, Philip, 69, 82–84, 103, 133, 174
gluteus medius, 332
gluteus minimus, 332
Gondwanaland, 53
Gordon, H. L., 204
gorillas, 5, 28, 45
cladists vs. evolutionary taxonomists on, 12–13
diet of, 86–88
phylogenetic relationships among Miocene hominoids and, 259–63, *262*
Gould, Stephen Jay, 15, 362–63
grabens, 187
gracile, 274–75
gracile mandibles, 207
gracilis, 160
grades, definition of, 18
Grand Coupure, 96
Granger, Walter, 145
greater palatine foramina, 263
greater trochanter, 115
greater tuberosity, 163
Greenfield, L., 266
greenhouse model, 50–51
Gregory, J. W., 187
groups, 367
Gulf of Mexico, 54–55
Guomde Formation, 298

habitats of Miocene hominoids, 255–56
Hadar Formation, 292–94
stratigraphic placement of hominids from, 292–93, *293*
Haile Selassie, Emperor of Ethiopia, 294
half-life, 24
hallucal metatarsal, 163
hallucal tarsometatarsal joint, 163
halluces, 6
in *Aegyptopithecus,* 163
in Notharctinae, 111
handaxes, 353
hands:
grasping, 39–41
prehensile, 40
see also forelimbs
Hanno, 3
Haplorhini, 45
nasal region of, 131, *131*
Harrison, T., 212–13, 246
Harvard University, 15
Museum of Comparative Zoology of, 290
Hay, Richard, 302
Hemiacodon, 123
Hershkovitz, P., 182–83
heterodonty, 378
heteromorphic, 162
Himalayas, uplifting of, 186, 193
hindlimbs:
of Adapidae, *114,* 128
of African Miocene dryomorphs, 225–27, *226*
of *Australopithecus afarensis,* 338
of lemurs, 113–15, *114*
of ramamorphs, 235–36
hip bones, 334–35, *335*
Hoffstetter, Robert, 132
home base, 349
Hominidae, 158
hominids:
association of alcelaphines and antilopines with, **289**
classification of, 357–65
from Hadar Formation, 292–93, *293*
locomotion of, 331–46
trends in pelvic morphology of, 344–46, *345*
Hominoidea, 160
from African Plio-Pleistocene, **274**
footprints of, 331, *331*
fossil record of, 274
impact of biomolecular data and recent fossil evidence on phylogeny of, 258–64
phylogeny and classification of, 256–67, 357–65
rise of, 200–248
Simons and Pilbeam's taxonomy of, 256–58, *257*
hominoids:
cladists vs. evolutionary taxonomists on, *13*
comparison of lower jaws of, *234*
first Eurasian appearance of, 191, 197
forelimbs of cercopithecoids vs. forelimbs of, 223–25, *224*
major lobes and sulci in brains of, *314*
mean endocranial capacity, estimated body mass, and EQs for, **325**
during Miocene epoch, *see* Miocene hominoids
hominoids, modern:
anatomy of lower face in Miocene hominoids and, *241*
craniodental comparison of African dryomorphs and, 217–20, *218*

hominoids (*continued*)
phylogenetic relationships among Miocene hominoids and, 259–64, *260, 262*
Homo, 4, 326–30
Australopithecus compared with, 271–72
bipedalism in, 344
craniodental skeleton of, 326–31, *327, 329, 330*
from East Africa, 289–90, 294, 296, 298, 300–302, 304
estimated age of, 19
evolutionary success of, 366
fossil record of, 275
opportunistic tools used by, 353–54, 356
phylogenetic relationships within, 363–65
Plio-Pleistocene appearance of, 271
species of, 275
from Sterkfontein, 282
from Swartkrans, 286
Homo erectus, 351
bipedalism in, 344
craniodental skeleton of, 328, *329*
from East Africa, 296–98, 300, 302, 304
phylogenetic relationships of, 363–64
from South Africa, 286, 288
tool use in, 354–57
Homo habilis, 351
australopithecine phylogenetic relationships and, 357–59
bipedalism in, 344
brain of, 326, 328–30
craniodental skeleton of, 326–28, *327*
from East Africa, 296–98, 301–2, 304
phylogenetic relationships of, 363
from South Africa, 282, 288
tool use in, 354–57
homologous, 10
homomorphic, 158
Homo sapiens, 325, 350
bipedalism in, 344
brain of, 326, 328–30
craniodental skeleton of, 328–30, *330*
cultural behavior in, 351
dominance of, 366
multiregional model of evolution of, 364, **365**
pelvis of, *341*
phylogenetic relationships of, 363–65, **365**
single-origin model of evolution of, 364, **365**
tool use in, 357
Homunculus, 253–54
honing facets, 103, 380
Hopkins, David, 53
Hopwood, A. T., 207
Howell, F. Clark, 294–96
Hughes, Alun, 281–82
Human Age, 366
humans, 45
anthropometric criteria distinguishing apes from, 4
cladists vs. evolutionary taxonomists on, 12–15
evolutionary link between apes and, 4–5
hip bones of, *335*
pentadactyl limbs of, 10
phyletogenetic relationships among Miocene hominoids and, 259–64, *262*
ramamorph affinities with, 264–66
surface-area-to-volume ratio in, 32
humans, archaic, behavioral and cultural trends in, 346–57
humans, modern:
bipedalism in, 349
center of gravity in, 336–37, *336, 337*
dental formula of, 36
maturation period in, 349, 351
pelvis in, 346
skeleton of, 373, *375*
upper dentition of, *321*
upper respiratory system of, 308
humerofemoral index, 311
humerus:
in *Aegyptopithecus,* 163
in African Miocene dryomorphs, 223
Huxley, Julian, 15
Hylobatidae, 158
hypocone, 381
hypoconid, 382
hypoconulid, 382

Ignacius, 64, 84–85
iliac pillar, 333
iliofemoral ligament, 335
ilium, 332–34
incisive canal, 239–40
incisors, 4, 378, 379
of Adapidae, 103, 105–6, 130
of African dryomorphs vs. modern African hominoids, 219
of Anthropoidea, 130
of *Australopithecus afarensis,* 309–11
of *Australopithecus boisei,* 321
of *Australopithecus robustus,* 318–19
of Carpolestidae, 79
of *Dendropithecus,* 215
of Fayum primates, 149
of gracile vs. robust *Australpoithecus,* 323–24
of *Homo habilis,* 328
of *Micropithecus,* 216

of Microsyopidae, 82
of New World monkeys, 254
of Old World monkeys, 249
of Omomyidae, 103, 119
of Paleocene primates, 88–89
of Parapithecidae, 152
of Paromomyidae, 84–85
of Picrodontidae, 86
of Plesiadapidae, 72–73, *73*
of Plesiadapiformes, 90
of *Pliopithecus*, 243–45
of *Proconsul*, 210
procumbent, 247
of Propliopithecidae, 159
of ramamorphs, 231, 233
of ramamorphs vs. dryomorphs, 239
of *Rangwapithecus*, 212
of Saxonellidae, 81
with spatulate crowns, 103
India, 186, 193
inferior transverse tori:
in *Aegyptopithecus*, 160
of *Amphipithecus*, 175
information content, cladograms vs. phylograms on, *14*, 15
infraorbital foramina, 121
infraorbital openings, 106
insectivorans, 40, 44*n*
insectivores, 44*n*
insectivorous primates:
dentition of, 88–89
effects of diet on dentition in, 34–36, *37*
in Eocene Europe, 98
relationship between body size and diet in, 32–34
Institute of Human Origins, 292
Institute of Paleontology, 132
intermembral index, 76, 113
internal carotid artery, 64, 105, 108
internal obliques, 332
International Omo Research Expedition, 294, 297
interorbital foramina, 183
intrabullar carotid circulation, 63–64
intraspecific dental variability, 234, *235*
inversion, 156
Iowa State University, 298
ischial callosities, 153
ischial tuberosities, 153
isochronous, 372
isometry, 28–29
isotopes, 21–24

Jarman-Bell principle, 33
jaws:
of Adapidae vs. Omomyidae, 103
of *Amphipithecus*, 174–75
of *Branisella*, 169
of *Oligopithecus*, 164, *165*
of *Plesiadapis gidleyi*, *69*
of *Pondaungia*, 173
of Siwalik hominoids, *234*
Jebel Qatrani Formation, *146*, 147, 154, 164
Johanson, Donald, 292, 309–11
Jolly, Clifford, 348
Jones, F. Wood, 39–40
Jungers, William, 337

Kada Hadar Member, 292
Kalb, John, 292
Kattwinkel, Professor, 301
Kay, Richard, 34, 36, 88–89, 134, 228, 233–34, 242, 264, 266
Kay's threshold, 34, 125, 130
KBS Tuff, 298, 301
Kent State University, 346
Kenya:
dryomorphs from, 204–7, 209–10, 212, 216
Miocene hominoids recently discovered in, 247
Old World monkeys from, 251
Plio-Pleistocene primates from, 274, 289–90, 297–301, 304
Kenyan Rift, 187
Kisingiri volcano:
hominoid fossils associated with, 194
Proconsul sites around, 211
Kitching, J. W., 284
Kleiber, M., 28
knuckle walking, 222
Kohl-Larsen, L., 291
Komba, 108
Koobi Fora Formation, stratigraphy of, 298–301, *299*
Kortlandt, A., 138
Koru, hominoid fossils associated with, 194–96
Kretzoi, M., 231
Kromdraai, 282–84
stratigraphy of, 283–84, *283*
Kubi Algi Formation, 298
Kurten, B., 158

Laccopithecus, 246
lacrimal bones, 373
of Adapidae, 105
of adapids vs. lemurs, 108
in Fayum primates, 149
in Plesiadapidae, 74
lacrimal foramina, 121
Laetoli, 290–92, 304

Lake Turkana, 297–301, *299*
 Miocene hominoids recently discovered near, 246–47
land-mammal ages, 372
 correlation of Tertiary epochs, standard stages and, *371*
 during Paleocene epoch, 56
 see also specific land-mammal ages
landmass movements:
 during Eocene epoch, 94–97
 during Miocene epoch, 191–94, *193*
 during Oligocene epoch, 139–41, *140, 141*
 during Paleocene epoch, 53–55
langur monkeys, 3
Lartet, Edouard, 201, 243
lateral sulcus, 121
Late Stone Age, tools used in, 356–57
Laurasia, 53
leaf floras, 93
Leakey, Jonathan, 302
Leakey, Louis S. B., 204–6, 301–2
Leakey, Mary Nicol, 207, 291, 301–2
Leakey, Richard, 247, 267, 294–98
leaping ability, 40, 156
Le Gros Clark, Sir Wilfred E., 5–6, 39, 206, 220
Lemur (genus), 4
Lemuriformes, 45
 Notharctinae skeleton compared to, 110, *112*, 113, *113, 114*
 Plesiadapiformes and Notharctinae fore-limb bones compared with, *112*
lemurophile hypothesis, 128–31, *129*
lemurs, 3–4, 28, 45*n*
 ectotympanic bone in, 64–65, 108
 geological origin of, 96–97
 hindlimbs of, 113–15, *114*
 internal carotid artery in, 108
 resemblance of adapids to, 107–9
Leptictidae, 59
Le Riche, W., 281
lesser trochanter, 115
Lewis, G. E., 233
Lewis, O. J., 221
Libypithecus, 253
Libypithecus markgrafi, skull of, *252*
life spans, 349–51
Limnopithecus, 213–14, *214*, 243
 in Simons and Pilbeam's taxonomy, 258
Limnopithecus legetet, 213–14, *214*
Limnopithecus macinnesi, 214
lingual, 378
Linnaean hierarchy, 44, **44**
Linnaeus, Carolus, 4, 44, 328
Lipotyphla, ties between primates and, 59
locomotion:
 bipedal, *see* bipedalism
 cursorial, 39
 of hominids, 331–46
 by knuckle walking, 222
 by leaping, 40, 156
 parasagittal, 223
 postcranial features associated with, 76–78, *77*
London, University of, 298, 350
lophs, 36, 381
lorises, 45*n*
 ectotympanic bone in, 108
 internal carotid artery in, 108
 orbital convergence in, 40
Lorisidae, 145
Lothagam Hill, Plio-Pleistocene primates from, 290, 304
Lovejoy, Owen, 340, 346, 348–49
Lucy, 292, 311, 325, 338
Lufeng:
 pliomorphs from, 246
 ramamorphs from, 236–37, *237*, **238, 239**
Lyell, Charles, 371–72

Maboko Island:
 Miocene hominoids recently discovered on, 246
 Old World monkeys from, 251
McHenry, Henry, 223–25, 290, 307, 347, 358
MacInnes, D. G., 206
McKenna, Malcolm, 182
Macrotarsius, 128
Madagascar, origin of lemurs in, 96–97
magnetic fields, 26–27
magnetostratigraphy, 372
Mahgarita, 105
Makapansgat, 284–86
 paleomagnetic record of, 285–86, *285*
mammae, 4
mammals:
 auditory bulla in, *61*
 dental formulas of, 36
 Eocene migration routes of, 94–97, *95*
 from Fayum Depression, 142
 Miocene migration of, 191–94, *193*
 Oligocene migration of, 139–41, *140, 141*
 of Paleocene epoch vs. Eocene epoch, 98
 Paleocene fossil record of, 56–57
 Paleocene migration of, 54–55
Mammals, Age of, 19, 49
mandibles, 378
 of *Aegyptopithecus*, 162

of African dryomorphs vs. modern African hominoids, 219
of *Amphipithecus,* 174, *176*
australopithecine phylogenetic relationships and, 358
of *Australopithecus africanus,* 312, 315
of *Australopithecus robustus,* 319
of *Branisella,* 169
coronoid and angular processes of, 179
of *Dendropithecus,* 214–15
gracile, 207
of New World monkeys, 254–55
of Old World monkeys, 251
of Parapithecidae, 152
of *Parapithecus,* 157–58
of *Pliopithecus,* 243
of *Proconsul,* 209–10
of Propliopithecidae, 158–59
of ramamorphs, 231–33, 236–37, 240–42
of ramamorphs vs. dryomorphs, 239
of Saxonellidae, *81*
mandibular fossae, 309
mandibular symphysis, 73, 378
of Adapidae, 103, 105, 108, 130
of Anthropoidea, 130
of *Apidium,* 154–55
of Fayum primates, 149
of lemurs, 108
of Notharctinae, 110
of Omomyidae, 103, 119
of Parapithecidae, 179
Mann, Alan, 349–51
manuports, 353
manus, 6
marine invertebrates, disappearance of, 50–51
Markgraf, Richard, 145, 154
Marsh, Othaniel C., 56
marsupials, 40, 98
Martin, Robert, 6, 231
masseter, 162
masticatory system, 360
mating behavior, 166
Matthew, William, 64
maturation patterns, 6
in *Australopithecus,* 349–51
in modern humans, 349, 351
Matuyama reversed polarity epoch, 286, 372
maxilla, 378
of *Dendropithecus,* 214
of *Proconsul,* 210
maxillary sinuses, 217, 373
Mayr, Ernst, 14, 19
medial carotid artery, loss of, 63–64
medial epicondyle, 156
medial pterygoid, 162
Mediterranean Sea, drying up of, 186, 272
megadont, 235
Melentis, J., 230–31
meniscus, 245
mental eminence, 330
mesial, 378
Mesopithecus pentelici, 252, *253*
skull of, *253*
mesostyle, 381
Mesozoic era, 50–51
geography of, 53
primate fossils of, 59
metabolic rate, correlation between body weight and, 29–32
metacone, 380
metaconid, 381
metaconule, 381
metacrista, 68
metastyle, 381
metopic suture:
in Fayum primates, 149
of Parapithecidae, 179
mice, surface-area-to-volume ratio in, 32
Michigan, University of, 69, 273, 350
Microchoerinae, 118, 120, 123
Microchoerus, 117
microfossils, 50
Micropithecus, 216–17, *216*
Micropithecus clarki, 216, *216*
Microsyopidae, 57, 81–84, 98–99
dentition of, 82, *83*
exclusion from order Primates of, 66
issues in classification of, 82–84
Middle Stone Age, 286–87
tools used in, 356–57
Minnesota, University of, 52
Miocene epoch, 20, 185–269
continental configuration during, 186
East Africa during, 186–91, *188, 189*
fossil record of, 194–200, *195, 199, 200*
fossils from Eurasia during, 197–99, *199*
land movements and dispersal routes during, 191–94, *193,* 197–98
monkeys of, *see* New World monkeys; Old World monkeys
paleoclimates and biogeography of, 185–94
Miocene hominoids:
age ranges of middle and late, *205*
anatomy of lower face in modern hominoids and, *241*
classification of, **202,** 256–68
dietary preferences, behavior, and habitats of, 255–56

Miocene hominoids (*continued*)
  dryomorph cluster of, *see* dryomorphs
  early, 201–2
  from East Africa, 201–3, **203**
  from Eurasia, 203–4
  from Lufeng, 236–37, *237*, **238, 239,** 246
  nomenclature for, 201
  phylogenetic relationships among living hominoids and, 259–64, *260, 262*
  pliomorph cluster of, *see* pliomorphs
  ramamorph cluster of, *see* ramamorphs
  recently discovered, 246–48
  rise of, 200–48
Mivart, St. George, 5
molariform, 36
molars, 30, 378, 380–82
  of Adapidae, 103, 106
  of *Aegyptopithecus*, 162
  of African dryomorphs vs. modern African hominoids, 217–20
  of *Afrotarsius*, 165
  of *Amphipithecus*, 174–75
  of *Australopithecus afarensis*, 309
  of *Australopithecus africanus*, 312
  of *Australopithecus boisei*, 321
  of *Australopithecus robustus*, 318–19
  of *Branisella*, 169–70, 182–83
  bunodont, 174
  of Carpolestidae, 79
  of *Dendropithecus*, 215
  of *Dolichocebus*, 172
  evolution of cusp pattern in, *67*
  of Fayum primates, 149
  of *Homo habilis*, 328
  of insectivorous primates, 88
  of *Micropithecus*, 216–17
  of New World monkeys, 254–55
  of Notharctinae, 110
  of Old World monkeys, 249–51
  of *Oligopithecus*, 164
  of Omomyidae, 103, 119–20
  of Parapithecidae, 152, 179
  of *Parapithecus*, 157, 168
  of Paromomyidae, 84–85
  of Picrodontidae, 86–87
  of Plesiadapidae, 73, 79
  of Plesiadapiformes, 91
  of *Plesiadapis*, 68
  of *Pliopithecus*, 245
  of *Pondaungia*, 173–74
  of *Proconsul*, 207, 209–10
  of Propliopithecidae, 158–59
  of *Qatrania*, 167–68
  quadritubercular, *66*, 68
  of ramamorphs, 231–34, 240–42
  or ramamorphs vs. dryomorphs, 239
  of *Rangwapithecus*, 212
  of Saxonellidae, 81
  of *Tremacebus*, 170, 183
  tritubercular, *66*, 68
molecular clocks, 259
moment of inertia, 76
monkeys, 3–5
  first appearance in Eurasia of, 198
  langur, 3
  of Miocene epoch, *see* New World monkeys; Old World monkeys
monophyletic groups, 10
Mons Basin, Paleocene fossils from, 56
morphoclines, 11
morphological characters, 11
morphology, skeletal, 373–89
  cranial, 373–78
  dentition, 378–82
  postcranial adaptations, 382–89
morphotypes, 101
multiregional model, 364, **365**
Musée de l'Homme, 292, 294–96

Nagri Formation, 198
Nannodectes, 71–73
Nannopithex, 123
Nannopithex fold, 68
Napier, John, 220
nasal fossae, 133
nassal incisive fossae, 239
nasal regions:
  in Haplorhini, 131, *131*
  in Strepsirhini, 131, *131*
nasoalveolar clivus, 240
nasolacrimal duct, 74
National Museums of Kenya, 291, 294–96, 301
Nature, 280
Navajovius, 89
navicular bone, 123
Neanderthal, 351
Necrolemur, 121–24, 134
negative allometric relationships, 30
  between body weight and nutritional requirements, 33
Neogene system, 20, 367
  correlation of time-stratigraphic units, geological time units, magnetic reversals, and environmental changes in Paleogene and, *369*
Neosaimiri, 255
Neudorf an der March, 230, 232
neurocranium, **180**
Newcastle upon Tyne, University of, 140
New World monkeys, 154, 160, 202, 253–55
  body weight ranges in, 125

classification of, **250,** 267–68
in Plio-Pleistocene epoch, 274
New World Oligocene anthropoids, classification of, 182–83
New World primates:
Branisella genus of, 169–70, 182–83
Dolichocebus genus of, *171,* 172, 183
from Miocene epoch, 199–200
from Oligocene epoch, 168–72, 182–83
origin of, 181–82
Tremacebus genus of, 170, *171,* 183
New York, City University of, 361–62
normal polarity, 27
North Africa, early Miocene primate sites in, 194, *195*
North America:
absence of higher primates from, 199–200
Adapidae from, 103–5, 130
dinosaurs from, 141
early Tertiary floras from, 94
Eocene fossil-bearing sites in, 98
fossils of Paleocene mammals from, 56–57
land bridge between Europe and, 95, 99
migration of New World primates from, 181
during Miocene epoch, 186
Oligocene dispersal route between South America and, 140–41
during Oligocene epoch, 138
Oligocene fossil record in, 142
Omomyidae from, 117–18, *127,* 130
plesiadapid fossils from, *60*
Northern Hemisphere:
during Oligocene epoch, 138
possible Eocene land mammal migration routes in, *95*
Notharctinae, 109–15
dental distinctions between Adapinae and, 116–17, *116*
dentition of, 110, *110,* 116–17, *116*
genera of, 109
North American extinction of, 105
Plesiadapiformes and Lemuriformes forelimb bones compared with, *112*
postcranial skeleton of, 110–15, *111, 112, 113, 114*
Notharctus, 106–7, 109–10, 128
Notharctus osborni, partially restored skeleton of, *111*
nuchal crest, 160
numerical taxonomy, 8
Nyanzapithecus, 246

Obik Sea, 55
obturator foramina, 113
occipital lobes, 133
occipital torus, 328
occlusal, 72, 378
Officer, Charles, 52
Oldowan tradition, 353, *354,* 356–57
Olduvai Gorge, 297, 301–4
stratigraphic sequence of, 302–4, *303*
Old World monkeys, 153, 202–3, 249–53
classification of, **250,** 267–68, *268*
forelimbs of hominoids vs. forelimbs of, 223–25, 224
in Plio-Pleistocene epoch, 273–74
relationship between Parapithecidae and, 177–79
Old World Oligocene primates, *see* Fayum primates
olecranon fossae, 113, 223–25
olecranon process, 223
olfactory bulbs:
in *Apidium,* 155
in Omomyidae and Anthropoidea, 133
Oligocene anthropoids, morphology of, **104**
Oligocene epoch, 20, 137–83
fossil record of, 142–43
geography of, 55
land movements and dispersal routes of, 139–41, *140, 141*
New World primates in, 168–72, 182–83
paleoclimates and biogeography of, 137–41
significant events in evolution of primates in, 183
Oligocene primates, 139, 242
Amphipithecus genus of, 104, 174–77, *176*
Branisella genus of, 169–70, 182–83
classification of, **144,** 172–83
Dolichocebus genus of, *171,* 172, 183
in New World, 168–72, 182–83
in Old World, *see* Fayum primates
Pondaungia genus of, 173–77
from South America, 168–69, *169*
Tremacebus genus of, 170, *171,* 183
Oligopithecus, 104, 164, *165,* 179
Oligopithecus savagei, 164
Olson, Todd, 361–62
Omo Basin, 294–97, *295*
Omomyidae, 117–25, 145
body size of, 125–28, *127*
canines of, 103, 119–20, 124
comparison of Adapidae and, 101–3
in comparison with later primates, 124, **125**
dental similarities linking Plesiadapidae and, 130
dentition of, *118,* 119–20, *120,* 124

Omomyidae (*continued*)
distinctive characteristics shared by, 119
ear region of, 134
ectotympanic bone in, 103, 108, 124, 130
genera of, 99
incisors of, 103, 119
jaw of, 103
mandibular symphysis in, 103, 119
molars of, 103, 119–20
North American and European stratigaphic distribution of, *127*
origins and distribution of, 117–19, 130
postcranial skeleton of, 123–24, *123*
premolars of, 103
skull of, *102*, 121–24, *122*, 132–33
snout of, 103
subfamilies of, 118
see also euprimates
Omomyinae, 118, 120
omomyophile hypothesis, *129*, 132–33
ontogenetic data, 11
opportunistic tools, 351–57
oral incisive fossae, 239–40
orangutans, 4
characters shared between *Sivapithecus* and, 264–66, *265*
cladists vs. evolutionary taxonomists on, 12–13
phylogenetic relationships among Miocene hominoids and, 259–64, *262*
ramamorph affinities with, 264–66
orbital convergence:
arboreal theory on, 39–41
relationship between leaping ability and, 40
visual predation theory on, 41
orbitosphenoid bone, 106
Oreopithecidae, 213, 246
Oreopithecus bambolii, 212–13, 246
orbito-temporal region, cranial bones forming, *75*
Origin of Species, The (Darwin), 5, 201
Orleanian age, 197
orthognathous, 239
Osborn, Henry Fairfield, 145
os planum, 74
Osteodontokeratic Culture, 284
Ouranopithecus macedoniensis, 230–31
owls, 41

Palaechthon, 82–84, 89
palates:
of African dryomorphs vs. modern African hominoids, 217
of *Australopithecus boisei*, 321
of ramamorphs vs. dryomorphs, 240
palatine bones, 373
Palenochtha, 82, 89
paleobotany, climates inferred from, 93–94
Paleocene epoch, 20, 49–91
extinctions at K/T boundary in, 49–50
mammals of Eocene epoch vs. mammals of, 98
movements of major landmasses and dispersal of species during, 53–55
paleoclimates and biogeography during, 49–55
Paleocene primates:
feeding adaptations of, 86–90
foreshortened faces of, 73
fossils of, 56–59
phylogeny and classification of, 90–91
similarities between bats and, 86
see also Plesiadapiformes
paleoclimates:
through Cenozoic, *272*
of Eocene epoch, 52–55, 93–97
inferred from paleobotany, 93–94
of Miocene epoch, 185–94
of Oligocene epoch, 137–41
of Paleocene epoch, 49–55
of Plio-Pleistocene, 272–73
Paleogene system, 20, 367
correlation of time-stratigraphic units, geological time units, magnetic reversals, and environmental changes in Neogene and, *369*
diversity of, *56*
paleomagnetic columns, 27
paleomagnetic measurement, 26–27
palmigrade, 222
palynological, 297
Pan:
pelvis of, *341*
Proconsul compared to, 209
Pangaea, 53
paracone, 380
paraconid, 68, 381
paracrista, 68
paraphyletic groups, 177
Parapithecidae, 144, 149–58, *155*
Apidium genus of, 154–57, *157*, 166, 168
dentition of, 152, 154, *154*, 177, 179
Parapithecus genus of, 157–58, 168
postcranial skeleton of, 153–54
Qatrania genus of, 158, 166–68
relationship between higher primates and, 177–81
species of, 149
Parapithecus, 157–58, 168
parasagittal movement, 223
parastyle, 381

Paromomyidae, 57, 84–85, 98–99
  craniodental skeleton of, *85*
  dentition of, 84–85
Paromomys, 84–85, 89–90
Partridge, Timothy, 281
parturition, pelvic dimensions and encephalization related to, *345,* 346
patella, 115
patellar groove, 115
Patterson, Bryan, 290
pelves:
  of *Australopithecus, 343,* 345–46
  of *Australopithecus afarensis,* 338–42, *341*
  of *Australopithecus afarensis, Pan,* and *Homo sapiens, 341*
  of *Australopithecus africanus,* 342
  of *Australopithecus robustus,* 342
  of *Homo erectus,* 344
  of Parapithecidae, 153–54
  of Plesiadapidae, Notharctinae, and Lemuriformes, *113*
  trends in hominid morphology of, 344–46, *345*
  wheel-and-axle design of, 332, *332*
Pelycodus, 109
  dental trends in, *110*
Pennsylvania, coal deposits of, 19
Pennsylvania, University of, 349
Periconodon, 128
periods, 20
peroneal tubercle, 163
peroneus longus, 123–24
pes, *see* feet
Petaurus, 89
petrosal bone, 61
Pettigrew, John, 86
phalanges, 76
Phenacolemur, 84–85, 88–90
phenetic taxonomy, 8
philtrum, 131
phyletic gradualism, 15–17
  punctuated equilibrium vs., *16,* 17, **17**
phylogenetic trees, *see* phylograms
phylogeny, 8, *see* classification
phylograms, 8
  cladograms vs., *14,* 15
Pickford, M., 242
Picrodontidae, 57, 98
  dentition of, 86–87, *87*
Picrodus, 86
Pilbeam, David, 158, 207, 228, 234, 236, 256–58, 266, 362–63
Pilgrim, Guy, 174, 233
Pinjor Formation, 198
plagiaulacoid dentition, 79–81
plantarflexion, 225
plate tectonics, 27
Platychoerops, 69, 72
Platyrrhini, 45, 155
  relationship between Parapithecidae and, 177, 181
Pleistocene epoch, 20
Plesiadapidae, 57, 68–78, 98–99
  dental similarities linking Omomyidae and, 130
  dentition of, 69–73, *71, 73,* 79
  evolution showing diversification in size and feeding behavior of, *89*
  fossils of, *60*
  locomotion mode of, 78
  Notharctinae and Lemuriformes pelves compared with, *113*
  Notharctinae skeleton compared with, 113–15, *114*
  phylogenetic relationships of, 69, *70*
  postcranial skeleton of, 75–78
  reduction in dental formulas of, 69–71, *71*
  skull of, 73–75
Plesiadapiformes, 45, 57–86, 129
  Carpolestidae family of, 57, 78–81, *80,* 98–99
  classification of, **58**
  dentition of, 68, 90–91
  emergence of, 49
  extinction of, 93, 98–99
  fossilized remains of, 57–64
  general morphology of, 63–68
  lack of predatory behavior in, 88–89
  Microsyopidae family of, 57, 66, 81–84, *83,* 98–99
  Notharctinae and Lemuriformes forelimb bones compared with forelimb bones of, *112*
  origins of, 59–63
  Paromomyidae family of, 57, 84–85, *85,* 98–99
  Picrodontidae family of, 57, 86, *87,* 98
  Plesiadapidae family of, *see* Plesiadapidae
  postcranial skeleton of, 90
  Saxonellidae family of, 57, 79–81, *81*
  as sister group of modern-looking primates, 90–91, *91*
  as sister group of Primates, 90, *91*
  skulls of, 63–68, *65*
  see also Paleocene primates
Plesiadapis, 69–73, 89–90
  cusp and crest terminology for upper incisors of, 72, *73*
  EQ of, 74
  molars of, 68
Plesiadapis gidleyi, lower jaw of, *69*
Plesiolestes, 84

plesiomorphic state, 10
Plesitarsiiformes, 90
plesitarsiiform-simiolemuriform classification, 128–31
Pliocene epoch, 20
pliomorphs, 242–46
  ancestral catarrhine features in, 245
  genera of, 201
  Pliopithecus genus of, *see Pliopithecus*
Pliopithecidae, 216, 242
Pliopithecus, 201, 203–4, 231
  catarrhine features in, 243–46, *244*
  in Simons and Pilbeam's taxonomy, 258
  species of, 243
Plio-Pleistocene, 20
  African hominids from, **274**
  fossil record of, 273–305
  paleoclimates and biogeography of, 272–73
  record of polarity reversals during, 285–86
Plio-Pleistocene primates, 271–366, **274**
  from East African Rift System, 275, *276, 278,* 288–305
  from South Africa, 274–88, *276, 278*
pneumatized, 308
Pocock, R. I., 131
polarity:
  normal, 27
  reversed, 27, 285–86
polarity epochs, 285–86, 372
polarity events, 286, 372
pollices, 111
polymorphic characters, 11
Pondaung Formation, 173
Pondaungia, 173–77
pongidae, 201
  pelvic dimensions, encephalization, and parturition in, *345*
populations, relict, 191
positive allometric relationships, 30
postcanines, 36, 378
  australopithecine phylogenetic relationships and, 358, 363
  of *Australopithecus,* 307–8
  of gracile vs. robust *Australopithecus,* 323–24
  of *Proconsul,* 209–10
  of Propliopithecidae, 158–59
  of *Rangwapithecus,* 212
postcranial skeletons, 382–89
  of adapids vs. lemurs, 109
  of *Aegyptopithecus,* 162–64
  of African Miocene dryomorphs, 220–27
  of *Apidium,* 155–56, *157*
  of *Australopithecus afarensis,* 311, 338
  of *Australopithecus africanus,* 342
  of *Australopithecus robustus,* 342–43
  of *Dendropithecus,* 214–15, 220
  different locomotor types associated with, 76–78, *77*
  of Fayum primates, 149
  of New World monkeys, 254–55
  of Notharctinae, 110–15, *111, 112, 113, 114*
  of Old World monkeys, 249–52
  of Omomyidae, 123–24, *123*
  of Parapithecidae, 153–54
  of Plesiadapidae, 75–78
  of Plesiadapiformes, 90
  of *Pliopithecus, 244,* 245
  of *Proconsul,* 207, 209, 220–27, *221, 222, 226*
  of Propliopithecidae, 160, 162–64, 181
  of ramamorphs, 235–36, 238
  of *Ramapithecus,* 238
  of *Sivapithecus,* 238
posterior iliac spine, 334
postorbital bar, 6, 373
postorbital closure, 6, 373
  in *Apidium,* 154
  of *Dolichocebus,* 172
  in Fayum primates, 149
  of Parapithecidae, 179
  of *Tremacebus,* 170
postorbital constriction, 160
postprotocingulum, 68
potassium-argon (K/A) dating, 25–26
prehallux, 163
prehension, 6
premaxilla, 378, 379
  of African dryomorphs vs. modern African hominoids, 217
  in Plesiadapidae, 74
premolariform canines, 103, 120
premolars, 378, 380
  of Adapidae, 103, 105, 108, 130
  of *Aegyptopithecus,* 162
  of African dryomorphs vs. modern African hominoids, 219
  of *Amphipithecus,* 174–75, *176*
  of Anthropoidea, 130
  australopithecine phylogenetic relationships and, 358, 360
  of *Australopithecus afarensis,* 309
  of *Australopithecus africanus,* 315
  of *Australopithecus boisei,* 321
  of *Australopithecus robustus,* 318–19
  of *Branisella,* 169–70
  of Carpolestidae, 79–81
  of *Dendropithecus,* 215
  of Fayum primates, 149
  of frugivorous primates, 88

heteromorphic, 162
of *Homo habilis,* 326–28
of lemurs, 108
of Microsyopidae, 82
of New World monkeys, 255
of Old World monkeys, 249
of *Oligopithecus,* 164
of Omomyidae, 103
of Parapithecidae, 152, 177–79
of Paromomyidae, 85
of Plesiadapidae, 72–73
of *Pliopithecus,* 243–45
of *Proconsul,* 207
of Propliopithecidae, 158–59, 180
of ramamorphs, 231–32, 235, 240
of Rangwapithecus, 212
of Saxonellidae, 81
sectorial, 103
of *Tremacebus,* 170
primate evolution:
major radiations in, 45, *46*
overview of, **22**
progressive loss of teeth in, *38*
significant events in, 183
primates:
definitions of, 4–7, 59
early descriptions of, 3–4
Primates (order):
classifications of, 393–403
exclusion of Microsyopidae from, 66
genera of, 4
Plesiadapiformes as sister group of, 90, *91*
traditional classification of, **42**
primates, extinct:
allometric analysis and, 28–32
effects of diet on dentition in, 34–36, *35, 37, 38*
making inferences about biology of, 27–38
relationship between size and diet in, 32–34
primates, modern:
Plesiadapiformes as sister group of, 90–91, *91*
relationship between body size and dietary preference in, *167*
skulls of, *6,* 65, *65*
Primates, origin of, 38–43
angiosperm radiation theory on, 41–43
arboreal theory on, 39–41
visual predation theory on, 41–42
Proconsul, 243
body weight of, 207–9
catarrhine features of, 207
dentition of, 210, *210, 211*
Dryopithecus compared with, 228–29
geographical distribution of, 207, 211
hindlimbs of, 225–27, *226*
Limnopithecus related to, 213–14
oral incisive fossa of, 239–40
postcranial skeleton of, 207, 209, 220–27, *221, 222, 226*
Rangwapithecus compared with, 211–12
in Simons and Pilbeam's taxonomy, 256, 258
sinuses in, 217–19
Sivapithecus vs., 232
species of, 206–11, 220
Proconsul africanus, 207–9, *208,* 213, 217, 220–27, *221, 222,* 242, 247, 257
Proconsulidae, 201, 206
Proconsul major, 209–10, *211,* 219, 230, 257
Proconsul nyanzae, 209–10, *210,* 226–27, 257
procumbent incisors, 247
prognathous, 207
Prohylobates, 251
promontory artery, 64
of Adapidae, 105
of Parapithecidae, 179
pronation, 221
Pronothodectes, 69–72
Pronycticebus, 105–6, 128
Propliopithecidae, 144, 158–64
Aegyptopithecus genus of, 108, 160–64, 166–68, 183, 219, 245–46
dentition of, 158–59, *159,* 162, 166
Propliopithecus genus of, 158–60, 166–68
relationship between higher primates and, 180–81, **180**
Propliopithecus, 158–60, 166–68
Prosimii, 5, 45
dental comb of, *12*
differences in cortical folding of Anthropoidea and, *133*
distinctions between Anthropoidea and, *133, 148,* 149
EQs of early and modern, **107**
relationship between Parapithecidae and, 181
skeleton of, 373, *374*
protocone, 380
protoconid, 381
protoconule, 381
proto-handaxes, 353
pseudohypocones, 116
pterion, 161
pterygoid plates, 323
Puget Group, leaf floras of, 93
punctuated equilibrium, 15–17
phyletic gradualism, *16,* 17, **17**
Punjab University, 97

Purgatorius, 82–84
Purgatorius ceratops, 59
pyroclastic rocks, K/A dating of, 25

Qatrania, 158, 166–68
quadritubercular molars, 381
  evolution of, *66,* 68
quadrupedalism, bipedalism vs., 347
Quaternary system, 20, 367
Queensland, University of, 86

radial notch, 223
Radinsky, L., 121
radioactive decay, principles of, 21–24
radioactive decay curve, *25*
radiometric dating, 20–27
  carbon-14 method of, 24
  fission track method of, 26
  K/A method of, 25–26
  methods of, 24–27
  paleomagnetic measurement method of, 26–27
  principle of, 21–24
  uranium-lead method of, 26
radius, 383
Rak, Yoel, 359–60
ramamorphs, 229–42
  from Africa, 240–42
  from China, 236–39, *237,* **238, 239**
  comparison of Asian dryomorphs and Asian, 239–40
  from Eurasia, 204, 229
  from Europe and Asia Minor, 230–33
  genera of, 201–2, 229, 236
  phylogeny of, 264–67
  from Siwalik Group, 233–36
Ramapithecinae, 229
Ramapithecus:
  from Africa, 240, 242
  from China, 236–38, *237,* **238**
  phyletic affinities of, 264–66
  in Simons and Pilbeam's taxonomy, 258
Ramapithecus punjabicus, 258
Ramapithecus wickeri, 232
Rangwapithecus:
  Proconsul compared with, 211–12
  species of, 211–13
Rangwapithecus gordoni, 212, *212,* 217–19
Rangwapithecus vancouveringi, 212–13, 217, 246
Rasmussen, Tab, 134
Raven, Peter, 86
relict populations, 191
Remane, A., 233
reversed polarity, 27
rhinarium, 131
Rift Valley Research Mission, 292
rift valleys:
  definition of, 187
  see also East African Rift System
Rio Gallegos, New World monkeys from, 253
Robinson, John T., 281, 286, 323–24, 342, 350, 359, 362
robust, 215, 274–75
rock units, 367
rodents, 40, 57, 59, 98
Rodman, Peter, 347
Rooneyia, 121–24, 134
Rose, K., 79, 81, 82–84
Rosenberger, A., 183
Rudabanya, 231
Rudapithecus, 231
Rusinga Island:
  dryomorphs from, 204–7, 209, 212–15, 222, 226
  hominoid fossils associated with, 194–96
Rutherford, Ernest, 21

sacroiliac joint, 334
sacrotuberous ligaments, 340
sacrum, 340
sagittal chest, 155
Sahni, Ashok, 97
St. Acheul, 356
Salmons, Josephine, 277
San Diego County, Calif., Eocene primate fossils from, 105
San Juan Basin, Paleocene fossils from, 56
Savage, Donald, 164
Saxonellidae, 57, 79–81
  mandible of, *81*
scapulae:
  of African Miocene dryomorphs, 223
  in *Apidium,* 155–56
Schepers, G., 281
Schimper, Wilhelm, 372
Schlosser, M., 158
Schweinfurth, Georg, 145
sciatic notch, 334
sclerophyllous vegetation, 186
sea-level changes through late Cretaceous and Tertiary, *55*
seals, 5
sectorial, 103, 380
seed-eating hypothesis, 348
semitendinosus, 160
series stages, 370
sexual-and-reproductive-strategy model, 348–49
sexually dimorphic, 103, 128, 166
sexual maturation, 6
shared derived characters, 9
Sheine, W., 36

Shoshonius, 121
Shungura Formation, 294–97
stratigraphy of, 294, *295*
siamangs, 12
Sidi Hakoma Member, 292, 294
sigmoid notch, 223–25
Simia, 4
simian shelf, 160, 175
Simiolemuriformes, 90
Simons, Elwyn, 99, 145, 159, 177, 213–14, 228, 233–34, 236, 256–58, 264, 266
Simons and Pilbeam's taxonomy, 256–58, *257*
Simpson, George Gaylord, 12, 15, 39
single-origin model, 364, **365**
sinuses:
of African dryomorphs vs. modern African hominoids, 217–19, *218*
in *Proconsul,* 217–19
sister groups, 9
Sivapithecus, 201–2, 204, 229, 258
from Africa, 240, 242
Afropithecus compared to, 247
characters shared between orangutans and, 264–66, *265*
from China, 236–39, *237,* **238, 239**
from Europe and Asia Minor, 230–32
morphological differences between *Ramapithecus* and, 238, **238**
phyletic affinities of, 264–67
from Siwalik Group, 233–36
Sivapithecus africanus, 235, 240, 242
Sivapithecus alpani, 232
Sivapithecus darwini, 230, 232, 235
Sivapithecus indicus, 231–32, 234–36, 238–39, **239,** 257
Sivapithecus lufengensis, 238–39, **239**
Sivapithecus meteai, 230–32, 235, 240, 264–66
Sivapithecus simonsi, 235
Sivapithecus sivalensis, 231, 236, 233–35, 257
Siwalik Group, 192–93, 198–99
comparison of intraspecific dental variability in modern chimpanzees and ramamorphs from, *235*
comparison of lower jaws of hominoids from, *234*
Old World monkeys from, 252
ramamorphs from, 233–36
skeletal morphology, *see* morphology, skeletal
skulls, 373
of Adapidae, *102,* 105–6, 128
of Adapinae, *115*
of *Aegyptopithecus,* 160–61, *161, 162*
of *Afropithecus turkanensis, 248*
of Anthropoidea, 132–33
of *Apidium,* 154–55
of *Australopithecus afarensis,* 309
of *Australopithecus robustus, 318*
of *Dolichocebus, 171,* 172
of euprimates, 106, *106*
of *Libypithecus markgrafi, 252*
of *Mesopithecus pentelici, 253*
of modern primates, *6,* 65, *65*
of New World monkeys, 254
of Omomyidae, *102,* 121–24, *122,* 132–33
of Parapithecidae, 152–55, *155*
of Plesiadapidae, 73–75
of Plesiadapiformes, 63–68, *65*
of Plesiadapiformes vs. modern primates, 65, *65*
of *Pliopithecus,* 243, *244,* 246
of *Proconsul,* 207–9, *208*
of ramamorphs, 236–39, **238, 239**
from Taung, 277–80
of *Tremacebus,* 170, *171*
of *Turkanapithecus,* 247
see also craniodental skeletons; craniofacial skeletons
Sloan, Robert, 52
smell, sense of, 41
Smilodectes, 107, 109
Smith, B. Holly, 350–51
Smith, G. Eliot, 39
Smith, William, 367
Smithsonian Institution, 105, 298
snouts, 103
Soan Formation, 198
Soddy, Frederick, 21
Songhor:
dryomorphs from, 204–7, 209–10, 212–14
hominoid fossils associated with, 194–96
Soriacebus, 254
South Africa:
association of alcelaphines and antilopines with early hominids in, **289**
caves of, 275–77
Kromdraai in, 282–84, *283*
Makapansgat in, 284–86, *285*
Plio-Pleistocene primates from, 274–88, *276, 278*
Sterkfontein in, 280–82, *281*
Swartkrans in, 286–87, *287*
Taung in, 277–80, 312, *313,* 315
South America:
appearance of higher primates in late Oligocene of, 168–69, *169*
dinosaurs in, 141
during Eocene epoch, 96
fossils of Paleocene mammals from, 56

South America *continued*)
migration of New World primates to, 181–82
during Miocene epoch, 186
Miocene primates from, 199–200, *200*
Oligocene fossil record in, 143
Oligocene geographical separation between Africa and, 139
Oligocene migration route between Africa and, 140
Oligocene migration route between North America and, 140–41
Oligocene primates from, 139
Plio-Pleistocene primates from, 274
Southern Hemisphere, Adapidae origin in, 103–4
Sparnacian land-mammal age, 98
spatulate, 103
speciation:
allopatric, 15–16
definition of, 15
models of, 15–17
phyletic gradualism model of, 15–17
punctuated equilibrium model of, 15–17
species dispersals, 181–82
during Eocene epoch, 94–97, *95*
during Miocene epoch, 191–94, *193*, 197–98
during Oligocene epoch, 139–41, *140*, *141*
during Paleocene epoch, 53–55
sphenoidal sinuses, 217, 373
squamosal suture, 323
standard stages, 370
correlation of land-mammal ages, Tertiary epochs and, *371*
stapedial artery, 149
of Adapidae, 105
of Parapithecidae, 179
stapedial canal, 149
Sterkfontein, 280–82
stratigraphic sequence of, 281–82, *281*
Stern, Jack, 338–40, 344
Stirton, R., 254–55
Stirtonia, 254
Stone Age, African tool use in, *352*
Stony Brook, State University of New York (SUNY) at, 337–38
strepsirhine-haplorhine classifications, 131–35
Strepsirhini, 45
brain in, 133
nasal region of, 131, *131*
stylar shelf, 381
subchrons, 286, 372
subnasal alveolar process, 240
subtalar joint, 225
superior transverse tori, 160
supination, 221
supraglabella region, 247
supraorbital tori, 207
supratoral sulcus, 236
surface-area-to-volume ratio, 29, 32
Susman, Randall, 338–40, 344
suspensory posture, 113
Sussman, Robert, 41–42, 86
sutures, 378
Swartkrans, stratigraphic sequence of, 286–87, *287*
sylvian sulcus, 121
symplesiomorphic characters, 10
synapomorphic characters, 10–11, 14
syndesmosis, 156
Systema Naturae (Linnaeus), 4
Szalay, F., 78, 82, 88, 133, 164, 174

Taieb, Maurice, 292
talocrural joint, 225
talon, 381
talonid, 68
talus, 163–64
taphonomic, 284
Tarling, D. H., 140
tarsiers, 3, 40
Tarsiidae, 144
Tarsiiformes, 45
features common to Anthropoidea and, **109**, 124, **125**
Plesiadapiformes as sister group of, 90, *91*
tarsiphile hypothesis, *129*, 132, 134–35
Tarsius, 89, 123
Omomyidae compared and contrasted with, 124, **125**
Tatrot Formation, 198
Taung, 277–80
Taung child, 312, *313*, 315
taxa, 9–10
diphyletic, 181
formal and anglicized names of, **45**
taxonomic ranks, 9
taxonomy:
alternate systems of, 45
evolutionary, *see* evolutionary taxonomy
limitations in, 18–19
of New World Oligocene anthropoids, 182–83
numerical, 8
phenetic, 8
of Simons and Pilbeam, 256–58, *257*
in text, 43–46
teeth:
enamel of, 203
evolutionary loss of, *38*
rows of, 119

size of, *30*
trenchant, 81
see also dentition; *specific kinds of teeth*
Teilhardina, 118–19, *120*
Tekkaya, I., 232–33
Tel Aviv University, 359
Teleki, Count Samuel, 187, 294
temporal fossae, **180**
temporalis muscle, 243
temporomandibular joint, 360
Terblanche, Gert, 282
terrestrial quadrupeds, postcranial adaptations in, 76–78
Tertiary period, 20
African fossil deposits from, 18
climate of, 52
correlation of land-mammal ages, standard stages and epochs of, *371*
epochs of, 371–72
geography of, 55
North American floras during, 94
sea-level changes during, *55*
Tethys Sea, 53, 55
regression of, 191, *193*
Tetonius, 121, 124
thermoluminescence dating, 287
third trochanter, 115
Thomas, D., 220
Thomson, Joseph, 187
Tibetan Plateau, uplifting of, 186, 193
tibia, 383
tibial malleolus, 225
Tiffanian-Clarkforkian mammal-age boundary, 57
time-stratigraphic units, *369,* 370
Tinderet volcano, hominoid fossils associated with, 194–96
Tobias, Phillip, 281
Tobien, H., 232
tools:
of Acheulian tradition, 354–56, *355,* 357
African Stone Age use of, *352*
basic forms of, 353
in cultural behavior, 351–57
of Developed Oldowan, 353, *355,* 356–57
of Oldowan tradition, 353, *354,* 356–57
opportunistic, 351–57
Torrejonia, 89–90
Transvaal Museum, 281–82, 286
transverse arch, 225
transverse tarsal joint, 235–36
tree shrews, olfactory signals in, 41
tree squirrels, olfactory signals in, 41
Tremacebus, 170, *171,* 183
trenchant teeth, 81
triceps, 223
trigone, 381
trigonid, 68, 381
trihedrals, 356
tritubercular molars, 380
evolution of, *66,* 68
trochlea, 115
trochlear ridges, 115
tuffs, 25, 275
Tulp, Claes Pieterzoon, 4
Tupaiidae, 59
Turkanapithecus, 247
Turkey, ramamorphs from, 231–32
Turolian age, 198, 204
tympanic membrane, 64
type localities, 370
type sections, 367

Uganda:
dryomorphs from, 204, 210, 216
Old World monkeys from, 251
ulna, 163, 383
unconformities, 298
ungulates, 98
upper respiratory systems, 308
uranium-lead dating, 26
Usno Formation, 294–97

Vallesian age, 198, 204
Van Couvering, John, 189–90
Vannier, Michael, 350
vegetation, 93–94, 143
of Equatorial Africa, *189*
sclerophyllous, 186
Vesalius, Andreas, 3
Vespertilio, 4
vicars, cercopithecoid, 178
Victoriapithecidae, 249
Victoriapithecus, 251
vision, binocular, 6
visual predation theory, 41–42
vocal tracts, 328–30
Vondra, Carl, 298
von Koenigswald, G. H. R., 231
Vrba, Elizabeth, 282–83

Wadi Moghara, Old World monkeys from, 251, 253
Walker, A., 209, 247
warm-bloodedness, 27–28
Wasatchian-Bridgerian land-mammal ages, 117–18
Wasatchian-Bridgerian mammal-age boundary, 110
Wasatchian land-mammal age, 98
Washington University, 41, 86, 350
White, T., 292, 309–11
Widan el Faras Basalt, 147
Wisconsin, University of, 89
Witwatersrand, University of the, 277, 281, 284

Wolfe, Jack, 52–53
Wolpoff, Milford, 263, 273
Wu Rukang, 238

Yale University, 56, 145, 233, 256
Y5 pattern, 309
Young, R. B., 279

Zanycteris, 86–87
Zinjanthropus boisei, 302
Zurich, University of, 6
zygomatic arches, 239
zygomatic bones, 373
zygomatic foramina, 263
zygomatic process, 240